Territorial Space and Transportation Coordinated Development

Territorial Space and Transportation Coordinated Development

Guest Editors

Gaoru Zhu
Linchuan Yang
Jing Chen

MDPI

Basel • Beijing • Wuhan • Barcelona • Belgrade • Novi Sad • Cluj • Manchester

Guest Editors

Gaoru Zhu
Department of Urban and Rural Planning
Beijing Jiaotong University
Beijing
China

Linchuan Yang
Department of Urban and Rural Planning
Southwest Jiaotong University
Chengdu
China

Jing Chen
Transport Planning and Research Institute
Ministry of Transport of China
Beijing
China

Editorial Office
MDPI AG
Grosspeteranlage 5
4052 Basel, Switzerland

This is a reprint of the Special Issue, published open access by the journal *Land* (ISSN 2073-445X), freely accessible at: https://www.mdpi.com/journal/land/special_issues/POTLWYLH5T.

For citation purposes, cite each article independently as indicated on the article page online and as indicated below:

Lastname, A.A.; Lastname, B.B. Article Title. *Journal Name* **Year**, *Volume Number*, Page Range.

ISBN 978-3-7258-4775-4 (Hbk)
ISBN 978-3-7258-4776-1 (PDF)
https://doi.org/10.3390/books978-3-7258-4776-1

Contents

About the Editors vii

Yinjie He, Biao Liu, Chengyou Xu and Dafang Wu
Literature Review on Public Transport and Land Use: Based on CiteSpace Statistical Analysis
Reprinted from: *Land* **2025**, *14*, 1096, https://doi.org/10.3390/land14051096 1

Haochen Yang, Nana Cui and Haishan Xia
The Evolution of the Interaction Between Urban Rail Transit and Land Use: A CiteSpace-Based Knowledge Mapping Approach
Reprinted from: *Land* **2025**, *14*, 1386, https://doi.org/10.3390/land14071386 22

Pengpeng Liang, Xu Cui, Jiexi Ma, Wen Song and Yao Xu
Exploring the Determinants of Spatial Vitality in High-Speed Rail Station Areas in China: A Multi-Source Data Analysis Using LightGBM
Reprinted from: *Land* **2025**, *14*, 1262, https://doi.org/10.3390/land14061262 51

Xixi Feng, Jixiao Li, Yadan Liu and Weidong Li
The Impact of High-Speed Rail on High-Quality Economic Development: Evidence from China
Reprinted from: *Land* **2025**, *14*, 1379, https://doi.org/10.3390/land14071379 80

Huanyu Yang, Wei Huang, Dong Yang and Ying Jiang
Innovation in Comprehensive Transportation Network Planning in the Context of National Spatial Development: Institutional Constraints and Policy Responses
Reprinted from: *Land* **2025**, *14*, 1046, https://doi.org/10.3390/land14051046 100

Lulin Liu, Xiaojie Ma and Yaolong Li
Does New Infrastructure Promote the Development of Rural Industries? A Nonlinear Analysis Based on Provincial Panel Data from China
Reprinted from: *Land* **2025**, *14*, 986, https://doi.org/10.3390/land14050986 120

Sangwan Lee, Jeongbae Jeon, Kuk Cho and Junhyuck Im
Does Intercity Transportation Accessibility Matter? Its Effects on Regional Network Centrality in South Korea
Reprinted from: *Land* **2025**, *14*, 873, https://doi.org/10.3390/land14040873 146

Jiexi Ma, Zhongwei Shen, Pengpeng Liang, Yu Zhao and Wen Song
Assessing Users' Satisfaction with the Urban Central Metro Station Area in Chengdu: An SEM-IPA Approach
Reprinted from: *Land* **2025**, *14*, 1023, https://doi.org/10.3390/land14051023 165

Zaiyu Fan, Weiyang Luo, Chang Liu and Mengyun Xie
The Role of Comprehensive Transportation in Shaping Spatial Expansion Patterns: A Case Study of the Yangtze River Middle Reaches Urban Agglomeration
Reprinted from: *Land* **2025**, *14*, 1064, https://doi.org/10.3390/land14051064 188

Yujian Gao, Long Bai, Shengqiang Liu, Hongjuan Zheng, Jie Liu, Jinxiang Cheng, et al.
Traffic–Tourism Spatial Interaction of Lai-Qu Expressway Based on the Traffic Flow Data
Reprinted from: *Land* **2025**, *14*, 1197, https://doi.org/10.3390/land14061197 212

Yiting Li, Jingwei Li, Ziyue Yu, Siying Li and Aoyong Li
Exploring the Relationship Between the Built Environment and Bike-Sharing Usage as a Feeder Mode Across Different Metro Station Types in Shenzhen
Reprinted from: *Land* **2025**, *14*, 1291, https://doi.org/10.3390/land14061291 228

Xu Lu, Mengqin Zhu, Zeting Li, Qingyu Li and Shan Huang
Urban Renewal Strategy Guided by Rail Transit Development Based on the "Node–Place–Revenue" Model: Case Study of Shenyang Metro Line 1
Reprinted from: *Land* **2025**, *14*, 1214, https://doi.org/10.3390/land14061214 . . . . **253**

Yichen Ruan, Xiaoyi Zhang, Shaohua Wang, Xiuxiu Chen and Qiuxiao Chen
Street View-Enabled Explainable Machine Learning for Spatial Optimization of Non-Motorized Transportation-Oriented Urban Design
Reprinted from: *Land* **2025**, *14*, 1347, https://doi.org/10.3390/land14071347 . . . . **275**

Yonggeng Xiong and Aibo Jin
Spatial Gradient Effects of Landscape Pattern on Ecological Quality Along the Grand Canal
Reprinted from: *Land* **2025**, *14*, 1310, https://doi.org/10.3390/land14061310 . . . . **297**

Honglin Feng, Zhefei Jin, Xiao Chu, Weihan Wang and Haiyuan Yao
Route Selection for the Extension Line of Pinglu Canal: A Comparative Study
Reprinted from: *Land* **2025**, *14*, 1266, https://doi.org/10.3390/land14040873 . . . . **320**

Xinyu Li, Gaoru Zhu, Jiaqi Sun, Leyao Wu and Yuting Peng
Research on the Coordination of Transportation Network and Ecological Corridors Based on Maxent Model and Circuit Theory in the Giant Panda National Park, China
Reprinted from: *Land* **2025**, *14*, 1465, https://doi.org/10.3390/land14071465 . . . . **340**

About the Editors

Gaoru Zhu

Gaoru Zhu is a professor of the Department of Urban and Rural Planning at Beijing Jiaotong University, with a doctoral degree from Peking University. He has previously served as a Chief Researcher at the Transport Planning & Research Institute of the Ministry of Transport of China and Deputy Director of the Transportation Bureau of Aba Prefecture in Sichuan Province. He has been selected as an Outstanding Young Science and Technology Talent by the Ministry of Transport. He focuses on the coordinated development between comprehensive transportation and territorial space. He serves as the Chair of the Territorial Space and Transportation Planning Technical Committee at the World Transportation Convention. He has led more than 40 projects, including National Natural Science Foundation, Humanities and Social Sciences Fund of Ministry of Education, and key scientific and technological project of the Ministry of Transport. He has published over 60 papers and has received 17 science and technology awards, including the outstanding Contributor to "Strong Transportation Country" Strategy Research of Ministry of Transport of China, the First Prize for Outstanding Engineering Consulting Achievements.

Linchuan Yang

Linchuan Yang is a professor and deputy head of the Department of Urban and Rural Planning at Southwest Jiaotong University, China. He obtained both PhD and MPhil degrees from The University of Hong Kong. He is the recipient of the Science Fund for Distinguished Young Scholars from the Science & Technology Department of Sichuan Province, China.

Jing Chen

Jing Chen is chief engineer of Transport Planning & Research Institute, Ministry of Transport of China. She is a nationally prominent expert in comprehensive transportation planning, focuses on integrated transport-corridor planning methodologies, regional transport-integration theory, and the optimal allocation of transport infrastructure within urban agglomerations. Recognized by the Ministry of Transport as a Leading Young and Mid-Career Scientific and Technological Innovator, she has led or participated in more than 100 national and provincial transport planning research projects as well as strategic studies on comprehensive transportation.

land

MDPI

Review

Literature Review on Public Transport and Land Use: Based on CiteSpace Statistical Analysis

Yinjie He [1], Biao Liu [1,*], Chengyou Xu [2] and Dafang Wu [2,*]

[1] Shenzhen Research Institute, The Chinese University of Hong Kong, Shenzhen 518057, China; yjhe@cuhkri.org.cn
[2] School of Geography and Remote Sensing, Guangzhou University, Guangzhou 510006, China; 2112401098@e.gzhu.edu.cn
* Correspondence: bliu@cuhkri.org.cn (B.L.); wudafang@gzhu.edu.cn (D.W.)

Abstract: With the growing demand for mobility fueled by global population expansion and rapid urbanization, the intricate interplay between public transport and land use, along with their economic, environmental, and social externalities, has emerged as a critical concern for policymakers and the public alike. This study assembles publicly available academic literature, including papers, reports, books, and news articles, to construct a comprehensive database. Using CiteSpace 5.8.R3 software, this study conducts a visualized analysis of 10,470 scholarly works on public transport and land use published since 1932, identifying and synthesizing the researcher, research stages, research theories, research models, and research hotspots. Findings reveal that since Mitchell and Rapkin first introduced the transport–land use relationship in 1954, research in this field has steadily gained traction, particularly after the 1973 oil embargo crisis. The Journal of Transport and Land Use and institutions such as the University of Minnesota's Transportation Research Center have played pivotal roles, particularly with the establishment of the World Society for Transport and Land Use Research (WSTLUR). In recent years, China's high-speed rail expansion has further revitalized interest in this field. Prominent scholars in this domain include Robert Cervero, Reid Ewing, Michael Duncan, and Peter Calthorpe. Major theoretical frameworks encompass utility theory, urban economic theory, and the human–land system theory. Key modeling approaches include the spatial interaction model, the stochastic utility model, and urban economic models. Current research hotspots center on safety and public health, equity and valuation, environmental sustainability and energy efficiency, as well as transit-oriented development (TOD) and accessibility. This systematic literature review offers valuable insights to inform land use planning, enhance spatial structure, guide transportation project decision making, and optimize transport infrastructure and service provision.

Keywords: public transport; land use; literature review; CiteSpace

1. Introduction

Sustainable development is recognized as one of the most pressing global challenges, with public transportation and land use serving as crucial links to achieving sustainability [1,2]. Both exert profound influences on economic growth, ecological sustainability, and social equity. Following World War II, rapid industrialization, urbanization, and the proliferation of automobiles led to widespread urban sprawl, exacerbating car dependency and raising concerns about inefficient land use and fragmented urban spatial structures [3]. As global population growth and rising income levels drive increasing demand for public transportation and land use, their interdependence strengthens [4,5]. Public transportation represents one of the most substantial infrastructure investments globally. The transit-oriented development (TOD) model [6] prioritizes accessibility enhancement, compact and mixed-use development, and multimodal transportation integration. This approach fosters higher land use intensity, improved connectivity, and a pedestrian-friendly urban structure. It reduces travel time for accessing essential services (e.g., work, shopping, entertainment), diversifies transportation options (including walking and cycling), and significantly lowers per capita vehicle ownership and usage [7]. This reduces overall transportation costs, encompassing both internal costs (direct consumer expenditures) and external costs (societal transportation burdens), while mitigating the secondary economic, environmental, and social consequences of urban sprawl. In the United States, TOD, alongside smart growth and new urbanism, is regarded as a crucial strategy by the Environmental Protection Agency (EPA) due to its profound influence on environmental sustainability, climate resilience, public health, historic preservation, and urban aesthetics and well-being.

Public transportation and land use are intrinsically linked, continuously interacting and exerting profound influences on social, economic, and environmental dimensions [8,9]. Their intricate relationship plays a pivotal role in shaping contemporary urban structures worldwide. Public transportation infrastructure profoundly shapes land use patterns, spatial structures, and urban layouts [10]. In turn, these elements influence public transportation demand, including its distribution, travel distances, routing, and modal choices [11]. The bidirectional and dynamic interplay between public transportation and land use is inherently complex, as it unfolds across multiple temporal and spatial scales and is often shaped by political, military, and cultural factors. Their interactions manifest in economic, environmental, and social dimensions as follows [12]:

1. Economic: Direct appreciation of nearby land value; increased land use intensity and density, or changes in land functions, leading to higher municipal tax revenues; reduced transportation infrastructure maintenance costs; allocation of public subsidies; enhanced productivity; and decreased transportation expenditures.
2. Environmental: Conservation of green spaces and wildlife habitats; mitigation of water pollution and the urban heat island effect; reduction in energy consumption and pollutant emissions; and enhancement of urban visual aesthetics.
3. Social: Advancement of social equity and expanded economic opportunities for low-income groups; enhanced accessibility and spatial cohesion; mitigation of housing affordability challenges; preservation of cultural heritage; promotion of public health through increased physical activity; and reduction in traffic-related incidents.

However, there are still many challenges in reality. The lack of forward-looking land use strategies and imprecise planning often result in economically inefficient and ecologically unsustainable public transportation routes [4]. Additionally, many projects focus solely on localized land use impacts while neglecting broader regional effects, potentially leading to biodiversity loss and degradation of critical ecosystem services due to transportation-induced pollution and habitat destruction. Consequently, a systematic review of the research progress in this domain holds significant theoretical and practical implications. The review systematically summarizes and categorizes the researchers, research stages, research theories, research models, and research hotspots. The purpose is not to introduce new empirical results, but to organize and systematize existing information, thereby helping researchers who are beginning to engage with or become familiar with this field to quickly access the information they seek.

2. Research Methods

In recent years, the body of literature on public transportation and land use has expanded exponentially, building upon a solid foundation established by previous studies. Given the extensive volume of existing studies, this research utilizes CiteSpace V.5.8.R3 software for literature analysis and synthesis. CiteSpace is an open source software developed by Chaomei Chen. He is a professor (tenured) in the School of Information Science and Technology at Drexel University, Philadelphia, PA, USA. This software is widely applied in literature reviews [13,14] and is particularly noted for its capability to identify high-impact clustering nodes and generate visual representations. Keyword searches for "public transportation" and "land use" were conducted across major international databases, including Web of Science, Scopus, Dimensions, and PubMed. These databases provide extensive coverage, encompassing nearly all publicly available academic resources, including peer-reviewed journal articles, reports, books, reviews, news, and patents. Through this comprehensive approach, a multi-source literature database spanning 1932 to 2023 was constructed, comprising a total of 10,470 records. Specifically, the dataset includes 3711 articles from Web of Science, 3901 from Scopus, 1858 from Dimensions, and 1000 from PubMed (see Table 1 for details). The clustering analysis in CiteSpace V.5.8.R3 was conducted using the Log-Likelihood Ratio (LLR) method, while all other parameters were set to the software's default settings, including slice length = 1, g-index (K = 25), LRF = 2, L/N = 10, LBY = 8, and e = 2.

Table 1. Search information.

Database	Website	Number of Articles	Search Method	Initial Year
Web of science	https://clarivate.com/webofsciencegroup/solutions/web-of-science/ (accessed on 7 March 2022)	3711	Full text	1989
Scopus	https://www.scopus.com/ (accessed on 7 March 2022)	3901	Title, abstract, keyword	1937
Dimensions	https://app.dimensions.ai/discover/publication (accessed on 7 March 2022)	1858	Title and abstract	1932
PubMed	https;//pubmed.ncbi.nlm.nih.gov/ (accessed on 7 March 2022)	1000	Best match	1972

3. Cluster Analysis Results

The CiteSpace V.5.8.R3 analysis results are presented in Figure 1 and Tables 2 and 3. Figure 1 visualizes the most influential research topics within the identified clusters, whereas Table 2 lists the top 10 most co-cited references along with their corresponding journals, as automatically generated by the software. A synthesis of highly co-cited literature suggests a strong correlation, and in some cases, causality, between the built environment and travel behavior. The built environment can actively promote walking and cycling behaviors. Therefore, experts advocate for land use policies that encourage moderate daily physical activity among residents by integrating land use planning, residential density, and urban design intersection density. Such policies facilitate the development of compact, mixed-use urban patterns [15,16], which enhance accessibility, yield public health benefits, and contribute to greenhouse gas emission reductions [17]. Meanwhile, researchers and policymakers are placing greater emphasis on equity and justice, advocating for the protection of individual rights, prioritization of vulnerable transportation groups, mitigation of inequality, and reduction of social exclusion effects.

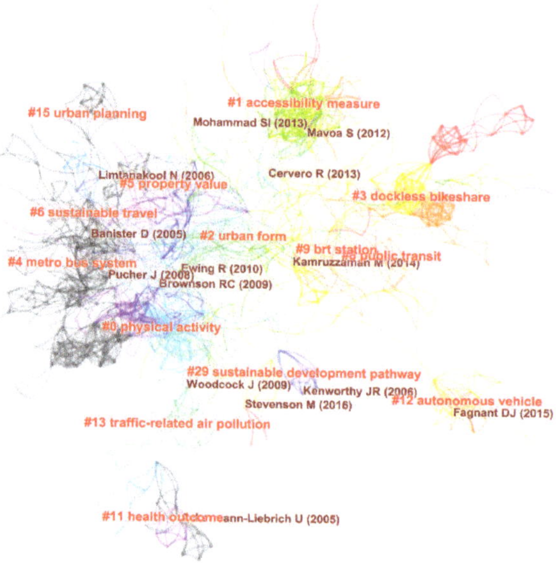

Figure 1. Cluster survey of 10,470 public transport and land use articles.

Table 2. Ten references with the highest co-citation intensity.

Author (Year)	Publication	Strength
Ewing R. (2010)	J AM PLANN ASSOC	37.48
Cao X.Y. (2009)	TRANSPORT REV	9.83
Saelens B.E. (2008)	MED SCI SPORT EXER	9.13
Woodcock J. (2009)	LANCET	8.68
Brownson R.C. (2009)	AM J PREV MED	8.6
Banister D. (2005)	UNSUSTAINABLE TRANSP	8.59
Pucher J. (2008)	TRANSPORT REV	8.16
Handy S. (2005)	TRANSPORT RES D-TR E	8
El-Assi W. (2017)	TRANSPORTATION	7.96
Frank L.D. (2010)	BRIT J SPORT MED	7.85

Table 3. Cluster analysis of public transportation and land use.

Cluster	Theme	Main Content
#0 physical activity	Active travel; environment feature; travel behavior; residential self-selection; environment characteristics; modal shift	Plan multifunctional land use combinations and design walkable communities and streets to reduce obesity risks (especially in the U.S.).
#1 accessibility measure	Public transport; urban area; transit accessibility; transportation sustainability; progressive development	Transportation policies should focus on the development of public transportation to improve accessibility to urban areas.
#2 urban form	Travel behavior; CO_2 emission; active transport; low-income population; national housing program	Urban development models should shift towards compact, low-carbon types, with a focus on low-income populations.
#3: dockless bikeshare	Bike-sharing usage; spatial variation travel mode; public bike sharing system demand	Bike-sharing systems meet spatially differentiated demands.
#4 metro bus system	Reliable transit service; transport mode choice; daily walking	Build a fast, reliable, and integrated public transportation system.
#5: property value	Town planning challenge; land use change; light rail	Public transportation leads to changes in land use and value.
#6 sustainable travel	Residential dissonance; transit use; neighborhood design characteristics structural equations model; transport modality	People's transportation choices; public transportation can reduce urban sprawl and promote the development of the TOD (transit-oriented development) model.
#8: public transit;	Spatial variation; ride-hailing service; smart card data; solving traffic congestion; dense Asian cities; traditional transport mean; complementarity pattern	The TOD development model addresses urban congestion, especially in densely populated Asian cities.
#13 traffic-related air pollution	Disease assessment; ambient nitrogen dioxide assessing greenhouse gas emission; health co-benefit; socioeconomic determinant	Evaluate the impact of vehicle emissions on health and society from a socioeconomic perspective.
#29 sustainable development pathway	Sustainable development planning; public participation	Advocate for energy conservation and environmental protection.

4. Researcher

The relationship between transportation and land use was first conceptualized by Mitchell and Rapkin (1954) in their seminal study [18]. Subsequently, Hansen (1959) examined Washington, D.C. as a case study and introduced the concept of reciprocal interactions between travel and location [19]. Since then, the interdependence of transportation and land use has garnered widespread academic recognition and sustained research interest. Over the past six decades, numerous interdisciplinary and cross-sectoral collaborations have sought to understand, integrate, and forecast the dynamics between transportation and land use. These efforts culminated in the establishment of the *Journal of Transport and Land Use*, a leading publication dedicated to this field. Scholars from the University of Minnesota's Transportation Research Center played a key role in co-founding the World Society for Transport and Land Use Research (WSTLUR).

In recent years, leading scholars specializing in public transportation and land use have included Robert Cervero, Reid Ewing, Michael Duncan, and Peter Calthorpe. Table 4 provides a summary of globally recognized experts whose contributions to transportation (including public transportation) and land use research are frequently cited in the literature. These experts come from diverse backgrounds, including urban planning, transportation research, statistics, public health, and governmental transportation agencies. Their contributions have been instrumental in advancing theories, analytical frameworks, evaluation methods, and modeling techniques in this field.

Table 4. Ten experts or scholars with outstanding contributions.

Name	Position	Contribution
William Alonso	Professor, Harvard University, United States	His book "*Location and Land Use*" is often considered the foundation of urban transportation economics.
Ira S. Lowry	Director, Policy Development and Research Office, U.S. Department of Housing and Urban Development	Proposed the Metropolis model, often regarded as the first to truly integrate land use and transportation.
Robert Cervero	Chair, Department of City and Regional Planning, UC Berkeley	Authority in urban public transportation and land use planning.
Reid Ewing	Distinguished Professor, City and Urban Land Research Institute, University of Utah	Advocate for transit-oriented design (TOD), authority in urban land and transportation research.
Peter Calthorpe	Architect, Urban Designer, and Urban Planner	Founding member of the Congress for the New Urbanism, and one of the advocates of TOD.
Kevin J. Krizek	Visiting Professor, Radboud University, Netherlands	Co-founder of the journal "*Transportation and Land Use*" and one of the founding members of the World Transportation and Land Use Research Society.
Lawrence D. Frank	Professor, University of California, San Diego	One of the first to quantify the links between the built environment, transportation, and health.
Robert L. Knight	Professor, Colorado State University	Highly cited scholar in the field of transportation and land use research.
Susan Handy	Distinguished Professor, University of California, Davis	Expert in the relationship between public transportation and land use, and strategies to reduce car dependency.
Michael Duncan	Professor, University of California	Expert in planning and design of residential, commercial, and public transportation projects.

5. Research Stages

An examination of the history of world transportation reveals that research on public transportation and land use has evolved through several distinct stages.

5.1. Pre-1950: The Pre-Research Stage

Before 1950, automobile ownership remained relatively low in most cities and regions worldwide, and travel primarily relied on walking, horse-drawn carriages, bicycles, and railroads. Consequently, academic attention to the interaction between public transportation and land use was minimal. The relatively slow modes of transportation also played a role in constraining disorderly urban expansion. London provides a representative example: in 1801, 87% of its 1.1 million residents lived in the inner city, which comprised only 20% of the total area (1500 square kilometers). Even by 1901, after the introduction of public transportation, 70% of London's 6.5 million residents still resided in the inner city.

5.2. 1950–1960: The Emergence of Public Transportation Research

Between 1950 and 1960, the first generation of research on public transportation and land use primarily examined changes in housing prices along transportation routes, relying largely on before-and-after comparative analyses. By the 1960s, researchers began incorporating various econometric models into their studies. Swedish scholar Alexander Paulsson analyzed how the expansion of Stockholm's metro system reshaped both local and regional transportation. He describes the gradual evolution of an integrated and collaborative planning process within Stockholm's public Metro system, which opened in 1950. His findings highlighted the role of metro expansion, intergovernmental cooperation, and state intervention in shaping the integration of public transportation and land use planning [20].

5.3. Post-1970: The Automobile Boom and the Shift in Urban Development

Following World War II, automobile ownership surged, and by the 1970s, private vehicles had become a dominant force in urbanization. At that time, public transportation was often perceived as a hindrance to economic development, as the convenience of automobiles facilitated mass suburbanization, accelerating urban sprawl. However, the 1980s marked a turning point. Amid rising environmental concerns and slowing economic growth, urban governance strategies shifted from managerialism to entrepreneurialism [21]. During this period, public transportation regained prominence in urban planning. Dejan Petkov analyzed the resurgence of tram systems in Western Europe during the 1980s, highlighting their role in shaping transit-oriented development policies [22].

5.4. 1973–1986: The Energy Crisis and the Rise of Integrated Research

The global oil crises between 1973 and 1986 triggered a surge of interest in the relationship between public transportation and urban form. Scholars such as Cervero, Knight, and Levinson conducted foundational research on household location choices, daily travel behavior, and transportation mode selection. These studies led to the development of predictive models and empirical analyses. A recurring finding was that public transportation played a crucial role in supporting low-income populations, as economically disadvantaged individuals living on urban peripheries were less reliant on private vehicles for daily commutes [20,23,24]. However, due to technological constraints, early research in this period lacked formal spatial data, limiting its analytical precision.

5.5. Post-2000: Advances in Empirical Research and Computational Methods

With the continued development of public transportation systems and the advent of sophisticated computational tools, post-2000 research benefited from the availability of extensive real-world datasets. Concurrently, econometric and spatial analysis methods matured, leading to a proliferation of studies utilizing panel data to examine the impact of public transportation on land development and property values (e.g., Boarnet and Haughwout [25]; Giuliano [26]).

During this stage, hedonic pricing models gained widespread adoption due to their simplicity and robustness in assessing the short- and medium-term effects of public transportation on real estate values (e.g., Debrezion et al. [27]; Mohammad and Graham [28]).

Empirical findings reinforced the notion of a strong bidirectional relationship between public transportation and land use, supported by case studies from multiple countries.

Boarnet and Haughwout found that metropolitan highways in the U.S. not only influenced land values but also reshaped population and employment distribution patterns [25]. Similarly, studies on Spain's high-speed rail system highlighted that accessibility and urban cohesion effects were more pronounced in major cities like Madrid and Seville compared to smaller urban centers. Albrechts et al. further emphasized high-speed rail's contribution to regional integration within Europe [29]. Notably, since 2008, China's rapid expansion of high-speed rail has generated a significant body of research on its socioeconomic and spatial impacts. Studies on the Chinese high-speed rail network have examined its role in housing price appreciation, economic growth, and regional connectivity. During this period, urban form variables, along with social and economic benefits, were increasingly incorporated into public transportation analyses.

5.6. Post-2010: The Maturity of Research and Emerging Critical Perspectives

As research continued to evolve, scholars began to identify unintended consequences of public transportation investments. For instance, Cook et al. observed that in some underdeveloped regions, transit development exacerbated environmental isolation, pollution, and noise, sometimes outweighing economic benefits [30]. This was particularly evident in economically disadvantaged areas of the United States and Latin America.

King conducted a landmark study on the causal relationship between transportation and land development, revealing that New York City's subway system did not necessarily lead to higher building densities in adjacent areas [31]. During this period, large-scale, long-term studies gained prominence, focusing on both the long-term benefits and potential drawbacks of public transportation networks. For example, Kasraian et al. analyzed public transportation projects across 22 U.S. cities, 21 European cities, and six Asian cities, revealing significant variations in outcomes across different urban contexts [32].

Simultaneously, a growing number of studies began shifting from macroeconomic factors to micro-level analyses of community residents' preferences, often employing surveys and interviews. Qualitative research in this domain flourished, particularly in Nordic countries. Scholars examined public attitudes toward transit-oriented development across cities of varying scales, from large metropolitan areas such as Oslo and Copenhagen to mid-sized cities like Aalborg (Denmark) and Gävle (Sweden), as well as small towns like Frederikshavn (Denmark) [33–35].

This period also saw a renewed emphasis on urban design principles, advocating for people-centric cities rather than merely viewing urban areas as economic hubs or transit corridors. Scholars such as McHarg and Deakin emphasized the importance of integrating transportation infrastructure with urban aesthetics to enhance the visual and functional quality of transit spaces [36,37]. The above simple summary is shown in Table 5.

Table 5. Characteristics of each stage of development.

Time Period	Characteristics	Keywords
1950–1986 Early Research Stage	Exploration of research theories, models, and methods	Urban expansion, public transport system, private bus system
1986–2010 Research Development Stage	Development of public transport providing empirical data and natural experimental schemes	Land use, land value, hedonic
2010–PresentResearch Maturity Stage	Focus not only on economics but also on reflecting the human, environmental, and health impacts of transportation development	Efficiency and effectiveness, ecology and environment
Future Research	Large-scale, long-time series research gaps need to be filled, with a greater focus on social equity and human issues	Spatialization, equity

6. Research Theories

The research field of public transportation and land use draws upon theories and concepts from a wide range of disciplines, including economics, psychology, geography, ecology, and engineering. A review of the literature reveals that micro-scale research mainly relies on utility theory (which explores how public transportation and land use affect individual travel choices); meso-scale research primarily draws on urban economic theory (which examines the impact of public transportation on land use, accessibility, and value); and macro-scale research largely employs human–earth system theory (which explores the coordinated development of public transportation and the Earth's environment).

6.1. Utility Theory

Utility refers to the decision-making process influenced by personal cognition and preferences, as well as social, economic, and environmental factors, wherein individuals assess the potential benefits and costs. In essence, utility reflects decision-makers' differing attitudes toward risk and reward, with higher risks generally accompanied by higher potential returns. In personal travel activities, decision-makers face uncertainty, such as choosing departure times, activities, destinations, transportation modes, and routes [38].

Utility theory can be applied to travel behavior and the benefits gained from selecting different transportation systems. Traditional utility theory assumes that decision making is based on absolute rationality and perfect information, yet it fails to account for psychological influences on travel decisions [39,40]. To address these limitations, some studies have incorporated theories of decision making under uncertainty, which treat such choices as analogous to a lottery-like process [41]. Under conditions of uncertainty, individual choices are probabilistic [38].

Both traditional utility theory and uncertainty-based decision-making theory focus on micro-level or individual-level analysis, utilizing mathematical models such as logistic regression. Subsequently, three standard decision theories under uncertain conditions were applied to public transportation and land use studies: expected utility theory [42], prospect theory [43], and regret theory [44]. The key principle underlying these theories is that decisions are based on the perceived value of potential gains and losses rather than final outcomes, with heuristic approaches employed to assess these gains and losses.

6.2. Urban Economic Theory

Emerging in the 1960s, urban economic theory occupies a middle ground between microeconomics, which focuses on individuals and markets, and macroeconomics, which examines national economies. It primarily seeks to understand the internal organization, spatial structure, and economic relationships within cities, addressing key urban economic issues such as housing, land use, transportation, and employment. This theory aims to enhance urban economic efficiency through both quantitative and qualitative analyses of overall economic benefits, land use benefits, and economies of scale [45].

Traditional economic theory initially used labor time as the sole measure of value, with the labor theory of value serving as the foundation of classical economics. It argues that economic savings can only be achieved through reductions in labor time. However, as economic development progressed, spatial factors such as location, mobility, and direction were recognized as integral to economic growth. Location theory subsequently became a key component of economic geography, regional science, and spatial economics.

In 1826, Johann Heinrich von Thünen, in his work *Der isolierte Staat* (*The Isolated State*) [46], introduced the Thünen model, which demonstrated how market accessibility influences the spatial organization of agricultural land. According to this model, transportation costs determine land rent: when transportation costs are low, location rent is high, and vice versa. This creates a rent gradient, where location rent decreases with distance from the market and eventually reaches zero.

In 1909, German location economist Alfred Weber, in his work *Über den Standort der Industrien* (*On the Location of Industries*) [47], developed the theory of industrial location, which he later refined in 1929. Weber's theory posits that the relative weight of raw materials and finished goods is a critical determinant of transportation costs and production site selection. When the production site is located near the source of raw materials and there is no significant quality loss during the production process, the total transportation cost will be lower when it is near the market.

Expanding on the Thünen model, William Alonso introduced accessibility theory in his 1964 work *Location and Land Use: Toward a General Theory of Land Rent* [48] to explain urban land use dynamics. He applied accessibility principles to different urban land uses (residential, commercial, and industrial). According to his theory, all households seek to maximize land occupancy while maintaining their required level of accessibility. Because land on the urban periphery is more affordable, wealthier households, with lower demand for central accessibility, tend to reside in suburban areas. In contrast, lower-income households require easier access to the city center to enhance their competitiveness in business or industry. While Von Thünen, Weber, and Alonso were not the sole contributors to location theory, they laid the foundation for its further development by geographers, economists, and regional scientists.

6.3. Human–Earth System Theory

Philosopher Immanuel Kant, adopting a dualistic philosophical approach, conceptualized geography as an analysis of human activities in time and space—an idea that later became fundamental to modern geographic thought. All human activities, including public transportation and land use, are shaped by various layers of the Earth's system, such as the atmosphere, hydrosphere, and biosphere.

In turn, human social and economic activities exert direct or indirect influences on the atmosphere (e.g., air pollution, greenhouse gas emissions, ozone depletion), lithosphere (e.g., mineral depletion, desertification, soil degradation), and biosphere (e.g., deforestation, biodiversity loss) [49,50]. As an extension of general systems theory [51], human–earth system theory provides a framework for understanding outcomes as complex adaptive systems, characterized by multiple interacting components and a state of perpetual disequilibrium.

7. Research Models

Transportation and land system models have long been central to research, leading to the development of various models over the years. Many of these models are applicable to studies on the relationship between public transportation and land use. Most are grounded in the aforementioned theories and are typically employed to simulate the interaction between transportation systems and land use, aiming to optimize travel efficiency and large-scale spatial development [52–56]. The land use system provides population and employment locations that generate travel demand in public transportation models, while the public transportation system, in turn, defines accessibility and travel times, which influence land use models in determining population and employment distributions.

The first generation of Land Use–Transportation Interaction (LUTI) models emerged in the late 1960s, primarily designed to enhance transportation models and mitigate congestion. Notable early models include Lowry's Metropolis model (1964), the PLUM model (1972), and the Lowry–Wilson model (1970). During the same period, the gravity/space interaction (SI) method gained popularity. Inspired by Newton's law of gravitation and empirical analyses of human spatial interactions, Lowry introduced the Metropolitan model for Pittsburgh, further advancing the SI method. Although Lowry's gravity-based approach was simple and easy to apply, it lacked a solid theoretical foundation [57]. Wilson addressed this limitation by providing a more rigorous theoretical framework for the SI method using entropy maximization [58]. Within this framework, interactions between activity zones could be computed using dual constraint, origin constraint, destination constraint, or unconstrained models. However, these models still exhibited several shortcomings, including mechanistic assumptions, theoretical gaps, high data requirements, and computational complexity [59].

By the mid-1990s, advancements in econometrics and behavioral economics, particularly random utility theory, enabled the prediction of choices among multiple discrete alternatives. This led to the development of the second generation of LUTI models [60,61]. These models emphasized theoretical soundness and methodological robustness, prioritizing flexibility, transparency, usability, reliability, scalability, and explanatory power. Over the past 60 years, interdisciplinary research and professional collaboration have contributed significantly to the evolution of operational LUTI models for assessing land use decisions and large-scale transportation investments. A major challenge in LUTI modeling, however, remains the integration of land use and transportation models [62,63]. Yin et al. [64] pointed out that LUTI models still struggle to fully capture the intricate interdependencies among land use, transportation, economic systems, and environmental factors, limiting their effectiveness in urban development planning and analysis.

In the past decade, fueled by advancements in both transportation and land use research, along with rapid progress in computing technologies, LUTI modeling has entered a new era of broader and more comprehensive applications [65–67]. A significant milestone in this evolution is UrbanSim, developed in 2012, which integrates urban simulation, 3D visualization, and open data-sharing functionalities. While UrbanSim extends the microsimulation tradition of land use/transportation modeling, it also builds upon the work of Wegener, Marsden, and Macett. As an open-source and modular system, UrbanSim leverages highly disaggregated data for dynamic simulations, marking a substantial leap forward from earlier LUTI models.

Beyond technological advancements, scholars have increasingly emphasized the role of governance policies in shaping land use and transportation outcomes [68]. However, most of the existing studies tend to focus on achieving singular policy objectives rather than exploring integrated multi-objective strategies [69]. Miller [70] argued that technological and behavioral paradigms are undergoing radical transformations, necessitating the development of Integrated Urban Models (IUMs) as policy analysis tools. He advocated for a shift beyond conventional LUTI frameworks to address potential disruptive forces, environmental challenges (e.g., climate change, natural disasters), public health crises (e.g., pandemics), and socio-demographic shifts (e.g., migration, aging populations).

Hawkins and Nurul Habib [71] further questioned whether the LUTI framework remains suitable for emerging trends such as peak car usage, shared mobility, and the impact of information and communication technology (ICT) on travel behavior. They noted that LUTI models were primarily developed during periods of simplified travel decision making and may struggle to accommodate the complexity of contemporary transportation dynamics. The next generation of LUTI models will need to incorporate factors such as declining population densities, the transformation of retail and service industries, ICT-driven mobility changes, and sustainable transportation policies (e.g., bicycle infrastructure investments). Current LUTI models are not yet fully equipped to analyze these trends comprehensively, nor do they provide sufficient empirical research on their long-term dynamic effects.

Given the diversity of LUTI models—each designed for different research purposes and embedded within distinct theoretical frameworks—their technical complexity and specialized terminology often pose challenges for interdisciplinary understanding [72–74]. As Donnelly et al. [75] pointed out, listing all LUTI models comprehensively is impractical. However, based on their underlying theories and scales, these models can be broadly categorized into four types (Table 6).

Table 6. Land use–transportation interaction (LUTI) model.

Model Category	Characteristics	Representative Models (Developer and Time)	Advantages
Spatial Interaction Model	Reflects the interactions between land use, transportation, economy, and environment; comprehensive, holistic, and hierarchical; supports urban development planning and analysis, but may oversimplify problems and fail to capture the essence of spatial interactions.	Lowry Model (Lowry, 1964)	Considered the first computer model to truly integrate land use and transportation.
		Projective Land Use Model (PLUM) (Goldner, 1972)	Based on the Lowry derivative model's sequential decision-making process.
		Lowry–Wilson Model (Wilson, 1970)	Extended the Lowry model using entropy maximization principles.
		Leeds Integrated Land Use and Transport (LILT) (Model Macett, 1983)	LILT incorporates Lowry–Wilson framework with travel demand model stages to explain traffic congestion effects.
		Integrated Transportation Land Use Package (ITLUP) (Putman, 1983)	One of the most widely used models.
		UrbanSim Model Paul (Waddell, 2012)	Extends Wegener and Macett models, integrating interactions between land use, transportation, economy, and environment, with 3D visualization for alternative planning.
		IUM Comprehensive Urban Model (Miller, 2018)	Focuses on governance policies for climate change, disasters, public health crises, and demographic changes.
Random Utility Model	Deals with the relationship between transportation and land use based on decision-makers' utility maximization; relates individual behavior to utility functions, effectively explaining the link between location features and traveler behavior.	Quigley Model (Quigley, 1976)	Early application of logit models to examine transportation and land use interactions.
		Lerman Model (Lerman, 1977)	Integrates and nests logit models.
		Anas Model (Anas, 1981)	Applies Random Utility Model to study combined travel modes and location choice.
Mathematical Programming Model	Aims to optimize urban spatial structure by minimizing costs or maximizing benefits; focuses on forecasting future outcomes rather than merely describing a spatial distribution process.	Technique for the Optimal Placement of Activities in Zones (TOPAZ) (Brotchie, 1969)	Developed as an overall planning tool, the model allocates activities to locations based on maximizing net benefits of spatial interactions and land use.
		Projective Optimisation Land Use Information System (POLIS) (Prastacos, 1985)	Redefines PLUM model from a mathematical programming perspective.
		Oppenheim Model (Oppenheim, 1993)	The location cost in the Oppenheim model is endogenously determined, directly related to opportunity costs, allowing for endogenous location and transport cost calculations.
Urban Economics Model	Considers the relationship between transportation and land use at the urban economic meso-scale; data and system parameter requirements are low, but the determination of coefficients can be subjective, influencing evaluation results.	Alonso Model (Alonso, 1965)	Addresses the relationship between location value and land use, offering a modified version of the classical consumer equilibrium theory, consisting of accessibility and location price.
		Rosen Model (Rosen, 1974)	Integrates hedonic theory, emphasizing the inherent heterogeneity of locations.
		Ellickson Model (Ellickson, 1981)	Direct specification and estimation of hedonic models.
		5-Stage Land Use Transport Model (5-LUT) (Martínez, 1992)	Links travel with socio-economic variables.

8. Research Hotspots

By analyzing the 29 research hotspot clusters identified through CiteSpace V.5.8.R3 software, several key themes frequently emerge in the literature: safety and health, equity and value, environment and energy conservation, and transit-oriented development (TOD) and accessibility.

8.1. Safety and Health (#0, #2, #8...)

Land use patterns significantly influence travel frequency, duration, destinations, and modes of transportation, all of which, in turn, impact safety and health [76–78]. Aditjandra et al. [79], among others, argue that higher land use intensity increases the likelihood of traffic accidents, as collisions are more frequent in urban areas compared to suburban or rural regions. However, due to lower speed limits and stricter regulations in urban settings, the severity of accidents is generally lower. Despite this, implementing road safety policies in urban development faces numerous challenges, including institutional inconsistencies, limited administrative capacity, weak law enforcement, and a lack of accountability. Consequently, essential pedestrian infrastructure such as sidewalks and plazas is often either missing or misappropriated as parking spaces, exposing pedestrians to hazardous traffic conditions. Designing pedestrian-friendly urban environments requires interdisciplinary improvements across fields such as urban aesthetics, spatial planning, public health, and psychology.

Ewing et al. [7] highlight that urban sprawl reduces walking and cycling, negatively affecting physical health. Additionally, air pollution and traffic noise further contribute to public health issues. The urban development model adopted by a city can have profound effects—both positive and negative—on public safety and health. This perspective is widely acknowledged in academia, and numerous models have been developed to analyze and predict individual travel behavior. However, some studies suggest that the causal link between land use and travel patterns remains weak (e.g., Cervero and Landis, [1]).

Streets, as one of the most critical forms of public space, not only support transportation but also serve as hubs for social interactions among urban residents. Strengthening road safety and improving accessibility are essential urban planning priorities. Research in this area increasingly focuses on reducing private vehicle dependency and mitigating public health concerns such as rising obesity rates.

8.2. Equity and Value (#2, #3, #5, #6. . .)

Equity is a complex and multifaceted concept. As Tomaney and Marques [80] point out, it is difficult to determine whether public transportation truly enhances equity within a region. There is no doubt that private transportation has disproportionately impacted economically disadvantaged communities and groups. Traditional transport planning has often prioritized improving conditions for private vehicles at the expense of pedestrian and bicycle infrastructure. Yet, the majority of people rely on walking, cycling, or public transportation for their daily commutes. More than 2 billion people worldwide use public transport daily, most of whom are low-income workers [81]. Urban transport policies should not sacrifice the interests of the majority for the convenience of a privileged few. The increasing number of private cars not only exacerbates traffic congestion but also reflects a widening social divide. Enhancing the convenience and accessibility of public transport and walkable environments remains one of the most effective strategies for providing economic opportunities to low-income groups at minimal cost [82,83].

Equity is inherently linked to geography, as public transport infrastructure and land use policies influence employment distribution and housing patterns, potentially leading to spatial segregation of wealth [84–86]. Investments in infrastructure, education, and healthcare often concentrate in high-income areas, creating disparate opportunity costs [87]. To promote equitable urban development, policymakers must work to mitigate these disparities.

Thomopoulos [88] used accessibility indicators to evaluate the effects of social exclusion, revealing that pedestrians and cyclists often derive few benefits from urban expansion while bearing its costs—such as the loss of green spaces, increased pollution, and heightened accident risks. Additionally, low-income communities face numerous disadvantages from urban sprawl, including reduced access to public services, employment centers, and social opportunities. Stevens [89] analyzed socioeconomic patterns in U.S. cities and found that older urban neighborhoods tend to experience higher poverty rates and related social issues, while newer, more spacious developments are generally wealthier, safer, and healthier. The concentration and isolation of vulnerable groups exacerbate social inequalities and limit economic mobility, reinforcing cycles of disadvantage. Government agencies should develop effective policy instruments to address the gap between the rich and the poor, reconcile the conflict between development and conservation, and coordinate resource allocation—ultimately achieving the goal of integrated development across economic, environmental, and social dimensions.

One of the most frequently debated questions in this field is: "Does public transport investment contribute to land value growth?" Henriksson and Summerton [90] reviewed literature on high-speed rail (HSR) and regional economic development, concluding that the impact of HSR remains inconclusive—its benefits depend on specific local contexts and require deliberate land use planning and policy interventions to maximize economic gains. Some studies suggest that public transit can increase surrounding land and real estate values, but the effect is gradual (e.g., Knight and Trygg, [23]; Boarnet and Haughwout, [25]; Giuliano, [26]; Cervero and Duncan, [91]; Deng, [92]). Conversely, in low-density areas, public transport may actually decrease property values due to noise, pollution, or other externalities.

Several studies from China highlight the economic benefits of high-speed rail connectivity. Guan and Peiser [93] applied principal component analysis (PCA) to assess neighborhood characteristics and subway accessibility in Shanghai's Pudong District, finding a positive correlation between metro infrastructure and property prices. Similarly, Chong et al. [94] used spatial econometrics on panel data from 268 Chinese cities (2008–2015), confirming that improved connectivity from HSR boosts GDP per capita. Pan [95] applied OLS, multilevel regression, and spatial econometric models to evaluate Houston's Metro Rail system, identifying significant positive effects on residential property values. Such studies frequently employ quantitative analysis, case studies, and regression modeling to examine the relationships between public transportation and economic outcomes [27,28].

8.3. Environment and Energy Saving (#8, #13, #29...)

For decades, integrating land use planning with public transport systems has been regarded as a key strategy for sustainable mobility. Transportation remains one of the largest contributors to carbon emissions and environmental degradation [96,97]. The International Energy Agency (IEA) warns that without strong and sustained mitigation policies, global transportation emissions could surpass all other energy sectors, potentially reaching 12 Gt CO_2-equivalent by 2050 [98]. The rapid increase in private vehicles has not only intensified urban air pollution and greenhouse gas emissions but has also contributed to rising traffic accidents and public health crises.

Private vehicle dominance reduces the external benefits of urban green spaces, such as wetlands, agricultural lands, and forest parks, which improve air and water quality, support biodiversity, and enhance food production [99–101]. Additionally, it heightens climate risks, increasing vulnerability to landslides, urban heat islands, water pollution, and flooding [65]. Therefore, countries such as the United States have implemented rigorous environmental policies to regulate the construction and operation of transportation infrastructure, aiming to balance human and ecological interests. Systematic environmental impact assessments (EIAs) are essential for ensuring that transport projects align with sustainable development goals.

The Tyndall Climate Change Research Centre in the UK employs GIS-based integrated land use and transport models to analyze climate risks under different spatial planning scenarios, urging urban planners to adopt balanced, climate-resilient strategies [102]. Ford and Airhihenbuwa [103] projected that future urban policies addressing climate change and rising energy prices—such as carbon taxes, emissions trading, and anti-sprawl regulations—will necessitate a shift in transportation investments toward public transit and non-motorized mobility solutions. Because public transportation remains the most efficient and sustainable mode for moving large numbers of people, it offers significant energy savings compared to private cars and aviation while also reducing congestion, improving urban air quality, and supporting compact city development through agglomeration effects.

In the face of expanding transportation infrastructure and mounting pressure to protect the ecological environment, proactive forecasting, scientific planning, and coordinated regulation may be the most effective means of promoting sustainable development and efficient land use. Such approaches can also help minimize the destruction of natural resources. Future decision making should take into account the high ecological externality costs and long-term maintenance expenses of projects, and should encourage the remediation, consolidation, and withdrawal of unnecessary land use, along with planned demolition and restoration to ecological land.

8.4. TOD and Accessibility (#2, #6, #8...)

Transit-oriented development (TOD) is a widely recognized approach in modern urban transportation and land use planning. TOD fosters livable cities and interconnected communities by strategically locating development around transit stations, thereby reducing car dependency. Public transport stations serve as key urban hubs for commercial, retail, social activities, and housing, promoting walkability and minimizing the need for private car ownership. Spaces previously designated for parking are increasingly being converted into green leisure areas. The expansion of public transport infrastructure is accelerating sustainable land use planning and strengthening the integration of transportation and land use policies. According to the United Nations Human Settlements Programme report, "A fair and sustainable city exhibits the following spatial characteristics: high-density, low-rise, mixed-use development, public transport-centered mobility, and integrated open space systems that prevent urban sprawl" [104]. TOD embodies these principles. For decades, urban planners and designers have advocated for better integration of transportation and land use to enhance sustainability and urban livability [6].

Peter Calthorpe, a prominent advocate of New Urbanism and TOD, was inspired by early studies on tram-based connections between suburbs and city centers. He argued that to mitigate the environmental impacts of urban expansion and foster a sense of community, cities should adopt an internal structure characterized by "a walkable environment with mixed housing, retail, offices, open spaces, and public facilities, enabling residents and workers to commute conveniently via bus, bicycle, walking, or car" [6]. Ian Canton provides a nuanced distinction between TOD and development-oriented transport (DOT): the former prioritizes urban development around transit hubs, while the latter emphasizes transportation planning as a means of guiding urban and regional expansion [105]. Report by Moore [106] highlight that TOD has gained significant traction in automobile-centric nations such as the United States, Australia, and Canada. Further, TOD extends to broader spatial planning concerns.In countries like Sweden, Denmark, and Finland, where robust transportation agencies coordinate urban development, TOD principles are seamlessly integrated into broader planning frameworks. Sweden's transport corridor planning and Denmark's "finger plan" share similarities with TOD, emphasizing high-quality public transit connections to central transport nodes. However, while Swedish corridor planning primarily focuses on public transportation.

As TOD gains prominence, studies assessing its benefits have emerged. Dittmar [107] evaluated TOD projects based on key factors such as planning, zoning, financing, and parking regulations, ultimately defining "high-benefit TOD" as developments characterized by accessibility, flexible residential–commercial land use integration, land value capture, and industrial agglomeration benefits. However, some scholars and practitioners caution that not all developments near transit stations qualify as TOD. Some may be poorly connected to transit networks, while others may inadvertently hinder accessibility—such as car-dependent retail centers, warehouses, and low-density gated communities.

The concept of accessibility lies at the intersection of transportation and land use and can be achieved through improved mobility or proximity. People seek accessibility in various domains, including workplaces, schools, commercial areas, and healthcare services. In practice, accessibility is assessed from multiple perspectives, including locational accessibility [108–110], personal accessibility [111,112], and economic accessibility. Regardless of the specific approach, four fundamental components define accessibility: land use, transportation, time, and the individual [113]. Accessibility measurement has been a critical focus in transportation and land use modeling since the 1950s. Scholars such as Levinson [114], Cervero et al. [115], Cervero and Duncan [116], Klonsky and Glenn [117], Melia et al. [118], and Steiner [119] have developed various methodologies to quantify accessibility. Rye and Hrelja [120] further argue that increasing urban density and enhancing public transportation infrastructure are essential strategies for improving accessibility.

9. Conclusions

The importance of public transportation and land use research has been widely recognized. Scholars have explored and practiced this issue from multiple perspectives, including economics, the environment, and society, making significant progress. The previous research has laid a solid foundation for future systematic studies. However, due to the complexity of the subject and current technological limitations, there are several areas that warrant further exploration.

9.1. Continuing to Strengthen the Fundamental Theoretical Research on Public Transportation and Land Use

Currently, much of the research on public transportation and land use focuses on economic aspects, such as regional housing prices, land values, or urban GDP. The interaction between public transportation and land use spans multiple disciplines, and further exploration from various perspectives is necessary. This would significantly enrich and refine the theoretical framework of public transportation and land use, thereby providing a solid theoretical foundation. Furthermore, traditional theoretical feedback on transportation and land use is insufficient to address emerging global challenges such as global warming, climate change, ecological destruction, social conflicts, and technological advancements. The balance between economic development and environmental protection requires responses from public transportation and land use. Future research should integrate insights from fields like public administration, engineering, geography, ecology, biology, psychology, and resource science, offering new perspectives to the field.

9.2. Focusing on Large-Scale Public Transportation and Land Use Research

Public transportation and land use exhibit varying characteristics and conclusions at different temporal and spatial scales. Some studies suggest a positive impact of public transportation, while others report weak or negative effects. These contradictions are often attributed to the political and economic conditions of the regions where case studies are conducted. Due to limitations in technology and data accessibility, most studies focus on medium-scale spatial and shorter temporal dimensions, with relatively simple and isolated cases. However, comprehensive research on large-scale and long-term series remains scarce, making it challenging to grasp the dynamic characteristics and evolution mechanisms of public transportation and land use as a whole. With advancements in high-resolution remote sensing technologies and big data analysis, future studies will benefit from large-area synchronized data, allowing for unified large-scale and long-term research.

9.3. Innovating Public Transportation and Land Use Evaluation Methods and Models

Public transportation and land use involve complex dimensions such as economics, the environment, society, politics, sports, culture, and even military factors, all of which influence one another. However, including too many dimensions can complicate models and make research work more difficult. Therefore, the focus should be on selecting the most representative and important factors to avoid redundancy. Developing mature, controllable, and replicable evaluation methods is crucial. The combination of multi-temporal remote sensing data, experimental analysis, and statistical survey data offers new ideas and pathways for future research. Geographic big data, Artificial Intelligence (AI), and 3S technologies provide spatial statistical analysis support, enabling the visualization of feedback loops between public transportation and land use. Therefore, future research should strengthen the integration of mathematical models and spatial technologies, incorporating new technological methods to support the related research between public transportation and land use, providing scientific evidence for decision making and planning.

9.4. Coordinating and Optimizing Public Transportation and Land Use

The development of public transportation has become the most important policy tool driving national economic growth in the post-industrial era. Research generally agrees that investing in transportation trunk lines and network integration can drive land space optimization and rehabilitation, leading to economic growth through labor market expansion. A systematic understanding of the relationship between public transportation and land use, as well as strategies to address future challenges, will provide decision-makers and investors with theoretical foundations for utilizing and protecting resources. For issues related to economic development, environmental protection, and social equity in public transportation and land use, future research must explore how to optimize and adjust existing public transportation routes from economic, environmental, and social perspectives. Especially in the current context of globalization, humanity is facing new transboundary challenges such as food security, ecological degradation, energy scarcity, and land degradation.

10. Discussion and Future Directions

Currently, a substantial body of literature has investigated the causal mechanisms between public transportation and land use. Future research, however, needs to place greater emphasis on the external effects of their interaction, particularly those related to economic benefits, environmental protection, and social equity. In line with the United Nations (IAEG-SDGs) "2030 Agenda for Sustainable Development" indicator framework, future efforts should aim to coordinate the economic, environmental, and social dimensions to promote integrated development.

10.1. Economic Dimension

The development of public transportation and land use generally exerts a positive economic impact. High passenger and freight throughput and improved accessibility encourage land use to become more intensive and efficient, thereby promoting economic growth. Negative economic impacts are mostly attributable to environmental degradation that results in economic losses. Developing high-quality, reliable, sustainable, and resilient infrastructure is a core task under Sustainable Development Goal (SDG) 9.1 of the 2030 Agenda.

10.2. Environmental Dimension

The environmental dimension primarily concerns the negative impacts of transport and land use interactions. The expansion of transportation infrastructure imposes considerable pressure on ecological conservation, especially in tropical and subtropical regions, which face severe challenges such as ecosystem fragmentation, illegal activities, land degradation, loss of biodiversity, water pollution, and increased carbon emissions. Protecting ecosystems and responding to climate-related disasters are key tasks under SDGs 13 and 15.

10.3. Social Dimension

From a social perspective, the focus lies on equity issues arising from imbalanced regional development. The essence of development should be to improve the quality of life and health standards for all individuals, including addressing the needs of vulnerable groups and promoting shared prosperity, which is a true reflection of equity. Ensuring equitable development is the central task of SDG 11.2.

In summary, studies on transportation and land use should not only focus on direct or localized impacts but must also consider indirect and cumulative effects. A development strategy that blindly seeks to promote economic growth through the expansion of transportation projects may encounter a series of economic, environmental, and social challenges. With the advancement of technology and the enrichment of geospatial big data, decision-making processes should comprehensively account for multidimensional impacts. Proactive strategic planning is necessary to ensure that transportation networks support both passenger and freight mobility, maximize public benefits, promote the efficient and intensive use of every piece of land, and minimize the destruction of natural resources, thereby achieving the goals of sustainable development.

Author Contributions: Conceptualization, B.L.; formal analysis, Y.H.; investigation, D.W.; resources, B.L. and D.W.; data curation, Y.H.; writing—original draft, Y.H.; writing—review and editing, C.X.; visualization, Y.H.; supervision, B.L.; funding acquisition, B.L., Y.H. and D.W. All authors have read and agreed to the published version of the manuscript.

Funding: This research study was supported by the project National Key Research and Development Program of China (2022YFB3903704), Guangdong Basic and Applied Basic Research Foundation (2025A1515011346).

Data Availability Statement: The original contributions presented in the study are included in the article, further inquiries can be directed to the corresponding author.

Conflicts of Interest: The authors declare no conflicts of interest.

References

1. Cervero, R.; Landis, J. Twenty years of the Bay Area Rapid Transit system: Land use and development impacts. *Transp. Res. Part A Policy Pract.* **1997**, *31*, 309–333. [CrossRef]
2. He, Y.; Wu, D.; Li, S.; Zhou, P. Ecological and Environmental Risk Warning Framework of Land Use/Cover Change for the Belt and Road Initiative. *Land* **2024**, *13*, 1281. [CrossRef]
3. Handy, S.; Cao, X.; Mokhtarian, P. Correlation or causality between the built environment and travel behavior? Evidence from Northern California. *Transp. Res. Part D Transp. Environ.* **2005**, *10*, 427–444. [CrossRef]
4. Laurance, W.F.; Clements, G.R.; Sloan, S.; O'connell, C.S.; Mueller, N.D.; Goosem, M.; Venter, O.; Edwards, D.P.; Phalan, B.; Balmford, A.; et al. A global strategy for road building. *Nature* **2014**, *513*, 229–232. [CrossRef] [PubMed]
5. He, Y.; Liu, X.; Sun, E.; Wu, D. Balancing economic growth and environmental conservation through the optimization of rail transit routes for sustainable development. *Sci. Rep.* **2024**, *14*, 31926. [CrossRef]
6. Calthorpe, P. *The Next American Metropolis: Ecology, Community, and the American Dream*; Princeton Architectural Press: Princeton, NJ, USA, 1993.
7. Ewing, R.; Meakins, G.; Hamidi, S.; Nelson, A.C. Relationship between urban sprawl and physical activity, obesity, and morbidity–Update and refinement. *Health Place* **2014**, *26*, 118–126. [CrossRef]

8. Sun, D.J.; Ding, X. Spatiotemporal evolution of ridesourcing markets under the new restriction policy: A case study in Shanghai. *Transp. Res. Part A Policy Pract.* **2019**, *130*, 227–239. [CrossRef]
9. Sun, D.J.; Zhang, K.; Shen, S. Analyzing spatiotemporal traffic line source emissions based on massive didi online car-hailing service data. *Transp. Res. Part D Transp. Environ.* **2018**, *62*, 699–714. [CrossRef]
10. Litman, T. Evaluating transportation land use impacts. *World Transp. Policy Pract.* **1995**, *1*, 9–16. [CrossRef]
11. Litman, T.; Steele, R. *Land Use Impacts on Transport*; Victoria Transport Policy Institute: Victoria, BC, Canada, 2017.
12. He, Y.; Wu, D.; Liu, Y. Review on the Impacts of Public Rail Transport on Urban Land Use Change: A Citespace-based Quantitative Analysis. *Adv. Earth Sci.* **2018**, *33*, 1259.
13. Chen, M.C.; Song, M. Visualizing a Field of Research: A Methodology of Systematic Scientometric Reviews. *PLoS ONE* **2019**, *14*, 1–25. [CrossRef] [PubMed]
14. Chen, M.C. Science Mapping: A Systematic Review of the Literature. *J. Data Inf. Sci.* **2017**, *2*, 1–40. [CrossRef]
15. Saelens, B.E.; Susan, L.; Handy, S. Built environment correlates of walking: A review. *Med. Sci. Sports Exerc.* **2008**, *40*, S550. [CrossRef]
16. Ewing, R.; Cervero, R. Travel and the built environment: A meta-analysis. *J. Am. Plan. Assoc.* **2010**, *76*, 265–294. [CrossRef]
17. Bassolas, A.; Barbosa-Filho, H.; Dickinson, B.; Dotiwalla, X.; Eastham, P.; Gallotti, R.; Ghoshal, G.; Gipson, B.; Hazarie, S.A.; Kautz, H.; et al. Hierarchical organization of urban mobility and its connection with city livability. *Nat. Commun.* **2019**, *10*, 4817. [CrossRef]
18. Mitchell, R.B.; Rapkin, C. *Urban Traffic*; Columbia University Press: New York, NY, USA, 1954.
19. Hansen, W.G. How accessibility shapes land use. *J. Am. Inst. Plan.* **1959**, *25*, 73–76. [CrossRef]
20. Paulsson, A. The city that the metro system built: Urban transformations and modalities of integrated planning in Stockholm. *Urban Studies* **2020**, *57*, 2936–2955. [CrossRef]
21. Harvey, D. Social processes and spatial form: An analysis of the conceptual problems of urban planning. *Pap. Reg. Sci. Assoc.* **1970**, *25*, 46–69. [CrossRef]
22. Petkov, D. The uneven development path of Bulgarian trolleybus transport–Leading back to the future? *Case Stud. Transp. Policy* **2020**, *8*, 1383–1392. [CrossRef]
23. Knight, R.L.; Trygg, L.L. Evidence of land use impacts of rapid transit systems. *Transportation* **1977**, *6*, 231–247. [CrossRef]
24. Cervero, R.; Seskin, S. *An Evaluation of the Relationships between Transit and Urban Form*; Transportation Research Board: Washington, DC, USA, 1995; p. 44.
25. Boarnet, M.G.; Haughwout, A.F. *Do Highways Matter? Evidence and Policy Implications of Highways' Influence on Metropolitan Development*; Working Papers; University of California Transportation Center: Berkeley, CA, USA, 2000.
26. Giuliano, G. Land use impacts of transportation investments. *Geogr. Urban Transp.* **2004**, *3*, 237–273.
27. Debrezion, G.; Pels, E.; Rietveld, P. The impact of railway stations on residential and commercial property value: A meta-analysis. *J. Real Estate Financ. Econ.* **2007**, *35*, 161–180. [CrossRef]
28. Mohammad, S.I.; Graham, D.J.; Melo, P.C.; Anderson, R.J. A meta-analysis of the impact of rail projects on land and property values. *Transp. Res. Part A Policy Pract.* **2013**, *50*, 158–170. [CrossRef]
29. Albrechts, L.; Healey, P.; Kunzmann, K.R. Strategic spatial planning and regional governance in Europe. *J. Am. Plan. Assoc.* **2003**, *69*, 113–129. [CrossRef]
30. Cook, D.A.; Levinson, A.J.; Garside, S.; Dupras, D.M.; Erwin, P.J.; Montori, V.M. Instructional design variations in internet-based learning for health professions education: A systematic review and meta-analysis. *Acad. Med.* **2010**, *85*, 909–922. [CrossRef]
31. King, D. Developing densely: Estimating the effect of subway growth on New York City land uses. *J. Transp. Land Use* **2011**, *4*, 19–32. [CrossRef]
32. Kasraian, D.; Maat, K.; Stead, D.; van Wee, B. Long-term impacts of transport infrastructure networks on land-use change: An international review of empirical studies. *Transp. Rev.* **2016**, *36*, 772–792. [CrossRef]
33. Rye, M.S.; Loiacono, D.M.; Folck, C.D.; Olszewski, B.T.; Heim, T.A.; Madia, B.P. Evaluation of the psychometric properties of two forgiveness scales. *Curr. Psychol.* **2001**, *20*, 260–277. [CrossRef]
34. Tillberg, K. *Families with Children's Daily Leisure Trips in the Car Society—A Time Puzzle with Geographical and Gender Variations*; Uppsala University: Uppsala, Sweden, 2001.
35. Riggs, W.; Shukla, S. Exploring the benefits of minimobility in the urban context: The case of central Stockholm. *J. Transp. Land Use* **2021**, *14*, 1019–1037. [CrossRef]
36. Mcharg, I.L. *Design with Nature*; American Museum of Natural History: New York, NY, USA, 1969.
37. Deakin, E. *Transportation, Land Use, and Environmental Planning*; Elsevier: Amsterdam, The Netherlands, 2019.
38. Rasouli, S.; Timmermans, H. Activity-based models of travel demand: Promises, progress and prospects. *Int. J. Urban Sci.* **2014**, *18*, 31–60. [CrossRef]
39. Ettema, D.F. *Activity-Based Travel Demand Modeling*; Cambridge University Press: Cambridge, UK, 1996.
40. Pinjari, A.R.; Bhat, C.R. *Activity-Based Travel Demand Analysis*; Edward Elgar Publishing: Cheltenham, UK, 2011.

41. Tversky, B. Pictorial encoding of sentences in sentence—Picture comparison. *Q. J. Exp. Psychol.* **1975**, *27*, 405–410. [CrossRef]
42. Savage, J.K. Economic Nature of the Cooperative Association. *J. Farm Econ.* **1954**, *36*, 529–534. [CrossRef]
43. Kahneman, D.; Tversky, A. On the interpretation of intuitive probability: A reply to Jonathan Cohen. *Cognition* **1979**, *7*, 409–411. [CrossRef]
44. Bell, D.E. Regret in decision making under uncertainty. *Oper. Res.* **1982**, *30*, 961–981. [CrossRef]
45. de la Barra, T. *Integrated Land Use and Transport Modelling. Decision Chains and Hierarchies*; Cambridge University Press: Cambridge, UK, 1989.
46. Von Thünen, J.H. *Isolated State*; Mohr Siebeck: Berlin, Germany, 1826.
47. Weber, A. *Regarding the Location of the Industry*; Mohr Siebeck: Berlin, Germany, 1909.
48. William, A. *Location and Land Use: Toward a General Theory of Land Rent*; Mohr Siebeck: Berlin, Germany, 1964.
49. Boulding, K.E. General systems theory—The skeleton of science. *Manag. Sci.* **1956**, *2*, 197–208. [CrossRef]
50. Forrester, J.W. *System Dynamics and the Lessons of 35 Years*; Springer: Boston, MA, USA, 1993.
51. Batty, M. *Planning Support Systems: Progress, Predictions, and Speculations on the Shape of Things to Come*; Citeseer: London, UK, 2007.
52. Felsenstein, D.; Axhausen, K.; Waddell, P. Land Use-Transportation Modeling with UrbanSim: Experiences and Progress Introduction to the Special Issue. *J. Transp. Land Use* **2010**, *3*, 1–3. [CrossRef]
53. Du, R.F.; Tan, Y. Assessing land-use and transport integration via a spatial composite indexing model. *Int. J. Environ. Sci. Technol.* **2015**, *12*, 803–816.
54. Ahmadzai, F. Analyses and modeling of urban land use and road network interactions using spatial-based disaggregate accessibility to land use. *J. Urban Manag.* **2020**, *9*, 298–315. [CrossRef]
55. Waldeck, L.; Holloway, J.; Van Heerden, Q. Integrated land use and transportation modelling and planning. *J. Transp. Land Use* **2020**, *13*, 227–254. [CrossRef]
56. Stiller, D.; Wurm, M.; Stark, T.; D'Angelo, P.; Stebner, K.; Dech, S.; Taubenböck, H. Spatial parameters for transportation: A multi-modal approach for modelling the urban spatial structure using deep learning and remote sensing. *J. Transp. Land Use* **2021**, *14*, 777–803. [CrossRef]
57. Lowry, I.S. *A Model of Metropolis*; Rand Corp Santa Monica Calif. Rand: Washington, DC, USA, 1964.
58. Wilson, A.G. The use of the concept of entropy in system modelling. *J. Oper. Res. Soc.* **1970**, *21*, 247–265. [CrossRef]
59. Lees, L. *The Geography of Gentrification: Thinking Through Comparative Urbanism*; SAGE Publications: Washington, CA, USA, 2012.
60. Lerman, S.; Borkman, R.F. *A Method for Detecting 8-Methoxypsoralen in the Ocular Lens*; Science: Washington, CA, USA, 1977.
61. Anas, A. The estimation of multinomial logit models of joint location and travel mode choice from aggregated data. *J. Reg. Sci.* **1981**, *21*, 223–242. [CrossRef]
62. Acheampong, R.A.; Silva, E.A. Land use–transport interaction modeling: A review of the literature and future research directions. *J. Transp. Land Use* **2015**, *8*, 11–38. [CrossRef]
63. Acheampong, R.A. *Understanding the Co-Emergence of Urban Location Choice and Mobility Patterns: Empirical Studies and an Integrated Geospatial and Agent-Based Model*; University of Cambridge: Cambridge, UK, 2017.
64. Yin, D.; Lopes, R.; Shlens, J.; Cubuk, E.D. A fourier perspective on model robustness in computer vision. *Adv. Neural Inf. Process. Syst.* **2019**, *32*, 1–11.
65. Wegener, M.; Fürst, F. *Land-Use Transport Interaction: State of the Art*; SSRN: Berlin, Germany, 2004.
66. Wegener, M. *Land-Use Transport Interaction Models*; Springer: Berlin, Germany, 2021.
67. Marsden, G.; Reardon, L. Questions of governance: Rethinking the study of transportation policy. *Transp. Res. Part A Policy Pract.* **2017**, *101*, 238–251. [CrossRef]
68. Johansson, F.; Tornberg, P.; Fernström, A. A function-oriented approach to transport planning in Sweden: Limits and possibilities from a policy perspective. *Transp. Policy* **2018**, *63*, 30–38. [CrossRef]
69. Howlett, M.; Rio, P. The parameters of policy portfolios: Verticality and horizontality in design spaces and their consequences for policy mix formulation. *Environ. Plan. C Gov. Policy* **2015**, *33*, 1233–1245. [CrossRef]
70. Miller, J.A.; Lacan, J. *The Four Fundamental Concepts of Psycho-Analysis*; Routledge: London, UK, 2018.
71. Hawkins, J.; Habib, K. Integrated models of land use and transportation for the autonomous vehicle revolution. *Transp. Rev.* **2019**, *39*, 66–83. [CrossRef]
72. Mingardo, G.; Van Wee, B.; Rye, T. Urban parking policy in Europe: A conceptualization of past and possible future trends. *Transp. Res. Part A Policy Pract.* **2015**, *74*, 268–281. [CrossRef]
73. Gerber, P.; Caruso, G.; Cornelis, E.; de Chardon, C.M. A multi-scale fine-grained LUTI model to simulate land-use scenarios in Luxembourg. *J. Transp. Land Use* **2018**, *11*, 255–272. [CrossRef]
74. Basu, R.; Ferreira, J.; Ponce-Lopez, R. A framework to generate virtual cities as sandboxes for land use-transport interaction models. *J. Transp. Land Use* **2021**, *14*, 303–323. [CrossRef]
75. Donnelly, R.; Upton, W.J.; Knudson, B. Oregon's Transportation and Land Use Model Integration Program. *J. Transp. Land Use* **2018**, *11*, 19–30. [CrossRef]

76. Meurs, H.; Haaijer, R. Spatial structure and mobility. *Transp. Res. Part D Transp. Environ.* **2001**, *6*, 429–446. [CrossRef]
77. Cao, X.; Mokhtarian, P.L. How do individuals adapt their personal travel? Objective and subjective influences on the consideration of travel-related strategies for San Francisco Bay Area commuters. *Transp. Policy* **2005**, *12*, 291–302. [CrossRef]
78. Grunfelder, J.; Nielsen, T.S. Commuting behaviour and urban form: A longitudinal study of a polycentric urban region in Denmark. *Geogr. Tidsskr.-Dan. J. Geogr.* **2012**, *112*, 2–14. [CrossRef]
79. Aditjandra, P.T.; Mulley, C.; Nelson, J.D. The influence of neighbourhood design on travel behaviour: Empirical evidence from North East England. *Transp. Policy* **2013**, *26*, 54–65. [CrossRef]
80. Tomaney, J.; Marques, P. Evidence, policy, and the politics of regional development: The case of high-speed rail in the United Kingdom. *Environ. Plan. C Gov. Policy* **2013**, *31*, 414–427. [CrossRef]
81. ITDP (Institute for Transportation and Development Policy). Rapid Transit Database. Available online: www.itdp.org/ (accessed on 7 March 2022).
82. Sanchez, T.W.; Stolz, R.; Ma, J.S. Inequitable effects of transportation policies on minorities. *Transp. Res. Rec.* **2004**, *1885*, 104–110. [CrossRef]
83. Bullard, R.D. *Highway Robbery: Transportation Racism & New Routes to Equity*; South End Press: Cambridge, UK, 2004.
84. Mollenkopf, J.H. *The Contested City*; Princeton University Press: Princeton, NJ, USA, 1983.
85. Logan, J.; Molotch, H. *The City as a Growth Machine*; Routledge: London, UK, 2012.
86. Liebs, C. *Main Street to Miracle Mile: American Roadside Architecture*; JHU Press: Baltimore, MD, USA, 1995.
87. Golub, A.; Marcantonio, R.A.; Sanchez, T.W. Race, space, and struggles for mobility: Transportation impacts on African Americans in Oakland and the East Bay. *Urban Geogr.* **2013**, *34*, 699–728. [CrossRef]
88. Thomopoulos, N.; Grant-Muller, S.; Tight, M.R. Incorporating equity considerations in transport infrastructure evaluation: Current practice and a proposed methodology. *Eval. Program Plan.* **2009**, *32*, 351–359. [CrossRef]
89. Stevens, B.L.; Lewis, F.L.; Johnson, E.N. *Aircraft Control and Simulation: Dynamics, Controls Design, and Autonomous Systems*; John Wiley & Sons: Hoboken, NJ, USA, 2005.
90. Henriksson, M.A.; Summerton, J.A. *High-Speed Trains: Effects on Local and Regional Development and Conditions for Integration with Public Transport. A Literature Review*; K2 Working Papers: Lund, Sweden, 2016.
91. Cervero, R.; Duncan, M. *Land Value Impacts of Rail Transit Services in Los Angeles County*; Report prepared for National Association of Realtors; Urban Land Institute: Washington, DC, USA, 2002.
92. Deng, T. *The Impact of Bus Rapid Transit on Land Development*; University of Aberdeen: Aberdeen, UK, 2010.
93. Guan, C.; Peiser, R. Accessibility, urban form, and property value: A study of Pudong, Shanghai. *J. Transp. Land Use* **2018**, *11*, 1057–1080. [CrossRef]
94. Chong, Z.; Qin, C.; Chen, Z. Estimating the economic benefits of high-speed rail in China: A new perspective from the connectivity improvement. *J. Transp. Land Use* **2019**, *12*, 287–302. [CrossRef]
95. Pan, Q. The impacts of light rail on residential property values in a non-zoning city. *J. Transp. Land Use* **2019**, *12*, 241–264. [CrossRef]
96. Bradshaw, C.J.; Warkentin, I.G.; Sodhi, N.S. Urgent preservation of boreal carbon stocks and biodiversity. *Trends Ecol. Evol.* **2009**, *24*, 541–548. [CrossRef]
97. Forman, R.T.T.; Sperling, D.; Bissonette, J.A.; Clevenger, A.P.; Cutshall, C.D.; Dale, V.H.; Fahrig, L.; France, R.L.; Goldman, C.R.; Heanue, K. *Road Ecology: Science and Solutions*; Island Press: Washington, DC, USA, 2003.
98. Brady, S.P.; Richardson, J.L.; Kunz, B.K. Incorporating evolutionary insights to improve ecotoxicology for freshwater species. *Evol. Appl.* **2017**, *10*, 829–838. [CrossRef] [PubMed]
99. IEA (International Energy Agency). Mobility Model. Available online: www.iea.org/etp/etpmodel/transport/ (accessed on 7 March 2022).
100. Campbell, W.B.; López Ortíz, S. *Integrating Agriculture, Ecotourism, and Conservation: Examples from the Field*; Springer: Berlin, Germany, 2011.
101. Lala, F.; Chiyo, P.I.; Omondi, P.; Okita-Ouma, B.; Kanga, E.; Koskei, M.; Tiller, L.; Morris, A.W.; Severud, W.J.; Bump, J.K. Influence of infrastructure, ecology, and underpass-dimensions on multi-year use of Standard Gauge Railway underpasses by mammals in Tsavo, Kenya. *Sci. Rep.* **2022**, *12*, 5698. [CrossRef]
102. Moeckel, R.; Garcia, C.L.; Chou, A.T.M.; Okrah, M.B. Trends in integrated land-use/transport modeling. *J. Transp. Land Use* **2018**, *11*, 463–476. [CrossRef]
103. Ford, C.L.; Airhihenbuwa, C.O. The public health critical race methodology: Praxis for antiracism research. *Soc. Sci. Med.* **2010**, *71*, 1390–1398. [CrossRef]
104. United Nations Human Settlements Programme. *Planning Sustainable Cities: Policy Directions, Global Report on Human Settlements*; UN-HABITAT: London, UK, 2009; pp. 1–17.
105. Ian, C. *Histories of Transit-Oriented Development: Perspectives on the Development of the TOD Concept*; Institute of Urban and Regional Development University of California: Berkeley, CA, USA, 2007; pp. 1–30.

106. Moore, T.; Throsnes, P. *The Transportation/Land Use Connection, American Planning Association*; Report; Planning Advisory Service: Chicago, IL, USA, 1999; pp. 448–449.
107. Dittmar, H. *The New Transit Town: Best Practices in Transit-Oriented Development*; Island Press: Washington, DC, USA, 2012.
108. He, Y.; Liu, X.; Wu, D.; Li, S.; Zhou, P. Ecological Risks and Patterns Associated With Land Use/Cover Changes Along the Belt and Road Initiative Routes. *Land Degrad. Dev.* **2025**, *28*, 5483. [CrossRef]
109. Wolf, J.P.; Song, C. *Finite-Element Modelling of Unbounded Media*; Wiley: Chichester, UK, 1996.
110. Geurs, K.T.; Bert, V.W. Accessibility evaluation of land-use and transport strategies: Review and research directions. *J. Transp. Geogr.* **2004**, *12*, 127–140. [CrossRef]
111. Pirie, G.H. Measuring accessibility: A review and proposal. *Environ. Plan. A* **1979**, *11*, 299–312. [CrossRef]
112. Kwan, M.P. Space-time and integral measures of individual accessibility: A comparative analysis using a point-based framework. *Geogr. Anal.* **1998**, *30*, 191–216. [CrossRef]
113. Farber, S.; Marino, M.G. Transit accessibility, land development and socioeconomic priority: A typology of planned station catchment areas in the Greater Toronto and Hamilton Area. *J. Transp. Land Use* **2017**, *10*, 879–902. [CrossRef]
114. Levinson, D.M. Accessibility and the journey to work. *J. Transp. Geogr.* **1998**, *6*, 11–21. [CrossRef]
115. Cervero, R.; Rood, T.; Appleyard, B. Tracking accessibility: Employment and housing opportunities in the San Francisco Bay Area. *Environ. Plan. A* **1999**, *31*, 1259–1278. [CrossRef]
116. Cervero, R.; Duncan, M. Walking, bicycling, and urban landscapes: Evidence from the San Francisco Bay Area. *Am. J. Public Health* **2003**, *93*, 1478–1483. [CrossRef]
117. Klonsky, E.D.; Glenn, C.R. Assessing the functions of non-suicidal self-injury: Psychometric properties of the Inventory of Statements About Self-injury (ISAS). *J. Psychopathol. Behav. Assess.* **2009**, *31*, 215–219. [CrossRef]
118. Melia, S.; Parkhurst, G.; Barton, H. The paradox of intensification. *Transp. Policy* **2011**, *18*, 46–52. [CrossRef]
119. Steiner, F.R. *The Living Landscape: An Ecological Approach to Landscape Planning*; Island Press: Washington, DC, USA, 2012.
120. Rye, T.; Hrelja, R. Policies for reducing car traffic and their problematisation. Lessons from the mobility strategies of British, Dutch, German and Swedish cities. *Sustainability* **2020**, *12*, 8170. [CrossRef]

land MDPI

Review

The Evolution of the Interaction Between Urban Rail Transit and Land Use: A CiteSpace-Based Knowledge Mapping Approach

Haochen Yang [1], Nana Cui [1,*] and Haishan Xia [1,2]

1 School of Architecture and Design, Beijing Jiaotong University, Beijing 100044, China; 23121611@bjtu.edu.cn (H.Y.); hshxia@bjtu.edu.cn (H.X.)
2 Transit Oriented Development Academy, Beijing Jiaotong University, Beijing 100044, China
* Correspondence: cuinn@bjtu.edu.cn

Abstract: Urban rail transit is a key enabler for optimizing urban spatial structures, and its interactive relationship with land use has long been a focus of attention. However, existing studies suffer from scattered methodologies, a lack of systematic analysis, and insufficient dynamic insights into global trends. This study comprehensively employs CiteSpace, VOSviewer, and Scimago Graphica to conduct bibliometric and knowledge map analysis on 1894 articles from the Web of Science database between 2004 and 2024, focusing on global research trends, collaboration networks, thematic evolution, and methodological advancements. Key findings include the following: (1) research on rail transit and land use has been steadily increasing, with a significant "US-China dual-core" distribution, where most studies are concentrated in the United States and China, with higher research density in Asia; (2) domestic and international research has primarily focused on themes such as the built environment, value capture, and public transportation, with a recent shift toward artificial intelligence and smart city technology applications; (3) research methods have evolved from foundational 3S technologies (GIS, GPS, RS) to spatial modeling tools (e.g., LUTI model, node-place model), and the current emergence of AI-driven analysis (e.g., machine learning, deep learning, digital twins). The study identifies three future research directions—technology integration, data governance, and institutional innovation—which provide guidance for the coordinated planning of transportation and land use in future smart city development.

Keywords: urban rail transit; land use; Web of Science; bibliometric analysis; visualization analysis; CiteSpace

1. Introduction

As the pace of global urbanization accelerates, urban rail transit (URT) plays a key role as an efficient and sustainable mode of transportation in optimizing urban spatial structures, alleviating traffic congestion, and promoting low-carbon development [1,2]. At the same time, the dynamic evolution of land use patterns directly impacts passenger distribution, operational efficiency, and socio-economic benefits of rail transit [3–5]. The interaction between the two has become a hot topic of research in fields such as urban planning, transportation engineering, and land management [6–8].

Despite a growing body of literature exploring this interaction, several key challenges persist in the current research landscape.

First, although early classical literature reviews have laid a systematic foundation for the disciplinary framework [9,10], the rapid evolution of transportation technology and urban spatial forms in recent years has made it imperative for the academic community to conduct a phased integration and reassessment of research findings and cutting-edge trends [11,12]. Current research still exhibits significant disciplinary fragmentation, with most studies focusing on specific regional cases (e.g., North American polycentric urban systems) [10] or micro-level issues (e.g., TOD [13], algorithm optimization of LUTI models [14], environmental assessment [15]), lacking an integrated analytical framework capable of synthesizing technological evolution logic, and spatial interaction effects [16,17].

Second, traditional literature reviews primarily rely on qualitative analysis, which limits their repeatability and objectivity when tracking the evolution of research topics, co-authorship networks, and methodological advancements [8,18]. As a result, they struggle to conduct a systematic assessment of the global evolution of core theoretical paradigms. In recent years, with the innovation of bibliometric tools, the academic community has gradually turned to using visualization analysis software to systematically mine knowledge from multiple sources of academic data [19,20]. Unlike existing studies that focus on statistical integration of general data, this paper focuses on the SCI-EXPANDED and SSCI journal databases and uses structured bibliometric methods to improve the original limited precision problem, thereby achieving an in-depth description of the cutting edge of this field and accurate tracking of theoretical developments [21].

Third, although emerging technologies such as AI, big data analysis, and digital twins have provided new avenues for URT-land use research, existing reviews suffer from timeliness lag [22]. Therefore, this study systematically reviews the latest advances in intelligent technologies such as machine learning [23] and deep learning [24] under the theoretical framework of coordinated development of rail transit and land use. Based on the existing literature review [25], this study provides a new theoretical foundation for further understanding the multi-level relationship between this field and new technologies.

This study conducts a comprehensive bibliometric and knowledge mapping analysis of 1894 academic publications from the Web of Science (WOS) Core Collection spanning 2004–2024. Leveraging CiteSpace, VOSviewer, and Scimago Graphica, the study investigates spatio-temporal publication trends, author collaboration networks, thematic clusters, and technical method adoption. The guiding research questions are as follows:

- RQ1: What are the global research trends in this field?
- RQ2: How have key research themes evolved over time?
- RQ3: How are emerging technologies changing research methods in this field?

This study makes significant contributions to the field of urban rail transit and land use interaction by conducting a systematic literature review and focusing on key research questions. Research reveals the dual-core spatial publication structure of China and the United States, clearly showing the distribution of global research forces. It identifies the evolving research theme trend from TOD construction to value capture and travel behavior, and reviews the development trends of methods such as machine learning and deep learning. These findings provide scholars with references for understanding research trends and innovating research methods, and integrate diverse content to establish a comprehensive research perspective. This offers urban planners and policymakers highly practical guidance for developing sustainable urban transport policies.

This study is structured to address the three core research questions through progressive analysis (Figure 1). Section 2 details the data and bibliometric methods. Section 3 responds to RQ1 and RQ2 by mapping global research trends. Section 4 further answers RQ2 and RQ3 through an interpretation of thematic evolution and technological progress based on keyword clustering. Each subsection corresponds to major clusters—built environment, value capture, travel behavior, and planning technologies—revealing their interconnections. Section 5 synthesizes insights and proposes future directions grounded in the analytical findings.

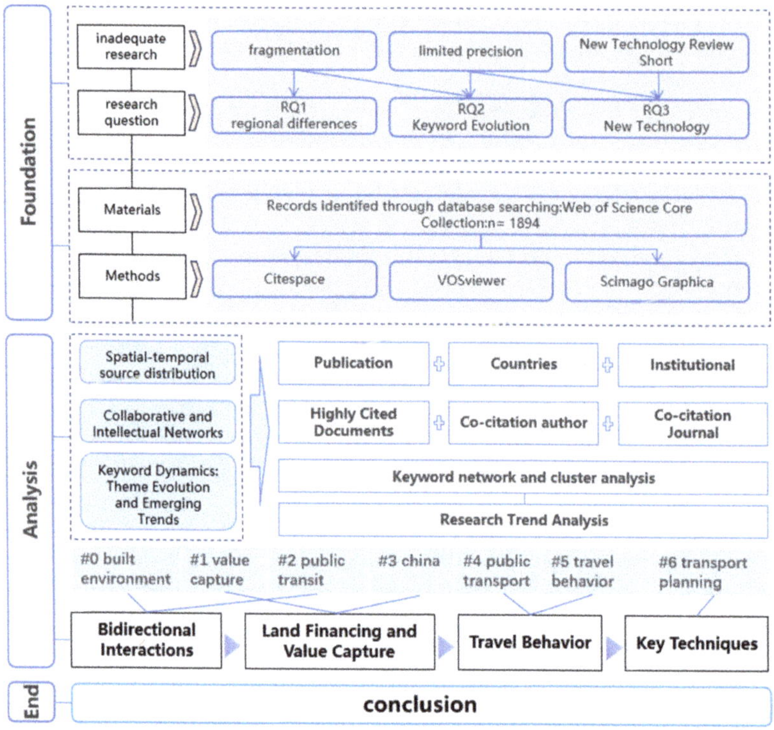

Figure 1. The framework of this research.

2. Materials and Methods

2.1. Data Collection

To examine the evolution of knowledge regarding the relationship between urban rail transit and land use, we compiled the pertinent literature from the Science Citation Index Expanded (SCI-EXPANDED) and Social Sciences Citation Index, both of which are integral databases within the Web of Science Core Collection. The literature search was conducted on 26 November 2024, using a targeted search method focused on the following keywords: (urban rail transit OR metro systems OR subway systems OR light rail transit OR urban public transportation) AND (land use OR land utilization OR land planning), with a time restriction from 1 January 2004, to 26 November 2024, to search the database content. Secondly, the initially obtained database literature was cleaned and deduplicated, resulting in 1894 articles. As an important component of the urban transportation system, urban rail transit systems play a crucial role in connecting various urban areas, while the evolution of land use patterns reflects the dynamic development of urban space. This study employs CiteSpace knowledge mapping technology to explore the research trajectory of the interaction between urban rail transit and land use.

2.2. Bibliometric Methods

The advent of the era of big data has placed researchers in a vast sea of literature. However, the continuous advancement of information technology and the growing capabilities of computer processing have opened new avenues for the refined and visual analysis of literature data. At present, more than 10 types of software are available for mapping foreign literature, each offering distinct advantages. For example, VOSviewer excels in topic clustering, providing clear and detailed visualizations, while HistCite organizes literature citations in a time-based network format. CiteSpace, renowned for its powerful co-citation analysis, has become a key tool in over 60 fields, including computer science, information science, and medicine, due to the ongoing refinement of its algorithms and functions. This study employed both CiteSpace 6.3.R1 and VOSviewer 1.6.20. to conduct a visual analysis of articles in the field, extracting cutting-edge research hotspots, connections, and trends. However, while CiteSpace is effective in mechanically calculating connections between articles and identifying prominent keywords and research topics, the specifics of the literature require manual reading and analysis for deeper understanding. In this study, CiteSpace 6.3.R1 was used to eliminate duplicates from the dataset, after which the author and keyword networks were explored. Additionally, Scimago Graphica **1.0.50.0** was utilized to visualize the national and regional distribution of research activity.

3. Performance Analysis

3.1. Spatial-Temporal and Source Distribution Characteristics

This study provides an overview of the global evolution and research landscape of the literature over the past two decades. It analyses the spatio-temporal dynamics of academic output (RQ1), including publication trends, national and institutional distribution patterns, and contributions from core journals. This reveals the key characteristics of the research field. This reveals the geographical and institutional diffusion pathways of related research. This lays the groundwork for subsequent studies on co-cited literature (RQ2 and RQ3).

3.1.1. Publication Trends

Figure 2 shows that the development of this research field can be illustrated by analyzing the number of papers published in recent years. The development of rail transit and land use research from 2004 to 2024 reflects distinct phases shaped by urbanization trends and policy shifts. There has been an overall upward trajectory in annual publications, rising from six to 209 papers.

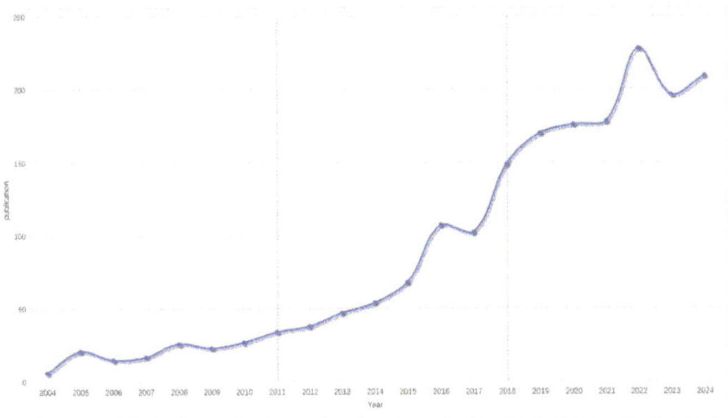

Figure 2. Number of publications each year from 2004 to 2024.

During the initial slow growth phase (2004–2011), limited data acquisition methods and an insufficient research foundation resulted in low awareness of land planning, particularly surrounding urban rail transit stations (the Transit-Oriented Development (TOD) concept had not yet been widely adopted), leading to limited related research. The subsequent rapid growth phase (2011–2018), driven by accelerated urbanization, particularly in East Asia, saw a dramatic increase in research activity, with studies examining themes such as destination accessibility and transport network proximity, and the global promotion of TOD models [10]. China's 2014 'New Urbanisation' strategy further stimulated this expansion by emphasizing coordinated rail–city development. The current high-quality growth phase (2018 to the present day) marks a strategic shift from quantity to sophistication. This is influenced by stricter Chinese regulations, such as the 2018 National Development and Reform Commission Document No. 52, which imposed financial safeguards and redirected the focus from construction scale to operational efficiency. Meanwhile, technological advancements have introduced big data analytics, machine learning and smart city concepts into research methodologies. This evolution clearly aligns research trends with policy priorities, progressing from initial technical explorations and mid-phase planning analyses to the current focus on smart, sustainable development.

3.1.2. Analysis of Countries

This study covers a wide range of the literature involving 56 countries across seven continents. As Figure 3 shows, the research results exhibit significant spatial distribution characteristics: they are concentrated in the Northern Hemisphere, primarily in countries with relatively favorable economic conditions, forming a dual-core structure centered on China and the United States. China and the United States stand out among these countries, with the highest research output: 737 and 626 papers were published by these two countries, respectively, accounting for 39% and 33% of the total literature. Canada and Australia follow with 134 and 126 papers, respectively. Although European countries have relatively few publications, they demonstrate strong interconnections, indicating high levels of academic collaboration within the region. Despite its small land area, Singapore has established extensive research collaboration networks with China, Indonesia, Malaysia and other countries thanks to its geographical location. Australia and New Zealand, despite their geographical distance from other continents, have continued to innovate in urban rail transit and land use research, maintaining close academic ties with European and American countries. By analyzing the spatial distribution patterns of literature and transnational cooperation networks around the world, this study reveals differences in academic activity levels and cooperation models between different regions.

Figure 3. Country collaboration map.

3.1.3. Institutional Analysis

Figure 4 shows the institutions with the highest number of new publications in the field of rail transit and land use from 2004 to 2024. The institution that has contributed the most to research in this field is Tongji University, which ranks first with 54 articles. Second is Wuhan University, with 52 papers published, followed closely by the University of Hong Kong and Southeast University, which have published 46 and 45 articles, respectively. In addition, Beijing Jiaotong University, with its advantages in transportation, has published 40 articles. Other institutions around the world, such as the University of Toronto, MIT, the University of California, Berkeley and the University of North Carolina, have continued to conduct interdisciplinary research in related fields, providing case studies for related research in developed countries around the world. Overall, research in this field is dominated by the US and China, which is consistent with the 'US-China dual-centre' pattern of global research output shown in Figure 3.

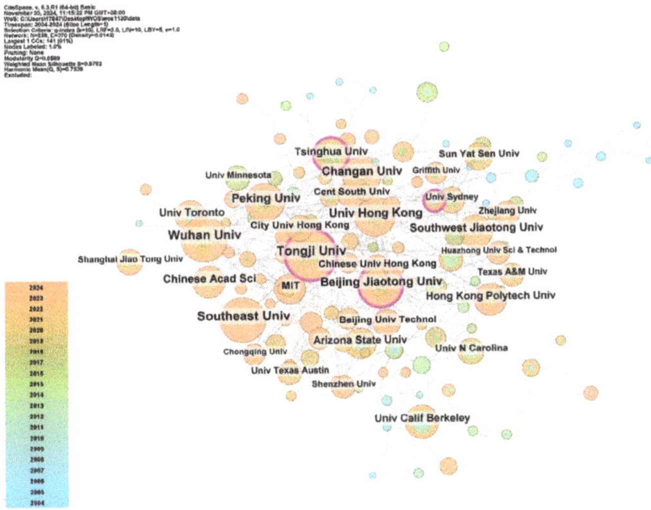

Figure 4. Institutional network.

3.2. *Collaborative and Intellectual Networks*

Building upon the macro-level spatio-temporal and institutional insights presented in Section 3.1, Section 3.2 delves deeper into the collaborative and intellectual structures that shape the field of urban rail transit and land use integration. The analysis is conducted from several dimensions, including highly co-cited documents and co-citation author networks. This allows us to trace the evolution of core research themes, identify seminal contributions, and map influential author clusters that have driven innovation in the field. These dimensions are directly related to the study's core research questions—particularly RQ2 and RQ3—as they reveal the methodological paradigms, disciplinary intersections, and knowledge diffusion patterns that influence how TOD-related challenges are conceptualized and addressed globally. The findings from this section lay the groundwork for understanding thematic clustering and knowledge frontiers in the next analytical stage.

3.2.1. Analysis of Co-Cited Documents

If two documents (or multiple papers) appear in a third document at the same time, then these two documents form a co-citation relationship. Figure 5 shows the network structure of co-cited documents, which shows the important role of these studies in the coordinated development of urban transportation. Through the analysis of co-cited documents, this study has discovered the knowledge structure in the field of sustainable urban transportation from a huge knowledge base, identified key knowledge bases in this field, and understood the development of this discipline. Accordingly, there were 429 co-cited documents in the sample data. In Table 1,the top ranked was Ibraeva A. (2020) with 53 co-citations [26], followed by An D.D. (2019) with 43 [27], Guo Y.Y. (2020) with 39 [28], Gan Z.X. (2020) with 38 [29], Ding C. (2019) with 36 [30], Ewing R. (2010) with 35 [10], Li S.Y. (2020) with 34 [31], Li Z.K. (2019) with 33 [32], Chen E.H. (2019) with 32 times [33], and the number of citations for the remaining articles is less than 30 [34–41]. The data shows that most of the literature with a high number of citations is concentrated in 2020, and as the frequency of citations decreases, the time of publication of the literature gradually becomes more diverse, with more of the literature from older years appearing.

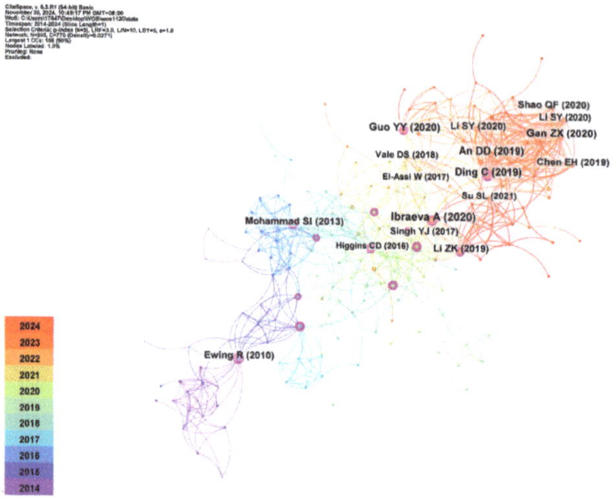

Figure 5. Co-cited documents network [34–41].

Table 1. Co-cited documents.

First Author	Year	Freq	Title
Ibraeva A.	2020	53	Transit-oriented development: A review of research achievements and challenges [26]
An D.D.	2019	43	Understanding the impact of built environment on metro ridership using open source in Shanghai [27]
Guo Y.Y.	2020	39	Built environment effects on the integration of dockless bike-sharing and the metro [28]
Gan Z.X.	2020	38	Examining the relationship between built environment and metro ridership at station-to-station level [29]
Ding C.	2019	36	How does the station-area built environment influence Metrorail ridership? Using gradient boosting decision trees to identify non-linear thresholds [30]
Ewing R.	2010	35	Travel and the Built Environment: A Meta-Analysis [10]
Li S.Y.	2020	34	Spatially varying impacts of built environment factors on rail transit ridership at station level: A case study in Guangzhou, China [31]
Li Z.K.	2019	33	Transit oriented development among metro station areas in Shanghai, China: Variations, typology, optimization and implications for land use planning [32]
Chen E.H.	2019	32	Discovering the spatio-temporal impacts of built environment on metro ridership using smart card data [33]

3.2.2. Analysis of Co-Citation Author Network

Cooperation between authors in the academic field plays a crucial role in advancing research, as collaborative efforts often lead to more comprehensive studies and innovative solutions. To analyze the productivity and collaborative networks of authors in the domain of urban rail transit and land use, this study utilized CiteSpace software to process and visualize the data. The resulting network of frequently cited authors provides a clear picture of the co-citation relationships that drive research in this field (Figure 6). Authors who are cited more frequently often serve as bridges between different research communities, fostering interdisciplinary dialog and enhancing the exchange of knowledge. According to the network analysis, the leading contributor is Cervero R., with his first collaboration dating back to 2005, and a total of 739 collaborative papers. He is closely followed by Ewing R., with 467 collaborative papers, and Frank L.D., who has co-authored 235 papers. Other notable contributors include Handy S. (166 collaborative papers), and Cao X.Y. (160 collaborative papers). The sixth to tenth positions are occupied by Litmante, Zhao P.J., Hess D.B., Handy S.L., and Debrezion, with 147, 145, 142, 142, and 139 co-authored publications, respectively. These authors have significantly influenced the field through their collaborative work, and their contributions have shaped the current understanding of urban rail transit and land use interactions. This collaborative network underscores the importance of interdisciplinary cooperation in advancing research in this domain.

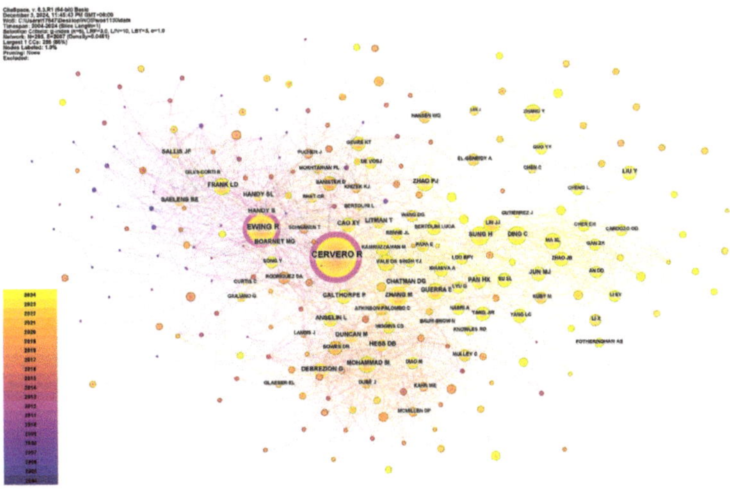

Figure 6. Co-citation author network.

3.2.3. Analysis of the Co-Cited Journal

The number of documents counted in this study is the total number of citations in the existing literature. It was found that 279 journals published 27,027 articles related to the interdisciplinary field of rail transit and land use. The top ten journals in this research field accounted for 29.74% of total publications, suggesting that academic achievements are concentrated in a few high-impact journals. Notably, *Transportation Research Part A: Policy and Practice* leads the list with 1029 articles, followed by *Journal of Transport Geography* with 1005 articles, and *Transportation Research Record* with 1003 articles. Other prominent journals include *Transportation Research Part D: Transport and Environment* with 815 articles, *Transport Policy* with 789 articles, and *Urban Studies* with 766 articles. The *Journal of the American Planning Association and Transportation* each contributed 722 and 694 articles, respectively, while *Cities and Sustainability* published 691 and 524 articles, respectively. Furthermore, journals publishing more than 10 articles in this field account for 75.27% of the total, suggesting that the dataset is highly representative of the journals actively contributing to the discourse on rail transit and land use (Figure 7).

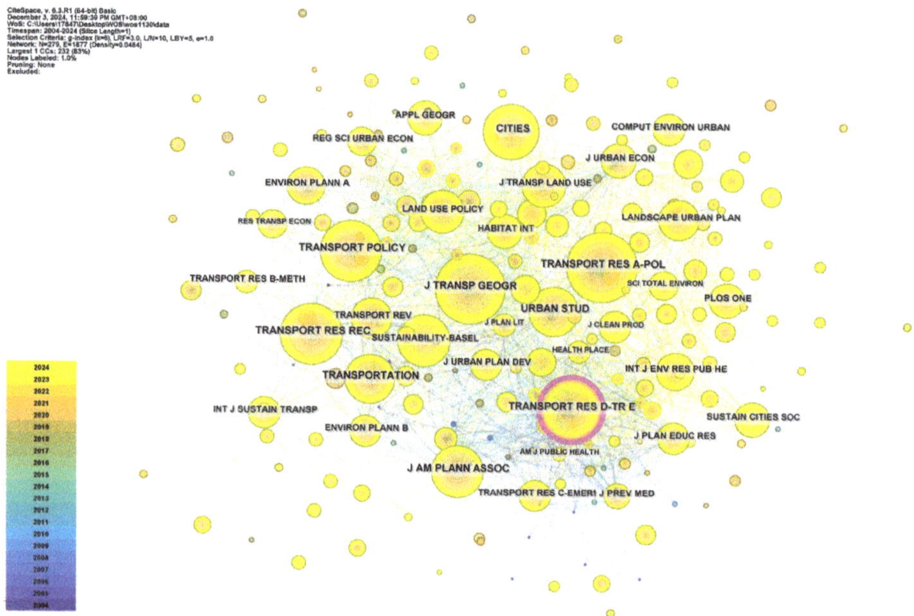

Figure 7. Co-citation journal network.

3.3. *Keyword Dynamics: Theme Evolution and Emerging Trends*

This section focuses on the core knowledge and thematic evolution in this field. Through clustering and analysis of high-frequency keywords, it reveals the stable core (such as built environment, land use, and transportation) and dynamic thematic shifts (such as value capture, gentrification, urban resilience, and machine learning) in the research. This section lays the groundwork for further deconstructing knowledge clusters and constructing a refined thematic structure in Section 4, while also delving into their intrinsic logical connections, interdisciplinary characteristics, and innovation-driven mechanisms.

3.3.1. Keyword Network and Cluster Analysis

A bibliometric analysis of keyword frequencies can reflect research frontiers, as keywords that appear more frequently often indicate widely concerned research priorities. To reflect the research focus over the past 20 years, annual slices were used, and "keywords" were used as the node type. In Figure 8, "land use" is the most important keyword, appearing 666 times, reflecting its central position in academic discussions. The second most frequently occurring keyword is "impact", which appears 437 times, highlighting the importance of understanding the interaction between urban rail transit and land use. The third most frequently occurring keyword was "built environment," which appeared 400 times, indicating that scholars generally pay attention to the physical and spatial environment of urban development. This was closely followed by "transportation" and "accessibility," which appeared 258 times each, indicating that as the built environment around stations improves, stations attract more and more passenger flow, and research on the role of transportation systems in improving urban accessibility has also received increasing attention. In addition, the word "city" appeared 248 times, indicating that research on the environment is mainly focused on large cities. Other keywords appeared less than 200 times, reflecting more specific aspects related to the research topic.

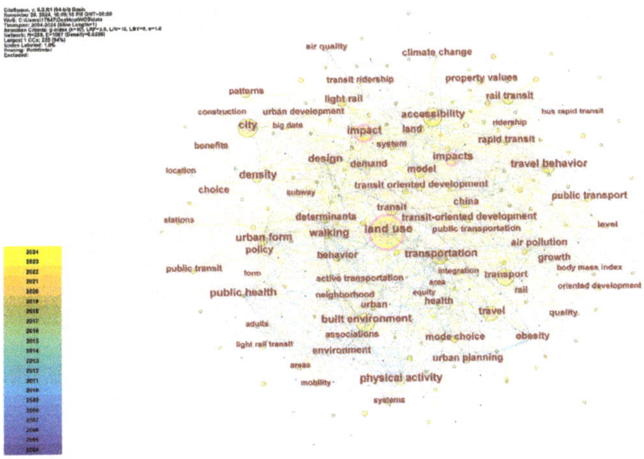

Figure 8. Keyword occurrence network.

By clustering the frequently occurring keywords, Figure 9 shows seven thematic clusters, and Table 2 summarizes the thematic clusters and their corresponding top five research focuses. '#0 Built environment' has the longest time span and the largest scale, running through the entire research period. Early terms in this category were relatively broad, covering general concepts such as urban development, urban ecology and land use planning, and included key terms such as 'physical activity', 'travel behaviour', 'walking' and 'environmental design'. These terms highlight the continued focus of scholars on the relationship between urban form and human mobility. '#1 Value Capture' reflects the land premium behavior brought about by rail transit stations, and includes keywords such as "non-linear effects", "spatial heterogeneity", "transit-oriented development" and "urban design" to reflect the increasing research on financing mechanisms with the development of rail transit stations. '#2 Public Transport' focuses on issues such as "air pollution", "Beijing", "urban transport" and "decentralisation", reflecting scholars' thinking about the problems of big cities in the process of urban construction, and indicating people's concerns about the traffic environment and governance level. '#3 China' is characterized by keywords such as "value capture", "housing prices" and "transport". China has attracted increasing attention from experts in the fields of urban planning and transport planning due to its rapid urban expansion and continuous construction of transport infrastructure in the past 20 years [42]. As a result, a large number of empirical studies have been carried out on China's rapid urbanization and transport investment. '#4 Public transport' is characterized by terms such as "service", "equity" and "urban vitality", indicating that with urban development and class differentiation, more and more scholars are concerned about how to coordinate transport resources to maintain social equity and the different impacts of public transport facilities on the travel of different social groups. It also further explores the factors that will affect the vitality and healthy growth of cities. '#5 Travel behaviour' is a recent research focus. This cluster includes terms such as 'vulnerability', 'resilience', 'COVID-19' [43] and 'climate change', which indicate that under the impact of the epidemic, global crises have an explosive impact on urban travel patterns, and urban resilience has received further attention. '#6 Transportation Planning' indicates that in the face of problems arising from the process of urban development, more and more scholars and government departments have carried out macro and micro adjustments to maintain the sustainable development of cities. This cluster includes terms such as "cellular automata", "urban sustainability" and "transport value capture", which reflect advanced modeling methods in transportation policy research.

Figure 9. Keyword clustering network.

Figure 9 presents a keyword clustering network where horizontal coloured lines divide the network into distinct thematic clusters, each representing a coherent group of research topics identified through keyword co-occurrence analysis. Within each cluster, curved lines illustrate the co-occurrence relationships between individual keywords. These arcs connect terms that frequently appear together in academic publications, indicating a strong conceptual link. The colour of each curved line corresponds to the cluster in which the connected keywords are located, visually reinforcing the internal consistency of a research theme. When a curved line links keywords across two different clusters, it reflects cross-thematic integration, highlighting the interdisciplinary nature of emerging research. By analysing the horizontal structure and curvature of the connections, it is possible to discern the internal cohesion of each cluster and trace the evolution and convergence of ideas across the broader research landscape. Thus, this dual-layered visualisation reveals both the stability of core themes and the dynamic interplay among them over time. Figure 9 shows that the keyword "United States" appeared early in the research, which is consistent with the results of our statistical analysis of the sources of the literature. During this period, the main focus was on the built environment and TOD, reflecting the initial emphasis on the spatial structure and planning strategies of urban areas. As research deepened, attention turned to urban construction-related issues, as evidenced by keywords such as "density", "pollution" and "climate change", which indicate a growing focus on environmental and sustainability issues. Keywords such as "investment," "housing prices" and "land value capture" emerged during the same period, indicating that the economic and financial impacts of construction and development were more pronounced in areas where rapid rail transit was being built. More recently, the scope of research has expanded to include social and health-related aspects. Keywords such as "older adult," "green space," and "COVID-19" indicate an emerging awareness of the role of urban rail transit in promoting public health, accessibility, and response to global crises. In the most recent period, keywords related to emerging technologies such as "big data", "travel mode choice", and "machine learning" have emerged, indicating a shift in research focus towards the use of advanced data analysis and AI to optimize transportation planning, predict commuting behavior, and improve decision-making processes.

Table 2. Top terms in clustering.

Cluster	Size	Top Terms (Log-Lihood, Ratio-Level)
#0 built environment	76	built environment (23.55, 0.0001); physical activity (12.47, 0.001); travel behavior (12.47, 0.001); walking (10.41, 0.005); environment design (8.83, 0.005);
#1 value capture	58	non-linear effects (11.36, 0.001); spatial heterogeneity (9.7, 0.005); transit-oriented development (9.68, 0.005); urban design (8.52, 0.005); sustainable mobility (8.52, 0.005);
#2 public transit	53	air pollution (10.3, 0.005); beijing (7.46, 0.01); urban transport (6.86, 0.01); decentralization (6.86, 0.01); urban growth (6.86, 0.01);
#3 china	51	value capture (12.13, 0.001); difference-in-differences (9.09, 0.005); built environment (8.88, 0.005); housing price (8.24, 0.005); transportation (7.94, 0.005);
#4 public transport	21	service (11.38, 0.001); equity (6.08, 0.05); gis (6.08, 0.05); urban vitality (5.68, 0.05); multiscale geographically weighted regression (5.68, 0.05);
#5 travel behavior	5	vulnerability (21.51, 0.0001); resilience (15.36, 0.0001); COVID-19 (15.36, 0.0001); climate change (7.76, 0.01); planning (6.56, 0.05);
#6 transport planning	5	cellular automata (20.87, 0.0001); urban sustainability (14.11, 0.001); system dynamics (11.9, 0.001); transit value capture (7.04, 0.01); difference-in-difference model (7.04, 0.01);

3.3.2. Research Trend Analysis

In this study, a comprehensive review of the interaction between urban rail transit and land use has been conducted, highlighting key trends and research hotspots over different time periods. Based on the analysis of the provided diagram (Figure 10), the keywords can be categorized into three distinct stages, each representing an evolving focus within the field. The progression of these stages illustrates the changing dynamics of urban rail transit and land use, influenced by technological advancements and contemporary socio-economic developments.

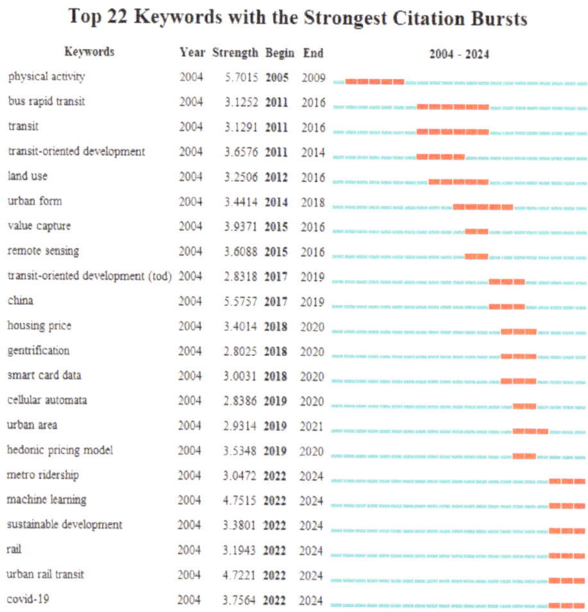

Figure 10. Keywords with the strongest citation bursts.

The early emerging keywords include terms that laid the foundation for the relationship between rail transit and urban development. Keywords such as physical activity, bus rapid transit, and TOD dominate this phase. This reflects the fact that the initial phase of research focused on exploring the concept of TOD and the use of high-precision remote sensing data.

In the mid-term, TOD remains an important theme, while at the same time China is receiving increasing attention, reflecting the region's rapid urbanization and the increase in the mileage of the rail network. The emergence of keywords such as house prices, gentrification and smart card data suggests that much attention is being paid to the socio-economic impacts of rail transport on property values and urban communities. Meanwhile, advanced modeling techniques such as cellular automata methods and feature pricing models have upgraded the original measurements as a tool to quantify the impact of urban rail transit on land use and real estate markets. With the continuous improvement of urbanization in East Asia, keywords such as, China, housing price, gentrification, smart card data, etc., gradually emerged in the middle of the research period. During this period, the level of urbanization in East Asia continued to increase, and the construction of rail transit exacerbated the heterogeneity of urban housing prices, causing more and more middle-class people to gather around rail transit stations. Using advanced modeling methods such as cellular automata, the impact of rail transit on housing prices can be more accurately quantified.

In recent years, major international events and the development of emerging technologies have had a significant impact on academic research priorities. The outbreak of COVID-19 has had a significant impact on urban rail transit operations for some time, prompting academia to focus on research on urban emergency response and resilience. At the same time, updates to machine learning technology have led to the emergence of research hotspots such as subway passenger volume forecasting, and empirical research on non-linear measures has increased significantly.

4. Discussion

Section 3 lays the foundation for Section 4 through keyword clustering analysis, clarifying the research theme and direction. In Section 4, Section 4.1 delves into the bidirectional interactive relationship between rail transit and land use based on the keywords "#0 Built Environment" and "#2 Public Transit" extracted from Section 3. Starting from foundational theory, it analyzes their interaction patterns and corresponding strategies. Section 4.2 combines keywords such as "#1 Value Capture" and "#3 China" to explore land value capture mechanisms and land development strategies. Through practical cases, particularly those in the context of China's urbanization, it further elucidates the practical application of value capture mechanisms. Section 4.3 corresponds to keywords such as "#4 Public Transport" and "#5 Travel Behavior," analyzing travel decision-making and commuting mode selection. Section 4.4 corresponds to "#6 Transport Planning" and "RQ3", reviewing recent technological evolution and key emerging technologies. Overall, a comprehensive research framework with a clear logical structure and seamless continuity is established.

4.1. Bidirectional Interactions Between Urban Rail Transit and Land Use

This section examines the bidirectional relationship between urban rail transit and land use. It focuses on three key aspects: theoretical frameworks, interactive mechanisms and strategic tools for integration. Models such as the Node-Place framework and coupling degree models are introduced to evaluate how transit infrastructure shapes land use patterns, and how land use influences transit performance in turn. Drawing on empirical case studies and emerging technologies such as GIS and AI, the section presents practical approaches to optimizing this coordination. These discussions establish a vital theoretical and empirical basis for subsequent chapters, particularly the analysis of land value capture (Section 4.2) and institutional mechanisms, by clarifying the interaction between physical infrastructure and spatial planning.

4.1.1. Theoretical Foundations of Urban Rail Transit and Land Use

Systematic research on rail transit and land use theory began in the 1960s, with its theoretical evolution following a trajectory from static analysis to dynamic synergy and from a single-perspective approach to multidisciplinary integration. Early studies, represented by Alonso's rent theory [44] and Hansen's accessibility model [45], revealed the one-way influence mechanism of transportation infrastructure on land value but failed to adequately consider the proactive role of planning interventions.

Following the widespread adoption of sustainable development concepts in the 1990s, Calthorpe proposed the TOD theory [46], Bertolini developed the node-place model [1], and Smolka [47] introduced the land value capture theory. These frameworks shifted the research focus toward proactive collaborative planning, emphasizing the mutual feedback development of transportation and land through high-density mixed-use development and functional alignment.

In the 21st century, research has further evolved into complex systems analysis, with scholars incorporating complex adaptive systems theory [48,49] and spatial syntax [50], combined with big data technology, to uncover the non-linear synergistic mechanisms underlying multi-stakeholder interactions. In recent years, with the emergence of carbon neutrality goals, low-carbon TOD and equity research have become new focal points, driving theoretical expansion toward sustainability and social inclusion [51]. Currently, how to integrate multidisciplinary perspectives and quantify dynamic synergistic effects remains a key issue requiring breakthroughs in future research.

4.1.2. Dynamic Interactions Between Urban Rail Transit Development and Land Use

There is a complex two-way interaction between rail transit and land use. On the one hand, rail transit stations enhance the accessibility of surrounding areas, promoting land appreciation and functional restructuring, and creating a "corridor effect" whereby commercial and residential land is concentrated around stations, thereby promoting sustainable transportation development [52]. On the other hand, the density and mix of land use influence the scale of passenger flow and operational efficiency at stations, creating a cyclical cumulative effect of "land use—passenger demand—rail service." For example, Canadian research indicates that the marginal benefits of relying solely on rail transit infrastructure development are limited, but when coordinated with land use policies, it can significantly promote economic growth and achieve TOD objectives [53].

This interactive mechanism is regulated by factors at different scales. At the macro level, urban spatial structure influences the layout efficiency of rail networks, while social factors (such as the concentration of minority groups in employment centers due to transportation restrictions in the United States) also affect passenger flow distribution [54]. At the meso level, the intensity of planning controls in station areas (e.g., zoning reforms and TOD incentives in New York and Toronto) affects the degree of development coordination, while the alignment of transportation corridor planning with regional density directly impacts rail transit utilization rates [55]. At the micro level, built environment design (e.g., "last-mile" connectivity experiences) and mixed-use land use integration [56] influence residents' mode choice, thereby affecting passenger flow generation.

The synergistic development of urban rail transit and land use exhibits significant spatial variations and dynamic changes. For example, cities in Europe and America (such as Barcelona [57] and Stockholm [58]) and Asian cities (such as Shanghai [59] and Tokyo [60]) exhibit different passenger flow-land use coupling patterns due to differing land policies. Additionally, technological advancements (such as spatial measurement methods and machine learning) can optimize transportation network design [61], but issues such as transfer system optimization [62] and social equity (e.g., accessibility for low-income groups) still require consideration. Therefore, future research should establish a more refined analytical framework, combining quantitative models (e.g., site selection simulation) [63], policy evaluation, and social equity analysis to comprehensively reveal their complex interplay.

4.1.3. Strategies for Enhancing Sustainable Rail Transit-Land Use Synergy

Recent developments in data-driven techniques offer fresh opportunities to improve the coordination between rail transit and land use. Geographic Information System (GIS)-based accessibility models refine spatial analysis by charting transit service areas and walkability measures, providing valuable, data-informed insights for policies promoting transit-oriented development [64].

In addition, analyzing mobile phone data helps to monitor passenger flows and travel demand in real time, which helps to strategically locate stations and design intermodal connections [65]. In addition, by integrating AI and machine learning into planning and design, non-linear passenger flow and land use prediction models are built to help urban planners manage potential development challenges [17].

Synergistic development between rail and urban land use can improve station accessibility, reduce congestion and support efficient land development. Key ways in which the two interact include mechanisms such as land value capture, land use regulations and governance reforms. Surplus land values from public transport investments are reinvested by refinancing rail infrastructure through tax improvements, tax incremental financing (TIF) and public–private partnerships (PPPs) [66]. The Urban Development Act encourages high-density and mixed-use development around stations by concentrating foot-traffic-concentrated public buildings as well as commercial areas around rail stations to increase the level of passenger supply on the one hand and reduce the demand for traveling by other modes on the other hand, slowing down urban sprawl [26]. Effective planning requires close coordination and integration between different sectors of government and society [67]. Linkages between stations and neighboring areas can be strengthened through the establishment of a comprehensive governance framework and improved financing models for land development. Future research should focus on AI-driven urban big model planning, integration of multiple types of intermodal transport systems, and summaries of empirical studies in other parts of the world.

4.2. Mechanisms of Land Financing and Value Capture

This section systematically explores the application logic of land value capture mechanisms in urban rail transit financing. By analyzing the relationship between urban rail transit investment and land value appreciation, the study reveals how governments can recover value through taxes, transfer of development rights (TDR) and joint development strategies. The final section focuses on the conflicting interests of governments, developers and the public, emphasizing the need to establish a fair value-sharing framework.

4.2.1. Value Capture

Value capture refers to the process by which governments recover land value gains generated by transport investments through mechanisms such as taxation and the TDR. The revenue collected is then reinvested to subsidize public transportation systems [68,69]. The core principle underlying this approach is that rail transit enhances land value by improving accessibility, creating a transmission chain of "accessibility → land value → capitalization" [70].

The expansion of urban rail transit networks has a multidimensional impact on urban land use patterns and spatial structures. Research indicates that rail infrastructure directly influences land development intensity and housing prices [71], as well as driving urban spatial restructuring by attracting the aggregation of commercial activity. For instance, the construction of the Sapporo subway system in Japan resulted in a substantial increase in the city's built-up area [72]. Within an 800-meter radius of newly constructed subway stations in the outskirts of Beijing, the catering and services sector exhibited a trend towards concentration, while also stimulating the demand for integrated development of underground commercial spaces, parking facilities, and public areas. Cities such as Tokyo, Hong Kong and Shanghai have improved both the travel experience of rail passengers and the efficiency of land utilization by constructing multifunctional underground systems. However, empirical research in Los Angeles suggests that incompatible zoning policies may restrict the development potential of land around subway stations, thereby suppressing the economic growth triggered by rail transit [73].

The conclusion that "rail transit significantly influences surrounding land values" has been widely validated by the academic community. Following the expansion of the metro system in Sunderland, UK, property prices around stations showed significant appreciation [68]; research in Wuhan, China, indicates that mixed-use areas exhibit the most pronounced property value appreciation effects [74]. Research based on complex network theory further reveals that the number of transportation nodes, network clustering coefficients, and land prices are positively correlated, while path length is negatively correlated with land prices [75]. These findings provide theoretical support for land value capture policies and reinforce the practical logic of "using rail transit investment to drive land appreciation gains that are reinvested in infrastructure development."

4.2.2. Mechanism Anatomy of Land Financing

Land financing mechanisms are a core tool for urban spatial design; governments around the world commonly adopt land value capture strategies to recover the spillover benefits of transport infrastructure and convert them into reinvestment capital for public projects. Among these, Transfer of Development Rights (TDRs), Tax Increment Financing (TIF), and Joint Development are three mechanisms that are widely used in countries around the world.

The TDR mechanism establishes a development rights trading market, allowing land rights holders in areas subject to development restrictions to transfer their unused floor area ratio to high-density development areas. This system not only protects historical heritage and ecological space, but also guides the orderly expansion of urban space. Taking the American zoning control system and the Italian historic urban renewal plan as examples, the TDR mechanism has successfully achieved the coordinated governance of land resource protection and urban density improvement [76]. On this basis, empirical studies have shown that the regulated operation of the TDR market can not only avoid the pressure of new tax burdens, but also effectively leverage private capital to participate in the urban development process, forming a financing model of co-governance by multiple entities [77].

The policy effectiveness of the incremental tax financing mechanism is mainly reflected in the fields of urban renewal and infrastructure financing. TIF enables municipalities to reinvest future tax revenues generated from increased property values into infrastructure improvements [78]. A case study from Fort Worth illustrates how TIF revenues were successfully employed to revitalize historic districts, transforming them into residential and commercial hubs [79].

Joint development is an essential strategy in TOD financing, fostering collaboration between governments and private developers to co-finance infrastructure projects. Under this model, governments provide land-use incentives (such as density bonuses or direct development rights) in exchange for private investment in public transit or urban renewal projects [80]. Case studies from Hong Kong's MTR and Tokyo's railway-led urban development demonstrate that aligning real estate projects with transport infrastructure investments enhances financial sustainability and urban connectivity [81]. However, researchers emphasize that clear governance frameworks and equitable revenue-sharing models are necessary to mitigate risks and ensure public benefits [77].

Future research on land mechanisms should focus on the dual-wheel drive of institutional innovation and technological empowerment. Mechanism adaptability should be deepened through cross-institutional environmental comparisons and multi-dimensional integrated design. AI dynamic models should be combined to achieve precise governance. At the same time, attention should be paid to social spatial justice, and a resilient urban development framework that balances the efficiency of land capitalization with social equity should be constructed. This will provide a systematic policy toolbox for addressing climate change and urban renewal.

4.2.3. Key Conflicts: Government–Market–Public Game in Value Distribution

The LVC mechanism is a crucial tool for financing urban development, yet it often generates conflicts among three major stakeholders: the government, the private sector, and the public. These conflicts are particularly pronounced in China's state-owned land system, where the government holds monopoly control over land supply and uses land-based financing to fund infrastructure and economic development [82]. While LVC strategies such as public land leasing, taxation, and joint development projects can generate substantial revenue, they also exacerbate social inequalities, leading to disputes over income distribution, property affordability, and urban redevelopment priorities.

From a government perspective, land sales and leasing are major revenue sources for local authorities, which depend on these funds for infrastructure expansion and debt servicing [83]. However, excessive reliance on land-based financing has led to soaring property prices, placing home ownership beyond the reach of middle- and low-income groups [84]. The government's dual role as both a regulator and a land seller creates conflicts of interest, as policies intended to curb speculation often contradict the need for sustained revenue generation.

The private sector, particularly real estate developers, advocates for reduced government intervention to facilitate market-driven development. However, developers tend to prioritize high-end housing projects, contributing to land supply distortions and reducing the availability of affordable housing [85]. In some cases, PPPs have attempted to balance market efficiency and public benefit, as seen in Hong Kong's MTR rail-plus-property model and Tokyo's railway-led urban development. However, without equitable benefit-sharing mechanisms, private developers often capture a disproportionate share of land appreciation gains, leading to public discontent [82].

The public, particularly lower-income residents, faces increasing housing affordability challenges due to rising land values. Public opposition to land expropriation, speculative investments, and urban renewal projects has become more vocal, especially in cities like Guangzhou and Shenzhen, where disputes over compensation for displaced residents have led to protests [86]. The need for inclusive development policies, such as inclusionary housing and social rental housing programs, has become more pressing.

To address the conflicts arising from LVC, it is essential to establish a comprehensive and transparent land value redistribution framework. Policymakers should focus on three key areas. First, strengthening LVC regulations by introducing progressive land taxation policies and equitable redistribution mechanisms can ensure that land value gains benefit governments, developers, and the public alike. Second, balancing market efficiency and social equity through inclusive housing policies, such as affordable housing quotas and community benefit agreements, can mitigate market-driven inequalities and enhance public well-being. Third, increasing public participation in urban governance by establishing community engagement platforms will allow local voices to influence planning decisions, fostering a more democratic and transparent land development process.

Future research should focus on resolving conflicts between government, market forces, and public interests through interdisciplinary approaches. Key areas of exploration include redefining the government's role by separating land ownership and regulatory functions, innovating market mechanisms through dynamic value-added income distribution models, and integrating digital public participation tools such as blockchain contracts and digital twin negotiation systems. Additionally, researchers should work on quantifying spatial justice using three-dimensional land price modeling and AI-driven site selection algorithms for affordable housing. Comparative studies across China's city policy database will contribute to a political economy framework tailored to China's unique land governance system. Finally, urban experimentation through policy laboratories will provide empirical insights into balancing efficiency and fairness in urbanization strategies.

4.3. The Impact of the Built Environment on Travel Behavior

This section focuses on the keywords '0# built environment' and '5# travel behaviour' in keyword clustering and establishes a three-layer analytical framework. Firstly, beginning with spatial attributes such as floor area ratio and land use patterns, we clarify how these attributes, alongside individual socioeconomic characteristics, influence commuting patterns. This verifies the correlation between the two themes. Secondly, within the TOD framework and incorporating behavioral economics theory, we analyze the impact of economic costs, spatial form and external shocks on travel decisions. Thirdly, we strengthen the logical connections through technical validation by integrating data from both sources using technologies such as big data and AI modeling. This lays the groundwork for further research on technology-enabled TOD models and sustainable travel strategies in future smart city development.

4.3.1. The Role of the Built Environment in Mobility Choices

Numerous empirical studies have shown that urban plot ratio, land use patterns, transport interchange patterns, and the distribution and quality of transport facilities play a crucial role in the use of transport modes [87]. Studies have shown that the socio-economic characteristics of individuals and households [88] and the spatial attributes of the built environment [89] together shape commuting patterns and overall travel demand. In particular, good neighborhood planning, high-density urban development and a well-developed street network can significantly promote the use of walking, cycling and public transport, thereby influencing residents' travel mode choices. Well-designed urban areas with integrated transport systems encourage active travel modes, while poorly planned environments discourage non-motorized travel and increase car dependence [90]. These findings emphasize the importance of linking urban design and transport policy for sustainable travel.

The extent of urban sprawl and the pedestrian environment are also key factors influencing travel. A study in Las Vegas found that unconstrained urban sprawl reduces the ease of walking [91]. Research in Wellington, New Zealand, found that land use and transport policies influence active commuting behavior, with GIS and WILUTE modeling demonstrating that increasing bus frequency and reducing fares significantly increased walking and cycling rates [92]. In Beijing, the large number of bike-sharing bikes around underground stations has improved the 'last mile' problem, expanding the underground traveling distance and the number of public transport riders [93]. A study using graphical Fourier transform theory found that bicycle-friendly transport corridors in Seoul reduced rail congestion, illustrating the complementary relationship between bicycles and public transport [94].

4.3.2. Travel Decisions and Commuter Mode Choices

Commuting mode choice is an important aspect of urban transport planning and is influenced by a combination of economic, spatial and behavioral factors. Within the framework of TOD, commuting decisions are influenced by multiple factors such as economic conditions, transport policies, and infrastructure investments [95].

Behavioral economics has highlighted the role of travel costs and economic incentives in commuting mode choice, especially among low-income populations [95]. The TPB research has shown that travel costs, convenience and personal attitudes directly affect travel mode preferences, with people relying less on public transport and using active transport or cars when fares increase. A case study in Wellington, New Zealand, showed that lower fares and increased frequency of service on public tools significantly increased ridership and active commuting behavior [92]. Urban form also plays an important role in influencing commuters' preferences and multi-modal transport switching. Studies have shown that the centrality of the metro network directly affects commuting patterns, and that transit nodes that integrate multi-modal transport minimize transfer costs and improve overall commuting efficiency [4]. A study of integrated transport nodes in Europe has shown that public transport utilization is higher in cities that have effectively combined selected multi-modal transport options such as rail, buses and microtransit [90]. In Germany, where a public transport fare-free scheme was carried out for a period of time, it was found that such policies could temporarily increase ridership but were not very effective in shifting transport choices in the long term and required more urban planning design interventions [96]. In the current phase of natural disasters and extreme events, the resilience of transport infrastructure is increasingly recognized as a key influence on commuting mode choice. When social or natural emergencies occur, disruptions in the public transport network can lead to a significant shift towards the use of private vehicles, increasing congestion and environmental burdens [97]. These findings highlight the importance of resilient public transport planning for sustainable urban transport systems. Technological advances such as mobile phone signaling data, AI-supported traffic modeling and multimodal integration strategies are gradually transforming transport planning [68]. Combining big data analytics and AI to predict commuting trends optimizes public transport pricing and infrastructure investment.

4.3.3. Technological Innovations in Mobility Planning

Advances in big data analytics, AI and Machine Learning can optimize the way cities plan transport and operate their public transport systems. New technologies such as real-time traffic monitoring, predictive analytics and demand response have greatly improved the efficiency and accessibility of traffic coordination. Artificial intelligence-driven multimodal transport macromodels have shown that these models are able to analyze complex travel behavior patterns and provide more adaptive and humane transport solutions [98]. A study of the Beijing Metro found that the application of mobile phone data and machine learning algorithms to analyze origin–destination (OD) passenger flows provided deeper insights into commuter mobility trends and improved the accuracy of transit demand forecasts [99].

TOD strategies have been improved through technological advances that have increased the level of coordination between land use and transit integration models. Research on GIS has shown that remote sensing technology is an artificial intelligence tool that can now be used to manually model the impact of accessibility and optimize the location of transit hubs through satellite imagery [100]. Research in digital twin technology further supports the implementation of TOD planning by creating real-time urban simulations that help decision makers predict the long-term effects of public transport investments [101]. These innovations contribute to evidence-based decision-making, reduce congestion, improve multimodal transport and encourage sustainable urban development.

The deployment of autonomous vehicles (AVs) and shared mobility is another emerging trend. A study on AV integration in Singapore found that strategically deploying AVs within first-mile and last-mile networks enhances public transit ridership and accessibility [102]. However, research suggests that unregulated AV expansion could undermine public transport systems, potentially increasing congestion and emissions unless carefully managed through policy interventions and smart urban planning [103].

Advancements in AI, ML, and big data analytics are reshaping mobility planning and TOD implementation, offering new opportunities for sustainable, data-driven urban transport solutions. Future research should explore AI-powered transit management, digital twin urban modeling, and climate-adaptive transport policies to ensure efficient and resilient urban mobility systems.

4.4. *Key Techniques: Categories, Applications, and Limitations*

4.4.1. History of Technological Development

In the early 2000s, urban rail transit and land use integration began with the rise of 3S technologies: GIS, GPS and RS, GIS provided the spatial foundation for land-use zoning, buffer analysis, and route planning [104,105]. RS allowed for large-scale monitoring of urban sprawl and detection of built environment changes through satellite imagery [106,107]. GPS data supported travel pattern studies and informed accessibility models by pinpointing real-time transit usage [108]. 3S technology has initially established a technical framework for research on the integration of urban rail transit and land use, addressing fundamental issues such as spatial analysis and data acquisition. However, it is difficult to explore the complex interactive relationship between the two. As research deepens and faces more complex interactions between land use and transportation, the data foundation provided by 3S technology alone is insufficient to meet the demand, leading to the emergence of spatial modeling tools.

As urban datasets expanded, spatial modeling tools evolved. LUTI models became increasingly influential, simulating how land development intensity affected ridership, congestion, and housing prices [14]. Researchers introduced Node–Place models to balance land intensity ("place") with transport connectivity ("node"), which informed optimal site selection for transit hubs [1,109,110]. With the digitization of transport systems, researchers began leveraging smart card data, mobile phone GPS, and POI data to assess urban mobility and land use on a granular level [111]. Transit authorities used this information to derive OD matrices and visualize passenger flows in real time [112]. Additionally, cellular automata, system dynamics models, and real-time geospatial platforms were deployed to simulate urban evolution, providing interactive decision tools for planners [113]. These innovations allowed for dynamic modeling of traffic, land value fluctuations, and housing demand near stations. Spatial modeling tools effectively simulate the mutual influence between land use and transportation by constructing various models, thereby optimizing the layout of transportation facilities.

The most recent phase has seen a surge in AI-powered urban analytics, with machine learning applied to forecast transit ridership, predict land values, and detect gentrification risks around station areas [114]. Deep learning models have further enhanced the accuracy of land-use classification and mode choice prediction [115]. At the infrastructure level, digital twins replicate entire cities—including rail lines, buildings, and demographic flows—within 3D, real-time models that support predictive simulation and crisis management [116]. These twins integrate data from IoT devices, BIM models, and transit APIs, enabling real-time planning for congestion, disaster response, and land policy optimization [117]. In parallel, smart city platforms are operationalizing these tools into municipal planning processes, making transit-land coordination more intelligent, equitable, and adaptive [118]. Artificial intelligence and digital twin technologies have enabled intelligent prediction and management of urban transportation and land use, greatly improving the scientific nature and foresight of decision-making, and promoting the integration of urban rail transit and land use to a higher level. Future research should focus on integrating AI ethics, open-source standards, and community engagement platforms to foster more inclusive and resilient cities.

4.4.2. Key New Technologies

In recent years, urban rail transit planning and land use coordination have increasingly relied on computational intelligence, driven by the rapid evolution of artificial intelligence, data science, and simulation modeling. These technologies not only provide analytical rigor but also enable predictive modeling, scenario simulation, and real-time decision support (Table 3).

Machine learning has emerged as a fundamental tool in urban predictive modeling, particularly for land use classification, mobility pattern recognition, and demand forecasting [119]. Supervised learning models, including Random Forest, XGBoost, and Support Vector Machines, demonstrate high accuracy in predicting land use categories and estimating transit hub passenger flows [120]. Unsupervised techniques such as K-means clustering and Latent Dirichlet Allocation effectively identify latent spatial patterns in Points of Interest datasets and geotagged social media data [121]. While ML enhances automation and scalability in urban analytics, its performance may decline in heterogeneous environments, and model interpretability remains a persistent challenge.

Deep learning excels in processing complex spatial data and high-resolution imagery. Convolutional Neural Networks extract fine-grained spatial features from satellite and street-view imagery, supporting zoning analysis and transit-oriented development mapping [122]. Generative Adversarial Networks, exemplified by LUCGAN+, simulate urban form evolution under diverse planning scenarios [123]. Graph Neural Networks model topological relationships between transit nodes, pedestrian networks, and land parcels [124]. Despite enabling high-precision spatial analysis, DL methods demand extensive labeled datasets and significant computational resources.

Evolutionary and swarm intelligence algorithms address complex trade-offs in land allocation. Genetic Algorithms and Particle Swarm Optimization optimize land use configurations under accessibility, ecological, and compactness constraints [125]. Advanced techniques like MOEA/D and NSGA-II enable high-resolution exploration of transit-oriented development layouts [126]. Hybrid approaches can effectively simulate the long-term interactions between land use and transport. Cellular automata and Markov chain models project urban expansion under different planning scenarios [127], while simulation results generated by the cellular automata model considering spatial interaction are significantly improved [128]. However, a key limitation is their inability to respond dynamically to real-time fluctuations in land markets, travel demand and zoning regulations.

Table 3. Key techniques: categories, applications, and limitations.

Technology	Key Methods	Applications	Strengths	Limitations
Machine Learning	Random Forest, XGBoost, Support Vector Machines, K-means Clustering, Latent Dirichlet Allocation	Land use classification, identifying spatial patterns	High accuracy, scalable automation, effective for classification and clustering tasks	Declines in heterogeneous environments, poor model interpretability
Deep Learning	Convolutional Neural Networks (CNNs) Generative Adversarial Networks, Graph Neural Networks (GNNs)	Zoning analysis, simulating urban form evolution, modeling topological relations	Handles complex and high-resolution spatial data, high-precision spatial analysis	Requires large labeled datasets, high computational resource demands
Evolutionary and Swarm Intelligence	Genetic Algorithms (GA), Particle Swarm Optimization (PSO), MOEA/D, NSGA-II	Land use configuration optimization, transit-oriented development layout planning	Good at solving multi-objective optimization, effective in exploring planning scenarios	Cannot dynamically respond to real-time changes in land markets, travel demand, and zoning regulations
Simulation Models	Cellular Automata Markov Chain Models	Urban expansion projection, land use–transport interaction simulation	Improved accuracy when considering spatial interaction	Lack of responsiveness to real-time fluctuations

5. Conclusions and Future Prospects

5.1. Research Progress and Existing Challenges

In the past two decades, with the acceleration of global urbanization and technological innovation in analysis, research on the collaborative mechanism between urban rail transit (URT) and land use has formed a systematic theoretical framework. This systematic review shows that (1) the research network shows significant geographical agglomeration characteristics, with China and the United States as the core nodes of knowledge production, forming a global academic cooperation network; (2) research mainly focuses on three major areas: the spatial feedback mechanism between the built environment and land use, the capture of rail transit premiums and innovation in financing models, multi-modal transportation network and travel decision behavior research; (3) the methodology has evolved through three stages: a transition from the neoclassical ground rent theory model to the spatial econometric model. In recent years, through deep integration with machine learning algorithms and smart city simulation technology, a comprehensive analysis method for multi-scale and multi-source data has been formed.

Current research still has three limitations: First, existing results are mostly concentrated on developed urban agglomerations in the northern hemisphere, with insufficient attention paid to urbanized areas in the southern hemisphere. Second, the dynamic complexity caused by the spatial expansion of megacities urgently requires the development of an integrated model that takes into account both macrostructural regulation and microbehavioral simulation. Third, there are institutional barriers to the data acquisition mechanism, and the phenomenon of data silos across departments and fields restricts the depth of research. A sustainable cross-institutional data collaboration platform needs to be established to ensure that data infrastructure keeps pace with technological development.

5.2. Future Research Directions and Implementation Paths

(1) Technology integration: Researchers should develop a decision support platform that integrates artificial intelligence and urban systems engineering. By integrating multi-source and heterogeneous data streams such as Internet of Things infrastructure data and social media behavior trajectories, an intelligent model with a dynamic feedback mechanism can be constructed. The focus is on breaking through the limitations of traditional machine learning methods in representing spatial heterogeneity and achieving a transition from static correlation analysis to dynamic analysis.

(2) Data governance: Establish a data collection framework for two-way calibration of "macro policies-micro behaviors". At the macro level, strengthen the standardized processing of new types of geospatial data such as satellite remote sensing and mobile phone signaling. At the micro level, the creation of a dynamic database, including behaviors such as commuting preferences and land transaction decisions, enables the secure integration of cross-domain data while ensuring that data security and privacy are effectively protected.

(3) Institutional innovation: Promote the formation of a collaborative research mechanism across disciplines and sectors, and establish a multi-actor participation mechanism covering government, businesses, and communities to find solutions to the challenges of spatial rights and interest allocation.

By addressing these issues, research in the field of rail transit and land use can transition from descriptive to normative and actionable, enabling cities to use urban transport systems as a catalyst for an equitable, sustainable, and resilient urban future.

Author Contributions: Conceptualization, N.C. and H.X.; methodology, H.Y.; writing—original draft preparation, H.Y. and N.C.; writing—review and editing, N.C. and H.X.; figures, H.X.; funding acquisition, N.C. All authors have read and agreed to the published version of the manuscript.

Funding: This research was supported by the National Natural Science Foundation of China (Grant No. 42401264) and the Humanities and Social Sciences Research Fund of the Ministry of Education (Grant No. 24YJCZH036).

Data Availability Statement: The dataset is available in Mendeley Data at [https://doi.org/10.17632/f37x36h5my.1].

Conflicts of Interest: The authors declare no conflicts of interest.

References

1. Bertolini, L. Nodes and Places: Complexities of Railway Station Redevelopment. *Eur. Plan. Stud.* **1996**, *4*, 331–345. [CrossRef]
2. Cervero, R. *The Transit Metropolis: A Global Inquiry*; Island Press: Washington, DC, USA, 1998.
3. Bertolini, L. Spatial Development Patterns and Public Transport: The Application of an Analytical Model in the Netherlands. *Plan. Pract. Res.* **1999**, *14*, 199–210. [CrossRef]
4. Peng, B.; Wang, T.; Zhang, Y.; Li, C.; Lu, C. Spatially Varying Effect Mechanism of Intermodal Connection on Metro Ridership: Evidence from a Polycentric Megacity with Multilevel Ring Roads. *ISPRS Int. J. Geo-Inf.* **2024**, *13*, 353. [CrossRef]
5. Sharifiasl, S.; Kharel, S.; Pan, Q.; Li, J. Assessing the Impact of Transit Accessibility on Employment Density: A Spatial Analysis of Gravity-Based Accessibility Incorporating Job Matching, Transit Service Types, and First/Last Mile Modes. *J. Transp. Geogr.* **2024**, *121*, 104053. [CrossRef]
6. Hu, N.; Legara, E.F.; Lee, K.K.; Hung, G.G.; Monterola, C. Impacts of Land Use and Amenities on Public Transport Use, Urban Planning and Design. *Land Use Policy* **2016**, *57*, 356–367. [CrossRef]
7. Abdi, M.H.; Lamiquiz-Dauden, P.J. Transit-Oriented Development in Developing Countries: A Qualitative Meta-Synthesis of Its Policy, Planning and Implementation Challenges. *Int. J. Sustain. Transp.* **2022**, *16*, 195–221. [CrossRef]
8. Jamme, H.-T.; Rodriguez, J.; Bahl, D.; Banerjee, T. A Twenty-Five-Year Biography of the TOD Concept: From Design to Policy, Planning, and Implementation. *J. Plan. Educ. Res.* **2019**, *39*, 409–428. [CrossRef]
9. Ewing, R.; Cervero, R. Travel and the Built Environment: A Synthesis. *Transp. Res. Rec. J. Transp. Res. Board* **2001**, *1780*, 87–114. [CrossRef]
10. Ewing, R.; Cervero, R. Travel and the Built Environment: A Meta-Analysis. *J. Am. Plan. Assoc.* **2010**, *76*, 265–294. [CrossRef]
11. Tian, G.; Danton, B.; Ewing, R.; Li, B. Varying Influences of the Built Environment on Household Travel in the United States—An Update with 36 Diverse Regions and Machine Learning. *Cities* **2024**, *155*, 105490. [CrossRef]
12. Luo, J.; Liu, P.; Xu, W.; Zhao, T.; Biljecki, F. A Perception-Powered Urban Digital Twin to Support Human-Centered Urban Planning and Sustainable City Development. *Cities* **2025**, *156*, 105473. [CrossRef]
13. Hess, D.B.; Lombardi, P.A. Policy Support for and Barriers to Transit-Oriented Development in the Inner City: Literature Review. *Transp. Res. Rec. J. Transp. Res. Board* **2004**, *1887*, 26–33. [CrossRef]
14. Acheampong, R.A.; Silva, E.A. Land Use–Transport Interaction Modeling: A Review of the Literature and Future Research Directions. *J. Transp. Land Use* **2015**, *8*, 11–38. [CrossRef]
15. Yigitcanlar, T.; Kamruzzaman, M. Investigating the Interplay between Transport, Land Use and the Environment: A Review of the Literature. *Int. J. Environ. Sci. Technol.* **2014**, *11*, 2121–2132. [CrossRef]
16. Xia, J.; Zhang, Y. Where Are Potential Areas for Transit-Oriented Development (TOD)-Exploring the Demands for Built Environment for TOD Planning. *Sustainability* **2022**, *14*, 8364. [CrossRef]
17. Rocco Di Torrepadula, F.; Napolitano, E.V.; Di Martino, S.; Mazzocca, N. Machine Learning for Public Transportation Demand Prediction: A Systematic Literature Review. *Eng. Appl. Artif. Intell.* **2024**, *137*, 109166. [CrossRef]
18. Geurs, K.T.; van Wee, B. Accessibility Evaluation of Land-Use and Transport Strategies: Review and Research Directions. *J. Transp. Geogr.* **2004**, *12*, 127–140. [CrossRef]
19. Yao, M.; Yao, B.; Cenci, J.; Liao, C.; Zhang, J. Visualisation of High-Density City Research Evolution, Trends, and Outlook in the 21st Century. *Land* **2023**, *12*, 485. [CrossRef]
20. Fang, Q.; Inoue, T.; Li, D.; Liu, Q.; Ma, J. Transit-Oriented Development and Sustainable Cities: A Visual Analysis of the Literature Based on CiteSpace and VOSviewer. *Sustainability* **2023**, *15*, 8223. [CrossRef]
21. He, Y.; Liu, B.; Xu, C.; Wu, D. Literature Review on Public Transport and Land Use: Based on CiteSpace Statistical Analysis. *Land* **2025**, *14*, 1096. [CrossRef]
22. Batty, M. The Size, Scale, and Shape of Cities. *Science* **2008**, *319*, 769–771. [CrossRef] [PubMed]

23. Tong, Z.; Zhang, Z.; An, R.; Liu, Y.; Chen, H.; Xu, J.; Fu, S. Detecting Anomalous Commuting Patterns: Mismatch between Urban Land Attractiveness and Commuting Activities. *J. Transp. Geogr.* **2024**, *116*, 103867. [CrossRef]
24. Li, Z.; Chen, B.; Wu, S.; Su, M.; Chen, J.M.; Xu, B. Deep Learning for Urban Land Use Category Classification: A Review and Experimental Assessment. *Remote Sens. Environ.* **2024**, *311*, 114290. [CrossRef]
25. Drici, H.; Carpio-Pinedo, J. Urban Land Use Mix and AI: A Systematic Review. *Cities* **2025**, *165*, 106102. [CrossRef]
26. Ibraeva, A.; de Almeida Correia, G.H.; Silva, C.; Antunes, A.P. Transit-Oriented Development: A Review of Research Achievements and Challenges. *Transp. Res. Part A Policy Pract.* **2020**, *132*, 110–130. [CrossRef]
27. An, D.; Tong, X.; Liu, K.; Chan, E.H.W. Understanding the Impact of Built Environment on Metro Ridership Using Open Source in Shanghai. *Cities* **2019**, *93*, 177–187. [CrossRef]
28. Guo, Y.; He, S.Y. Built Environment Effects on the Integration of Dockless Bike-Sharing and the Metro. *Transp. Res. Part D Transp. Environ.* **2020**, *83*, 102335. [CrossRef]
29. Gan, Z.; Yang, M.; Feng, T.; Timmermans, H.J.P. Examining the Relationship between Built Environment and Metro Ridership at Station-to-Station Level. *Transport. Res. Part D Transport. Environ.* **2020**, *82*, 102332. [CrossRef]
30. Ding, C.; Cao, X.; Liu, C. How Does the Station-Area Built Environment Influence Metrorail Ridership? Using Gradient Boosting Decision Trees to Identify Non-Linear Thresholds. *J. Transp. Geogr.* **2019**, *77*, 70–78. [CrossRef]
31. Li, S.; Lyu, D.; Huang, G.; Zhang, X.; Gao, F.; Chen, Y.; Liu, X. Spatially Varying Impacts of Built Environment Factors on Rail Transit Ridership at Station Level: A Case Study in Guangzhou, China. *J. Transp. Geogr.* **2020**, *82*, 102631. [CrossRef]
32. Li, Z.; Han, Z.; Xin, J.; Luo, X.; Su, S.; Weng, M. Transit Oriented Development among Metro Station Areas in Shanghai, China: Variations, Typology, Optimization and Implications for Land Use Planning. *Land Use Policy* **2019**, *82*, 269–282. [CrossRef]
33. Chen, E.; Ye, Z.; Wang, C.; Zhang, W. Discovering the Spatio-Temporal Impacts of Built Environment on Metro Ridership Using Smart Card Data. *Cities* **2019**, *95*, 102359. [CrossRef]
34. Higgins, C.D.; Kanaroglou, P.S. A Latent Class Method for Classifying and Evaluating the Performance of Station Area Transit-Oriented Development in the Toronto Region. *J. Transp. Geogr.* **2016**, *52*, 61–72. [CrossRef]
35. Yang, L.; Chau, K.W.; Szeto, W.Y.; Cui, X.; Wang, X. Accessibility to Transit, by Transit, and Property Prices: Spatially Varying Relationships. *Transp. Res. Part D: Transp. Environ.* **2020**, *85*, 102387. [CrossRef]
36. El-Assi, W.; Salah Mahmoud, M.; Nurul Habib, K. Effects of Built Environment and Weather on Bike Sharing Demand: A Station Level Analysis of Commercial Bike Sharing in Toronto. *Transportation* **2017**, *44*, 589–613. [CrossRef]
37. Higgins, C.D.; Kanaroglou, P.S. Forty Years of Modelling Rapid Transit's Land Value Uplift in North America: Moving beyond the Tip of the Iceberg. *Transp. Rev.* **2016**, *36*, 610–634. [CrossRef]
38. Singh, Y.J.; Lukman, A.; Flacke, J.; Zuidgeest, M.; Van Maarseveen, M.F.A.M. Measuring TOD around Transit Nodes—Towards TOD Policy. *Transp. Policy* **2017**, *56*, 96–111. [CrossRef]
39. Vale, D.S.; Viana, C.M.; Pereira, M. The Extended Node-Place Model at the Local Scale: Evaluating the Integration of Land Use and Transport for Lisbon's Subway Network. *J. Transp. Geogr.* **2018**, *69*, 282–293. [CrossRef]
40. Li, S.; Lyu, D.; Liu, X.; Tan, Z.; Gao, F.; Huang, G.; Wu, Z. The Varying Patterns of Rail Transit Ridership and Their Relationships with Fine-Scale Built Environment Factors: Big Data Analytics from Guangzhou. *Cities* **2020**, *99*, 102580. [CrossRef]
41. Shao, Q.; Zhang, W.; Cao, X.; Yang, J.; Yin, J. Threshold and Moderating Effects of Land Use on Metro Ridership in Shenzhen: Implications for TOD Planning. *J. Transp. Geogr.* **2020**, *89*, 102878. [CrossRef]
42. Lu, K.; Zhang, L.; Li, S.; Huang, Y.; Ding, X.; Hao, J.; Huang, S.; Li, X.; Lu, F.; Zhang, H. Urban Rail Transit in China: Progress Report and Analysis (2015–2023). *Urban Rail Transit* **2025**, *11*, 1–27. [CrossRef]
43. Wang, Z.; Guo, R.; Zou, L.; Li, T.; Yao, X. Impact Evaluation of COVID-19 on Transit Ridership: A Case Study of the Beijing Subway. *Urban Rail Transit* **2024**, *10*, 297–316. [CrossRef]
44. Alonso, W. *Location and Land Use: Toward a General Theory of Land Rent*; Harvard University Press: Cambridge, MA, USA, 1964; ISBN 978-0-674-73085-4.
45. Hansen, W.G. How Accessibility Shapes Land Use. *J. Am. Inst. Plan.* **1959**, *25*, 73–76. [CrossRef]
46. Calthorpe, P. *The Next American Metropolis—Ecology, Community, and the American Dream*; Princeton Architectural Press: New York, NY, USA; Abingdon, UK, 1993; Volume 60.
47. Smolka, M.O. A New Look at Value Capture in Latin America. Available online: https://www.lincolninst.edu/app/uploads/2024/04/2099_1420_New_Look_at_Value_Capture_Latin_America_0712LL.pdf (accessed on 12 March 2025).
48. Nel, D.; Du Plessis, C.; Landman, K. Planning for Dynamic Cities: Introducing a Framework to Understand Urban Change from a Complex Adaptive Systems Approach. *Int. Plan. Stud.* **2018**, *23*, 250–263. [CrossRef]
49. Rhodes, M.L.; Murray, J. Collaborative Decision Making in Urban Regeneration: A Complex Adaptive Systems Perspective. *Int. Public Manag. J.* **2007**, *10*, 79–101. [CrossRef]
50. Hillier, B. Studying Cities to Learn about Minds: Some Possible Implications of Space Syntax for Spatial Cognition. *Environ. Plan. B Plan. Des* **2012**, *39*, 12–32. [CrossRef]

51. Pezeshkriejad, P.; Monajem, S.; Mozafari, H. Evaluating Sustainability and Land Use Integration of BRT Stations via Extended Node Place Model, an Application on BRT Stations of Tehran. *J. Transp. Geogr.* **2020**, *82*, 102626. [CrossRef]
52. Cervero, R.; Sarmiento, O.L.; Jacoby, E.; Fernando Gomez, L.; Neiman, A. Influences of Built Environments on Walking and Cycling: Lessons from Bogota. *Int. J. Sustain. Transp.* **2009**, *3*, 203–226. [CrossRef]
53. Lavery, T.A.; Kanaroglou, P. Rediscovering Light Rail: Assessing the Potential Impacts of a Light Rail Transit Line on Transit Oriented Development and Transit Ridership. *Transp. Lett.* **2012**, *4*, 211–226. [CrossRef]
54. Grengs, J. Equity and the Social Distribution of Job Accessibility in Detroit. *Environ. Plan. B Plan. Des.* **2012**, *39*, 785–800. [CrossRef]
55. Antanaitis, A. Developing Up and Not Out: Understanding the Barriers to and Opportunities for Reurbanization along Waterloo's Central Transit Corridor. Master's Thesis, University of Waterloo, Waterloo, ON, Canada, 2015.
56. Cervero, R.; Kockelman, K. Travel Demand and the 3Ds: Density, Diversity, and Design. *Transport. Res. Part D Transport. Environ.* **1997**, *2*, 199–219. [CrossRef]
57. MartÍn, J.C.; Gutiérrez, J.; Román, C. Data Envelopment Analysis (DEA) Index to Measure the Accessibility Impacts of New Infrastructure Investments: The Case of the High-Speed Train Corridor Madrid-Barcelona-French Border. *Reg. Stud.* **2004**, *38*, 697–712. [CrossRef]
58. Paulsson, A. The City That the Metro System Built: Urban Transformations and Modalities of Integrated Planning in Stockholm. *Urban Stud.* **2020**, *57*, 2936–2955. [CrossRef]
59. Tian, L.; Li, Y.; Yan, Y.; Wang, B. Measuring Urban Sprawl and Exploring the Role Planning Plays: A Shanghai Case Study. *Land Use Policy* **2017**, *67*, 426–435. [CrossRef]
60. Sani Roychansyah, M.; Ishizaka, K.; Omi, T. Considerations of Regional Characteristics for Delivering City Compactness: Case of Studies of Cities in the Greater Tokyo Area and Tohoku Region, Japan. *J. Asian Archit. Build. Eng.* **2005**, *4*, 339–346. [CrossRef]
61. Cats, O.; Gkioulou, Z. Modeling the Impacts of Public Transport Reliability and Travel Information on Passengers' Waiting-Time Uncertainty. *EURO J. Transp. Logist.* **2017**, *6*, 247–270. [CrossRef]
62. Vuchic, V.R. *Urban Transit Systems and Technology*; Wiley: Hoboken, NJ, USA, 2007.
63. Chai, S.; Liang, Q.; Zhong, S. Design of Urban Rail Transit Network Constrained by Urban Road Network, Trips and Land-Use Characteristics. *Sustainability* **2019**, *11*, 6128. [CrossRef]
64. Langford, M.; Fry, R.; Higgs, G. Measuring Transit System Accessibility Using a Modified Two-Step Floating Catchment Technique. *Int. J. Geogr. Inf. Sci.* **2012**, *26*, 193–214. [CrossRef]
65. Borole, N.; Rout, D.; Goel, N.; Vedagiri, P.; Mathew, T.V. Multimodal Public Transit Trip Planner with Real-Time Transit Data. *Procedia-Soc. Behav. Sci.* **2013**, *104*, 775–784. [CrossRef]
66. Debrezion, G.; Pels, E.; Rietveld, P. The Impact of Railway Stations on Residential and Commercial Property Value. *J. Real Estate Finance Econ.* **2007**, *35*, 161–180. [CrossRef]
67. Wu, F. *Planning for Growth: Urban and Regional Planning in China*; Routledge: New York, NY, USA, 2015.
68. Du, H.; Mulley, C. The Short-Term Land Value Impacts of Urban Rail Transit: Quantitative Evidence from Sunderland, UK. *Land Use Policy* **2007**, *24*, 223–233. [CrossRef]
69. Zhu, X.; Liu, S. Analysis of the Impact of the MRT System on Accessibility in Singapore Using an Integrated GIS Tool. *J. Transp. Geogr.* **2004**, *12*, 89–101. [CrossRef]
70. Saxe, S.; Miller, E.J. Transit and Land Value Uplift: An Introduction. 2016. Available online: https://uttri.utoronto.ca/wp-content/uploads/sites/19/2017/10/16-02-04-02-Transit-and-Land-Value-Uplift-An-Introduction.pdf (accessed on 12 March 2025).
71. Efthymiou, D.; Antoniou, C. How Do Transport Infrastructure and Policies Affect House Prices and Rents? Evidence from Athens, Greece. *Transp. Res. Part A Policy Pract.* **2013**, *52*, 1–22. [CrossRef]
72. Lin, D.; Broere, W.; Cui, J. Metro Systems and Urban Development: Impacts and Implications. *Tunn. Undergr. Space Technol.* **2022**, *125*, 104509. [CrossRef]
73. Schuetz, J.; Giuliano, G.; Shin, E.J. Does Zoning Help or Hinder Transit-Oriented (Re)Development? *Urban Stud.* **2018**, *55*, 1672–1689. [CrossRef]
74. Li, J.; Huang, H. Effects of Transit-Oriented Development (TOD) on Housing Prices: A Case Study in Wuhan, China. *Res. Transp. Econ.* **2020**, *80*, 100813. [CrossRef]
75. Wen, S.; Shi, J.; Zhang, W. Impact of Urban Rail Transit Network on Residential and Commercial Land Values in China: A Complex Network Perspective. *Complexity* **2021**, *2021*, 8849066. [CrossRef]
76. Colavitti, A.M.; Serra, S. The Transfer of Development Rights as a Tool for the Urban Growth Containment: A Comparison between the United States and Italy. *Pap. Reg. Sci.* **2018**, *97*, 1247–1266. [CrossRef]
77. St Lewis, J. Revitalizing Private Intercity Rail in the U.S. A Case Study of Brightline's Use of Development-Based Land Value Capture. Master's Thesis, Lund University, Lund, Sweden, 2024. Available online: https://lup.lub.lu.se/student-papers/search/publication/9171362 (accessed on 12 March 2025).

78. Weber, R.; Goddeeris, L. Tax Increment Financing: A Property-Based Tool for Local Regeneration? *J. Urban Regen. Renew.* **2007**, *1*, 55–68. [CrossRef]
79. Newman, G.; Lee, J.; Berke, P. Using the Land Transformation Model to Forecast Vacant Land. *J. Land Use Sci.* **2016**, *11*, 450–475. [CrossRef]
80. Chang, Z.; Phang, S.-Y. Urban Rail Transit PPPs: Lessons from East Asian Cities. *Transp. Res. Part A Policy Pract.* **2017**, *105*, 106–122. [CrossRef]
81. Aydın, E. Financing an Urban Express Railway Line via Real Estate Development: Hızray (Istanbul) Case. 2023. Available online: https://polen.itu.edu.tr/items/6850e32e-55f7-4166-bcb1-b79a1cebb960 (accessed on 12 March 2025).
82. Wu, F. Land Financialisation and the Financing of Urban Development in China. *Land Use Policy* **2022**, *112*, 104412. [CrossRef]
83. Tsui, K.Y. China's Infrastructure Investment Boom and Local Debt Crisis. *Eurasian Geogr. Econ.* **2011**, *52*, 686–711. [CrossRef]
84. El-Nagdy, M.; El-borombaly, H.; Khodeir, L. Threats and Root Causes of Using Publicly-Owned Lands as Assets for Urban Infrastructure Financing. *Alex. Eng. J.* **2018**, *57*, 3907–3919. [CrossRef]
85. Nzau, B.; Trillo, C. Harnessing the Real Estate Market for Equitable Affordable Housing Provision through Land Value Capture: Insights from San Francisco City, California. *Sustainability* **2019**, *11*, 3649. [CrossRef]
86. Huang, X.; He, D.; Tang, S.; Li, X. Compensation, Housing Situation and Residents' Satisfaction with the Outcome of Forced Relocation: Evidence from Urban China. *Cities* **2020**, *96*, 102436. [CrossRef]
87. Zhang, M. The Role of Land Use in Travel Mode Choice: Evidence from Boston and Hong Kong. *J. Am. Plan. Assoc.* **2004**, *70*, 344–360. [CrossRef]
88. Hanson, S.; Pratt, G. Reconceptualizing the Links between Home and Work in Urban Geography. *Econ. Geogr.* **1988**, *64*, 299–321. [CrossRef]
89. Sohn, J. Are Commuting Patterns a Good Indicator of Urban Spatial Structure? *J. Transp. Geogr.* **2005**, *13*, 306–317. [CrossRef]
90. Celbiş, M.G.; Bouzouina, L. To What Extent Walking and Biking Are Substitutes or Complements to Public Transport? Interpretable Machine Learning Findings from the University of Lyon, France. *J. Transp. Geogr.* **2025**, *123*, 104071. [CrossRef]
91. Bartshe, M.; Coughenour, C.; Pharr, J. Perceived Walkability, Social Capital, and Self-Reported Physical Activity in Las Vegas College Students. *Sustainability* **2018**, *10*, 3023. [CrossRef]
92. Mackenbach, J.D.; Randal, E.; Zhao, P.; Howden-Chapman, P. The Influence of Urban Land-Use and Public Transport Facilities on Active Commuting in Wellington, New Zealand: Active Transport Forecasting Using the WILUTE Model. *Sustainability* **2016**, *8*, 242. [CrossRef]
93. Fan, A.; Chen, X.; Wan, T. How Have Travelers Changed Mode Choices for First/Last Mile Trips after the Introduction of Bicycle-Sharing Systems: An Empirical Study in Beijing, China. *J. Adv. Transp.* **2019**, *2019*, 5426080. [CrossRef]
94. Li, C.-X.; Yoon, C.-J. Analysis of Urban Rail Public Transport Space Congestion Using Graph Fourier Transform Theory: A Focus on Seoul. *Sustainability* **2025**, *17*, 598. [CrossRef]
95. Cheng, L.; Bi, X.; Chen, X.; Li, L. Travel Behavior of the Urban Low-Income in China: Case Study of Huzhou City. In *Proceedings of the Intelligent and Integrated Sustainable Multimodal Transportation Systems Proceedings from the 13th Cota International Conference of Transportation Professionals (Cictp2013)*; Zhang, L., Wei, H., Li, Z., Zhang, Y., Li, M., Eds.; Elsevier Science B.V.: Amsterdam, The Netherlands, 2013; Volume 96, pp. 231–242.
96. Guajardo Ortega, M.F.; Link, H. Mode Choice Inertia and Shock: Three Months of Almost Fare-Free Public Transport in Germany. *Econ. Transp.* **2025**, *41*, 100382. [CrossRef]
97. Barnes, M. Transit Systems and Ridership under Extreme Weather and Climate Change Stress: An Urban Transportation Agenda for Hazards Geography. *Geogr. Compass* **2015**, *9*, 604–616. [CrossRef]
98. Lukic Vujadinovic, V.; Damnjanovic, A.; Cakic, A.; Petkovic, D.R.; Prelevic, M.; Pantovic, V.; Stojanovic, M.; Vidojevic, D.; Vranjes, D.; Bodolo, I. AI-Driven Approach for Enhancing Sustainability in Urban Public Transportation. *Sustainability* **2024**, *16*, 7763. [CrossRef]
99. Zou, L.; Wang, Z.; Guo, R. Real-Time Prediction of Transit Origin–Destination Flows during Underground Incidents. *Transp. Res. Part C Emerg. Technol.* **2024**, *163*, 104622. [CrossRef]
100. Rodriguez Baquero, L.M.; Quiroga, J.S.; Angulo, M.G.; Gómez Caicedo, M.I.; Costa, L.V.; Ballesteros, V.; Alarcon Villamil, N.O. Bibliometric Analysis on the Application of Artificial Intelligence, Big Data and the Internet of Things in Traffic Management Systems in Latin America: Advances, Challenges and Opportunities. *Preprints* **2024**.
101. Bellini, P.; Collini, E.; Fanfani, M.; Palesi, L.A.I.; Nesi, P. Smart City Digital Twin Platform Architecture for Mobility and Transport Decision Support Systems. In Proceedings of the 2024 IEEE International Conference on Big Data (BigData), Washington, DC, USA, 15–18 December 2024; pp. 5486–5495.
102. Ng, V.; Kim, H.M. Autonomous Vehicles and Smart Cities: A Case Study of Singapore. In *Smart Cities for Technological and Social Innovation*; Kim, H.M., Sabri, S., Kent, A., Eds.; Academic Press: Cambridge, MA, USA, 2021; pp. 265–287.
103. Fagnant, D.J.; Kockelman, K. Preparing a Nation for Autonomous Vehicles: Opportunities, Barriers and Policy Recommendations. *Transp. Res. Part A Policy Pract.* **2015**, *77*, 167–181. [CrossRef]

104. Zhang, L.; Cui, X.; Wang, M.; Hu, Q.; Yu, Y. The Development and Status of 3S Technology in China. In Proceedings of the 2010 2nd International Conference on Advanced Computer Control, Shenyang, China, 27–29 March 2010; IEEE: Piscataway, NJ, USA, 2010; pp. 214–217.
105. Wu, W. Digital Urban Planning Based on "3S" Technology. *J. Anhui Agric. Sci.* **2007**, *35*, 8035–8037.
106. Lemenkova, P. Urban-Rural Environmental Gradient in a Developing City: Testing ENVI GIS Functionality. *arXiv* **2018**, arXiv:1812.10378.
107. Das, S.; Angadi, D.P. Land Use Land Cover Change Detection and Monitoring of Urban Growth Using Remote Sensing and GIS Techniques: A Micro-Level Study. *GeoJournal* **2022**, *87*, 2101–2123. [CrossRef]
108. Alkhazraji, H. Predicting Traffic Speed and Recommending Optimal Routes Using GPS Data: A Smart Navigation System for Commuters. Available online: https://pure.northampton.ac.uk/en/studentTheses/predicting-traffic-speed-and-recommending-optimal-routes-using-gp#:~:text=This%20thesis%20introduces%20a%20novel%20framework%20designed%20to,hour%20ahead%20and%20recommend%20optimal%20routes%20for%20commuters (accessed on 12 March 2025).
109. Baghestani, A.; Najafabadi, S.; Salem, A.; Jiang, Z.; Tayarani, M.; Gao, O. An Application of the Node–Place Model to Explore the Land Use–Transport Development Dynamics of the I-287 Corridor. *Urban Sci.* **2023**, *7*, 21. [CrossRef]
110. Cao, Z.; Asakura, Y.; Tan, Z. Coordination between Node, Place, and Ridership: Comparing Three Transit Operators in Tokyo. *Transp. Res. Part D Transp. Environ.* **2020**, *87*, 102518. [CrossRef]
111. Li, Z.; Tang, J.; Feng, T.; Liu, B.; Cao, J.; Yu, T.; Ji, Y. Investigating Urban Mobility through Multi-Source Public Transportation Data: A Multiplex Network Perspective. *Appl. Geogr.* **2024**, *169*, 103337. [CrossRef]
112. Zeng, W.; Fu, C.-W.; Arisona, S.M.; Erath, A.; Qu, H. Visualizing Mobility of Public Transportation System. *IEEE Trans. Vis. Comput. Graph.* **2014**, *20*, 1833–1842. [CrossRef]
113. Guzman, L.A.; Escobar, F.; Peña, J.; Cardona, R. A Cellular Automata-Based Land-Use Model as an Integrated Spatial Decision Support System for Urban Planning in Developing Cities: The Case of the Bogotá Region. *Land Use Policy* **2020**, *92*, 104445. [CrossRef]
114. Mazzetto, S. A Review of Urban Digital Twins Integration, Challenges, and Future Directions in Smart City Development. *Sustainability* **2024**, *16*, 8337. [CrossRef]
115. Bibri, S.E.; Huang, J.; Jagatheesaperumal, S.K.; Krogstie, J. The Synergistic Interplay of Artificial Intelligence and Digital Twin in Environmentally Planning Sustainable Smart Cities: A Comprehensive Systematic Review. *Environ. Sci. Ecotechnology* **2024**, *20*, 100433. [CrossRef]
116. Lehtola, V.V.; Koeva, M.; Elberink, S.O.; Raposo, P.; Virtanen, J.-P.; Vahdatikhaki, F.; Borsci, S. Digital Twin of a City: Review of Technology Serving City Needs. *Int. J. Appl. Earth Obs. Geoinf.* **2022**, *114*, 102915. [CrossRef]
117. Allam, Z.; Dhunny, Z.A. On Big Data, Artificial Intelligence and Smart Cities. *Cities* **2019**, *89*, 80–91. [CrossRef]
118. Deng, T.; Zhang, K.; Shen, Z.-J. (Max) A Systematic Review of a Digital Twin City: A New Pattern of Urban Governance toward Smart Cities. *J. Manag. Sci. Eng.* **2021**, *6*, 125–134.
119. Mirzakhani, N.; Ali Madad Soltani, A.; Hedayatifard, M. Application of Machine Learning in Understanding Urban Neighborhood Gentrification: A Meta-Synthesis Review. *Cities* **2025**, *162*, 105973. [CrossRef]
120. Peng, C.; Yang, S.; Zhang, P.; Hu, S. Exploring Nonlinear and Interaction Effects of TOD on Housing Rents Using XGBoost. *Cities* **2025**, *158*, 105728. [CrossRef]
121. Wang, J.; Biljecki, F. Unsupervised Machine Learning in Urban Studies: A Systematic Review of Applications. *Cities* **2022**, *129*, 103925. [CrossRef]
122. Xu, L.; Luo, S.; O'Hern, S.; Delbosc, A.; Chen, Z.; Fu, X. Do Protected Cycle Lanes Make Cities More Bike-Friendly? Integrating Street View Images with Deep Learning Techniques. *Cities* **2025**, *161*, 105890. [CrossRef]
123. Yu, B.; Lee, Y.; Sohn, K. Forecasting Road Traffic Speeds by Considering Area-Wide Spatio-Temporal Dependencies Based on a Graph Convolutional Neural Network (GCN). *Transp. Res. Part C Emerg. Technol.* **2020**, *114*, 189–204. [CrossRef]
124. Shanthappa, N.K.; Mulangi, R.H.; Manjunath, H.M. Origin-Destination Demand Prediction of Public Transit Using Graph Convolutional Neural Network. *Case Stud. Transp. Policy* **2024**, *17*, 101230. [CrossRef]
125. Porta, J.; Parapar, J.; Doallo, R.; Rivera, F.F.; Santé, I.; Crecente, R. High Performance Genetic Algorithm for Land Use Planning. *Comput. Environ. Urban Syst.* **2013**, *37*, 45–58. [CrossRef]
126. Masoumi, Z.; van Genderen, J. Artificial Intelligence for Sustainable Development of Smart Cities and Urban Land-Use Management. *Geo-Spat. Inf. Sci.* **2024**, *27*, 1212–1236. [CrossRef]

127. Arsanjani, J.J.; Kainz, W.; Mousivand, A.J. Tracking Dynamic Land-Use Change Using Spatially Explicit Markov Chain Based on Cellular Automata: The Case of Tehran. *Int. J. Image Data Fusion* **2011**, *2*, 329–345. [CrossRef]
128. Xia, C.; Zhang, A.; Wang, H.; Zhang, B. Modeling Urban Growth in a Metropolitan Area Based on Bidirectional Flows, an Improved Gravitational Field Model, and Partitioned Cellular Automata. *Int. J. Geogr. Inf. Sci.* **2019**, *33*, 877–899. [CrossRef]

land

MDPI

Article

Exploring the Determinants of Spatial Vitality in High-Speed Rail Station Areas in China: A Multi-Source Data Analysis Using LightGBM

Pengpeng Liang, Xu Cui *, Jiexi Ma *, Wen Song and Yao Xu

School of Architecture, Southwest Jiaotong University, 999 Xi'an Road, Chengdu 611756, China; liang_pengpeng@my.swjtu.edu.cn (P.L.); songwen@my.swjtu.edu.cn (W.S.); xuyao@my.swjtu.edu.cn (Y.X.)
* Correspondence: cuixu@swjtu.edu.cn (X.C.); mjx@my.swjtu.edu.cn (J.M.)

Abstract: High-speed rail (HSR) station areas play a vital role in shaping urban form, stimulating economic activity, and enhancing spatial vitality. Understanding the factors that influence this vitality is key to supporting sustainable urban development and transit-oriented planning. This study investigates 66 HSR station areas in 35 Chinese cities by integrating multi-source data—Sina Weibo check-in records, urban support indicators, station attributes, and built environment variables—within a city–node–place analytical framework. Using Multiple Linear Regression (MLR) and Light Gradient Boosting Machine (LightGBM) models, we identify key drivers of spatial vitality, while SHAP analysis reveals nonlinear and interaction effects. The results show that city population size, urbanization level, commercial land use, transit accessibility, and parking facilities significantly enhance station area vitality. However, diminishing returns are observed when commercial land and bus stop densities exceed certain thresholds. The station location index shows a negative correlation with spatial vitality. The analysis of interaction effects highlights strong synergies between urban development and functional configuration, as well as between accessibility and service infrastructure. Different station types exhibit varied spatial patterns and require differentiated strategies. This study offers empirical insights for aligning transport infrastructure and land use planning, supporting the development of vibrant, accessible, and sustainable HSR station areas.

Keywords: high-speed rail; station area spatial vitality; multi-source data; Light Gradient Boosting Machine (LightGBM)

1. Introduction

In recent decades, the rapid expansion of high-speed rail (HSR) networks has become a transformative force reshaping the spatial organization of cities and regions [1,2]. As an efficient, high-capacity mode of intercity transport, HSR enhances regional connectivity, compresses travel time, and drives urban integration and spatial restructuring across scales [3–5]. In China, the development of the world's largest HSR system has not only fulfilled key transportation demands but has also profoundly impacted urban form, land use patterns, and population distribution [6–8].

HSR stations, serving as both transport nodes and urban centers, play a pivotal role in promoting economic vitality and guiding development direction [9–11]. However, the success of these stations extends beyond their transport functionality—it is increasingly reflected in the intensity, diversity, and persistence of human activities they attract. This has given rise to the concept of spatial vitality, which refers to the dynamic state of human presence and interaction within urban space over time. Spatial vitality is a multidimensional construct encompassing economic vitality (e.g., business activity, consumption), social vibrancy (e.g., pedestrian flows, diversity of social interactions), and environmental engagement (e.g., the use of public and green spaces). In the context of HSR station areas, spatial vitality serves as a key indicator of how effectively transportation infrastructure integrates with urban life [12–16].

To explore what drives spatial vitality in HSR station areas, this study focuses on three categories of influencing factors—urban support capacity, station attributes, and built environment characteristics—based on a city–node–place analytical framework. These dimensions are selected because they collectively reflect the macro-level development background, the meso-level transportation functions of the station, and the micro-level spatial qualities experienced by users. Specifically, urban support indicators such as population size and urbanization level represent the demand base and economic potential of the host city. Station-level variables, including service frequency, location accessibility, and connectivity, determine how well the station functions as a mobility hub. Built environment indicators—such as commercial land use ratio, transportation facilities, and functional diversity—describe the spatial configuration and service provision in the immediate station area. By jointly analyzing these three levels, the study aims to reveal not only which factors significantly affect spatial vitality, but also why and how they do so under different conditions and in combination.

Understanding what drives this vitality is essential for enhancing land use efficiency, optimizing functional configurations, and promoting transit-oriented, sustainable development. More importantly, spatial vitality plays a fundamental role in improving the broader quality of urban life. Economically, vibrant station areas can stimulate commercial development and attract investment. Socially, they facilitate inclusiveness, accessibility, and human interaction. Environmentally, vital station areas can promote compact development, reduce vehicle dependence, and encourage the use of public and green spaces. Therefore, enhancing spatial vitality around HSR stations is not only a matter of functional efficiency, but also a strategic pathway toward inclusive, resilient, and environmentally sustainable cities.

Despite growing academic attention, current research on HSR station area vitality faces several notable limitations [17–19]. First, many studies adopt a static perspective, relying on instantaneous indicators such as heatmaps (e.g., Baidu thermal maps), which capture short-term activity snapshots but fail to reflect the cumulative vitality that develops over time—a key aspect for assessing the long-term spatial dynamics of station areas. Second, traditional statistical models typically assume linear relationships and struggle to capture complex nonlinear interactions or threshold effects among influencing variables, limiting their explanatory power in multifactorial urban contexts. Third, conventional data sources often involve survey-based or aggregated statistics, which lack spatial precision and temporal continuity. Such data mask local variations and individual behaviors within station areas. In contrast, emerging location-based big data, like social media check-ins, provide fine-grained, real-time insights, enabling more accurate mapping of human activity patterns and overcoming the coarse granularity and static nature of prior data.

With the emergence of geospatial big data and social media-based human mobility records, there is a new opportunity to conduct high-resolution analyses of spatial vitality [20–22]. In particular, the use of location-based check-in data provides valuable insights into real-time activity patterns, enabling more dynamic assessments of how people interact with station environments. Moreover, combining classical regression models with machine learning algorithms offers a robust framework to capture both the interpretable and nonlinear aspects of population behavior [23–27].

Recent academic efforts have increasingly leveraged multi-source data and advanced analytical methods to study the dynamics of HSR station areas. For instance, Wang et al. focused on the Yangtze River Delta, developing an integrated framework to assess urban vitality through indicators such as concentration, accessibility, livability, and functional diversity [6]. Yue et al. analyzed mobile phone signaling data in Jiangsu Province, clustering 71 HSR stations by passenger flow time series and employing geographically weighted multinomial logit models to examine the roles of entertainment POI density, population, GDP, and built area in shaping station classifications [9]. Additionally, recent research by Doan et al. applied machine learning models combined with SHAP (Shapley Additive Explanations) to explore the nonlinear and threshold effects of built environment, traffic, and air quality factors on urban vitality in Manhattan, using street-view pedestrian presence data [28]. Building on our earlier exploratory work, which employed panel data and a difference-in-differences (DID) approach to examine the temporal effects of HSR on commercial agglomeration, this study makes a significant conceptual and methodological advancement. We shift from a temporal to a spatial comparative perspective, analyzing the vitality mechanisms of 66 HSR station areas across 35 cities. Through interpretable machine learning and a city–node–place framework, we aim to uncover the multiscale drivers of station area vitality and provide actionable planning insights [29]. Together, these studies highlight the growing potential of integrating diverse datasets and interpretable computational models to uncover complex spatial patterns.

Therefore, this study aims to comprehensively examine the spatial vitality of HSR station areas across China by integrating multi-source data and applying both multiple linear regression (MLR) and Light Gradient Boosting Machine (LightGBM) models. Drawing on check-in data from Sina Weibo, urban statistical indicators, and detailed built environment metrics, we analyze 66 representative HSR stations in 35 cities to identify key drivers of population clustering. We further explore the threshold and interaction effects of influencing factors using SHAP (Shapley Additive Explanations) analysis [30,31].

To comprehensively understand the factors shaping spatial vitality in HSR station areas, this study adopts a city–node–place analytical framework and combines both traditional statistical models and advanced machine learning techniques. By integrating multi-source data including social media check-ins, statistical indicators, and built environment features, we aim to explore the nonlinear, threshold, and interaction effects underlying population clustering. The following sections detail the study area, data sources, variable design, modeling methods, and empirical results, culminating in theoretical discussions and policy implications.

2. Study Area and Data

2.1. Study Area

As of now, HSR services have been introduced in 269 cities across China, covering approximately 72.3% of all cities nationwide. To ensure the representativeness of our research, we selected 35 cities through a stratified sampling framework. The selection considered five key criteria: (1) regional diversity (eastern, central, western, and ethnic minority regions); (2) administrative level (including municipalities, provincial capitals, and sub-provincial cities); (3) urban scale and population density; (4) level of economic development and functional importance; and (5) the maturity and intensity of HSR operations. This approach ensures that the sample reflects China's diverse urban contexts and developmental stages.

From these cities, 66 representative HSR stations were selected based on comprehensive criteria, including station classification, operational scale, geographic location, service capacity, and data availability [11,29,32,33]. In order to avoid bias and maintain consistency, stations co-located with airports or lacking adequate data—such as Chengdu Xipu Station and Guangzhou Baiyun Airport North Station—were excluded. We also avoided stations with weak service capabilities or low integration with the surrounding urban fabric. This filtering process ensures the analytical validity and robustness of our sample, making it suitable for comparative analysis.

The final sample includes a wide range of station types (e.g., central city hubs, regional gateways, peripheral nodes), covering diverse geographic and functional contexts. These stations span metropolitan cores such as Beijing and Shanghai, as well as key inland centers like Xi'an and Chengdu. This distribution enables a comprehensive understanding of how HSR integrates with varying urban structures [34,35]. Overall, the selected 66 stations in 35 cities provide a robust empirical foundation for analyzing spatial vitality in different socioeconomic and planning environments across China (Figure 1).

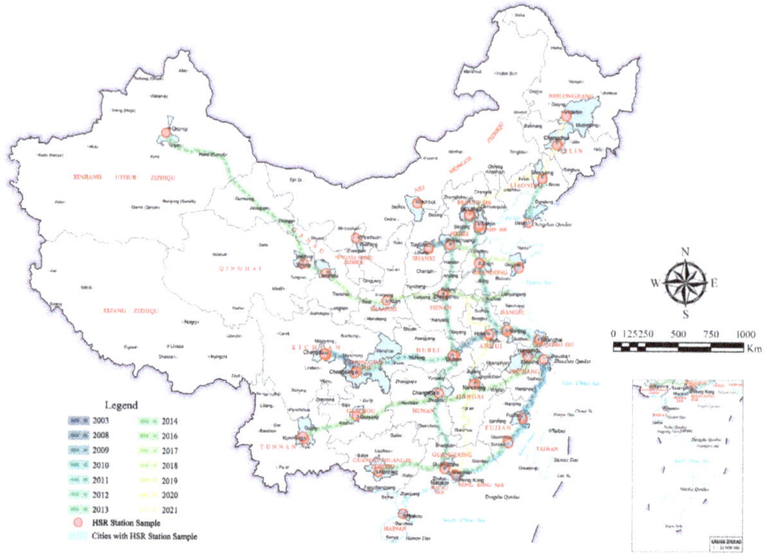

Figure 1. Study on the spatial distribution of HSR stations. Note: The original map (No. GS(2019)1686) is from the China National Natural Resources Standard Map Service website (http://bzdt.ch.mnr.gov.cn/download.html?searchText=1686, accessed on 5 June 2025). The authors mapped the spatial distribution of HSR stations based on it.

2.2. HSR Station Areas

In the existing literature, scholars have adopted various approaches to delineate the spatial extent of areas influenced by HSR stations, depending on their specific research objectives [36–42]. Among these, distance-based and time-based criteria are widely applied. For example, Schütz (1998) classified the surrounding areas into three development zones based on station accessibility: a 5–10 min access zone, a zone accessible within 15 min, and a zone beyond 15 min [43]. Building on this, Zheng et al. (2024) investigated development patterns and multi-level spatial interactions within a 1500 m radius around HSR stations, shedding light on the mechanisms of station area development and spatial structure [1]. Similarly, Wang Lan et al. (2014) proposed a concentric zoning approach based on spatial distance, dividing the station's influence area into a core zone (2000 m), an impact zone (4000 m), and a peripheral zone (8000 m) [44].

Findings from these studies indicate that the functional intensity of facilities around HSR stations generally decreases with increasing distance from the station, forming a layered, centripetal spatial structure [45–47]. Based on this insight, the present study adopts a 1500 m fixed-radius buffer centered on each HSR station as the primary area of investigation. This decision is supported by both existing literature and the practical focus of this study. On one hand, the 1–2 km range is widely used in the existing literature as a typical zone of transit influence, ensuring methodological consistency. On the other hand, since this research focuses on spatial vitality, which is typically concentrated within walkable distances from the station—especially during the early and middle stages of development—a 1500 m range effectively captures the core activity zone.

Additionally, using a uniform fixed-radius buffer allows for standardized comparison across 66 HSR stations in 35 cities, enhancing the robustness of cross-sectional analysis. While network-based travel distances may offer a more realistic representation of accessibility, such data are often unavailable or inconsistent across large samples due to differences in local infrastructure quality and data availability.

2.3. Data Sources and Processing

Population agglomeration is a critical indicator for evaluating the vibrancy and economic consumption potential of HSR station areas [48,49]. Meanwhile, users' check-in behaviors at various spatial locations reflect their preferences for different types of activity spaces. Therefore, to gain deeper insights into the spatial behavior patterns of individuals within HSR station areas, this study employed a web crawler to collect check-in data from Sina Weibo within defined station boundaries.

The acquired check-in data were first matched with corresponding spatial coordinates. Subsequently, the spatial locations were categorized based on a point-of-interest (POI) classification system. This processing enabled the identification of the spatial distribution patterns of active individuals and the characterization of activity hotspots within the station areas.

Notably, Sina Weibo initially provided an open API for check-in data access in 2012 [50–52]. However, the original interface was later discontinued due to overwhelming access traffic. In response to this limitation, this study improved upon previous methods by developing a targeted data acquisition strategy. Specifically, we conducted a detailed search of the Sina Weibo source pages, applied precise filtering techniques, and utilized both the location service API and user service API in an integrated manner. This optimized workflow significantly enhanced both the accuracy and efficiency of data collection, thereby laying a solid foundation for subsequent analysis.

To ensure the integrity of the dataset, we implemented a multi-stage preprocessing workflow. First, we removed duplicate entries based on user ID, timestamp, and spatial coordinates. Second, we filtered out invalid check-ins lacking essential attributes such as geolocation or content. Third, we eliminated advertising-related records using keyword-based content filtering to exclude promotional or irrelevant posts. These steps collectively ensured that only high-quality, user-generated activity records were retained for analysis.

The data acquisition process involved several key steps (Figure 2): identifying available interface paths, constructing query addresses for HSR station areas, sending requests to retrieve POI collections, iteratively accessing URLs and parsing the returned JSON files, and finally, performing data validation and storage. The collected check-in data covered the period up to 31 December 2024, and each entry was precisely matched with the latitude and longitude of its corresponding HSR station area (Table 1).

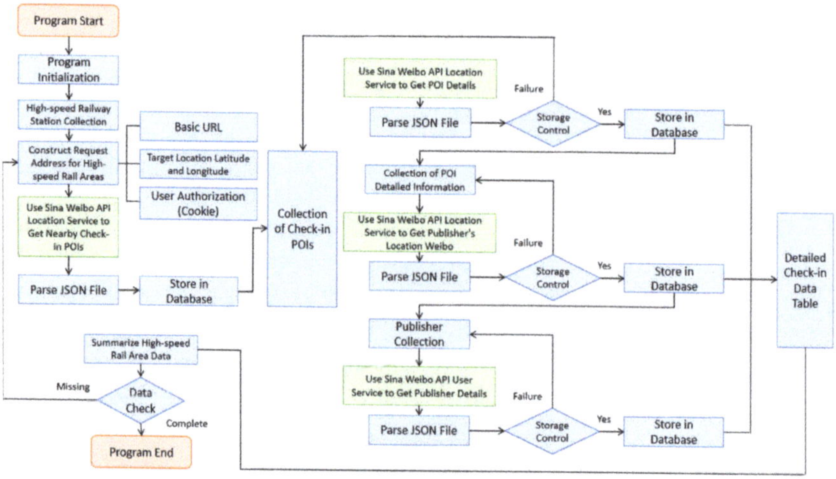

Figure 2. Workflow for acquiring check-in data from Sina Weibo.

Table 1. Sample records from Weibo check-in dataset.

Field Name	Data Type	Description	Example
record_index	int	Sequential index of the data record	1
station_name	string	Name of HSR station	Shanghai Hongqiao Station
poi_name	string	Name of the point of interest (POI)	Haidilao Hotpot (Zhouhong Road Branch)
poi_address	string	Full address of the POI	799 Shenchang Road, Minhang District, Shanghai
checkin_frequency	int	Number of check-in occurrences	126
poi_category	string	Category/type of the POI	Hotpot Restaurant
longitude	double	Geographic longitude of the POI	121.30929749345
latitude	double	Geographic latitude of the POI	31.193352869547752

2.4. Variables

The influencing factors of population agglomeration represent a multidimensional and complex research domain, involving the interaction of various spatial and functional components. At the urban level, the capacity of a city to support population concentration serves as a fundamental condition. This capacity encompasses urban scale, economic development, transportation network structure, and the orientation of planning policies (Table 2). Together, these factors determine the city's attractiveness and its ability to accommodate large populations [53–55].

In selecting urban support variables, we prioritized those with solid theoretical foundations and empirical support in the field of urban vitality research. Specifically, we identified "Population Scale," "Economic Development," and "Transportation Network" as core dimensions, as they comprehensively reflect a city's development stage and population attraction capacity. These macro-level factors play a fundamental role in determining the intensity of population activity and the spatial vitality of HSR station areas.

First, the scale and level of economic development define the availability of employment, education, and living resources, which directly enhance the city's capacity to attract people. Second, the layout and efficiency of the transportation network facilitate the flow of people and goods, making the city a key node in regional mobility. Additionally, urban planning strategies and policy support optimize resource allocation, thus further reinforcing population concentration. The data for these indicators are primarily derived from municipal statistical yearbooks, government bulletins, and open-access databases.

Table 2. Urban support variables.

Category	Dimension	Variables
Urban Support (C)	Population Scale (C-POP)	Population of the city where the HSR station is located (C1), population of the district/county (C2)
	Economic Development (C-ED)	City GDP (C3), district GDP (C4), tertiary industry GDP of the city (C5) and district (C6), urbanization rate of the city (C7) and district (C8), general public budget revenue of the city (C9) and district (C10)
	Transportation Network (C-TN)	Road network density of the city (C11) and district (C12), city-level HSR connection frequency index (C13), one-stop reachable cities index (C14), direct HSR link index (C15), city-level degree centrality (C16), city-level closeness centrality (C17), city-level betweenness centrality (C18), city-level eigenvector centrality (C19)

At the node level, the characteristics of HSR stations also exert a significant influence on the degree of population aggregation. Factors such as the rationality of station location, the suitability of its design and scale, and the completeness of supporting facilities jointly determine the attractiveness of the station as a transportation hub [56,57]. Stations located in central urban areas or major traffic corridors are more likely to attract large flows of people. A well-organized spatial layout and functional zoning can enhance user experience, while high-quality supporting facilities—including waiting areas, commercial amenities, and accessibility features—encourage frequent use and promote further agglomeration. Data on HSR station characteristics are sourced from official railway reports, open databases, and transportation planning documents. Table 3 presents the variables used to assess the HSR station conditions.

Table 3. HSR station condition variables.

Category	Dimension	Variables
HSR Station Condition (S)	Opening Time and Level (S-TL)	HSR service start time (S1), station classification level (S2)
	Station Location (S-Z)	Number of HSR connections (S3), number of one-stop connections (S4), number of direct links (S5), station-level degree centrality (S6), station-level closeness centrality (S7), station-level betweenness centrality (S8), station-level eigenvector centrality (S9), distance to the city center (S10), location index (S11), public transit time (S12), driving time (S13), total train services (S14), originating/terminating services (S15), percentage of originating/terminating services (S16), percentage of through services (S17)
	Station Scale and Design (S-SD)	Number of platforms (S18), number of rail lines (S19), station building area (S20), square area in front of the station (S21)
	Connecting Transport (S-JT)	Number of metro entrances (S22), metro lines (S23), bus stops (S24), bus lines (S25), long-distance bus stations (S26)
	Commercial and Leisure Facilities (S-CF)	Number of commercial and leisure facilities (S27)

At the place level, the built environment of the station area serves as the immediate spatial carrier for population activity and thus plays a key role in influencing agglomeration [16,58–61]. Core dimensions of the built environment include spatial density, functional diversity, design quality, transportation accessibility, and facility availability. Appropriate population density ensures urban vitality; functional diversity refers to the mix of land uses, commercial types, and social activities; and high-quality design enhances visual appeal and usability. Furthermore, the accessibility of the station area to the broader urban context—through both physical connections and service coverage—affects whether people choose it as a destination for travel or daily activity. Relevant data are collected from urban planning documents, remote sensing imagery, and POI data from Gaode Maps. Table 4 presents the built environment variables of station areas.

Table 4. Station areas built environment variables.

Category	Dimension	Variables
Station Area Built Environment (E)	Development Scale (E-DS)	Building density (E1), floor area ratio (E2), development proportion (E3), underground space density (E4)
	Land Use (E-LU)	Proportion of residential land (E5), commercial/service land (E6), public administration/service land (E7), green space (E8), transport infrastructure (E9), public utility land (E10), land use mix index (E11)
	Spatial Design (E-SD)	Road network density (E12), number of intersections (E13), building proximity index (E14), green view index (E15), sky openness (E16), enclosure index (E17)
	Transportation Connectivity (E-TC)	Number of metro entrances (E18), number of metro lines (E19), number of bus stops (E20), number of bus lines (E21), number of parking facilities (E22)
	Functional Facilities (E-FF)	Number of commercial service facilities (E23), residential facilities (E24), public facilities (E25), business facilities (E26)

3. Methods

3.1. Multiple Linear Regression Model

The Multiple Linear Regression (MLR) model is a widely used statistical method that explores the linear relationship between a dependent variable and multiple independent variables. In the context of high-speed rail (HSR) station area studies, this model is particularly valuable as it enables the quantification of how various factors—such as urban scale, transportation networks, and station-area facilities—affect spatial vitality.

Specifically, the MLR model helps identify and evaluate the extent to which different urban and infrastructural variables contribute to population agglomeration and activity intensity around HSR stations. By establishing a regression equation, it becomes possible to measure the individual contribution of each explanatory variable to the dependent variable and to reveal the underlying mechanisms that drive spatial dynamics.

In this study, we adopt the MLR approach by using spatial vitality in HSR station areas as the dependent variable and a set of multidimensional influencing factors as independent variables. The basic mathematical form of the model is expressed as follows:

$$\text{Vitality}\left(\text{check}-\text{ins}/\text{km}^2\right) = \beta_0 + \beta_1 X_1 + \beta_2 X_2 + \cdots + \beta_n X_n + \epsilon \quad (1)$$

In this equation, Vitality (check-ins/km^2) denotes the dependent variable, representing the intensity of human activity per unit area within the 1500 m buffer zone of each HSR station during its operational period. The variables $X_1, X_2, \ldots, X_n$ are the independent variables, corresponding to influencing factors such as transportation accessibility, functional diversity, and station-area infrastructure. β_0 is the intercept term, and $\beta_1, \beta_2, \ldots, \beta_n$ are the regression coefficients, each representing the marginal effect of a corresponding independent variable on the dependent variable. The term ϵ captures the random error, accounting for unexplained variability in the model.

Each regression coefficient has a clear interpretive meaning. For instance, the coefficient β_1 indicates the magnitude and direction of change in spatial vitality Y when the corresponding factor X_1 increases by one unit, holding all other variables constant. This allows for a detailed understanding of the relative importance of different influencing factors.

3.2. LightGBM Model

Unlike traditional multiple linear regression models, LightGBM (Light Gradient Boosting Machine) is an ensemble learning method based on the gradient boosting framework. It builds multiple decision trees to model data and iteratively optimizes the objective function to enhance prediction accuracy. As a result, LightGBM demonstrates significant advantages in handling complex nonlinear relationships and large-scale datasets, making it an effective tool for uncovering hidden patterns and influential features in data.

The core algorithm underlying LightGBM is the Gradient Boosting Decision Tree (GBDT). The main idea is to construct a strong predictive model by sequentially training a series of weak learners, typically shallow decision trees, and minimizing the loss function in each iteration.

The general form of the loss function is defined as

$$L = \sum_{i=1}^{n} l(y_i, \hat{y}_i) + \Omega(T) \tag{2}$$

In this expression, L represents the total loss, y_i is the actual value of the *i*-th sample, $\hat{y}_i$ is the corresponding predicted value, $l(\cdot)$ denotes the loss function for an individual sample (e.g., squared error or log loss), and $\Omega(f)$ is a regularization term that controls the complexity of the model, such as tree depth or the number of leaf nodes.

During each iteration, LightGBM leverages both the first-order gradient (i.e., the derivative of the loss function with respect to the prediction) and the second-order gradient (i.e., the second derivative) to update the model. This use of gradient information enables LightGBM to improve computational efficiency and convergence speed while maintaining high prediction accuracy.

To further improve training speed and reduce memory consumption, LightGBM introduces a histogram-based algorithm to construct decision trees (Figure 3). Instead of evaluating every possible split point, LightGBM first buckets continuous feature values into discrete bins (i.e., histograms). Then it finds the optimal split based on these bins, significantly reducing the number of comparisons needed during training.

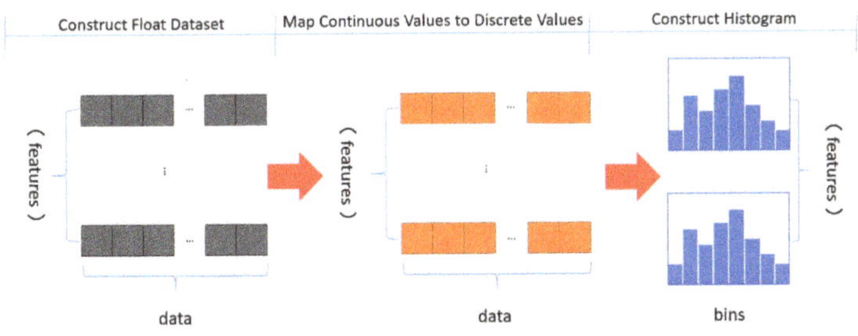

Figure 3. Illustration of LightGBM histogram optimization principle.

3.3. Shapley Additive Explanation (SHAP)

Although LightGBM and other machine learning methods can produce highly accurate predictions, their "black-box" nature often makes it difficult to understand how the models arrive at specific decisions. To address this issue, SHAP (Shapley Additive Explanations) was developed as a model-agnostic interpretability framework that offers transparent explanations for complex machine learning models.

The SHAP model is theoretically grounded in the Shapley value from cooperative game theory. It quantifies the contribution of each feature to the prediction by computing its marginal effect across all possible feature combinations. Therefore, SHAP not only maintains the predictive power of advanced models but also reveals the underlying mechanisms driving their outputs.

$$\phi_i = \sum_{S \subseteq N/\{i\}} \frac{|S|!(|N|-|S|-1)!}{|N|!}[f(S \cup \{i\}) - f(S)] \tag{3}$$

In this equation, ϕ_i denotes the Shapley value of feature i, representing its contribution to the model's prediction. N is the set of all features, and S is any subset that does not include feature i. The function f(S) refers to the model output using only the features in subset S, while $f(S \cup \{i\})$ refers to the output when feature i is added to the subset.

By averaging the marginal contributions over all possible subsets, the Shapley value provides a fair and comprehensive measure of feature importance. As a result, SHAP serves as a powerful tool for enhancing the interpretability of machine learning models without sacrificing predictive accuracy.

3.4. Model Selection and Comparative Justification

Given the heterogeneous and high-dimensional nature of the data used in this study—including structured socioeconomic indicators, spatial POI distributions, and user-generated check-in behaviors—we adopt a dual-modeling strategy using both Multiple Linear Regression (MLR) and Light Gradient Boosting Machine (LightGBM).

The MLR model provides robust interpretability and is well suited for structured variables with assumed linear effects. It supports inference of marginal impacts and is ideal for testing policy-relevant hypotheses. However, it falls short in capturing complex nonlinearity and variable interactions.

LightGBM, on the other hand, excels in handling nonlinearities, feature interactions, and sparse behavioral data—common in social sensing applications. It is especially valuable for uncovering threshold effects and high-order dependencies, albeit at the cost of interpretability. This is addressed by incorporating SHAP.

Therefore, the combined use of MLR and LightGBM allows us to bridge explanatory clarity and analytical depth, leading to a more comprehensive understanding of what drives spatial vitality in HSR station areas.

4. Results

4.1. Spatiotemporal Variations in the Vitality of HSR Station Areas

From the cumulative check-in bar chart, it is evident that human activity in HSR station areas exhibits a significant spatial clustering pattern, particularly concentrated in economically developed metropolitan regions (Figure 4). For instance, Shanghai Hongqiao Station, Futian Station, Beijing Station, and Shenzhen Station rank among the top nationwide in terms of cumulative check-ins, reaching 327,000, 224,000, 183,000, and 151,000 check-ins, respectively. Notably, Shanghai Hongqiao Station recorded the highest number of check-ins, highlighting its strong attractiveness as a core transportation hub.

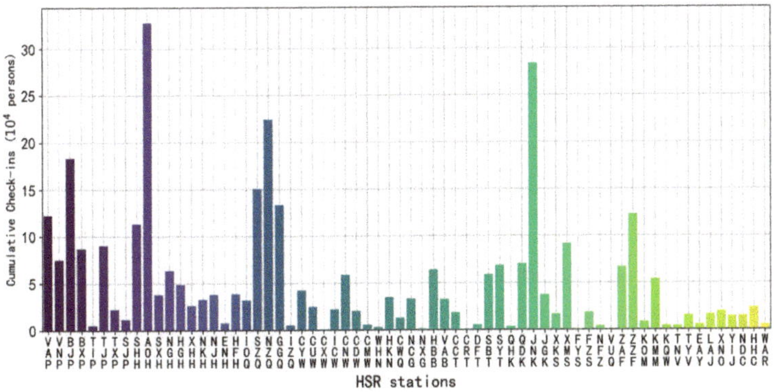

Figure 4. Cumulative check-in volume at HSR station areas. Note: For legibility, abbreviated names are used. See Appendix A for the full station name list.

In addition, stations such as Jinan Station (284,000), Guangzhou East Station (134,000), Zhengzhou Station (123,000), and Shanghai Station (113,000) also recorded high levels of activity. These stations are typically located near city centers and benefit from convenient transportation links, which effectively attract large crowds. It is also worth noting that popular tourist destinations like Xiamen Station have drawn substantial numbers of active users, indicating that HSR station areas also possess strong tourism appeal.

Figure 4 illustrates the distribution of cumulative check-ins across different HSR station areas. Based on functional attributes, the active spaces surrounding the stations can be categorized into commercial service space, business space, public space, leisure space, transportation space, residential space, green space, industrial space, and administrative space. The activity levels in each space type were analyzed (Figure 5). Among them, commercial service spaces exhibited the highest level of activity, with approximately 1.37 million check-ins, reflecting their strong commercial attractiveness. Transportation and residential spaces followed, with 610,000 and 600,000 check-ins, respectively. This pattern not only underscores the critical role of HSR stations as transportation hubs but also demonstrates the effective development of residential functions in adjacent areas.

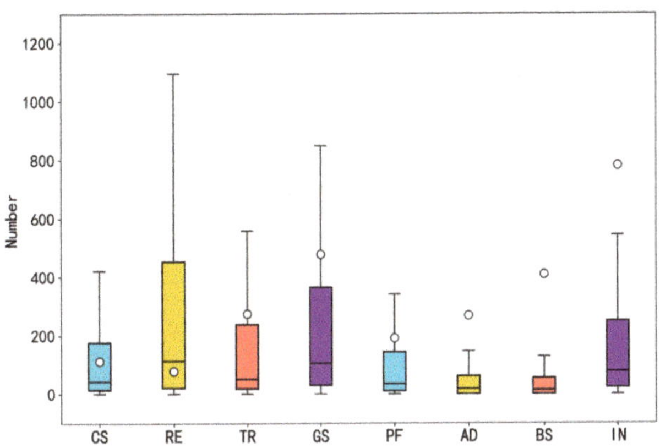

Figure 5. Number of check-ins in different types of spaces around high-speed rail stations. Note: For clarity and brevity, space types are represented using abbreviations: GS stands for Green Space, CS for Commercial Services, PF for Public Facilities, TR for Transportation, RE for Residential, AD for Administrative, BS for Business, and IN for Industrial. These abbreviations correspond to major urban functional zones.

Furthermore, public and green spaces recorded 420,000 and 160,000 check-ins, respectively, showing a certain degree of vitality. This suggests that in addition to transportation and commercial services, HSR station areas also provide public amenities and green leisure environments that meet the diverse needs of residents and travelers.

According to the spatial distribution of check-in data, the activity patterns of populations in HSR station areas can be generally categorized into three types: balanced distribution, fan-shaped clustering, and star-shaped dispersion (Figure 6). First, stations with a balanced distribution are typically located near city centers. These areas tend to develop evenly, with commercial and business functions highly concentrated, which generates strong attractiveness and leads to relatively uniform population aggregation. Typical examples include Beijing Station, Tianjin Station, Shanghai Station, Futian Station, Guangzhou East Station, Chengdu South Station, and Zhengzhou Station. These stations represent the typical development pattern of central urban HSR stations.

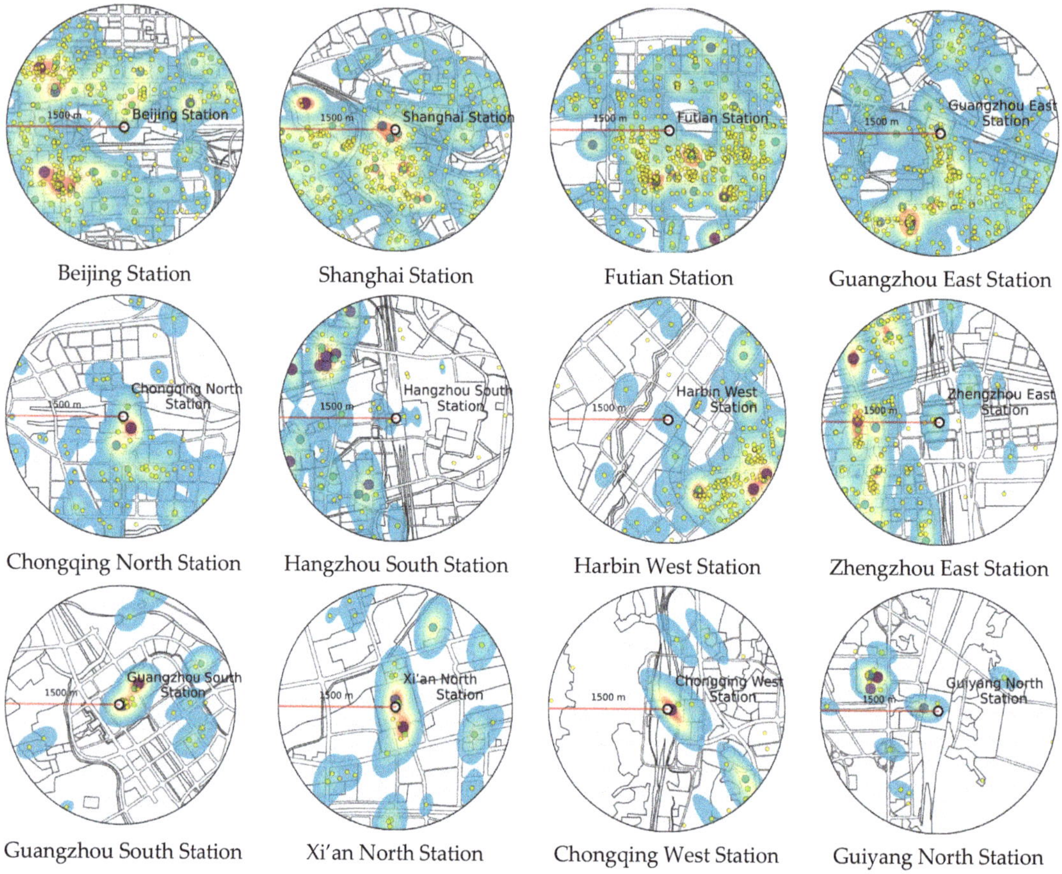

Figure 6. Spatial distribution map of crowd check-in activities in HSR station areas.

Second, fan-shaped clustering occurs in HSR stations that expand along existing urban development axes. Influenced by the built environment, these station areas extend in a directional pattern that mirrors urban spatial growth. Stations such as Chongqing North, Hangzhou South, Harbin West, and Zhengzhou East fall into this category, and their development aligns closely with the city's expansion trajectory.

Lastly, star-shaped dispersion is observed in HSR stations located farther from urban centers. These stations often feature undeveloped land or lack adequate supporting facilities, resulting in weaker integration with the city and a more scattered pattern of population activity. Despite their current lower levels of aggregation, these stations hold the potential to become key nodes in future urban development. Representative examples include Guangzhou South, Xi'an North, Chongqing West, Guiyang North, Kunming South, and Changchun West stations.

4.2. Results of Linear Regression

In this study, all explanatory variables were subjected to a multicollinearity test. The results showed that the Variance Inflation Factors (VIFs) for all variables were below 5, indicating that multicollinearity among the variables was not significant and the model estimates are statistically reliable. The MLR model achieved an R-squared value of 0.3975, a Root Mean Squared Error (RMSE) of 5.014, and a Mean Absolute Error (MAE) of 3.712. Table 5 presents the results of the MLR model.

Table 5. Results of the multiple linear regression model.

Variable Code	Variable Description	Coef.	Std Err.	t	*p*-Value
C1	Population size of the HSR station city	7.0798	3.017	2.346	0.022 *
C7	Urbanization level of the HSR station city	8.0828	3.646	2.217	0.03 **
S11	Station location index	−9.5416	2.821	−3.383	0.001 ***
S15	Number of originating and terminating train services	6.5358	2.996	2.181	0.033 **
S24	Number of bus stops near the station	12.4662	3.202	3.893	0.001 ***
E6	Land area for commercial service facilities	9.9389	4.107	2.42	0.018 **
E22	Number of parking facilities in the station area	8.7181	3.224	2.704	0.009 ***

Note: Significance levels—* $p < 0.05$, ** $p < 0.01$, *** $p < 0.001$.

Several factors related to urban support capacity exhibited significant effects on the spatial clustering of active populations. Specifically, both the population size of the city (C1) and the level of urbanization (C7) were strongly correlated with the degree of population aggregation, showing statistically significant positive relationships. The regression coefficient for city population size (C1) was 7.0798 with a *p*-value of 0.022, suggesting that larger urban populations tend to attract more active individuals to gather around HSR station areas. Similarly, the urbanization level (C7) had a coefficient of 8.0828 and a *p*-value of 0.03, indicating that more urbanized environments promote higher levels of population clustering.

Regarding station-level characteristics, the station location index (S11) had a significant negative effect on population aggregation, with a coefficient of −9.5416 and a *p*-value of 0.001. This result implies that stations with lower location index values—i.e., more favorable or central locations—tend to attract higher levels of population activity. In contrast, the number of originating and terminating train services (S15) positively influenced population clustering, with a coefficient of 6.5358 and a *p*-value of 0.033, demonstrating that increased service frequency enhances the station's ability to attract people. Additionally, the number of bus stops near the station (S24) also had a statistically significant positive effect, with a coefficient of 12.4662 and a *p*-value of 0.001, underscoring the importance of public transportation connectivity in reinforcing population gathering.

Furthermore, several elements of the built environment around the station area were found to significantly influence the level of population activity. The area designated for commercial service facilities (E6) had a positive and significant impact, with a coefficient of 9.9389 and a *p*-value of 0.018. This suggests that expanding commercial land use around stations can substantially boost their attractiveness and crowd-gathering capacity. Similarly, the number of parking facilities (E22) had a positive and significant effect, with a coefficient of 8.7181 and a p-value of 0.009, indicating that adequate parking infrastructure supports both accessibility and functional completeness, thereby contributing to greater population clustering.

These findings align with urban economic theory, which suggests that larger urban populations create stronger consumption bases and attract higher levels of pedestrian activity. The positive effect of commercial land use reflects the principle of agglomeration economies, where mixed-use developments increase land efficiency and social interaction.

In contrast, the negative effect of the station location index suggests that stations located in peripheral or poorly integrated zones suffer from low spatial synergy, as predicted by urban spatial mismatch theory. The accessibility factor, such as the number of bus stops, supports the notion that multi-modal connections enhance last-mile connectivity, which is a cornerstone of transit-oriented development (TOD).

4.3. Results of the LightGBM Regression Model and SHAP Analysis

According to the convergence evaluation plot of the active population prediction model, the model's loss value showed a significant decrease during the initial phase of training (Figure 7). In particular, within the first 50 iterations, the loss dropped sharply from a high value near 7.0 to approximately 5.75. This indicates that the optimization algorithm quickly adjusted the hyperparameters in the early training stage and substantially reduced the prediction error. As the number of iterations increased, the rate of loss reduction gradually slowed, and the curve began to flatten. After around 250 iterations, the loss value stabilized at approximately 5.2, suggesting that further training did not lead to noticeable improvement, and the model reached convergence with stable performance.

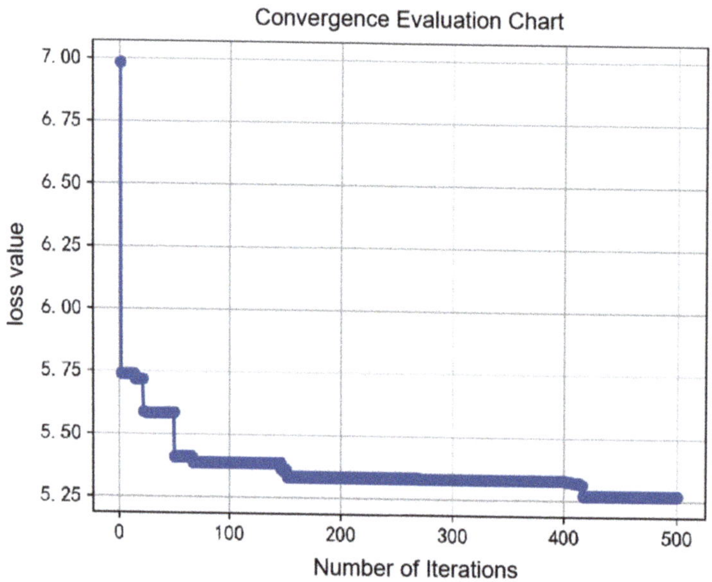

Figure 7. Convergence evaluation chart of the crowd aggregation model in HSR station areas.

Based on the hyperparameter tuning results, a scatterplot matrix was used to visualize the interactions between parameters and their impact on model performance (Figure 8). The selected optimal parameter configuration achieved a well-balanced trade-off between model fitting and generalization. Specifically, the learning rate was set to 0.01 to ensure stable optimization and reduce the risk of overfitting. The number of trees was set to 1041 to improve accuracy without introducing excessive computational burden. The maximum tree depth was limited to 10, effectively controlling model complexity and enhancing generalization capability. Each tree was allowed a maximum of 15 leaf nodes, increasing fitting capacity while mitigating overfitting. The subsample ratio was set at 0.648, helping introduce randomness and reduce overfitting. Additionally, the feature subset ratio was set to 0.624, which further enhanced model diversity. Lastly, the minimum number of samples per leaf was set to 10, ensuring node stability and robust decision splits. Collectively, these hyperparameter settings enabled the model to achieve strong and stable performance in predicting active population clustering. The LightGBM model achieved an R-squared value of 0.7347, a Root Mean Squared Error (RMSE) of 3.3273, and a Mean Absolute Error (MAE) of 1.9621. Figure 9. Learning curve of the LightGBM model showing the loss trajectories for both the training and validation sets. As the number of iterations increases, both losses decrease and gradually stabilize, with no upward trend in validation loss—indicating good model convergence and no significant overfitting.

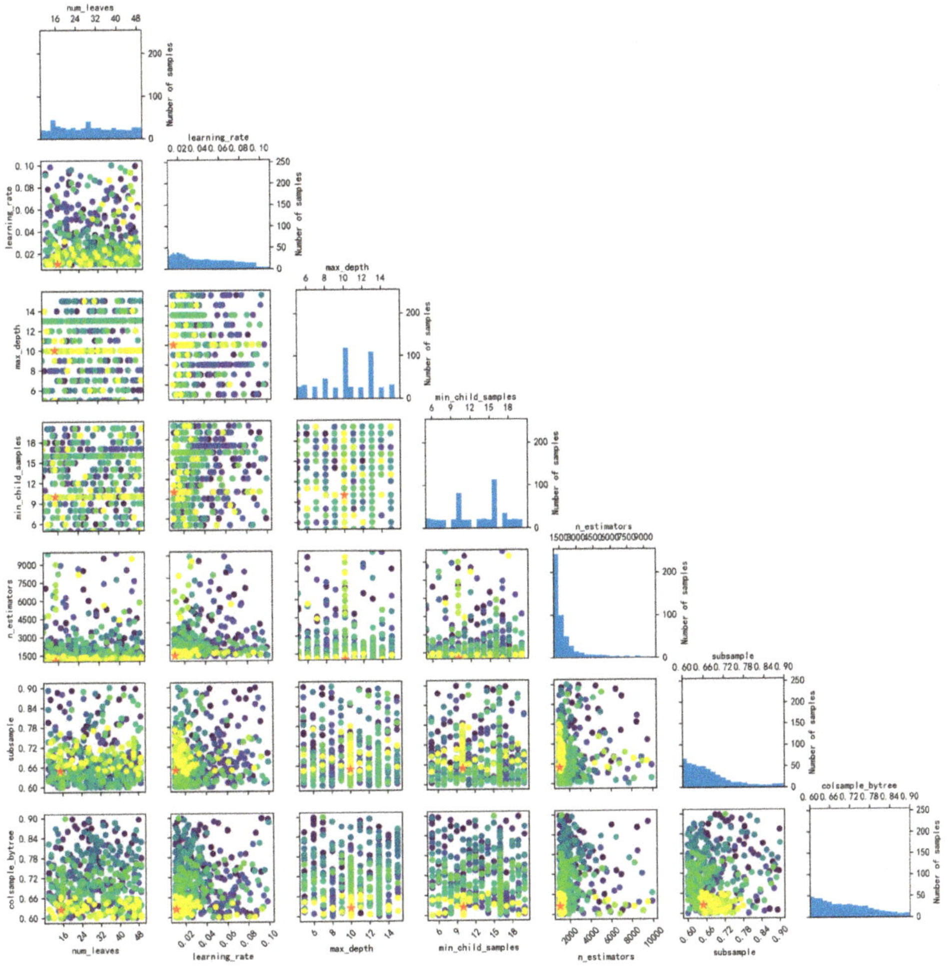

Figure 8. Model parameter optimization.

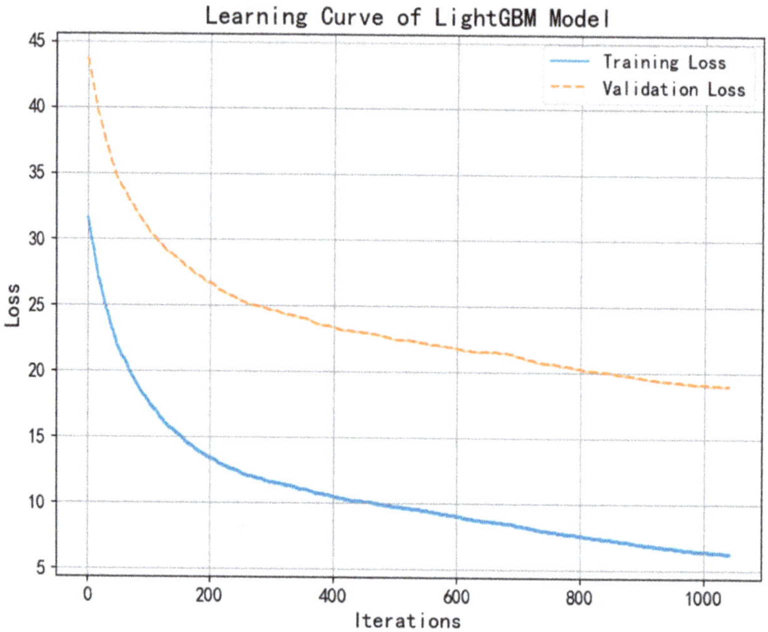

Figure 9. Learning curve of the LightGBM model.

After training the model, SHAP analysis was conducted to interpret the contribution of each input variable to the prediction outcomes. The resulting variable importance ranking (Figure 10) and SHAP value distribution plots (Figure 11) revealed that the factors influencing active population clustering differed from those affecting passenger flow or dwellers. In particular, active population distribution more strongly reflected perceptions of spatial design, economic vitality, and local consumption potential in HSR station areas.

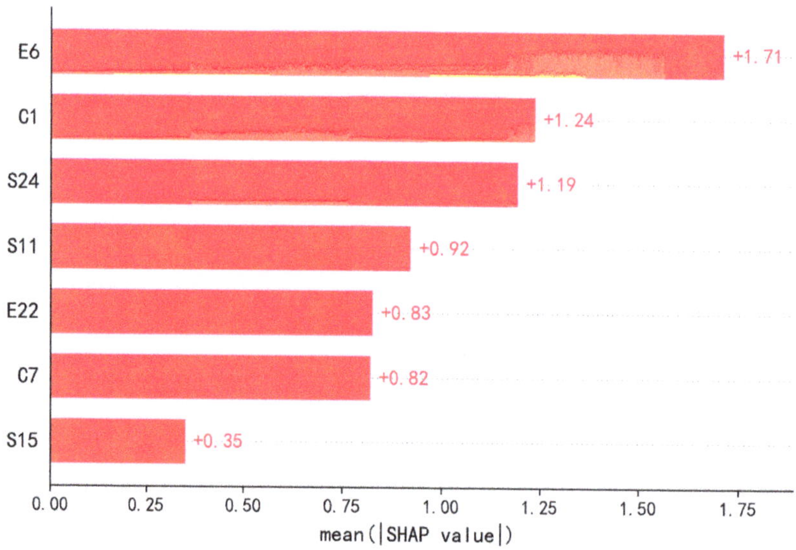

Figure 10. Feature importance ranking based on mean SHAP values.

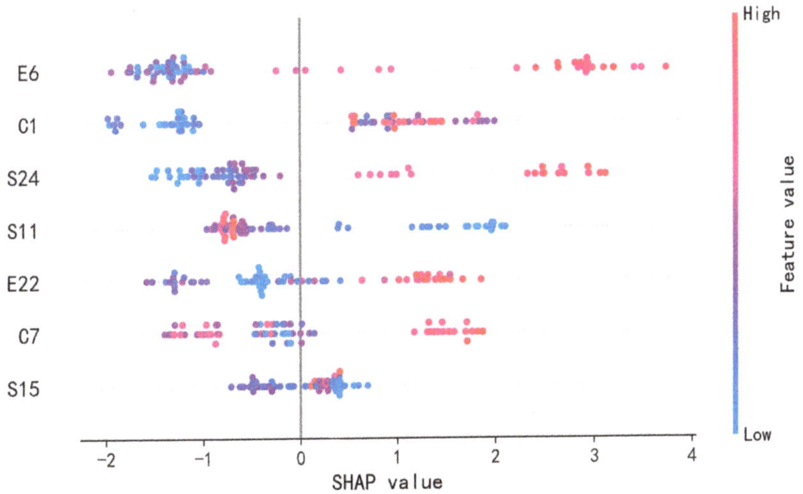

Figure 11. SHAP summary plot showing feature effects and value distributions.

Among all features, the proportion of land used for commercial service facilities within the station area (E6) exhibited the highest SHAP value of +1.71, indicating it had the most substantial positive impact on active population clustering. In addition, the population size (C1) and urbanization level (C7) of the HSR station's host city had SHAP values of +1.24 and +0.92, respectively, suggesting that both city scale and development level significantly promote human activity near the stations. The station location index (S11) also showed a notable SHAP value of +0.92, confirming that favorable geographic positioning and transit hub advantages play a key role in attracting people. Moreover, the number of bus stops near the station (S24) and the number of parking facilities in the station area (E22) had SHAP values of +0.83 and +0.82, respectively, underscoring the importance of well-developed transit infrastructure in enhancing population aggregation.

4.4. Threshold Effects

The primary purpose of the univariate driving model is to examine the independent effect of a single variable on population clustering, thereby identifying and understanding the specific degree and pattern of influence each factor has on high-speed rail (HSR) station areas. Through the quantitative assessment of each variable, the model reveals how these factors operate under different conditions and how their effects may change across value ranges.

The results indicate a strong relationship between commercial facilities and population clustering in HSR station areas. Specifically, the partial dependence plot for the proportion of commercial service land use in the station area (E6) shows that when the share increases from a low baseline to around 12%, the predicted value of active population rises sharply. However, once the proportion exceeds 15%, the predicted value tends to stabilize (Figure 12). This pattern reflects the classical economic law of diminishing marginal utility and suggests that while moderate commercial land allocation enhances attractiveness and pedestrian density, excessive concentration may cause functional redundancy, spatial congestion, or underutilization—consistent with the "resource saturation" theory in urban land use planning. This finding suggests that a moderate increase in commercial land use can significantly enhance the vibrancy of the station area by attracting more consumers and stayers. Therefore, in planning areas surrounding HSR stations, commercial, cultural, and office functions should be strategically allocated to match actual demand, thus strengthening the overall attractiveness and functional diversity of the district.

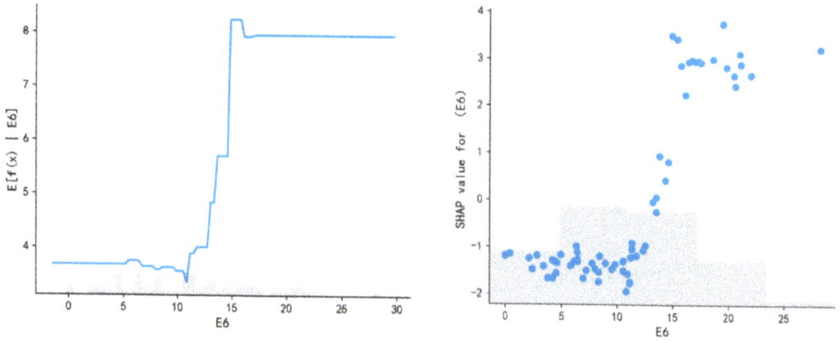

Figure 12. Partial dependence plot for feature E6.

In addition, the station location index (S11) exhibits a significant negative correlation with the predicted value of active population clustering (Figure 13). The partial dependence plot shows that when S11 is low—indicating a superior spatial location within the city—the predicted population level is high. As S11 increases, the predicted value drops rapidly and levels off, particularly after the value exceeds 0.3. This result highlights the critical role of spatial positioning in attracting human activity. And then levels off, reinforcing the spatial mismatch theory which states that urban facilities located far from core areas often suffer from reduced accessibility and weaker socio-economic integration. Consequently, the planning and development of HSR station areas should prioritize their integration within the broader urban spatial framework. Strengthening connectivity with central urban zones and major functional areas, improving transport accessibility, and ensuring the availability of high-quality public services can collectively enhance the locational advantages of HSR stations, thereby promoting greater population clustering and urban vitality.

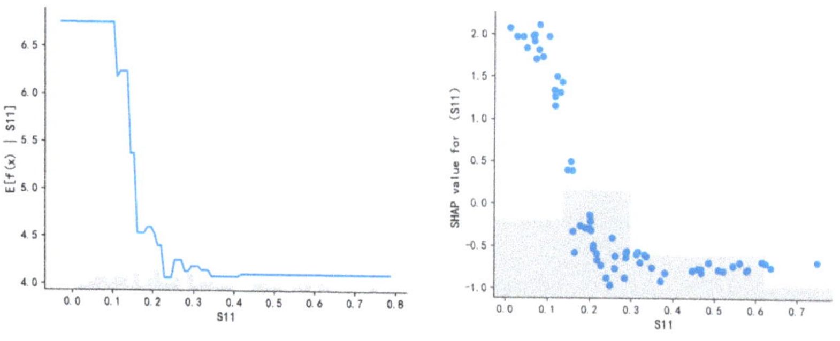

Figure 13. Partial dependence plot for feature S11.

Furthermore, the number of bus stops (S24) near the station has a clear and positive effect on population activity (Figure 14). According to the partial dependence plot, as the number of bus stops increases, the predicted value of active population rises notably—especially after the number surpasses 10, at which point the prediction exhibits a step-like increase and gradually stabilizes. This pattern confirms that a well-developed bus transfer system is a key factor in enhancing the vitality of HSR station areas. Increasing the number of bus stops not only improves transportation accessibility but also expands the service coverage of the station area, thus attracting more pedestrian flow and encouraging longer stays. Accordingly, future planning of station-area transit infrastructure should prioritize the density and coverage of the bus network, support multimodal integration, and promote the construction of efficient and convenient public transportation systems to better support human activity and sustainable station-area development.

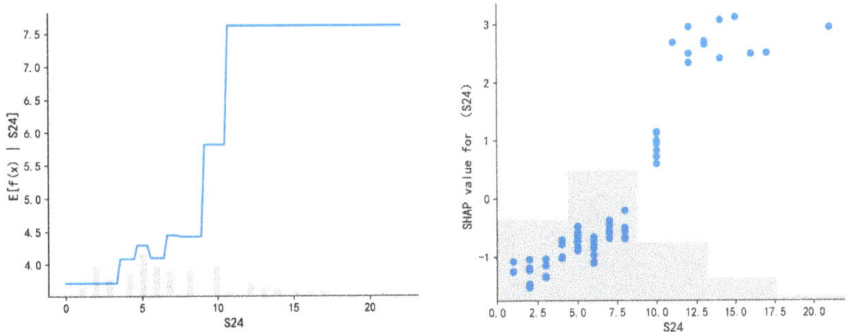

Figure 14. Partial dependence plot for feature S24.

These patterns reflect the classical economic concept of diminishing marginal returns, which states that beyond a certain threshold, the additional benefit of increasing an input (e.g., commercial land ratio or bus stop density) decreases.

From a theoretical standpoint, the observed turning points align with threshold theory, which emphasizes the presence of critical values beyond which system responses change nonlinearly. Additionally, the phenomenon is related to resource saturation theory, where excessive infrastructure or functional allocation may exceed local demand or spatial capacity, leading to diminishing efficiency or even congestion effects.

These findings suggest that urban planning around HSR stations must consider not only increasing supply but also optimizing it to avoid over-concentration and spatial redundancy.

4.5. Interaction Effects

The main objective of the multi-factor interaction model is to assess the combined influence of multiple variables on population clustering. This model emphasizes the interaction and synergy among different variables, revealing how their joint effects can significantly enhance clustering outcomes. By analyzing the composite influence of variable combinations, the model helps identify which interactions are most impactful in shaping human activity around HSR station areas. Figure 15 shows the interaction effects of influencing factors on spatial vitality in HSR station areas.

To quantify interaction effects, we used SHAP's built-in method, which separates each prediction into the effect of individual features and the added effect from feature pairs. For any two features, SHAP calculates how much their combined influence differs from the total of their separate effects. If the combined effect is stronger, this indicates a synergy—what we call a "synergy bonus." These observed synergies are entirely data-driven and reflect the nonlinear relationships captured by the LightGBM model. We did not assign any fixed values or assumptions; instead, the SHAP interaction values reflect the complex relationships automatically learned by the LightGBM model.

The interaction plot reveals that the urbanization level of the HSR station's host city (C7) and the proportion of commercial service land use in the station area (E6) jointly affect the predicted intensity of population activity (Figure 16). The results show that when the urbanization level is high (above approximately 85%) and the proportion of commercial service land use exceeds 15%, the predicted value of population activity rises significantly, peaking at around 8.9. This indicates a strong clustering effect. Conversely, when either the urbanization level is low or the proportion of commercial land is insufficient, the predicted values remain at a relatively low level. This supports the "urban function–infrastructure synergy" theory, where a mature urban context provides sufficient population density and mobility demand, while commercial land facilitates consumption and social interaction.

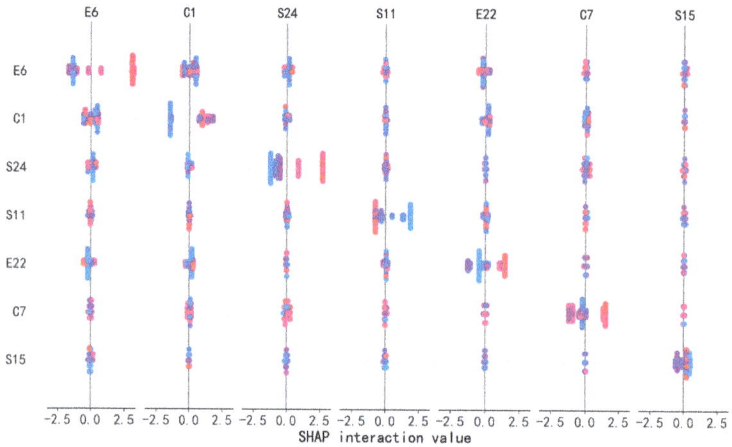

Figure 15. SHAP interaction values for spatial vitality in HSR station areas.

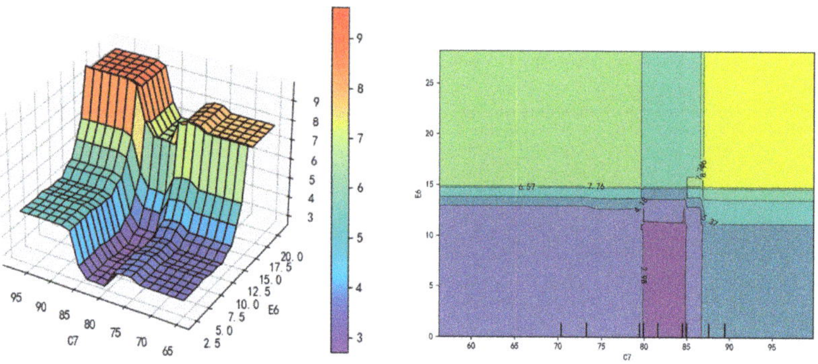

Figure 16. Three-dimensional and two-dimensional joint partial dependence plots for features C7 and E6.

This suggests a clear synergistic relationship between a city's development level and the functional capacity of the station area. A high level of urbanization typically brings higher population density and better infrastructure, while commercial services offer spaces for consumption and social engagement. The combination of these two factors maximizes the vitality of HSR station areas. Therefore, improving station-area vibrancy depends not only on local planning and design but also on aligning with the broader developmental context of the city. In cities with high urbanization levels, special attention should be paid to enhancing functional diversity within the station area—particularly by increasing the supply and quality of commercial service facilities—to ensure efficient land use and foster a positive cycle of population clustering.

Another interaction plot shows the joint effect of the number of bus stops (S24) and the proportion of commercial service land use (E6) on population activity (Figure 17). The plot reveals that when the number of bus stops is relatively high and the commercial land use ratio is moderate to high (between approximately 10% and 20%), the predicted value of population activity increases significantly, peaking at around 10.3. This demonstrates a strong positive synergy. In contrast, if the number of bus stops is low or the commercial land share is insufficient, the predicted activity level drops considerably. This aligns with transit-oriented development (TOD) principles: multimodal connectivity enhances last-mile accessibility, while commercial functions support destination attractiveness.

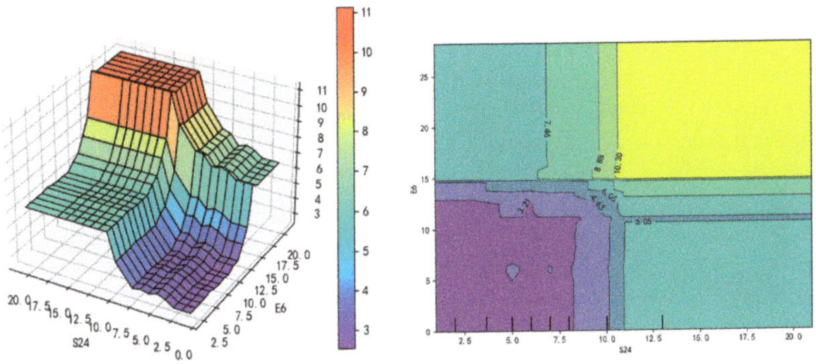

Figure 17. Three-dimensional and two-dimensional joint partial dependence plots for features S24 and E6.

Additionally, the interaction between the station location index (S11) and the number of parking facilities (E22) also influences both population clustering and mobility within the station area. As shown in Figure 18, when the station location index is low (approximately 0.1) and the number of parking facilities increases substantially (e.g., up to 400 spaces), the predicted value of population activity rises to 7.87, reflecting a pronounced clustering effect. A higher location index typically means the station is farther from the city center, leading to weaker transport connections. In such cases, sufficient parking becomes critical—especially for car users—by lowering access costs and improving station reachability. This interaction illustrates a compensatory mechanism: parking facilities can mitigate the disadvantage of poor location by reducing access costs for car users, thus broadening the station's functional catchment. This finding resonates with access–cost theory and supports the importance of integrated planning in peripheral station areas.

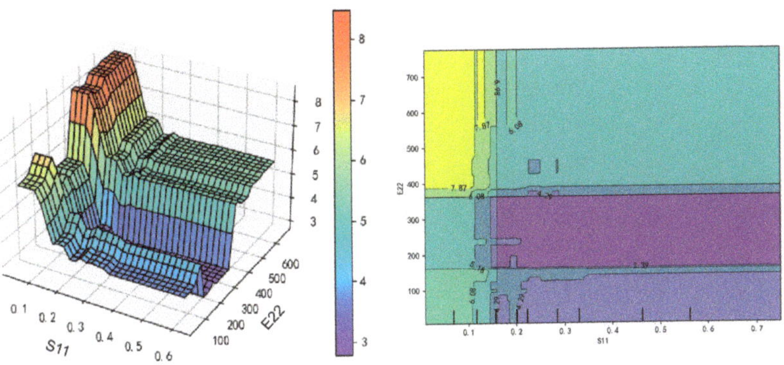

Figure 18. Three-dimensional and two-dimensional joint partial dependence plots for features S11 and E22.

As the number of parking facilities increases, the area's transportation support capacity improves, which in turn promotes population aggregation. Particularly when the station's spatial location is less advantageous, optimizing parking infrastructure can help offset locational disadvantages and attract more users. The interaction between station location and parking capacity highlights the importance of coordinated planning. Enhancing transportation linkages and parking management can significantly improve accessibility and appeal, contributing to sustainable human activity and functional development in HSR station areas.

The observed synergies between urbanization and commercial land use suggest that spatial vitality is a product of both macro-level urban development and micro-level spatial design. This supports theories from urban systems and complexity science, which emphasize the interdependence between urban form and function. The enhanced activity observed at well-connected, well-equipped station areas validates the role of integrated planning in maximizing land value and urban accessibility.

4.6. Individual-Level Influences

Due to differences in land use types, development intensity, and spatial configuration patterns, high-speed rail (HSR) station areas exhibit diverse developmental characteristics within the urban spatial structure. Different types of station areas show distinct patterns in how they attract populations, which factors most influence clustering, and what development strategies are most effective. Therefore, analyzing individual station areas helps identify key influencing factors and enables the formulation of tailored planning approaches.

For HSR station areas dominated by residential functions, population clustering is primarily driven by the availability of lifestyle amenities, commercial services, and transportation accessibility (Figure 19). For example, at Beijing South Station, several variables were found to positively influence population aggregation, including the number of commercial and leisure facilities within the station area (S27), general public budget revenue of the host city (C10), centrality within the urban network (C16), and the number of directly connected stations via one-transfer routes (S4). These factors collectively indicate that the station attracts people by offering robust commercial amenities and strong connectivity. However, certain constraints were also identified, such as limited sky openness (E16), high road density (C12), and a lower number of metro lines (S23), which may hinder further population clustering. Therefore, the development strategy for such residential-oriented station areas should focus on enhancing everyday service facilities, strengthening metro and bus integration, and improving land use efficiency to increase the convenience and livability of the area, thereby enhancing its overall attractiveness.

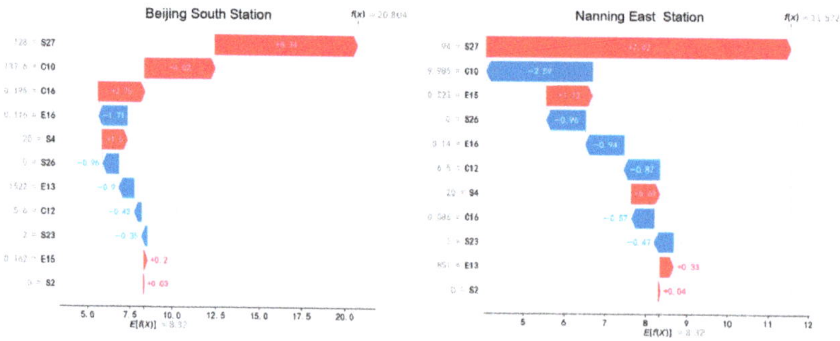

Figure 19. SHAP feature contribution for Beijing South Station and Nanning East Station.

In commercially oriented station areas, such as Zhengzhou Station, population clustering is mainly influenced by the quality of commercial facilities, the radiation effect of the city center, and transportation connectivity (Figure 17). Key promoting factors include urban centrality (C16), the number of station-based commercial and leisure facilities (S27), and the number of directly connected stations (S4). These variables demonstrate that Zhengzhou Station effectively attracts passengers and dwellers through its commercial vitality and strong transport hub functions, enhancing its clustering effect (Figure 20). However, several limiting factors were also identified, including general public budget revenue (C10), sky openness (E16), road network density (C12), number of road intersections (E13), and the number of metro lines (S23). These constraints indicate that although Zhengzhou Station performs well in commercial and transportation dimensions, it still has room for improvement in environmental comfort, traffic network optimization, and vertical spatial design. To further promote clustering effects in this station area, it is recommended to integrate surrounding resources through comprehensive and future-oriented planning and renewal efforts.

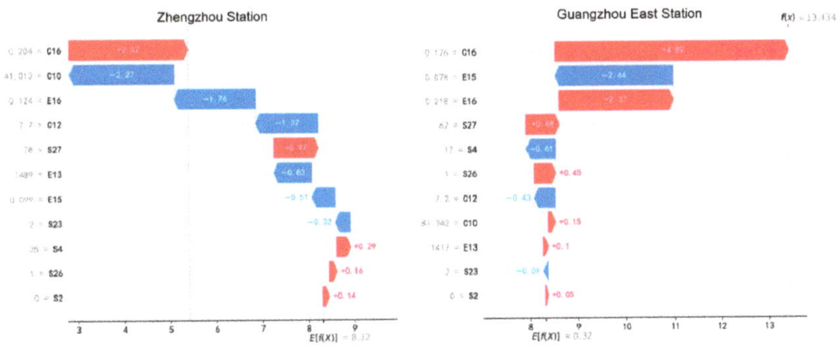

Figure 20. SHAP feature contribution for Zhengzhou Station and Guangzhou East Station.

Overall, these station-level differences reaffirm the necessity of differentiated planning strategies. Different types of HSR station areas exhibit their own unique spatial characteristics and development potential. Residential-, transport-, commercial-, and public service-oriented station areas each have their own advantages and challenges. Therefore, development strategies should be tailored according to the station's functional orientation, geographic context, infrastructure quality, and socioeconomic background. For each station type, enhancing infrastructure, optimizing spatial layout, and reinforcing inter-functional connectivity are key strategies for fostering population aggregation and achieving sustainable development of HSR station areas.

5. Discussion

5.1. Rethinking Station Area Vitality Through Nonlinear Urban Dynamics

This study's findings challenge the conventional linear assumptions in urban spatial analysis and contribute to a more nuanced understanding of how high-speed rail (HSR) station areas evolve. The revealed nonlinear relationships—particularly the saturation effects of commercial land and bus stop density—indicate that vitality does not increase indefinitely with infrastructural inputs. This echoes recent shifts in urban theory emphasizing complex systems thinking, where urban space responds to incremental interventions in nonlinear, often threshold-dependent ways.

In comparison with existing literature, which primarily views commercial land and transit accessibility as linear enablers of urban vitality, our results add empirical weight to the idea that "more" is not always "better." The discovery of diminishing marginal returns in facility input highlights the importance of optimal functional balance in station-area development. From a theoretical standpoint, this finding extends the urban spatial equilibrium framework by introducing nonlinear feedback effects triggered by over-saturation or spatial crowding.

5.2. Spatial Thresholds and the Redefinition of Planning Efficiency

The identification of clear spatial thresholds—such as the optimal commercial land ratio (10–15%) and ideal bus stop density—offers a critical redefinition of what constitutes planning efficiency in station area contexts. Rather than maximizing infrastructure, planners should aim for functionally calibrated interventions based on local capacity and contextual needs. This perspective moves beyond traditional engineering-driven logic and embraces performance-based spatial planning.

Moreover, these thresholds provide actionable tools for resource prioritization in cities facing financial or spatial constraints. For instance, peripheral stations may not benefit from excessive commercial zoning or transit lines, whereas moderate-scale, context-sensitive development could yield more sustainable outcomes. This insight has practical value for transit-oriented development (TOD), supporting phased, evidence-based investment models that align with evolving population flows and spatial usage.

5.3. Theoretical and Policy Implications for Node–Place Synergy

The findings reinforce and refine the "node–place" framework by revealing that spatial vitality emerges not from the node (HSR infrastructure) or the place (urban environment) alone, but from their synergy, particularly when modulated by city-level conditions (e.g., urbanization rate, centrality). The interaction effects between macro-structural variables (e.g., C7, urbanization) and micro-spatial attributes (e.g., E6, commercial land) suggest that infrastructure performance is contingent upon urban context compatibility.

This insight has clear policy relevance. For national and provincial planning bodies, it calls for differentiated station development models based on regional typologies. For instance, stations in highly urbanized zones should prioritize function mixing and pedestrian-centric design, while those in transition zones may benefit from mobility-enhancing infrastructure like park-and-ride systems. Furthermore, integrating social sensing data (e.g., Weibo check-ins) into urban monitoring can facilitate adaptive governance and promote real-time evaluation frameworks for major transport infrastructure.

To further contextualize the findings, we compare two representative cases: Chengdu and Beijing. Chengdu, as a rapidly expanding city, features HSR stations like Chengdu East and Chengdu South that are still in the process of urban integration. Here, planning should emphasize balanced commercial land development, robust multimodal access, and staged infrastructure investment to prevent premature saturation. In contrast, mature hubs such as Beijing South or Beijing West require strategies that alleviate congestion, optimize transfer systems, and elevate service quality. These differentiated approaches underscore that the node–place synergy must be adapted to each city's growth stage and spatial characteristics, reaffirming the importance of localized, context-aware planning in HSR station area development.

6. Conclusions

This study systematically examined the spatial vitality of high-speed rail (HSR) station areas in China by integrating multi-source data and applying both statistical and machine learning methods. Based on a city–node–place framework, we evaluated 66 HSR station areas across 35 cities using Sina Weibo check-in data, urban support capacity, station attributes, and built environment indicators.

Importantly, this study considered both the broader urban context in which each station is embedded and the detailed environmental features within the surrounding area. City-level factors such as population size, urbanization level, and transit network configuration form the underlying support conditions, while local-scale features—including commercial land use ratio, bus stop density, and parking facilities—shape the functional performance of each station area. Our use of a 1500 m buffer allowed for consistent comparison across cities, capturing meaningful differences in how both urban structure and immediate surroundings influence vitality.

The results show that city population size, urbanization level, commercial land use ratio, transit accessibility, and parking facilities are key positive contributors to station area vitality. SHAP analysis revealed nonlinear threshold effects and interaction relationships—such as diminishing returns in commercial land use and bus stop density, and strong synergies between urban development and functional configuration.

These findings have several policy implications. First, land use and transit planning can be integrated to enhance functional synergy around HSR stations. Second, moderate commercial land allocation (around 10–15%) is most effective in attracting population activity. Third, improving multimodal transport infrastructure—including bus stops and parking—can strengthen last-mile connectivity. Fourth, tailored development strategies are needed for different station types (e.g., residential vs. commercial). Lastly, data-driven evaluation tools like social media analytics and interpretable machine learning can support real-time monitoring and adaptive policy formulation.

Nonetheless, this study has some limitations. The use of Sina Weibo check-in data, while providing high spatial-temporal granularity, may introduce demographic and geographic biases. Younger users and residents of more developed cities are overrepresented, while the behavior of older adults and residents in less developed or peripheral areas may be underrepresented. These biases could influence the comprehensiveness of vitality assessment. Future research should further integrate diverse datasets—such as mobile phone signaling, smart card records, and public Wi-Fi logs—to improve representativeness. Additionally, the present study focuses on a single temporal snapshot. Future studies could explore how spatial vitality evolves over time using longitudinal data sources. Such efforts may reveal seasonal rhythms, lifecycle stages, or long-term shifts in station activity patterns, thereby enriching the understanding of station area development dynamics. Overall, this study offers empirical insights and methodological innovations to guide the sustainable, accessible, and vibrant development of HSR station areas in China and beyond.

Author Contributions: Conceptualization, P.L. and X.C.; Investigation, W.S. and Y.X.; Methodology, P.L. and J.M.; Software, P.L. and J.M.; Writing—original draft, P.L.; Writing—review and editing, X.C., J.M. and W.S. All authors have read and agreed to the published version of the manuscript.

Funding: This research was funded by the National Natural Science Foundation of China, grant number 52472331, 52278081 and U20A20330.

Data Availability Statement: The data presented in this study are available on request from the corresponding author. The data are not publicly available due to privacy.

Conflicts of Interest: The authors declare no conflicts of interest.

Appendix A

Table A1. HSR Stations, Corresponding Codes, and Key Independent Variable Values.

No.	Station Name	Station Code	C1 (Million)	C7 (%)	S11	S15 (Count)	S24 (Count)	E6 (%)	E22 (Count)
1	Beijing North	VAP	21.84	87.6	0.1	109	5	11.3	590
2	Beijing South	VNP	21.84	87.6	0.08	577	3	9.5	399

Table A1. *Cont.*

No.	Station Name	Station Code	C1 (Million)	C7 (%)	S11	S15 (Count)	S24 (Count)	E6 (%)	E22 (Count)
3	Beijing	BJP	21.84	87.6	0.03	81	8	21	719
4	Beijing West	BXP	21.84	87.6	0.12	289	2	17.3	562
5	Tianjin South	TIP	13.63	84.5	0.29	1	2	4.5	41
6	Tianjin	TJP	13.63	84.5	0.13	143	7	13.5	631
7	Tianjin West	TXP	13.63	84.5	0.2	158	5	7.7	261
8	Shijiazhuang	SJP	11.22	71.44	0.19	106	8	13.5	188
9	Shanghai	SHH	24.76	89.46	0.21	295	12	14.9	659
10	Shanghai Hongqiao	AOH	24.76	89.46	0.25	736	14	19.5	158
11	Shanghai West	SXH	24.76	89.46	0.24	0	4	11.3	247
12	Ningbo	NGH	9.62	79.9	0.2	94	10	10.9	427
13	Hangzhou East	HGH	12.38	84	0.05	195	6	14.6	294
14	Hangzhou South	XHH	12.38	84	0.58	0	3	12.5	198
15	Nanjing South	NKH	9.49	87.01	0.32	226	2	18.6	192
16	Nanjing	NJH	9.49	87.01	0.07	155	8	8	238
17	Hefei South	ENH	12.04	84.6	0.23	229	5	8.3	115
18	Hefei	HFH	12.04	84.6	0.22	5	10	21.1	315
19	Shenzhen North	IOQ	17.66	99.8	0.16	562	10	8.3	224
20	Shenzhen	SZQ	17.66	99.8	0.12	225	10	13.8	621
21	Futian	NZQ	17.66	99.8	0.01	53	15	28.2	668
22	Guangzhou East	GGQ	18.73	86.48	0.21	356	5	16.4	600
23	Guangzhou South	IZQ	18.73	86.48	0.62	804	5	11.7	103
24	Shapingba	CYW	32.13	71	0.37	57	12	11.1	451
25	Chongqing North	CUW	32.13	71	0.07	237	2	7.2	243
26	Chongqing West	CXW	32.13	71	0.55	164	3	3.8	72
27	Chengdu East	ICW	21.27	79.89	0.16	657	5	5.8	165
28	Chengdu South	CNW	21.27	79.89	0.12	103	10	16.7	549
29	Chengdu	CDW	21.27	79.89	0.09	4	6	17	248
30	Chengdu West	CMW	21.27	79.89	0.18	40	10	4.2	133
31	Wuhan	WHN	13.74	84.66	0.33	248	1	6.9	55
32	Hankou	HKN	13.74	84.66	0.16	287	2	17.5	376
33	Changsha South	CWQ	7.68	83.27	0.32	256	7	4.2	97
34	Nanchang	NCG	6.54	79.58	0.22	73	14	10.5	362
35	Nanchang West	NXG	6.54	79.58	0.51	260	12	3.4	56
36	Harbin	HBB	9.4	56.3	0.07	138	13	22	571
37	Harbin West	VAB	9.4	56.3	0.2	117	7	12.3	133
38	Changchun	CCT	9.07	67.1	0.26	166	16	14.3	464
39	Changchun West	CRT	9.07	67.1	0.46	27	2	0	7
40	Dalian North	DFT	7.53	73	0.45	137	3	11.2	149
41	Shenyang North	SBT	9.15	84.99	0.13	66	6	16.1	478
42	Shenyang	SYT	9.15	84.99	0.04	142	8	15.7	775
43	Qingdao North	QHK	10.34	77.32	0.29	157	4	2.4	67
44	Qingdao	QDK	10.34	77.32	0.16	156	17	6.4	363
45	Jinan	JNK	9.42	74.3	0.07	42	21	15.4	421
46	Jinan West	JGK	9.42	74.3	0.34	126	6	6.1	128
47	Xiamen North	XKS	5.31	90.2	0.47	155	6	4.9	61
48	Xiamen	XMS	5.31	90.2	0.08	162	11	10.5	495
49	Fuzhou South	FYS	8.45	73.27	0.58	120	7	8.9	49
50	Fuzhou	FZS	8.45	73.27	0.47	220	10	4.6	152
51	Nanning East	NFZ	8.89	70.36	0.49	306	5	2.1	46
52	Haikou	VUQ	8.89	82.7	0.64	17	1	0.4	9
53	Zhengzhou East	ZAF	12.83	79.4	0.56	345	5	13.2	231
54	Zhengzhou	ZZF	12.83	79.4	0.15	29	4	19.8	374
55	Kunming South	KOM	8.5	80.5	0.75	273	4	2.8	45
56	Kunming	KMM	8.5	80.5	0.28	90	13	20.5	376
57	Guiyang North	KQW	6.22	80.3	0.29	263	6	6.4	49
58	Taiyuan South	TNV	5.44	89.3	0.35	269	6	10.5	123
59	Taiyuan	TYV	5.44	89.3	0.25	7	5	20.6	393
60	Xi'an North	EAY	13	79.6	0.62	474	2	10.8	51

Table A1. *Cont.*

No.	Station Name	Station Code	C1 (Million)	C7 (%)	S11	S15 (Count)	S24 (Count)	E6 (%)	E22 (Count)
61	Lanzhou West	LAJ	4.42	84.07	0.2	133	4	6.3	75
62	Xining	XNO	5.95	61.43	0.52	55	7	4.4	100
63	Yinchuan	YIJ	7.28	66.34	0.38	88	8	6.4	102
64	Hohhot East	NDC	3.55	79.8	0.31	98	5	9.8	67
65	Hohhot	HHC	3.55	79.8	0.12	2	8	8.4	226
66	Urumqi	WAR	4.08	96.5	0.26	77	7	11.1	87

References

1. Zheng, W.; Wei, S. A 'node-place-network-city' framework to examine HSR station area development dynamics: Station typologies and development strategies. *J. Transp. Geogr.* **2024**, *120*, 103993. [CrossRef]
2. Li, J.; Sun, X.; Cong, W.; Miyoshi, C.; Ying, L.C.; Wandelt, S. On the air-HSR mode substitution in China: From the carbon intensity reduction perspective. *Transp. Res. A Policy Pract.* **2024**, *180*, 103977. [CrossRef]
3. Shao, S.; Tian, Z.; Yang, L. High speed rail and urban service industry agglomeration: Evidence from China's Yangtze River Delta region. *J. Transp. Geogr.* **2017**, *64*, 174–183. [CrossRef]
4. Chen, Z.; Haynes, K.E. Impact of high-speed rail on regional economic disparity in China. *J. Transp. Geogr.* **2017**, *65*, 80–91. [CrossRef]
5. Xu, W.A.; Zhou, J.; Yang, L.; Li, L. The implications of high-speed rail for Chinese cities: Connectivity and accessibility. *Transp. Res. A Policy Pract.* **2018**, *116*, 308–326. [CrossRef]
6. Wang, F.; Wei, X.; Liu, J.; He, L.; Gao, M. Impact of high-speed rail on population mobility and urbanisation: A case study on Yangtze River Delta urban agglomeration, China. *Transp. Res. A Policy Pract.* **2019**, *127*, 99–114. [CrossRef]
7. Jia, S.; Zhou, C.; Qin, C. No difference in effect of high-speed rail on regional economic growth based on match effect perspective? *Transp. Res. A Policy Pract.* **2017**, *106*, 144–157. [CrossRef]
8. Shaw, S.; Fang, Z.; Lu, S.; Tao, R. Impacts of high speed rail on railroad network accessibility in China. *J. Transp. Geogr.* **2014**, *40*, 112–122. [CrossRef]
9. Yue, Y.; Chen, J.; Feng, T.; Ma, X.; Wang, W.; Bai, H. Classification and determinants of high-speed rail stations using multi-source data: A case study in Jiangsu Province, China. *Sustain. Cities Soc.* **2023**, *96*, 104640. [CrossRef]
10. Deng, T.; Gan, C.; Perl, A.; Wang, D. What caused differential impacts on high-speed railway station area development? Evidence from global nighttime light data. *Cities* **2020**, *97*, 102568. [CrossRef]
11. Diao, M.; Zhu, Y.; Zhu, J. Intra-city access to inter-city transport nodes: The implications of high-speed-rail station locations for the urban development of Chinese cities. *Urban Stud.* **2017**, *54*, 2249–2267. [CrossRef]
12. Wei, S.; Wang, L. Classifying High-Speed Rail Stations in the Yangtze River Delta, China: The Node-Place Modelling Approach. *Appl. Spat. Anal. Policy* **2022**, *16*, 625–646. [CrossRef]
13. Wang, L.; Zheng, W.; He, S.; Wei, S. Assessing urban vitality and its determinants in high-speed rail station areas in the Yangtze River Delta, China. *J. Transp. Land Use* **2022**, *15*, 333–354. [CrossRef]
14. Lan, F.; Gong, X.; Da, H.; Wen, H. How do population inflow and social infrastructure affect urban vitality? Evidence from 35 large- and medium-sized cities in China. *Cities* **2020**, *100*, 102454. [CrossRef]
15. Chen, Y.; Yu, B.; Shu, B.; Yang, L.; Wang, R. Exploring the spatiotemporal patterns and correlates of urban vitality: Temporal and spatial heterogeneity. *Sustain. Cities Soc.* **2023**, *91*, 104440. [CrossRef]
16. Li, Q.; Cui, C.; Liu, F.; Wu, Q.; Run, Y.; Han, Z. Multidimensional Urban Vitality on Streets: Spatial Patterns and Influence Factor Identification Using Multisource Urban Data. *ISPRS Int. J. Geo-Inf.* **2021**, *11*, 2. [CrossRef]
17. Wang, B.; de Jong, M.; van Bueren, E.; Ersoy, A.; Meng, Y. Transit-Oriented Development in China: A Comparative Content Analysis of the Spatial Plans of High-Speed Railway Station Areas. *Land* **2023**, *12*, 1818. [CrossRef]
18. Zhang, X.; Wang, L.; Wei, S.; Cui, Y. Multifunctional development patterns and driving mechanisms of high-speed rail station areas in the Yangtze River Delta, China. *Transp. Policy* **2025**, *167*, 264–275. [CrossRef]
19. Yao, Y.; Liu, X.; Li, X.; Zhang, J.; Liang, Z.; Mai, K.; Zhang, Y. Mapping fine-scale population distributions at the building level by integrating multisource geospatial big data. *Int. J. Geogr. Inf. Sci.* **2017**, *31*, 1220–1224. [CrossRef]
20. Kim, S.; Lee, K.Y.; Shin, S.I.; Yang, S. Effects of tourism information quality in social media on destination image formation: The case of Sina Weibo. *Inf. Manag.* **2017**, *54*, 687–702. [CrossRef]
21. Huang, H.; Long, R.; Chen, H.; Sun, K.; Li, Q. Exploring public attention about green consumption on Sina Weibo: Using text mining and deep learning. *Sustain. Prod. Consum.* **2022**, *30*, 674–685. [CrossRef]

22. Li, S.; Dragicevic, S.; Castro, F.A.; Sester, M.; Winter, S.; Coltekin, A.; Pettit, C.; Jiang, B.; Haworth, J.; Stein, A.; et al. Geospatial big data handling theory and methods: A review and research challenges. *ISPRS J. Photogramm. Remote Sens.* **2016**, *115*, 119–133. [CrossRef]
23. Yang, L.; Lu, Y.; Cao, M.; Wang, R.; Chen, J. Assessing accessibility to peri-urban parks considering supply, demand, and traffic conditions. *Landsc. Urban Plan.* **2025**, *257*, 105313. [CrossRef]
24. Yang, L.; Yang, H.; Yu, B.; Lu, Y.; Cui, J.; Lin, D. Exploring non-linear and synergistic effects of green spaces on active travel using crowdsourced data and interpretable machine learning. *Travel Behav. Soc.* **2024**, *34*, 100673. [CrossRef]
25. Shafiq, S.M.; Tian, Z.; Bashir, A.K.; Jolfaei, A.; Yu, X. Data mining and machine learning methods for sustainable smart cities traffic classification: A survey. *Sustain. Cities Soc.* **2020**, *60*, 102177. [CrossRef]
26. Liu, L.; Silva, E.A.; Wu, C.; Wang, H. A machine learning-based method for the large-scale evaluation of the qualities of the urban environment. *Comput. Environ. Urban Syst.* **2017**, *65*, 113–125. [CrossRef]
27. Jing, Y.; Hu, H.; Guo, S.; Wang, X.; Chen, F. Short-Term prediction of urban rail transit passenger flow in external passenger transport hub based on LSTM-LGB-DRS. *IEEE Trans. Intell. Transp. Syst.* **2020**, *22*, 4611–4621. [CrossRef]
28. Doan, Q.C.; Ma, J.; Chen, S.; Zhang, X. Nonlinear and threshold effects of the built environment, road vehicles and air pollution on urban vitality. *Landsc. Urban Plan.* **2025**, *253*, 105204. [CrossRef]
29. Liang, P.; Cui, X.; Lin, M.; Yang, T.; Wu, B. High-speed rail effects on station area-level business commercial agglomeration: Evidence from 110 stations in China. *Front. Environ. Sci.* **2022**, *10*, 1045959. [CrossRef]
30. Yin, H.; Xiao, R.; Fei, X.; Zhang, Z.; Gao, Z.; Wan, Y.; Tan, W.; Jiang, X.; Cao, W.; Guo, Y. Analyzing "economy-society-environment" sustainability from the perspective of urban spatial structure: A case study of the Yangtze River delta urban agglomeration. *Sustain. Cities Soc.* **2023**, *96*, 104691. [CrossRef]
31. Li, X.; Shi, L.; Shi, Y.; Tang, J.; Zhao, P.; Wang, Y.; Chen, J. Exploring interactive and nonlinear effects of key factors on intercity travel mode choice using XGBoost. *Appl. Geogr.* **2024**, *166*, 103264. [CrossRef]
32. Loo, B.P.Y.; Huang, Z. Location matters: High-speed railway (HSR) stations in city evolution. *Cities* **2023**, *139*, 104380. [CrossRef]
33. Deng, Y.; Bai, Y.; Cui, L.; He, R. Travel Mode Choice Behavior for High-Speed Railway Stations Based on Multi-Source Data. *Transp. Res. Rec.* **2023**, *2677*, 525–540. [CrossRef]
34. Liu, Y.; Xu, S.; Tian, J.; Liu, T.; Dong, T. What matters in promoting new town by High-Speed Railway station? Evidence from China. *Transp. Policy* **2024**, *159*, 241–253. [CrossRef]
35. Deng, T.; Wang, D.; Yang, Y.; Yang, H. Shrinking cities in growing China: Did high speed rail further aggravate urban shrinkage? *Cities* **2019**, *86*, 210–219. [CrossRef]
36. Lin, C. Research of impact factors of High-speed railways Hub area development in China. *Urban Plan. Int.* **2011**, *26*, 72–77.
37. Wang, X.; Pan, H. Land Change Pattern in High-Speed Rail Station Area: Empirical Research on Yangtze River Delta Region in China from 2010 to 2020. In *Socioeconomic Impacts of High-Speed Rail Systems, Proceedings of the 2nd International Workshop on High-Speed Rail Socioeconomic Impacts, online, 12–13 September 2023*; Springer: Cham, Switzerland, 2023; pp. 315–334. [CrossRef]
38. Niu, B.; Yin, P.; Shen, P. Industrial Spatio-Temporal Distribution of High-Speed Rail Station Area from the Accommodation Facilities Perspective: A Multi-City Comparison. *Land* **2023**, *12*, 332. [CrossRef]
39. Sugie, L.; Sung, H.; Joong, C.M. Analysis of Users' Facility Preference for Mixed Land Use Development in the High-Speed Rail Station Areas. *J. Korea Plan. Assoc.* **2014**, *49*, 211–224.
40. Ribalaygua, C.; Perez-Del-Caño, S. Assessing spatial planning strategy in high-speed rail station areas in Spain (1992–2018): Towards a sustainable model. *Eur. Plan. Stud.* **2019**, *27*, 595–617. [CrossRef]
41. Wang, B.; Ersoy, A.; van Bueren, E.; de Jong, M. Rules for the Governance of Transport and Land use Integration in High-speed Railway Station Areas in China: The Case of Lanzhou. *Urban Policy Res.* **2022**, *40*, 122–141. [CrossRef]
42. Doronzo, D.; Caggianelli, A.; Dellino, P. A ghost eruption behind "Le vulcanoclastiti di Craco" ghost town, Basilicata, Italy. *Int. J. Earth Sci.* **2019**, *108*, 545–546. [CrossRef]
43. Schütz, E. Stadtentwicklung durch Hochgeschwindigkeitsverkehr: Konzeptionelle und methodische Ansätze zum Umgang mit den Raumwirkungen des schienengebundenen Personen-Hochgeschwindigkeitsverkehrs (HGV) als Beitrag zur Lösung von Problemen der Stadtentwicklung. *Informationen Zur Raumentwickl.* **1998**, *6*, 369–383.
44. Lan, W.; Can, W.; Chen, C.; Hao, G. Development and Planning of the Surrounding Area of High-Speed Rail Stations: Based on Empirical Study of Beijing-Shanghai Line. *Urban Plan. Forum* **2014**, *4*, 31–37.
45. Zou, Z.; Tang, Y. Evaluation of Sustainable Development Potential of High-Speed Railway Station Areas Based on "Node-Place-Industry" Model. *ISPRS Int. J. Geo-Inf.* **2023**, *12*, 349. [CrossRef]
46. Du, J.; Druta, O.; van Wesemael, P. Place quality in high-speed rail station areas: Concept definition. *J. Transp. Land Use* **2021**, *14*, 1165–1186. [CrossRef]
47. Wang, L.; Yuan, F.; Duan, X. How high-speed rail service development influenced commercial land market dynamics: A case study of Jiangsu province, China. *J. Transp. Geogr.* **2018**, *72*, 248–257. [CrossRef]

48. Jiao, J.; Zhang, Q.; Jiang, R.; Lyu, G. Does high-speed rail contribute to cross-boundary agglomeration of migrant workers? Evidence from China. *J. Transp. Geogr.* **2025**, *125*, 104170. [CrossRef]
49. Wang, F.; Liu, Z.; Xue, P.; Dang, A. High-speed railway development and its impact on urban economy and population: A case study of nine provinces along the Yellow River, China. *Sustain. Cities Soc.* **2022**, *87*, 104172. [CrossRef]
50. Jendryke, M.; Balz, T.; Liao, M. Big location-based social media messages from China's Sina Weibo network: Collection, storage, visualization, and potential ways of analysis. *Trans. GIS* **2017**, *21*, 825–834. [CrossRef]
51. Lu, Y.; Li, J.; Yang, Z.; Ou, X.; Xie, W. Data mining and social networks processing method based on support vector machine and k-nearest neighbor. *J. Comput. Methods Sci. Eng.* **2021**, *21*, 435–447. [CrossRef]
52. Zhou, M.; Wang, M.; Hu, Q. A POI Data Update Approach Based on Weibo Check-in Data. In Proceedings of the 2013 21st International Conference on Geoinformatics (Geoinformatics), Henan University, Kaifeng, China, 20–22 June 2013.
53. Deng, T.; Wang, D.; Hu, Y.; Liu, S. Did high-speed railway cause urban space expansion?-Empirical evidence from China's prefecture-level cities. *Res. Transp. Econ.* **2020**, *80*, 100840. [CrossRef]
54. Diao, M. Does growth follow the rail? The potential impact of high-speed rail on the economic geography of China. *Transp. Res. A Policy Pract.* **2018**, *113*, 279–290. [CrossRef]
55. Guo, Y.; Cao, L.; Song, Y.; Wang, Y.; Li, Y. Understanding the formation of City-HSR network: A case study of Yangtze River Delta, China. *Transp. Policy* **2022**, *116*, 315–326. [CrossRef]
56. Xu, M.; Shuai, B.; Wang, X.; Liu, H.; Zhou, H. Analysis of the accessibility of connecting transport at High-speed rail stations from the perspective of departing passengers. *Transp. Res. A Policy Pract.* **2023**, *173*, 103714. [CrossRef]
57. Zhang, X.; Loo, B.P.Y.; Wang, L.; Wei, S. Disparities in high-speed rail accessibility in the Yangtze River Delta, China: A door-to-door travel time perspective. *Res. Transp. Bus. Manag.* **2025**, *59*, 101308. [CrossRef]
58. Yang, J.; Cao, J.; Zhou, Y. Elaborating non-linear associations and synergies of subway access and land uses with urban vitality in Shenzhen. *Transp. Res. A Policy Pract.* **2021**, *144*, 74–88. [CrossRef]
59. Mouratidis, K.; Poortinga, W. Built environment, urban vitality and social cohesion: Do vibrant neighborhoods foster strong communities? *Landsc. Urban Plan.* **2020**, *204*, 103951. [CrossRef]
60. Li, X.; Li, Y.; Jia, T.; Zhou, L.; Hijazi, I. The six dimensions of built environment on urban vitality: Fusion evidence from multi-source data. *Cities* **2022**, *121*, 103482. [CrossRef]
61. He, Q.; He, W.; Song, Y.; Wu, J.; Yin, C.; Mou, Y. The impact of urban growth patterns on urban vitality in newly built-up areas based on an association rules analysis using geographical 'big data'. *Land Use Policy* **2018**, *78*, 726–738. [CrossRef]

land

MDPI

Article

The Impact of High-Speed Rail on High-Quality Economic Development: Evidence from China

Xixi Feng [1], Jixiao Li [2], Yadan Liu [1] and Weidong Li [1,*]

[1] School of Economics and Management, Beijing Jiaotong University, Beijing 100044, China; 22110147@bjtu.edu.cn (X.F.); 23111152@bjtu.edu.cn (Y.L.)
[2] School of Finance and Taxation, Hebei University of Economics and Business, Shijiazhuang 050061, China; lijixiao2002@163.com
* Correspondence: wdli@bjtu.edu.cn

Abstract: Utilizing data from 282 prefecture-level cities in China from 2005 to 2021, this study constructs an evaluation index system for high-quality economic development across the following five dimensions: innovation, coordination, green, openness, and sharing. A continuous difference-in-differences approach is employed for regression analysis to empirically examine the impact of high-speed rail on high-quality economic development, further exploring its mechanisms and spatial spillover effects. The findings reveal that (1) HSR significantly promotes high-quality economic development; (2) with the development of HSR, from 2005 to 2021, China's high-quality economic development showed an evolutionary trend of overall improvement, with a gradual optimization of spatial patterns; (3) it facilitates high-quality economic development by enhancing capital and labor mobility, strengthening industrial chain resilience, and advancing industrial structure upgrading; (4) high-speed rail development in neighboring regions generates positive spatial spillover effects on local urban economic quality; and (5) the impact of high-speed rail on high-quality economic development exhibits significant heterogeneity across cities with different regions, tiers, scales, and resource endowments. These results confirm the positive role of high-speed rail in fostering high-quality economic development.

Keywords: high-speed rail (HSR); heterogeneity analysis; robustness check; mechanism analysis; high-quality economic development (HQE); continuous difference-in-differences

1. Introduction

The 2022 National Congress of the Communist Party of China, convened from October 16 to 22, highlighted that the contemporary socialist nation's overall building requires prioritizing high-quality development. HQE essentially entails the harmonious integration of quantity and quality, encompassing economic, social, environmental, and innovative dimensions [1]. Economic growth does not equate to economic development [2]. In the four decades since the initiation of its reform and opening-up policy, China has witnessed remarkable economic expansion, but increasing resource and environmental constraints have rendered the traditional extensive development model unsustainable [3]. The pressure for green transformation is substantial, making it imperative to promote HQE [4].

Since the launch of the Beijing–Tianjin Intercity Railway in 2008, China's HSR infrastructure has experienced swift expansion, amassing the most extensive operational mileage globally and accounting for nearly 67% of the international total [5]. This extensive rail network has bolstered regional interconnectivity [6]. With the implementation

https://doi.org/10.3390/land14071379

of the "Eight Vertical and Eight Horizontal" plan, China has fully embraced the "high-speed rail era" [7]. The increasing density of the HSR network has significantly boosted regional connectivity, effectively promoting the movement of people, goods, information, and capital along its routes [8]. This has injected new momentum into regional economic development, driving transformative improvements in economic activity quality, efficiency, and green transformation [9]. Beyond stimulating regional economic growth, HSR also curbs urban pollutants such as sulfur dioxide and wastewater, reduces carbon emissions, and enhances urban ecological efficiency [10], thereby influencing undesirable outputs in economic development and delivering the economic effects of both emission reduction and efficiency enhancement [11]. As a result, the HSR expansion has become a crucial driver for high-quality urban development [12].

Has more than a decade of HSR development spurred HQE? What are the mechanisms through which HSR drives HQE? This study explores several key questions, as follows: Does HSR still have a positive effect on HQE after accounting for other policies and confounding factors? Do regional differences exist in the impact of HSR on HQE? Does HQE create spatial spillover effects, and if so, what are the boundaries of these effects? To address these questions, this research constructs an evaluation system for HQE in China, covering the following five dimensions: innovation, coordination, green development, openness, and sharing. The HQE index for 282 cities from 2005 to 2021 is calculated using the entropy-weighted TOPSIS method. A continuous difference-in-differences model is employed to empirically examine the role of HSR in promoting HQE. By reducing spatial and temporal barriers, HSR accelerates the flow of labor and capital, enhances industrial chain resilience, promotes industrial structure upgrading, and improves green innovation levels, all of which contribute to HQE. The impact of HSR shows significant heterogeneity across regions, city tiers, transportation hub status, and resource endowments. It has a significantly positive effect on high-quality development in non-transportation hub cities, peripheral cities, and eastern and western regions, as well as growth-stage and mature resource-based cities. However, it shows no significant impact on growth-stage and renewable resource-based cities, central cities, and transportation hub cities, while exerting a significantly negative effect on declining resource-based cities. Additionally, HQE generates positive spatial spillover effects, with an attenuation boundary of 1000 km.

The primary contributions of this paper are mainly reflected in the following aspects: (1) innovatively selecting 33 effective indicators and employing the entropy-weighted TOPSIS method to measure the economic quality indices of 282 prefecture-level cities; (2) comprehensively validating the significantly positive impact of HSR on various mechanism variables, including labor mobility, capital mobility, industrial structure upgrading, and industrial chain resilience; (3) from the perspective of urban spatial connectivity, utilizing a continuous difference-in-differences model and a spatial Durbin model to comprehensively analyze the impact of HSR on urban HQE and its spatial spillover effects, with the spatial spillover boundary estimated at 1000 km, providing valuable insights for future large-scale HSR construction; and (4) to some extent, it enriches the research on HSR and HQE and provides an in-depth analysis of the spatial evolution characteristics of high-quality urban development.

The rest of the paper is structured as follows: the second section presents the literature review; the third section outlines the research method; the fourth section analyzes the empirical analysis; the fifth section presents the discussion of findings; and the sixth section concludes the paper.

2. Literature Review

The measurement of HQE indices in the existing literature is primarily divided into the following two approaches: single indicators and comprehensive evaluation index systems. Among these, single indicators often employ total factor productivity [13] and green total factor productivity [14] as proxy variables, while comprehensive evaluation index systems construct measurement frameworks from multiple dimensions. According to research, HSR's introduction dramatically improves urban high-quality development [15]. The development of evaluation index systems has been examined by academics from some of the following angles: creating metrics for high-quality urban development that take into account infrastructure, social, ecological, and economic factors [16]; developing indicators for high-quality economic growth that are grounded in industrial restructuring, overall factor productivity, technological innovation, ecological environment, and enhancements in the standard of living for residents [17]; creating a HQE index based on the dimensions of development momentum, structural improvement, methodological strategies, and development results [18]; proposing a framework encompassing economic structure, economic scale, economic efficiency, and coordination [19]; assessing HQE through economic efficiency, stability, and sustainability; and innovatively constructing a HQE index based on the five development concepts of innovation, coordination, green development, openness, and sharing [20].

China's interregional population mobility has been spatially constrained by natural geographical factors, leading to labor market segmentation and hindering the free allocation of high-skilled labor [21]. Geographical, cultural, and specialization differences across regions have resulted in heterogeneous labor under market segmentation [22]. The dual characteristics of HSR, namely "time–space compression" and "boundary breakthrough", spatially dismantle labor segmentation caused by natural factors and expand the matching radius between firms and labor [23]. On the supply side, HSR lowers the expense of moving between regions for people [24] and increases the accessible distance for job seekers [25]. The integration and complementarity of heterogeneous labor with diverse knowledge and skills significantly optimize urban labor resource allocation and reduce frictional costs [26]. On the demand side, HSR expands urban boundaries, creating more job opportunities and increasing urban labor demand, thereby attracting higher-quality human capital [27]. Thus, HSR accelerates interregional population mobility [28], enhances labor matching efficiency [29], and fosters HQE through its resource allocation effects [30].

Green development necessitates stringent control of environmental pollution and the restoration of damaged ecosystems, objectives that rely heavily on green technologies [31]. Following the opening of HSR, the enhanced accessibility of cities has increased the mobility of innovation resources [11]. The cross-regional flow of high-level human capital induced by HSR [32] facilitates green innovation through technological advancements, knowledge spillovers, and efficiency improvements [33]. Furthermore, HSR strengthens interregional learning and collaboration, promoting the diffusion and sharing of intangible elements [34]. By elevating regional green innovation levels, it enhances urban GTFP [35] and accelerates the flow of innovation factors [36], amplifying communication and scale effects, thereby fostering regional green innovation and green transformation [37] and propelling high-quality economic growth in cities [38].

Following the inauguration of HSR, listed companies within cities have exhibited a greater propensity to establish subsidiaries in other regions, thereby facilitating cross-regional capital mobility [39]. The establishment of these subsidiaries not only increases fixed asset investments [40], but also drives local economic development through optimized resource allocation [41]. Transportation infrastructure induces the spatial redistribution of capital, altering the geographic distribution of economic activities and subsequently influencing the reduction in regional economic disparities [42]. Given that the per capita capital stock in developed regions significantly exceeds that in less developed areas, the law of diminishing marginal returns in neoclassical economic theory suggests that the return on capital in developed regions will be lower than that in less developed regions [43]. As improved transportation conditions reduce barriers to cross-regional capital flows, they promote the movement of capital from developed to less developed regions, thereby narrowing regional economic disparities [44].

The industrial chain is a linear structure composed of multiple enterprises at different production stages, interconnected through nodes, with various chains interwoven into a network. Damage to any node can have a cascading impact on the entire chain [45]. As a fundamental conduit for the circulation of products across industries, HSR directly facilitates intra-regional industrial division and collaboration by reducing the costs of product, factor, and technology flows, thereby enhancing regional production efficiency [46]. As a significant investment activity, building HSR encourages the growth of associated industries and increases regional industrial diversification [47]. Given that different industries exhibit varying demand elasticities and trade orientations [48], a diversified industrial structure can effectively mitigate and disperse external risks when faced with shocks, thereby strengthening the resilience of the industrial chain [49].

HSR primarily focuses on passenger transport, accelerating labor mobility in cities along its routes, particularly benefiting industries heavily reliant on human resources. Simultaneously, the expansion of HSR networks tends to increase rental costs and property prices [50], thereby reducing the proportion of manufacturing sectors sensitive to land prices and production costs and consequently lowering the share of the secondary sector in the economy while promoting the growth of the service industry. Employment in the service industry is greatly increased with the advent of HSR, such as tourism, accommodation, and catering within cities, but has a limited impact on non-service employment [51]. HSR notably boosts employment in the hospitality industry [52], and sectors that are intensive in information and knowledge, like finance, education, and tourism, gain more from the connectivity that HSR offers [53–55]. It is anticipated that HSR will facilitate industrial transformation in relevant cities.

3. Research Method

3.1. Variable Selection and Explanation

(1) Dependent variable: As shown in Table 1, the HQE index, which is derived through the entropy-weighted TOPSIS method, encompasses the following five key dimensions: innovation, coordination, green development, openness, and sharing.

(2) Core explanatory variable: The HSR variable is quantified by the count of HSR lines.

(3) Control variables: The level of human capital is indicated by the proportion of college and university students relative to the total year-end population; urban economic density is assessed by the ratio of regional GDP to the administrative land area; passenger turnover is determined by taking the natural logarithm of the aggregate passenger volumes from highways, waterways, and civil aviation; the level of transportation infrastructure is reflected by the natural logarithm of highway mileage; and the level of economic development is shown by the natural logarithm of regional GDP.

(4) Mechanism variables: Labor mobility is assessed using the natural logarithm of the number of employed individuals in urban units at year-end; capital mobility is represented by the natural logarithm of total fixed asset investment; industrial structure upgrading is calculated as the ratio of the added value of the tertiary sector to that of the secondary sector; industrial chain resilience is obtained from a composite index computed via the entropy weight method based on resistance recovery and transformation renewal indices; and green

innovation is measured by the natural logarithm of the quantity of green invention patents issued in the current year.

Table 1. Construction of high-quality economic development indicators.

Primary Indicators	Second-Class Indicators	Third-Class Indicators	Basic Indicators	Unit	Type of Indicator
Innovation	Innovation input	Innovation input	R&D expenditure/GDP	%	+
			Science and technology expenditure/local general public budget expenditure	%	+
	Innovation output	Innovation output	Urban Innovation Index	-	+
		Patent applications granted per 10,000 persons	Number of patents granted/population of the region		+
	Innovative foundations	Scale of education	The count of students attending general higher education programs	Tens of thousands	+
		Investment in education	Share of education expenditure in general budget expenditure of local finance	%	+
	Innovation efficiency	Labor productivity	GDP/average annual number of employees	%	+
		Capital productivity	GDP/total investment in fixed assets	%	+
Coordination	Industrial structure	Rationalization of industrial structure	Percentage of tertiary sector output	%	+
		Advanced industrial structure	Tertiary sector output/secondary sector output	%	+
	Urban and rural coordination	The ratio of disposable income per capita for urban and rural residents	Urban disposable income per capita/rural disposable income per capita	%	-
		Urbanization rate	Permanent urban population/(permanent urban population + permanent rural population)	%	+
	Financial structure	Financial risk	Balance of deposits and loans/GDP	%	-
Greenness	Energy consumption	Electricity consumption per unit of output	Industrial Electricity Consumption/GDP	Kilowatt-hours/billion dollars	-
		Wastewater discharge per unit of output	Industrial wastewater discharge/GDP	Tons/billion dollars	-
		Exhaust emissions per unit of output	Smoke emissions per unit of output	Tons/billion dollars	-
		Emissions of smoke and dust per unit of output	Industrial soot and dust discharges/GDP	Tons/billion dollars	-
	Pollution emission	Haze pollution	Annual average PM2.5 concentration	μg/m^3	-
Openness	Level of foreign trade	Degree of openness	Total Trade Imports and Exports/GDP	-	+
	Introduction of foreign capital	Effectiveness of openness	Total utilized foreign capital/GDP	-	+
	Tourism openness		International tourism revenue/GDP	-	+
	Tourism openness	International tourism revenue/GDP	Domestic tourism revenue/GDP	-	+
		Domestic tourism revenue/GDP	The number of inbound tourists received	10,000 people	+

Table 1. *Cont.*

Primary Indicators	Second-Class Indicators	Third-Class Indicators	Basic Indicators	Unit	Type of Indicator
Sharing	Income distribution	Per capita income	Real GDP per capita	-	+
		Remuneration for labor	Average wages of employees	Yuan	+
	Consumption level	Share of consumption	Social Retail Consumption/GDP	%	+
	Urban–rural sharing	Engel coefficient of urban households	Household food expenditure accounts for the proportion of urban consumption expenditure	%	-
		Engel coefficient of rural households	Household food expenditure accounts for the proportion of rural consumption expenditure	%	-
	Public services	Per capita expenditure on education	Education expenditures/ total population	%	+
		Cultural resources	Public library collection per capita	Number of books/million people	+
		Health resources	Hospital bed count per 10,000 individuals	Number	+
		Employment effect	Registered urban unemployment rate	%	-
	Data sharing	Internet penetration	The number of internet broadband access users among 100 people	%	+

3.2. Data

This study analyzes 282 prefecture-level cities from 2005 to 2021, encompassing 4794 observations, taking into account data availability and dependability. The China Statistical Yearbook and the City Statistical Yearbook are the sources of the variable data. Table 2 offers definitions of the variables, while Table 3 displays descriptive statistics.

Table 2. Description of variables.

Variable Types	Variable Symbol	Variable Name	Description and Measurement Method
Dependent variable	HQE	High-quality economic development	Comprehensive index calculation
Explanatory variable	HSR	High-speed rail	Number of HSR lines opened
Control variables	HCI	Level of human capital	Logarithm of the number of employees at year end
	UED	Urban economic density	Logarithm of total fixed asset investment
	PTK	Passenger turnover	Logarithm of the sum of road passenger traffic, water passenger traffic, and civil aviation passenger traffic
	Road	Level of transportation infrastructure	Logarithm of highway mileage
	Economy	Level of economic development	Logarithm of regional gross domestic product
Mediating variables	Labor	Labor mobility	Logarithm of the number of employees at year end
	Capital	Capital mobility	Logarithm of total fixed asset investment
	Industrial structure	Industrial structure upgrading	Ratio of the added value of the tertiary industry to the added value of the secondary industry
	Resilience	Industrial chain resilience	Composite index of industrial chain resilience
	Green innovation	Green innovation	Logarithm of the number of green invention patents

Table 3. The primary variables' descriptive statistics.

	N	Mean	Standard Deviation	Minimum	Maximum
HQE	4794	2.1865	1.8612	0.5477	12.4517
HSR	4794	0.7795	1.2323	0	9
HCI	4794	0.0201	0.0264	0	0.1238
UED	4794	0.2445	0.4190	0.004	2.7076
PTK	4794	12.897	1.8406	8.3087	17.8353
Road	4794	6.0009	0.5870	4.4886	7.1892
Economy	4794	16.2875	1.0383	13.9946	19.0017

3.3. *Construction of the Continuous Difference-in-Differences Model*

Traditional difference-in-differences models ignore the diverse levels of influence brought about by different levels of HQE and only take into account the binary effects of HSR on cities along its lines. In order to examine the effect of HSR on the HQE of cities along its routes, this study uses the number of HSR lines as a continuous proxy variable for the HSR effect. It carries this out by using a continuous difference-in-differences model. Equation (1) represents the baseline DID regression model, as follows:

$$hqed_{it} = \lambda_0 + \lambda_1 hsr_{it} + X'_{it}\gamma + f_i + \delta_t + \epsilon_{it} \tag{1}$$

In the equation, $hqed_{it}$ reflects the degree of HQE in city i during period t; hsr_{it} is a dummy variable representing the number of HSR lines in city i in year *t*, serving as the core explanatory variable to capture the rail effect. If λ_1 is significant, it indicates that HQE substantially influences the high-quality economic growth in cities along its routes. X'_{it} denotes the set of control variables; f_i and δ_t represent individual and time effects, respectively; and ϵ_{it} represents the random error term. This study employs city–time clustering to adjust the standard errors of the estimated coefficients, addressing potential heteroskedasticity and spatial autocorrelation issues that may affect the parameter estimates.

3.4. *Spatial Evolution Characteristics of High-Quality Economic Development*

To perform a thorough examination of the spatial evolution features of urban HQE, this research visualizes the regional distribution of HQE between 2005 and 2021 using ArcGIS 10.8 software, utilizing the natural breaks classification method to categorize development levels, thereby facilitating year-specific analyses. As illustrated in Figure 1, from 2005 to 2021, China's HQE exhibited an evolutionary trend characterized by "overall advancement and enhanced regional coordination", with a progressively optimized spatial pattern. In 2005, the vast majority of regions across the country were in a low-level development stage, with HQE indices converging between 0.547 and 1.665. The geographic dispersion of HQE was highly uneven, with central and western regions predominantly occupying low-value zones. There were only a handful of developed cities, forming isolated high-value "islands". By 2021, with the sustained implementation of national strategies promoting innovation-driven growth, green development, and integrated development, the HQE index experienced widespread improvement. Medium–high-value regions (above 3.088) increased significantly, indicating that most areas had transitioned from quantitative growth to qualitative growth. Simultaneously, the developmental gap between western and central regions narrowed, reflecting a shift from an "east—high, west—low" pattern to a "gradient progression and regional synergy" trend, indicating the start of a new stage of HQE.

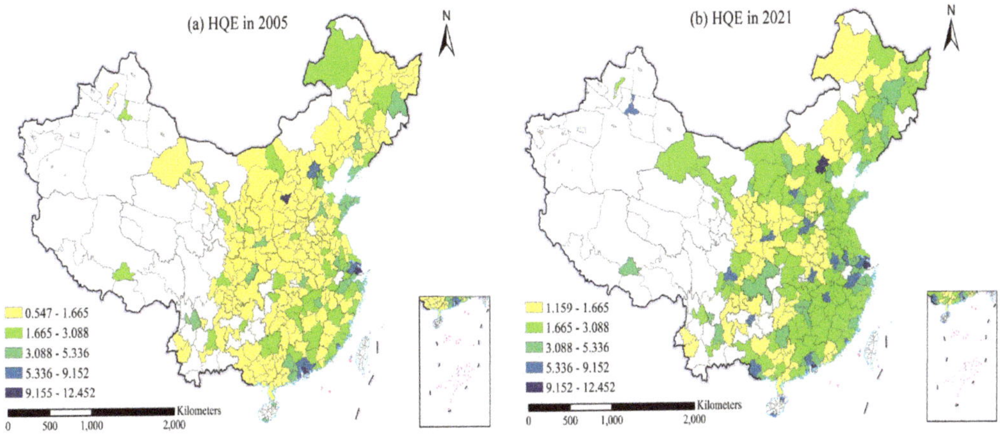

Figure 1. Features of high-quality urban economic growth in terms of spatial development between 2005 and 2021.

From a spatial perspective, the HQE index in 2005 displayed a distinct "east—strong, west—weak, point-based distribution" pattern. High-value zones were primarily concentrated in a few first-tier cities and provincial capitals along the eastern coast, forming limited high-quality development "core nodes", while vast central and western regions, along with some border areas, remained low-value zones, highlighting pronounced spatial disparities. By 2021, the geographical allocation of HQE underwent significant transformation, manifesting as "areal diffusion and multi-core-driven growth". Regional disparities tended to diminish and spatial equilibrium strengthened, indicating that the spatial organizational structure of China's HQE is evolving toward greater rationality, efficiency, and sustainability.

4. Results

4.1. Baseline Regression

In the baseline regression model, HSR is the primary explanatory factor, while HQE is the dependent variable. Model (1) presents regression outcomes without considering control variables, while Models (2) and (3) progressively incorporate control variables. The positive and statistically significant coefficients of HSR service intensity in Models (1) to (4) suggest that the development of HSR has a significantly positive effect on urban HQE. Detailed regression results are provided in Table 4.

Table 4. Baseline regression.

	(1)	(2)	(3)	(4)
	HQE	**HQE**	**HQE**	**HQE**
HSR	0.0422 *** (0.0123)	0.0779 *** (0.0135)	0.0748 *** (0.0135)	0.0742 *** (0.0135)
HCI		2.3578 ** (1.0903)	2.1194 * (1.0931)	1.9819 * (1.0923)
UED		−0.4401 *** (0.0668)	−0.4678 *** (0.0672)	−0.4989 *** (0.0677)

Table 4. *Cont.*

	(1)	(2)	(3)	(4)
	HQE	**HQE**	**HQE**	**HQE**
PTK			−0.0742 *** (0.0182)	−0.0771 *** (0.0182)
Road			0.0982 (0.1111)	0.0713 (0.1112)
Economy				0.2005 *** (0.0556)
Constant	1.7866 *** (0.0358)	1.7748 *** (0.0428)	2.0476 *** (0.7011)	−0.8141 (1.0587)
id	YES	YES	YES	YES
Year	YES	YES	YES	YES
N	4794	4794	4794	4794
R^2	0.2582	0.2659	0.2688	0.2709

Standard errors in parentheses. * $p < 0.1$, ** $p < 0.05$, and *** $p < 0.01$.

4.2. Parallel Trend Test for High-Speed Rail

This research adopts the continuous difference-in-differences model with city–year double fixed effects. The premise of this paper's application of the model is to comply with the parallel trend, i.e., before the policy occurs, where the trend of change is the same for both the realization group and the control group, and in this paper, if there are variations between the experimental and control groups before the HSR's opening, then it shows that the variation in HQE is not attributed to the establishment of HSR. In this research, if the control group and the experimental group show inconsistent trends before the HSR's opening, then it indicates that the change in HQE is not induced by the construction of HSR. The findings reveal no noticeable distinction between the treated and untreated groups before the commencement of HQE, while in the later stage of building HSR, as shown in Figure 2, the outcome is affirmative and notable, signifying compliance with the parallel trend assessment.

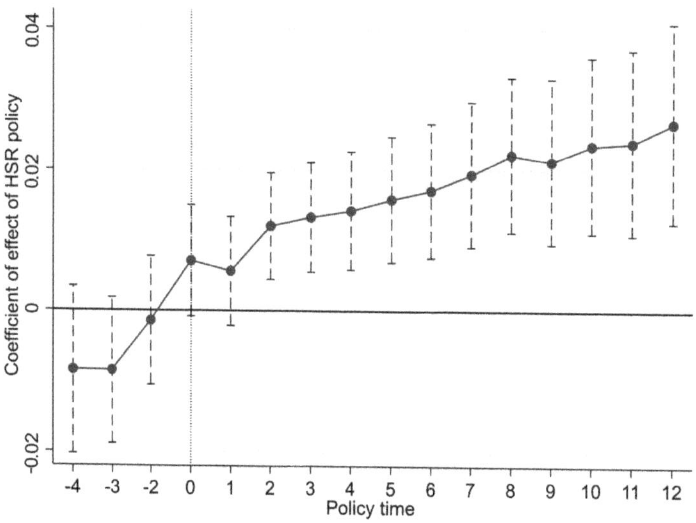

Figure 2. Parallel trend test graph.

4.3. Robustness Checks

As shown in Table 5, (1) the regression results further validate the robustness of the baseline regression conclusions by demonstrating a considerably favorable influence on HQE when HSR opening is used as the independent variable. (2) After excluding the special years of the 2008 financial crisis and the 2020 COVID-19 pandemic, the regression results remain significant, indicating that the positive effect of HSR on HQE is not disrupted by extreme economic events. (3) The findings indicate that HSR continues to considerably support high-quality economic growth in ordinary prefecture-level cities, even when municipalities are removed from direct central government control, eliminating the additional influence of policy priority and resource allocation. (4) One period behind the primary explanatory variable, the findings remain robust, demonstrating that HSR opening not only has an immediate positive effect on HQE, but also exhibits a significant lagged effect, continuously driving HQE in the long term. (5) Using green total factor productivity as a substitute for the HQE index, the regression results remain significant.

Table 5. Robustness test results I.

	(1)	(2)	(3)	(4)	(5)
	HQE	**HQE**	**HQE**	**HQE**	**GTFP**
HSR	0.0733 *** (0.0155)	0.0667 *** (0.0147)	0.0670 *** (0.0131)	0.0664 *** (0.0139)	0.0118 *** (0.0017)
Control variables	YES	YES	YES	YES	YES
Constant	−0.6575 (1.0615)	−0.0834 (1.1662)	−1.9733 * (1.0219)	−0.1753 (1.0902)	0.1399 (0.1324)
id	YES	YES	YES	YES	YES
Year	YES	YES	YES	YES	YES
N	4794	4230	4726	4512	4794
R^2	0.2696	0.2718	0.2881	0.2756	0.2751

Standard errors in parentheses. * $p < 0.1$ and *** $p < 0.01$.

When assessing the effect of HSR on HQE, regression results are presented to exclude potential policy interferences. As shown in Table 6, Model (1) excludes the effect of the smart city pilot policy, and the empirical results demonstrate that the positive effect of HSR on HQE is still statistically significant. Model (2) excludes the low-carbon city policy, and the regression analysis continues to reveal a favorable effect of HSR. Model (3) excludes the Broadband China pilot policy, and the positive effect of HSR remains unchanged and significant. Model (4) excludes the new energy demonstration city policy, and the regression results continue to demonstrate that HSR significantly facilitates HQE.

To confirm the credibility of the research results, this research addresses the potential influences of terrain fluctuation and resident population size on HSR construction by introducing interaction terms between terrain fluctuation and time, as well as resident population and time, as control variables to mitigate sample selection bias. As shown in Table 7, Column (1) presents regression results with the terrain fluctuation–time interaction term fixed, showing that HSR continues to significantly promote HQE. Column (2) reports regression results with the resident population–time interaction term fixed, indicating that the favorable influence of HSR is not significantly affected by population dynamics. Column (3) simultaneously fixes the interaction terms of terrain fluctuation time and resident population time, demonstrating that the effect of HSR on HQE remains robust. Columns (4) and (5) show instrumental variable regression results, with Column (4) confirming the validity of the instrumental variable and passing the weak instrument test and Column (5) showing that HSR construction exerts a positive impact on HQE at the 1% significance level, consistent with the results in Column (1), further validating the robustness of the findings.

Table 6. Robustness test results II.

	(1)	(2)	(3)	(4)
	HQE	**HQE**	**HQE**	**HQE**
HSR	0.0746 *** (0.0135)	0.0744 *** (0.0135)	0.0709 *** (0.0135)	0.0736 *** (0.0135)
Smart City Pilot	−0.0225 (0.0352)			
Low-Carbon City Policy		−0.0080 (0.0352)		
Broadband China Pilot Program			0.1112 *** (0.0365)	
New Energy Demonstration City				0.0482 (0.0420)
Control variables	YES	YES	YES	YES
Constant	−0.8472 (1.0601)	−0.8188 (1.0591)	−0.9774 (1.0591)	−0.7406 (1.0606)
id	YES	YES	YES	YES
Year	YES	YES	YES	YES
N	4794	4794	4794	4794
R^2	0.2709	0.2709	0.2724	0.2711

Standard errors in parentheses. *** $p < 0.01$.

Table 7. Robustness test results III.

Description	**Alleviate Sample Selection Bias**			**Instrumental Variable Method**	
Model	**(1)**	**(2)**	**(3)**	**(4)**	**(5)**
Variables	**HQE**	**HQE**	**HQE**	**HQE**	**HQE**
HSR	0.065 *** (0.016)	0.036 ** (0.017)	0.040 ** (0.017)		0.0787 *** (0.0199)
IV core explanatory variable lag 1				0.8432 *** (0.009)	
Constant term	−10.002 *** (3.246)	−21.282 *** (5.220)	−32.281 *** (6.190)	-	-
Terrain × Time	YES	NO	YES	NO	NO
Population × Time	NO	YES	YES	NO	NO
Control variables	YES	YES	YES	YES	YES
id	YES	YES	YES	YES	YES
Year	YES	YES	YES	YES	YES
F-test	-	-	-	4132.87	-
R-square	0.271	0.272	0.274	-	0.275
N	4794	4794	4794	4512	4512

Standard errors in parentheses. ** $p < 0.05$ and *** $p < 0.01$.

4.4. Mechanism Analysis

Table 8 reports the results of the mediating effect test of HSR opening on the quality of economic growth, with the following model constructed:

$$M_{it} = \beta_0 + \beta_1 hsr_{it} + \beta_2 control_{it} + \theta_i + \gamma_t + \varepsilon_{it} \quad (2)$$

Equation (2) in Model (3) bears resemblance to the specification of the benchmark regression model in Model (1). In this paper, five indicators are selected as the dependent variables, denoted as Mit.

Table 8. Mechanism analysis.

	(1)	(2)	(3)	(4)	(5)
	Asset Liquidity	**Labor Mobility**	**Industrial Structure Upgrading**	**Industrial Chain Resilience**	**Green Innovation**
HSR	0.3442 *** (0.0157)	0.4716 *** (0.0671)	0.0230 *** (0.0050)	0.0084 *** (0.0006)	0.0702 *** (0.0141)
Control variables	YES	YES	YES	YES	YES
Constant	−19.2800 ***	−0.9734	9.0130 ***	0.4505 ***	−7.1813 ***
	(1.2327)	(5.2587)	(0.3950)	(0.0489)	(1.1079)
id	YES	YES	YES	YES	YES
Year	YES	YES	YES	YES	YES
N	4794	4794	4794	4794	4794
r2	0.6273	0.2018	0.5764	0.4242	0.7019

Standard errors in parentheses. *** $p < 0.01$.

Column (1) reveals that the regression coefficient for HQE is 0.3442, which passes the 1% significance level test, indicating that HSR significantly facilitates labor mobility and enhances the quality of economic development by optimizing human capital allocation. Column (2) demonstrates that the regression coefficient for HQE is 0.4716, passing the 1% significance level test, suggesting that HSR promotes capital mobility, providing financial support for economic growth. Column (3) shows that the regression coefficient for HQE is 0.0230, which also passes the 1% significance level test, indicating that HSR advances industrial structuring and optimizes the economic structure. Column (4) indicates that the regression coefficient for HQE is 0.0084, passing the 1% significance level test, suggesting that HSR strengthens the resilience of industrial chains, thereby ensuring economic stability and competitiveness. Column (5) reveals that the regression coefficient for HQE is 0.0702, passing the 1% significance level test, demonstrating that HSR fosters green innovation, driving green transformation and sustainable development.

4.5. Heterogeneity Analysis

This study uncovers notable distinctions in the promotional impact of HSR on the HQE of the eastern, central, western, and northeastern regions. As shown in Table 9, Column (1) shows that the impact of HSR on the HQE of the eastern region is positively significant at the 5% level, potentially benefiting from its solid economic foundation and optimized industrial structure. Column (3) demonstrates that the effect of HSR on the western region is positively significant at the 1% level, likely due to its underdeveloped infrastructure, as HSR has markedly improved transportation conditions. Columns (2) and (4) reveal that while the impacts of HSR on the central and northeastern regions are positive, they are not statistically significant, potentially related to their economic foundations, structures used in industry, and the extent of HSR network coverage, which have yet to completely realize the economic benefits of HSR. Thus, there exists significant diversity in the effect of HSR on the HQE of various regions.

Table 9. Heterogeneity regression results I.

	(1)	(2)	(3)	(4)
	Eastern Region	**Central Region**	**Western Region**	**Northeastern Region**
HSR	0.0560 ** (0.0243) (2.5680)	0.0107 (0.0261) (2.5571)	0.1053 *** (0.0257) (1.7817)	0.0628 (0.0505) (4.0934)
id	YES	YES	YES	YES
Year	YES	YES	YES	YES
N	1445	1343	1428	578
R^2	0.2191	0.3572	0.3341	0.2245

Standard errors in parentheses. ** $p < 0.05$ and *** $p < 0.01$.

There are two categories for the sample cities, as follows: hub cities and non-hub cities. According to Table 10, Model (1) shows that the effect of HQE on the HQE of hub cities is positive but not statistically significant, potentially due to their already advanced infrastructure and economic development levels, resulting in smaller implications of HSR on the margins. Model (2) reveals that HSR exerts a significantly positive influence on the HQE of non-hub cities at the 1% significance level, likely because their infrastructure is relatively weaker, and HSR notably enhances transportation convenience and resource mobility. When cities are categorized into central cities and peripheral cities, Model (3) demonstrates that HSR positively impacts the HQE of peripheral cities, with statistical significance at the 1% level, potentially because HSR improves transportation accessibility, facilitating factor mobility and industrial agglomeration. Model (4) shows that HSR has a positive but insignificant effect on central cities, possibly because they possess strong economic foundations and resource agglomeration capabilities, with HSR primarily manifesting as a resource agglomeration factor rather than a new growth driver.

Table 10. Heterogeneity regression results II.

	(1)	(2)	(3)	(4)
	Transportation Hub City	**Non-Transportation Hub City**	**Central City**	**Peripheral City**
HSR	0.0420 (0.0576)	0.0679 *** (0.0141)	0.0412 (0.0446)	0.0962 *** (0.0145)
Constant	4.9098 (4.4179)	−2.1140 * (1.1102)	−1.0299 (3.8566)	−1.5988 (1.0852)
id	YES	YES	YES	YES
Year	YES	YES	YES	YES
N	323	4471	612	4182
R^2	0.3129	0.2872	0.1878	0.3192

Standard errors in parentheses. * $p < 0.1$ and *** $p < 0.01$.

The samples are classified into resource-based and non-resource-based cities, with resource-based cities further divided into growing, mature, declining, and regenerative types. The results in Table 11 indicate that Models (1) and (2) show that the impact of HSR on the HQE of both resource-based and non-resource-based cities is significantly positive at the 1% level. Model (3) reveals that the influence of HSR on growing cities is significantly positive at the 1% level, potentially due to its acceleration of resource exploitation, industrial optimization, and market expansion. Model (4) demonstrates that the effect of HSR on mature cities is significantly positive at the 5% level. Model (5) shows that the impact of HSR on declining cities is significantly negative at the 1% level, likely because it has fails to effectively address resource depletion and industrial transformation issues. Model (6) indicates that the effect of HSR on regenerative cities is positive but not significant, possibly because they are in the early stages of transformation, and the positive effects of HSR have not yet fully materialized.

Table 11. Heterogeneity regression results III.

	Resource-Based City					
	(1)	(2)	(3)	(4)	(5)	(6)
	Resource-Based City	Non-Resource-Based City	Growing	Mature	Declining	Rejuvenating
HSR	0.0659 *** (0.0254)	0.0770 *** (0.0168)	0.1875 *** (0.0659)	0.0744 ** (0.0368)	−0.1341 *** (0.0259)	0.0433 (0.1322)
Control variables	YES	YES	YES	YES	YES	YES
Constant	2.7866 * (1.4884)	−2.7713 * (1.4809)	6.4608 (8.6322)	2.4413 (2.0037)	1.4723 (2.9810)	25.9972 ** (11.0512)
Urban FE	YES	YES	YES	YES	YES	YES
Year FE	YES	YES	YES	YES	YES	YES
N	1904	2890	238	1020	391	255
R^2	0.3388	0.2501	0.6914	0.3427	0.6350	0.2916

Standard errors in parentheses. * $p < 0.1$, ** $p < 0.05$, and *** $p < 0.01$.

4.6. Analysis of Spatial Spillover Effects

Inter-city interactions indicate that the HQE of one city may be influenced by other cities, suggesting the presence of spatial correlation. Therefore, this study constructs a spatial difference-in-differences model to further examine the spatial impact effects of HSR on HQE.

4.7. Spatial Correlation Test

If the global Moran's I index is greater than zero and the p-value is statistically significant, this indicates the presence of positive spatial effects. As shown in Table 12, under the economic geography weight matrix, spatial correlation is significant, which suggests that the use of a spatial econometric model is appropriate.

In the localized Moran scatter plot, the first quadrant represents regions with high levels of HQE surrounded by similar regions, while the third quadrant denotes low-level regions surrounded by analogous areas, both indicating positive spatial autocorrelation. The second quadrant illustrates low-level regions encircled by high-level regions, and the fourth quadrant depicts high-level regions surrounded by low-level regions, both reflecting negative spatial autocorrelation. The results demonstrate a significant spatial dependence between HSR and HQE, with high-level regions adjacent to other high-level regions and low-level regions neighboring other low-level regions. As shown in Figures 3 and 4, the fitted regression line suggests a positive relationship.

Table 12. The global Moran's I index results for HQE.

Economic Geography Weight Matrix					
Year	Moran' I	*p*-Value	Year	Moran' I	*p*-Value
2005	0.2966	0.0000	2013	0.3147	0.0000
2006	0.3105	0.0000	2014	0.3035	0.0000
2007	0.3391	0.0000	2015	0.3004	0.0000
2008	0.3420	0.0000	2016	0.2869	0.0000
2009	0.2907	0.0000	2017	0.2931	0.0000
2010	0.3419	0.0000	2018	0.2554	0.0000
2011	0.3292	0.0000	2019	0.2324	0.0000
2012	0.3241	0.0000	2020	0.3021	0.0000
			2021	0.3305	0.0000

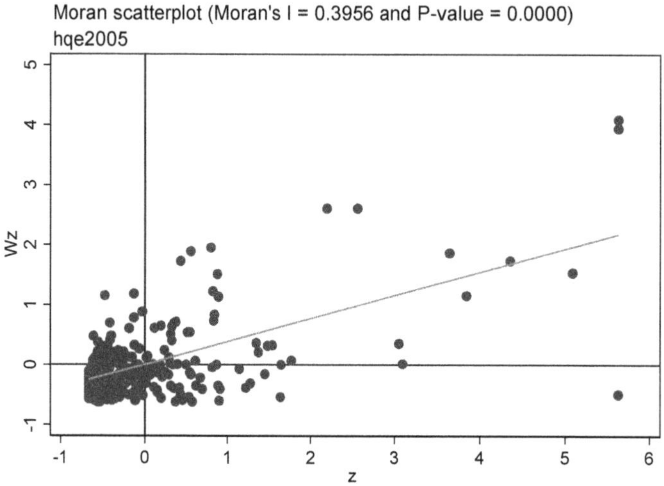

Figure 3. Localized Moran scatterplot of high-quality economic development in 2005.

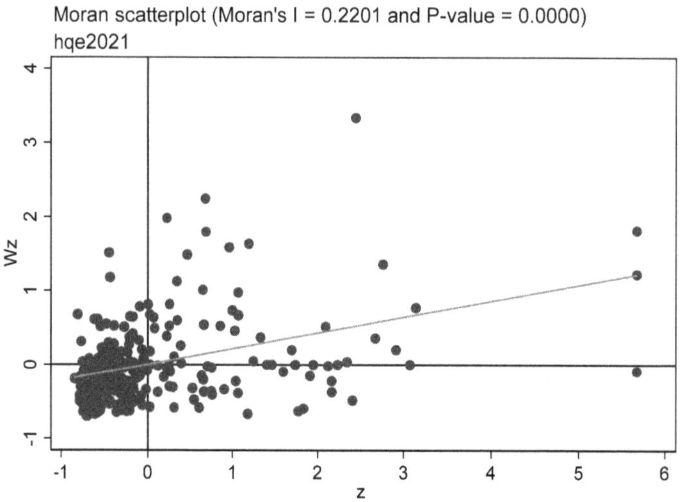

Figure 4. Localized Moran scatterplot of high-quality economic development in 2021.

This research uses a spatiotemporal fixed-effects Durbin model to examine the spatial correlation between HSR and HQE. By utilizing partial differentiation, the regression results are decomposed into direct, indirect, and total effects, as presented in the table. As shown in Table 13, the direct effect of the economic–geographical weight matrix is positive and statistically significant at the 1% level, in accordance with the baseline regression results, indicating that the introduction of HSR fosters HQE in local areas. Furthermore, the indirect effect is also highly significant and positive at the 1% level, suggesting that HQE exerts a substantial beneficial influence on the high-quality economic growth of neighboring cities.

Table 13. Estimation results of the spatial Durbin difference-in-differences model effects.

Variable	Main	Direct	Indirect	Total
HSR	0.0746 *** (0.0131)	0.0767 *** (0.0135)	0.0944 ** (0.0444)	0.1711 *** (0.0475)
rho	0.1373 *** (0.0271)			
Control variables	YES	YES	YES	YES
Urban FE	YES	YES	YES	YES
Year FE	YES	YES	YES	YES
N	4794	4794	4794	4794

Standard errors in parentheses. ** $p < 0.05$ and *** $p < 0.01$.

4.8. Spatial Distance Decay Effect

Having established the significant spatial spillover effects of HSR on urban HQE, this study employs a spatial Durbin difference-in-differences model with a geographically weighted matrix to further analyze the spatial decay boundary of HSR's impact on HQE. The spatial spillover coefficients of HSR service intensity are examined by setting varying distance thresholds. If the geographical distance, dij, between city i and city j exceeds the specified threshold, the corresponding spatial weight matrix element is assigned a value of 1/ij d; otherwise, it is set to 0. The following is the precise calculation method:

$$f(x) = \begin{cases} 1/d_{ij}, \text{ When } d_{ij} \text{ is outside the distance threshold} \\ 0\,, \text{ When } d_{ij} \text{ is inside the distance threshold} \end{cases} \quad (3)$$

First, the initial distance threshold for $1/d_{ij}$ is set at the minimum inter-city distance of 100 km, incrementing by 100 km, with the step size reduced near the peak. Second, regression analysis is conducted using the spatial Durbin difference-in-differences model.

Figure 5 illustrates the spillover effect coefficients and their 95% confidence intervals across varying geographical distances. As shown in Figure 5, the spatial spillover coefficients exhibit a "U-shaped" distribution. In particular, there are four distinct intervals in the connection between the geographical distance threshold and the HSR spillover impact coefficient, as follows: (1) HQE in nearby regions favorably helps local high-quality economic growth when the geographical distance threshold falls between 100 and 200 km, demonstrating a positive spatial spillover effect; (2) when the threshold lies between 200 and 300 km, the spillover effect is not statistically significant; (3) when the threshold spans from 300 to 740 km, the development of HSR in surrounding areas exerts a restraining effect on local high-quality economic growth; and (4) when the threshold ranges from 740 to 1000 km, the development of HSR in surrounding areas again promotes local high-quality economic growth. This phenomenon may arise because, within this distance range, HSR begins to demonstrate its advantages as a long-distance rapid transportation mode.

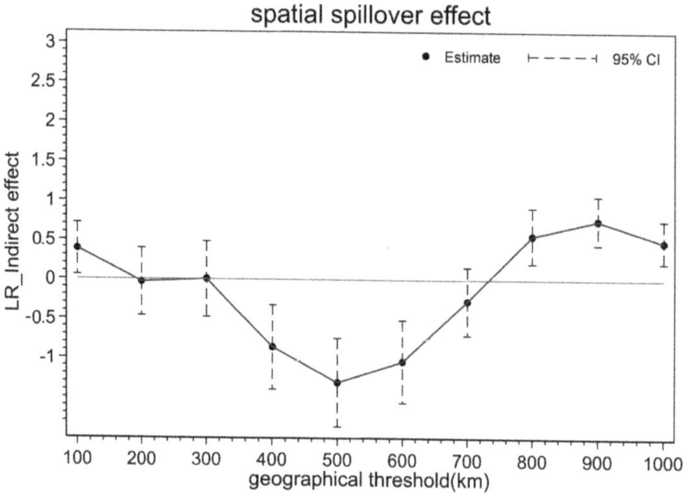

Figure 5. Spatial distance decay effect.

5. Discussion

Our empirical results demonstrate that HSR significantly promotes HQE in China, with mechanisms including enhanced labor and capital mobility, industrial structure upgrading, and green innovation. However, the heterogeneity in effects across regions and city types suggests that the benefits of HSR are not uniformly distributed. For instance, the lack of a significant impact in central and northeastern regions may reflect infrastructural or policy gaps that need addressing. Additionally, the spatial spillover effects, while positive, exhibit a decay boundary of 1000 km, indicating that the economic benefits of HSR are geographically bounded. These findings underscore the importance of tailored regional policies to maximize HSR's potential and ensure equitable development. Future research could explore the long-term dynamics of HSR's impact and the role of complementary policies in amplifying its benefits.

6. Conclusions

This research, relying on city-level panel data from 2005 to 2021, employs a continuous difference-in-differences approach to empirically analyze the impact of HSR on urban HQE, further analyzing its mechanisms and spatial spillover effects. The findings reveal the following: First, HQE significantly enhances high-quality economic growth. Second, HSR enhances urban HQE by facilitating labor mobility, capital flow, industrial structure upgrading, supply chain resilience, and green innovation. Third, HQE generates favorable spatial spillover effects on the high-quality economic growth of neighboring cities, with a distinct spatial decay boundary of 1000 km. Fourth, the effect of HSR on urban energy efficiency exhibits significant heterogeneity across regions, city tiers, hub statuses, and resource endowments. Specifically, HSR positively and significantly influences HQE in eastern regions at the 5% level and in western regions at the 1% level, while its impact on central and northeastern regions, though positive, is not statistically significant. HQE positively affects hub cities, but not significantly, while its impact on non-hub cities and peripheral cities exhibits a significantly positive impact at the 1% level. Its influence on core cities is positive but not significant. In addition, this study reveals a remarkable spatial evolution of China's HQE from 2005 to 2021, transitioning from an uneven "point-based distribution" to a more coordinated "multi-core driven" pattern.

In light of these findings, this study puts forward the following policy implications: First, given the significant role of HSR in driving HQE, the optimization of HSR network layouts should be prioritized, with accelerated construction in central, western, and remote regions to enhance regional accessibility and foster coordinated economic development. Second, as HSR boosts high-quality urban economic development by facilitating labor and capital mobility, advancing industrial upgrading, strengthening supply chain resilience, and stimulating green innovation, local governments should strengthen industrial park development along HSR routes, attract high-end manufacturing and modern service industries, and encourage green technological innovation to drive economic green transformation. Third, while HSR exhibits positive spatial spillover effects on neighboring cities within a 1000 km boundary, regional economic collaboration along HSR corridors should be strengthened to establish a coordinated industrial division system, enabling resource sharing and mutual benefits. Fourth, given the heterogeneous impacts of HSR across regions and city types, with more pronounced effects in eastern and western regions and greater benefits for non-hub and peripheral cities, policies should be more targeted. For instance, HSR resource allocation in eastern and western areas can be optimized to amplify economic benefits while exploring distinctive development pathways for HSR economies in central and northeastern regions based on regional strengths.

Author Contributions: Writing—original draft, X.F.; writing—review & editing, X.F.; data curation, J.L.; software, J.L.; validation, Y.L.; methodology, Y.L.; supervision, W.L.; project administration, W.L.; resources, W.L. All authors have read and agreed to the published version of the manuscript.

Funding: This research is funded by Humanities and Social Science Planning Project [2023JBW8006].

Data Availability Statement: Data is unavailable due to confidentiality restrictions.

Conflicts of Interest: The author declares no conflicts of interest.

References

1. Wang, Y.; Miao, Z.; Lu, Y.; Zhu, Y. The impact of economic development on urban livability: Evidence from 40 large and medium-sized cities of China. *J. Geogr. Sci.* **2023**, *33*, 1123–1139. [CrossRef]
2. Rodrik, D. The past, present, and future of economic growth. *Challenge* **2014**, *57*, 5–39. [CrossRef]
3. Ren, X.; Tong, Z.; Sun, X.; Yan, C. Dynamic impacts of energy consumption on economic growth in China: Evidence from a non-parametric panel data model. *Energy Econ.* **2022**, *107*, 105855. [CrossRef]
4. Long, F.; Zheng, L.; Song, Z. High-speed rail and urban expansion: An empirical study using a time series of nighttime light satellite data in China. *J. Transp. Geogr.* **2018**, *72*, 106–118. [CrossRef]
5. Zheng, S.; Kahn, M.E. China's bullet trains facilitate market integration and mitigate the cost of megacity growth. *Proc. Natl. Acad. Sci. USA* **2013**, *110*, E1248–E1253. [CrossRef]
6. Qin, Y. 'No county left behind?' The distributional impact of high-speed rail upgrades in China. *J. Econ. Geogr.* **2017**, *17*, 489–520. [CrossRef]
7. Ke, X.; Chen, H.; Hong, Y.; Hsiao, C. Do China's high-speed-rail projects promote local economy?—New evidence from a panel data approach. *China Econ. Rev.* **2017**, *44*, 203–226. [CrossRef]
8. Lin, Y. Travel costs and urban specialization patterns: Evidence from China's high speed railway system. *J. Urban Econ.* **2017**, *98*, 98–123. [CrossRef]
9. Vickerman, R. Can high-speed rail have a transformative effect on the economy? *Transp. Policy* **2017**, *62*, 31–37. [CrossRef]
10. Tang, Z.; Wang, L.; Wu, W. The impact of high-speed rail on urban carbon emissions: Evidence from the Yangtze River Delta. *J. Transp. Geogr.* **2023**, *110*, 103641. [CrossRef]
11. Sun, Y.; Razzaq, A.; Kizys, R.; Bao, Q. High-speed rail and urban green productivity: The mediating role of climatic conditions in China. *Technol. Forecast. Soc. Change* **2022**, *185*, 122055. [CrossRef]
12. Chen, Z.; Haynes, K.E. Impact of high-speed rail on regional economic disparity in China. *J. Transp. Geogr.* **2017**, *65*, 80–91. [CrossRef]
13. Saleem, H.; Shahzad, M.; Khan, M.B.; Khilji, B.A. Innovation, total factor productivity and economic growth in Pakistan: A policy perspective. *J. Econ. Struct.* **2019**, *8*, 7. [CrossRef]
14. Amp, Z.J.; Amp, W.G.; Jipeng, Z. The Estimation of China's provincial capital stock: 1952–2000. *Econ. Res. J.* **2004**, *10*, 35–44.

15. Hu, Z.; Wu, G.; Han, Y.; Niu, Y. Unraveling the dynamic changes of high-speed rail network with urban development: Evidence from China. *Socio-Econ. Plan. Sci.* **2023**, *85*, 101380. [CrossRef]
16. Lu, Y.; Zeng, L. How do high-speed railways facilitate high-quality urban development: Evidence from China. *Land* **2022**, *11*, 1596. [CrossRef]
17. An, J. High-Speed Rail Network, Entrepreneurial Activity and High-Quality Development: Empirical Data from Chinese Cities. *Theor. Econ. Lett.* **2024**, *14*, 2382–2406. [CrossRef]
18. Kong, Q.; Li, R.; Jiang, X.; Sun, P.; Peng, D. Has transportation infrastructure development improved the quality of economic growth in China's cities? A quasi-natural experiment based on the introduction of high-speed rail. *Res. Int. Bus. Financ.* **2022**, *62*, 101726. [CrossRef]
19. Qi, J. Fiscal expenditure incentives, spatial correlation and quality of economic growth: Evidence from a Chinese province. *Int. J. Bus. Manag.* **2016**, *11*, 191–201. [CrossRef]
20. Kong, Q.; Peng, D.; Ni, Y.; Jiang, X.; Wang, Z. Trade openness and economic growth quality of China: Empirical analysis using ARDL model. *Financ. Res. Lett.* **2021**, *38*, 101488. [CrossRef]
21. Feng, Q.; Chen, Z.; Cheng, C.; Chang, H. Impact of high-speed rail on high-skilled labor mobility in China. *Transp. Policy* **2023**, *133*, 64–74. [CrossRef]
22. Kang, S.K.; DeCelles, K.A.; Tilcsik, A.; Jun, S. Whitened Resumes: Race and Self-Presentation in the Labor Market. *Adm. Sci. Q.* **2016**, *61*, 469–502. [CrossRef]
23. Li, B.; Zhao, Q.; Shahab, Y.; Kumar, S. High-speed rail construction and labor investment efficiency: Evidence from an emerging market. *Res. Int. Bus. Financ.* **2022**, *64*, 101848. [CrossRef]
24. Baum-Snow, N.; Brandt, L.; Henderson, J.V.; Turner, M.A.; Zhang, Q. Roads, railroads, and decentralization of Chinese cities. *Rev. Econ. Stat.* **2017**, *99*, 435–448. [CrossRef]
25. Zhang, A.; Wan, Y.; Yang, H. Impacts of high-speed rail on airlines, airports and regional economies: A survey of recent research. *Transp. Policy* **2019**, *81*, A1–A19. [CrossRef]
26. Yan, L.; Tu, M.; Chagas, A.L.S.; Tai, L. The impact of high-speed railway on labor spatial misallocation: Based on spatial difference-in-differences analysis. *Transp. Res. Part A Policy Pract.* **2022**, *164*, 82–97. [CrossRef]
27. Shi, Y.; Guan, C.; Chang, J.-W.; Liu, H. High-Speed Rail: A Fast Lane for Information Delivery of China's IPOs. *Emerg. Mark. Financ. Trade* **2025**, *61*, 273–288. [CrossRef]
28. Wang, Y.; Guo, J.; Zhang, S. Market Access, Population Mobility and Labor Force Structure: Evidence from China's High-Speed Railway. *SAGE Open* **2024**, *14*, 21582440241285750. [CrossRef]
29. Wang, F.; Wei, X.; Liu, J.; He, L.; Gao, M. Impact of high-speed rail on population mobility and urbanisation: A case study on Yangtze River Delta urban agglomeration, China. *Transp. Res. Part A Policy Pract.* **2019**, *127*, 99–114. [CrossRef]
30. Guo, Y.; Xie, W.; Yang, Y. Dual green innovation capability, environmental regulation intensity, and high-quality economic development in China: Can green and growth go together? *Financ. Res. Lett.* **2024**, *63*, 105275. [CrossRef]
31. Clements, K.; Pawlowski, J.; Manouselis, N. Open educational resources repositories literature review—Towards a comprehensive quality approaches framework. *Comput. Hum. Behav.* **2015**, *51*, 1098–1106. [CrossRef]
32. Guirao, B.; Lara-Galera, A.; Campa, J.L. High Speed Rail commuting impacts on labour migration: The case of the concentration of metropolis in the Madrid functional area. *Land Use Policy* **2017**, *66*, 131–140. [CrossRef]
33. Zhang, F.; Wang, F.; Yao, S. Does government-driven infrastructure boost green innovation? Evidence of new infrastructure plan in China. *J. Asian Econ.* **2024**, *95*, 101828. [CrossRef]
34. Wu, Y.; Lee, C.C.; Lee, C.C.; Peng, D. Geographic proximity and corporate investment efficiency: Evidence from high-speed rail construction in China. *J. Bank. Financ.* **2022**, *140*, 106510. [CrossRef]
35. Zhang, F.; Yao, S.; Wang, F. The role of high-speed rail on green total factor productivity: Evidence from Chinese cities. *Environ. Sci. Pollut. Res.* **2023**, *30*, 23314–23327. [CrossRef]
36. Li, S.; Liu, Q.; Shao, Z. Connect to green: High-speed rail network and firms' environmental performance. *Transp. Policy* **2024**, *159*, 297–313. [CrossRef]
37. Zhou, Y.; Xu, X.; Tao, L. The impact mechanism of high-speed railway on regional green innovation spillover under multi-dimensional paths. *Environ. Impact Assess. Rev.* **2022**, *95*, 106795. [CrossRef]
38. Chen, X.; Di, Q.; Jia, W.; Hou, Z. Spatial correlation network of pollution and carbon emission reductions coupled with high-quality economic development in three Chinese urban agglomerations. *Sustain. Cities Soc.* **2023**, *94*, 104552. [CrossRef]
39. Chen, Z.; Xue, J.; Rose, A.Z.; Haynes, K.E. The impact of high-speed rail investment on economic and environmental change in China: A dynamic CGE analysis. *Transp. Res. Part A Policy Pract.* **2016**, *92*, 232–245. [CrossRef]
40. Shao, S.; Tian, Z.; Yang, L. High-speed rail and urban service industry agglomeration: Evidence from China's Yangtze River Delta region. *J. Transp. Geogr.* **2017**, *64*, 102–112. [CrossRef]
41. Acemoglu, D.; Johnson, S.; Robinson, J.A. Institutions as a fundamental cause of long-run growth. *Handb. Econ. Growth* **2005**, *1*, 385–472.

42. Donaldson, D. Railroads of the Raj: Estimating the impact of transportation infrastructure. *Am. Econ. Rev.* **2018**, *108*, 899–934. [CrossRef]
43. Alfaro, L.; Kalemli-Ozcan, S.; Volosovych, V. Why doesn't capital flow from rich to poor countries? An empirical investigation. *Rev. Econ. Stat.* **2008**, *90*, 347–362. [CrossRef]
44. Dong, X.; Zheng, S.; Kahn, M.E. The role of transportation speed in facilitating high-skilled teamwork across cities. *J. Urban Econ.* **2020**, *115*, 103212. [CrossRef]
45. Li, L.; Lu, L.; Xu, Y.; Sun, X. Influence of high-speed rail on tourist flow network in typical tourist cities: An empirical study based on the Hefei–Fuzhou high-speed rail in China. *Asia Pacific J. Tour. Res.* **2020**, *25*, 1215–1231. [CrossRef]
46. Wang, D.; Niu, Y.; Qian, J. Evolution and optimization of China's urban tourism spatial structure: A high-speed rail perspective. *Tour. Manag.* **2018**, *64*, 218–232. [CrossRef]
47. Lin, J.J.; Xie, Z.X. The associations of newly launched high-speed rail stations with industrial gentrification. *J. Transp. Geogr.* **2020**, *83*, 102662. [CrossRef]
48. Chen, Z.; Haynes, K.E. Impact of high-speed rail on housing values: An observation from the Beijing–Shanghai line. *J. Transp. Geogr.* **2015**, *43*, 91–100. [CrossRef]
49. Dong, X. High-speed railway and urban sectoral employment in China. *Transp. Res. Part A Policy Pract.* **2018**, *116*, 603–621. [CrossRef]
50. Meng, X.; Qu, Y.; Xu, X. The Sustainable Impact of High-Speed Rail Connection on the Local and Neighboring Regions' Employment: Evidence from China. *Sustainability* **2024**, *16*, 16178. [CrossRef]
51. Gao, Y.; Su, W.; Wang, K. Does high-speed rail boost tourism growth? New evidence from China. *Tour. Manag.* **2019**, *72*, 220–231. [CrossRef]
52. Xu, W.A.; Zhou, J.; Yang, L.; Li, L. The implications of high-speed rail for Chinese cities: Connectivity and accessibility. *Transp. Res. Part A Policy Pract.* **2018**, *116*, 308–326. [CrossRef]
53. Jiao, J.; Wang, J.; Zhang, F.; Jin, F.; Liu, W. Roles of accessibility, connectivity and spatial interdependence in realizing the economic impact of high-speed rail: Evidence from China. *Transp. Policy* **2020**, *91*, 1–15. [CrossRef]
54. Ren, X.; Zeng, G.; Dong, K.; Wang, K. How does high-speed rail affect tourism development? The case of the Sichuan-Chongqing Economic Circle. *Transp. Res. Part A Policy Pract.* **2023**, *169*, 103588. [CrossRef]
55. Bhatt, A.; Kato, H.; Hayashi, Y. High-speed rails and knowledge productivity: A global perspective. *Transp. Policy* **2021**, *101*, 174–186. [CrossRef]

land

Article

Innovation in Comprehensive Transportation Network Planning in the Context of National Spatial Development: Institutional Constraints and Policy Responses

Huanyu Yang [1], Wei Huang [2], Dong Yang [1] and Ying Jiang [3,*]

[1] China Academy of Transportation Sciences, Beijing 100029, China; yanghy@motcats.ac.cn (H.Y.); yangdong@motcats.ac.cn (D.Y.)
[2] CCCC Water Transport Planning and Design Institute Co., Ltd., Beijing 100008, China; huangwei1@pdiwt.com.cn
[3] School of Transportation, Southeast University, Nanjing 211189, China
* Correspondence: yingjiang@seu.edu.cn

Abstract: This study investigates the institutional innovation pathways for integrating comprehensive transportation networks into China's territorial spatial planning system, with a focus on resolving the conflicts between ecological conservation and infrastructure development. By proposing a 'constraint-coupling-innovation' framework, this research addresses the gaps in existing spatial governance mechanisms, particularly the insufficient alignment between transportation planning and the 'three zones and three lines' (ecological conservation, agricultural production, and urban development zones with binding redline) system. The study employs mixed-method approaches, including geospatial conflict analysis (GIS), AI-driven policy coordination tools, and case studies from the Yangtze River Economic Belt. It demonstrates that rigid ecological constraints (e.g., ecological sensitivity veto power) can reduce planning conflicts effectively, while adaptive governance models enhance land use efficiency and stakeholder collaboration. Key findings reveal a significant negative correlation ($R^2 = 0.75$) between ecological protection redline (EPR) coverage and transportation network density, underscoring the necessity for differentiated governance strategies in high-conflict regions. A comparative analysis with the EU's Natura 2000 sites and TEN-T networks further highlights China's unique hierarchical governance model, which integrates top-down ecological mandates with localized technological innovations, such as digital twins and polycentric decision making. This study contributes to global debates on sustainable spatial planning by offering actionable pathways for balancing infrastructure expansion with ecological resilience, while also proposing institutional reforms, such as a National Transportation Spatial Governance Index (NTSGI), to standardize ecological compliance. These insights provide both theoretical advancements in spatial institutionalism and practical tools for policymakers navigating the dual challenges of urbanization and climate resilience.

Keywords: comprehensive transportation network; territorial spatial planning; institutional constraints; policy coupling; technological innovation

1. Introduction

The integration of transportation infrastructure with ecological conservation goals has emerged as a critical challenge in spatial planning worldwide, particularly in rapidly urbanizing regions where infrastructure expansion often conflicts with biodiversity protection. Globally, this tension manifests in frameworks such as the EU's Natura 2000 sites and TEN-T networks [1], where habitat fragmentation and carbon-intensive development persist despite regulatory efforts [2,3]. China's unique 'three zones and three lines' system provides a novel institutional context to explore spatial governance innovations. China's integration of comprehensive transportation planning into its multi-plan system under the territorial spatial planning (TSP) framework signifies a strategic shift toward coordinated and sustainable urbanization. Since 2020, this approach has prioritized the alignment of transportation infrastructure with land use efficiency, ecological conservation, and socio-economic objectives, aiming to mitigate urban sprawl and reduce carbon emissions [4]. By embedding transportation strategies within broader developmental agendas—such as economic connectivity and climate resilience—the multi-plan integration system seeks to optimize resource allocation and institutionalize low-carbon mobility solutions, including green public transit [5]. However, the implementation of this system encounters institutional barriers, including fragmented departmental priorities, weak enforcement of EPR, persistent vehicle-centric planning paradigms, which are often overlooked in current studies [6].

The urgency of addressing these issues is heightened by China's dual carbon goals—achieving peak emissions by 2030 and carbon neutrality by 2060—which necessitate systemic reforms in transportation infrastructure investment and land use governance. These reforms are essential for facilitating a transition toward green technologies, including electric vehicles, smart logistics, renewable energy-integrated transit systems, digital innovation, and green finance mechanisms [7–9]. While existing studies have examined polycentric governance [10], digital planning tools [11–13] and innovative planning [14–18] in isolation, few have investigated how hierarchical governance systems, such as China's, can incorporate ecological constraints into multi-level decision making. This gap is significant, as traditional planning models tend to prioritize economic growth over spatial carrying capacity, resulting in irreversible ecological damage [6].

This study addresses three underexplored dimensions through the following aims:

1. Institutional constraints: To analyze how China's 'three zones and three lines' territorial spatial governance system can institutionalize ecological constraints (e.g., EPR veto power) within transportation planning, thereby reconciling infrastructure expansion with land resource sustainability.
2. Policy coupling mechanisms: To explore adaptive governance pathways that vertically align national carbon neutrality goals with local land use priorities, focusing on the role of multi-level stakeholder engagement in resolving conflicts between transportation corridors and protected ecological zones.
3. Spatial decision support: To demonstrate how geospatial tools (e.g., GIS-based conflict mapping) and polycentric governance models can enhance the transparency and equity of land allocation for transportation projects, particularly in ecologically sensitive regions.

By integrating spatial institutionalism theory with empirical case studies, this study advances a hybrid 'constraint-coupling-innovation' framework. This framework diverges from Ostrom's polycentric governance by introducing rigid ecological thresholds (e.g., veto mechanisms) while leveraging digital twins for dynamic land use simulations—an approach particularly suited to hierarchical governance contexts. Comparative analysis with EU and Japanese models further contextualizes China's innovations within global spatial planning debates, offering actionable pathways for balancing infrastructure expansion with ecological resilience. Methodologically, this research employs a mixed-methods approach that combines geospatial analytics and qualitative case studies. Quantitative GIS tools are utilized to map conflicts between transportation corridors and ecological redlines, while in-depth case analyses of four provinces (Yunnan, Hebei, Jiangsu, Shandong) reveal institutional barriers and innovation pathways. These provinces were selected for their representative ecological–transportation conflict profiles, which range from biodiversity hotspots to high-density urban corridors.

2. Theoretical Framework

2.1. Spatial Governance Theory

The theory of spatial governance, particularly in relation to territorial space, can be comprehensively understood through several critical dimensions derived from contemporary governance frameworks. The framework proposed by Douglass North introduces three lines as entities that redefine territorial spaces. This paradigm emphasizes the significance of both formal and informal boundaries within governance structures. North's conceptualization suggests that governance is shaped by the interplay of property rights, political authority, and social norms that define interactions within a given space. Consequently, institutional arrangements should not be viewed in isolation but rather as dynamic entities that exist within specific territorial contexts. Such delineations are crucial for understanding how policies are crafted and implemented based on geographical and institutional parameters, thereby creating a new institutional environment where governance can more effectively respond to localized needs and circumstances [19]. Elinor Ostrom's work on polycentric governance advocates for a multi-level decision making framework that transcends traditional hierarchical governmental structures. Ostrom posited that polycentric governance systems comprise multiple, overlapping institutions operating at various scales, involving a range of stakeholders, including governmental bodies, nonprofits, community-based organizations, and the private sector. This governance model fosters collaboration and coordination among different sectors to address complex social and environmental issues. The flexibility inherent in polycentric governance facilitates adaptive management strategies capable of responding to the specific challenges faced by diverse communities [20]. The success of governance systems significantly relies on the capacity to foster effective communication and collaboration among all relevant stakeholders. Polycentric frameworks are particularly adept at managing resources that span multiple jurisdictions, necessitating collaborative actions that transcend political and social boundaries. By acknowledging the interconnectedness of various sectors and the imperative for cooperative governance, Ostrom underscored that knowledge sharing, mutual understanding, and interdisciplinary approaches can yield more effective and resilient outcomes in spatial governance [21]. In summary, spatial governance as articulated through North's three lines and Ostrom's polycentric governance addresses the need for innovative institutional arrangements and proactive collaboration across multiple levels and sectors. By establishing well-defined boundaries for governance and encouraging a pluralistic approach to decision making, these theories foster a navigable path for policy implementation that is responsive to both local and global challenges. The integration of these concepts into governance practices thus cultivates an environment conducive to fostering sustainable development and equitable resource management [22,23].

2.2. Analytical Dimensions

The relationship between spatial planning documents, transportation networks, and local policy innovation is multifaceted and critical for achieving sustainable urban development. Furthermore, the regulatory constraints associated with spatial planning emphasize the legally binding nature of planning documents, which fundamentally influence land use and development activities. These legally binding documents establish the frameworks through which land use is regulated, thereby providing clear guidelines and restrictions for development projects. The robustness of these regulations is essential for ensuring compliance and facilitating orderly urban growth. Notably, in regions where these documents are strictly enforced, conflicts arising from land use decisions are often mitigated, which enhances predictability and stability in urban planning processes [24]. When examining the responsiveness of transportation networks to land use zoning, it is essential to recognize that the effective integration of transport and land use planning is crucial for optimizing urban mobility. Well-structured land use zoning should inherently reflect the accessibility needs of the population alongside the corresponding transport infrastructure that supports them. The quality and configuration of transportation networks directly influence urban development; therefore, a responsive transport network enhances the efficacy of spatial planning. For instance, when land use zones are established with an emphasis on mixed-use development and transit-oriented design, transportation systems can efficiently facilitate movement, reduce reliance on private vehicles, and promote sustainable commuting patterns. The interplay between land use planning and transportation directly impacts mobility and accessibility, thereby enriching the overall urban experience [4]. Regarding the scalability of local policy experiments, innovative expansion in urban planning is paramount. Flexible and amenable local policies that allow for experimentation can lead to significant urban improvements. Pilot programs and experimental initiatives enable municipalities to test new strategies for urban development, assess their effectiveness, and evaluate their scalability. A crucial aspect of this innovative approach is the incorporation of stakeholder engagement and community input, which fosters local support and responsiveness. Successful local policy experiments not only yield immediate benefits but also serve as models for broader application in similar urban contexts. Therefore, careful evaluation and documentation of these experiments can inform future planning practices and regulatory structures across different regions [25]. In conclusion, the effectiveness of spatial planning is significantly influenced by the legal framework established by regulatory constraints, the adaptability of transportation networks to land use zoning, and the capacity for local policy experiments to scale up innovative practices. Each of these elements plays a critical role in fostering sustainable and efficient urban environments that are well integrated and responsive to community needs.

To further clarify the theoretical innovation of this study, Table 1 contrasts the proposed 'constraint-coupling-innovation' framework with Elinor Ostrom's polycentric governance theory. While Ostrom's approach emphasizes decentralized negotiation among equal actors [10], the framework integrates top-down ecological constraints (e.g., redline veto) with bottom-up technological innovation, thereby addressing the unique challenges of land use conflicts in territorial spatial planning. This hybrid model is particularly well-suited to China's hierarchical governance context, where centralized ecological goals must coexist with localized development needs.

Table 1. Comparative analysis: Ostrom's polycentric governance vs. the proposed 'constraint-coupling-innovation' framework.

Dimension	Ostrom's Polycentric Governance Theory	Proposed Constraint-Coupling-Innovation Framework	Theoretical Advancement and Land Science Contribution
Governance Structure	Polycentric decentralization: multiple autonomous decision centers (e.g., communities, local governments) coordinate resource allocation through self-organization.	Multi-level synergy: central policy constraints (e.g., EPR) coupled with local innovations, forming a dynamic 'vertical constraint–horizontal collaboration' network.	Introduces rigid ecological thresholds as central constraints, transcending traditional polycentric theory's 'equal negotiation' limitation, better suited to land resource regulation in territorial spatial planning.

Table 1. *Cont.*

Dimension	Ostrom's Polycentric Governance Theory	Proposed Constraint-Coupling-Innovation Framework	Theoretical Advancement and Land Science Contribution
Institutional Constraints	Relies on community consensus and soft rules (e.g., social norms, trust mechanisms).	Legalized rigid constraints via ecological veto power, mandating project compliance with EPR.	Upgrades ecological protection from 'flexible negotiation' to institutional veto instruments, resolving power asymmetries in land resource development–conservation conflicts.
Policy Instruments	Self-governance tools (e.g., community covenants, collective action rules).	Policy coupling tools (e.g., dynamic adjustment between national carbon goals and local land quotas) and technological innovations (digital twins, AI conflict detection).	Integrates vertical policy coordination with horizontal technological empowerment, achieving tripartite 'institution–technology–data-' driven land governance.
Conflict Resolution	Incremental negotiation through repeated games (e.g., common-pool resource management).	Data-driven preemptive conflict identification and rigid intervention (e.g., ecological veto halting projects, supplemented by alternative solutions generation).	Replaces ex-post negotiation with spatial simulation-aided preemption (e.g., GIS overlay analysis), addressing irreversible land development impacts and minimizing ecological damage.
Application Context	Small-scale common-pool resources (e.g., forests, fisheries) under community self-governance.	Macro-scale territorial governance (e.g., interprovincial transportation networks vs. EPR), emphasizing land multifunctionality (production living–ecological space balance).	Expands beyond Ostrom's 'local autonomy' to provide institutional paradigms for regional composite land use.
Theoretical Goal	Avoids 'tragedy of the commons' through sustainable resource use.	Balances land development (e.g., transport land expansion) and conservation (e.g., ecological protection redline) to achieve spatial justice and intergenerational equity.	Extends land ethics from 'resource utilization' to spatial equity (e.g., ecological compensation), addressing normative imperatives in territorial spatial planning.

3. Institutional Impacts of Spatial Planning

3.1. Planning Paradigm Shift

The paradigm shift in spatial planning prioritizes ecological sensitivity over traditional sectoral optimization, addressing the historical fragmentation of decision making across departments like transportation and urban development. Earlier approaches often overlooked ecological interdependencies, resulting in habitat loss and unsustainable resource use [26]. The emerging framework institutionalizes an ecological sensitivity veto—a mechanism requiring rigorous assessments of project impacts on biodiversity, landscape integrity, and ecosystem resilience before approval. This policy redefines ecological thresholds as non-negotiable constraints, superseding short-term economic imperatives to align development with spatial carrying capacities [27]. By integrating geospatial tools and transdisciplinary collaboration, planners quantify ecological trade-offs, enforce adaptive strategies, and foster institutional coordination across sectors [28].

The Yangtze River Economic Belt's high-speed rail (Figure 1) planning exemplifies this transition. Here, the ecological sensitivity veto mandates cross-regional cooperation to define shared ecological limits and prioritize low-impact corridor designs, counterbalancing past industrialization-driven degradation. Advanced spatial analytics map critical habitats and vulnerability zones, guiding data-driven decisions that align with China's dual carbon goals [29]. The veto also spurs innovation, such as eco-engineering solutions minimizing land disturbance, while embedding green infrastructure principles into transport systems. This approach demonstrates how ecological constraints can reconcile infrastructure demands with biodiversity conservation, offering scalable models for regions grappling with urbanization and climate resilience [30]. Institutionalizing such mechanisms underscores a broader commitment to ecological–economic synergy, where ecological health becomes foundational to sustainable progress [31].

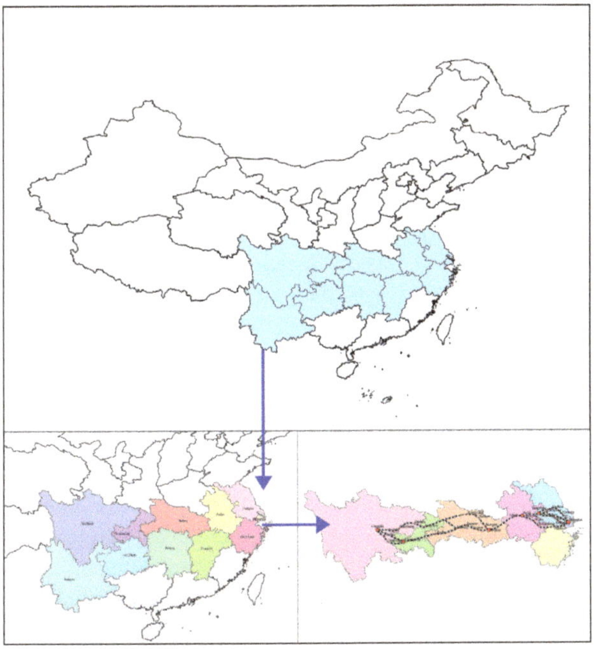

Figure 1. The Yangtze River Economic Belt area.

The ecological veto mechanism's application in the Yangtze River Economic Belt (Section 3.1) was further validated through digital twin simulations (Section 4.2), demonstrating its efficacy in conflict resolution.

3.2. Decision Making Restructuring

Environmental governance reforms increasingly emphasize reconciling infrastructure demands with ecological preservation through two interlinked mechanisms: empowering natural resource departments with veto authority and integrating environmental impact assessments (EIAs) with spatial suitability analyses. Strengthened veto power enables these departments to enforce ecological thresholds during project approvals, countering historical tendencies to prioritize economic growth over environmental safeguards [32]. Simultaneously, combining EIAs—which evaluate ecosystem risks—with spatial suitability assessments ensures that development aligns with land use compatibility and minimizes habitat disruption [33,34]. This dual approach institutionalizes science-based decision making, mitigating degradation from unchecked urbanization by embedding ecological viability as a non-negotiable criterion for infrastructure planning [35].

The synergy of these measures fosters adaptive governance frameworks that balance growth with sustainability. Rigorous assessments provide policymakers with data to avoid ecologically fragile zones, while veto authority ensures compliance with long-term sustainability benchmarks—critical as urban expansion intensifies pressures on natural systems [36,37]. For instance, spatial analyses guide transport networks away from biodiversity hotspots, and veto interventions halt projects exceeding carrying capacities. Such reforms exemplify a shift toward viewing ecosystems as foundational rather than expendable, offering scalable models to address global ecological–economic tensions without stifling development.

3.3. Resource Allocation Reconfiguration

The allocation of transportation land quotas is intricately linked to the establishment of urban development boundaries, ensuring that land resources are utilized efficiently in alignment with urban planning objectives and developmental needs. By implementing clearly defined boundaries, local authorities can effectively regulate land use, prioritize transportation infrastructure, and manage urban sprawl, thereby enhancing overall urban sustainability and livability [38]. Furthermore, establishing a market-oriented mechanism, particularly through Transferable Development Rights (TDR), is essential for promoting the trade of development rights within corridor planning frameworks. TDR allow property owners to transfer their development rights to designated areas for higher density development, thereby facilitating a more flexible land use strategy. This mechanism not only incentivizes conservation in less developed regions but also enables urban developers to address the demands of growing populations by concentrating development in areas equipped with existing infrastructure [24]. The integration of TDR within urban planning has been shown to facilitate both economic efficiency and environmental sustainability. By reallocating development rights, cities can mitigate the spatial injustices commonly associated with land use transitions and the reduction in construction land (CLR). This approach promotes a balance between developmental aspirations and ecological conservation, thereby enhancing the acceptability of urban expansion while reinforcing the principles of spatial equity [39]. Consequently, focusing on TDR can lead to better-designed urban corridors that support economic growth while preserving the critical ecological functions of surrounding areas. Furthermore, the market-oriented nature of TDR programs can enhance transparency and trust in urban development processes, as rights are traded in an open market, reflecting their real-world value based on demand and supply dynamics. This can catalyze investments in infrastructure and improve land utilization, providing a contemporary avenue to address the challenges associated with rapid urbanization [40]. Additionally, the successful application of both transportation land quota allocation and TDR mechanisms requires a comprehensive understanding of local contexts, adherence to regulatory frameworks, and a commitment to stakeholder engagement in order to optimize urban spatial planning outcomes. Such strategic planning maximizes the potential of urban spaces while minimizing the conflicts arising from competing land use interests [41]. In summary, leveraging transportation land quotas in conjunction with TDR frameworks in corridor planning can significantly enhance the efficiency and sustainability of urban development, paving the way for resilient urban environments that can adapt to future challenges and demands.

4. Innovation Frontiers

4.1. Institutional Innovations

Institutional innovation is essential for addressing the challenges posed by the need for efficient governance of traffic spaces. Establishing a cross-departmental negative list system for traffic space governance represents a significant advancement toward the effective regulation and management of urban traffic. This system delineates specific activities or entities that are prohibited or restricted in traffic space usage, thereby providing a clearer framework within which traffic governance can operate. Such a framework encourages collaboration among various governmental departments, necessitating input from stakeholders, including urban planning, transportation, and environmental agencies, to effectively align their policies and operational guidelines. This coordination can lead to improved regulatory compliance and the more streamlined enforcement of traffic regulations, ultimately enhancing the overall traffic environment [42]. Additionally, the proposal to create a dynamic adjustment mechanism for traffic corridors, accompanied by a ±10% buffer zone, could effectively address fluctuations in traffic flow and enhance flexibility in traffic management. This buffer allows for real-time adjustments based on traffic conditions, enabling authorities to respond promptly to congestion, accidents, or other disruptions. For instance, during peak hours, traffic corridors can be adjusted to accommodate higher volumes, while during less busy times, the adjustments can prioritize alternate routes or modes of transportation. This dynamism contributes to optimizing road usage and enhancing the efficiency of traffic flows, minimizing delays and potentially reducing emissions through smoother traffic management [43]. Implementing such strategies could alleviate many issues associated with transportation systems, including congestion, air pollution, and inefficiencies in resource allocation, while fostering a stronger commitment to sustainable urban development. The introduction of a negative list and a dynamic corridor adjustment mechanism is likely to enhance the quality of life for residents by creating a more robust, efficient, and sustainable traffic environment that can adapt to changing conditions and demands.

4.2. Technological Breakthroughs

Technological innovations are reshaping the integration of transportation planning with territorial spatial governance, particularly in resolving conflicts between infrastructure demands and ecological conservation. Two key tools demonstrate this potential:

1. Digital twins for dynamic land use simulation. Digital twin platforms synthesize multi-source geospatial data (e.g., EPR boundaries, transportation corridors, land use zoning maps) to simulate real-time interactions between infrastructure expansion and protected areas [44]. For instance, in the Yangtze River Economic Belt, a digital twin model identified 15 high-risk zones where proposed railway routes intersected wetlands designated as EPR. This simulation prompted planners to reroute 32% of the railway network, reducing potential habitat fragmentation by 18%. Data sources for the digital twin models included: (1) EPR boundaries from the Ministry of Natural Resources (2023); (2) Transportation corridor plans from provincial GIS databases; (3) Land use zoning maps at 1:50,000 scale. Spatial conflicts were quantified using the Overlap Analysis Tool of ArcGIS Pro 3.4.3.

These models also support scenario testing: by adjusting parameters such as buffer zone widths (±10%) around redlines, planners can dynamically evaluate trade-offs between connectivity gains and ecological losses [45].

2. AI-Driven conflict detection in multi-plan coordination. Artificial intelligence algorithms are deployed to detect spatial conflicts across planning documents. For example, a system trained on China's 'three zones and three lines' territorial spatial plans analyzed 766 provincial-level transportation projects. The system prioritized conflicts based on ecological sensitivity indices (e.g., species richness, landscape connectivity), enabling policymakers to enforce the 'ecological veto' mechanism more efficiently [46]. Data source: provincial transportation departments and the Ministry of Natural Resources.

Table 2 summarizes AI-detected conflicts and resolution outcomes across four provinces, revealing significant variations in both conflict rates and governance efficacy.

Provinces with high ecological sensitivity (e.g., Yunnan) exhibited the highest conflict rates (27.6%), while economically developed regions (e.g., Jiangsu) demonstrated stronger resolution capacities (81.8%).

These technologies not only enhance planning precision but also institutionalize ecological thresholds into land use decision making. By converting abstract regulatory constraints into actionable spatial data, they bridge the gap between policy intent and on-ground implementation.

According to Figure 2, significant differences exist in the degree of conflict and spatial distribution among the four provinces: Hebei, Jiangsu, Shandong, and Yunnan. Hebei Province exhibits a maximum buffer zone value range of 0.001 to 269.375, indicating a high potential conflict intensity between its railway network and nature reserves. Notably, the high-value range of 242.439 to 269.375 suggests the presence of serious conflict areas, although the relative range is small. In contrast, Jiangsu and Shandong provinces demonstrate a relatively narrow numerical range (Jiangsu: 0.001–2.464; Shandong: 0.001–2.04), indicating a lower overall degree of conflict. This can be attributed to the railway planning in the wetlands of northern Jiangsu and the ecologically sensitive areas of the Shandong Peninsula. Yunnan Province's numerical range (0.001–2.04) is comparable to that of Shandong and Jiangsu. However, due to its mountainous terrain and the scattered distribution of protected areas, conflict areas in Yunnan are concentrated in river valleys or low-altitude transportation corridors. This situation underscores the necessity for a refined balance between railway construction and ecological protection.

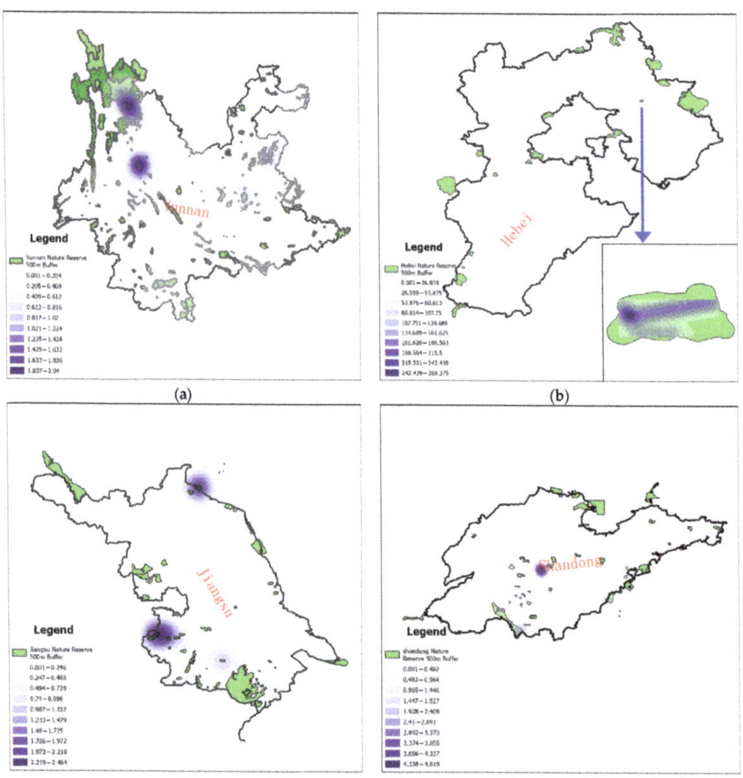

Figure 2. Conflict zones between Yunnan (**a**), Hebei (**b**), Jiangsu (**c**), and Shandong (**d**) railway networks and national nature reserves.

Table 2. AI-detected conflicts in provincial transportation projects.

Province	Total Projects	Redline Conflicts	Conflict Rate (%)	Resolved Through Ecological Veto	Resolution Rate (%)
Yunnan	185	51	27.6	42	82.4
Hebei	176	34	19.3	26	76.5
Jiangsu	210	22	10.5	18	81.8
Shandong	195	29	14.9	25	86.2
Total	766	223	19.3	181	81.2

4.3. Governance Experiments

In the context of the deep integration of territorial spatial planning and comprehensive transportation networks, governance experiments focused on ecology and sustainability are essential. They provide innovative frameworks to address complex environmental challenges effectively.

The Transportation-Oriented Ecological Restoration (TOER) pilot project in China's national parks serves as an innovative vehicle for governance experimentation. This project explores a new model for the coordinated development of transportation infrastructure and ecological conservation, driven by the dual forces of institutional constraints and policy responses. Central to the TOER pilot project is the concept of "the 'life community' concept integrating mountains, rivers, forests, and grasslands" which effectively bridges the traditional disconnect between transportation planning and ecological conservation. It emphasizes systematic restoration and spatial integration, achieving physical fusion and functional complementarity between transportation facilities and ecological restoration through innovative technologies such as wildlife-friendly passages and intelligent management platforms. Focused on resolving the 'protection–development' dilemma, the project has established multi-level systems, including the 'Land Space Ecological Protection and Restoration Project Acceptance Specification', to mitigate the inertia of 'development first' through institutional constraints and activate multi-stakeholder participation through policy responses. Ultimately, it aims to achieve the synergistic goals of enhancing transportation functionality while increasing ecological value.

The cross-border planning alliance has transitioned from an 'administrative region economy' to a 'functional zone economy' through institutional and tool innovations, thereby offering a replicable governance paradigm for transportation network planning within the context of territorial spatial dynamics. This study examines the transportation coordination practices of the Guangdong–Hong Kong–Macao Greater Bay Area as a case study. The approach includes the establishment of independent legal entities, such as the Guangdong–Hong Kong–Macao Transportation Coordination Committee, which coordinates the formulation of cross-regional transportation special plans. Furthermore, the creation of data-sharing platforms and mutual recognition systems, exemplified by the Greater Bay Area Transportation Data Center, the promotion of aligned technical standards, the clarification of dispute resolution mechanisms through legislation, and the incorporation of blockchain and AI technologies to enhance governance efficiency collectively provide an innovative paradigm for cross-regional transportation network planning, shifting the focus from 'physical connectivity' to 'institutional integration'.

4.4. Quantitative Analysis of Provincial Conflicts

To systematically evaluate the spatial trade-offs between ecological preservation and transportation development, this study conducted a province-level comparative analysis using two key indicators: EPR area (as a percentage of provincial territory) and comprehensive transportation network (railways, highways, and inland waterways) density (km per 100 km^2). Data were extracted from *China Statistical Yearbook* (2024) and the website of the Ministry of Natural Resources of China, covering five strategically significant regions: Hebei, Jiangsu, Shandong, and Yunnan (Table 3).

Table 3. Key indicators of province-level comparative analysis.

Province	EPR Area (10^4 km^2)	Comprehensive Transportation Network Density (10^4 km/(10^4 km^2))	Railway Mileage (10^4 km)	Highway Mileage (10^4 km)	Inland Waterway Mileage (10^4 km)	Area (10^4 km^2)
Yunnan	11.84	0.86	0.52	32.93	0.47	39.41
Hebei	4.05	1.16	0.85	21.11	0	18.88
Jiangsu	1.82	1.75	0.47	15.87	2.44	10.72
Shandong	2.07	1.91	0.77	29.34	0.11	15.81

Key observations from Figure 3 reveal that Yunnan has the largest EPR area, but the lowest comprehensive transportation network density, which is in line with its rich ecological resources but complex terrain and difficult transportation construction. In contrast, Jiangsu possesses both the minimum EPR area but the second highest density of its comprehensive transportation network, suggesting a balance between development and protection. These quantified disparities underscore the necessity for differentiated governance: high-conflict provinces, such as Yunnan, require stringent ecological sensitivity vetoes in transport planning, while low-density regions like Hebei may prioritize the conservation of ecological connectivity when expanding their comprehensive transportation networks.

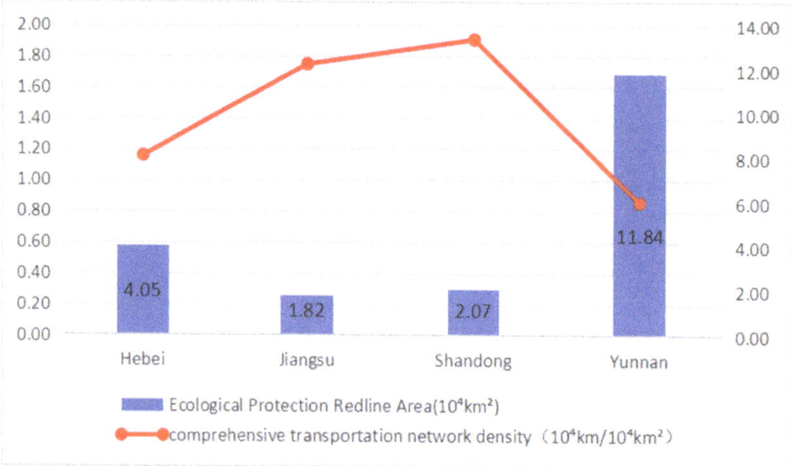

Figure 3. Comparison of EPR area and comprehensive transportation network density in some provinces of China (2023).

Figure 4 shows a significant negative correlation ($R^2 = 0.75$) between the area of EPR and the density of transportation networks, indicating that the strict protection of ecologically sensitive areas may limit the expansion of transportation infrastructure. This finding supports the necessity of ecological sensitivity rejection mechanisms. Pearson correlation analysis was conducted using SPSS 28.0, with significance thresholds set at $p < 0.05$.

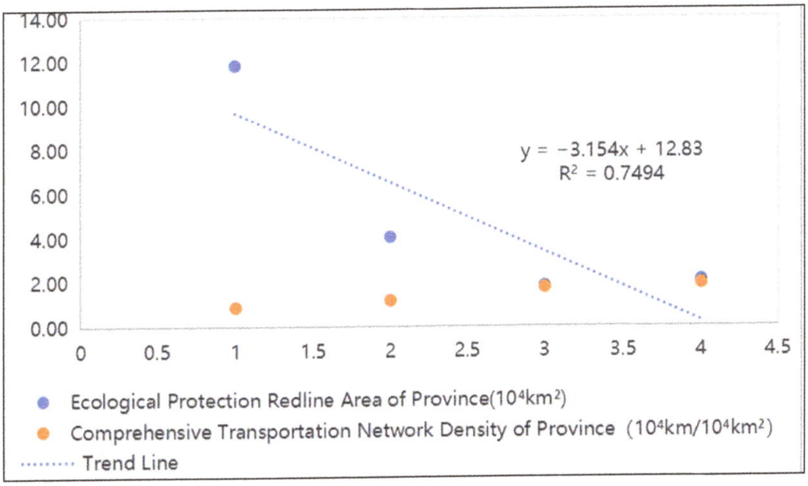

Figure 4. Correlation analysis between EPR Area and Comprehensive Transportation Network Density.

5. International Comparative Perspectives

5.1. EU Experience: Conflict Management

Natura 2000 sites and TEN-T networks integration exemplifies EU efforts to reconcile transport and biodiversity goals. The TEN-T aims to enhance connectivity across EU member states through an extensive transport system, while the Natura 2000 sites are designed to protect valuable habitats and species under the Birds and Habitats Directives. The interplay between these two networks often leads to conflicts, since the urban and infrastructure development associated with the TEN-T may encroach upon Natura 2000 areas, which are designated to conserve biodiversity and maintain ecological integrity. An analysis of urbanization trends indicates that urban cover within Natura 2000 sites is significantly lower than that outside these areas; however, urban growth rates within Natura 2000 sites are increasing, suggesting an incipient urban sprawl that could jeopardize conservation objectives [3]. To address these challenges, spatial planning must prioritize eco-friendly integration strategies. This involves assessing and mitigating the potential adverse effects of the Trans-European Transport Network (TEN-T) on Natura 2000 sites. For example, 34% of TEN-T corridors within Natura 2000 sites exhibit high conflict levels (Figure 5), according to data from the European Commission (2023) and EEA (2023). Policies should ensure that the development of transport infrastructure takes into account ecological connectivity and functions, thereby fostering resilience in protected areas. Effective land management approaches could employ comprehensive planning tools that consider ecological factors, aiding in achieving a balance between transport infrastructure and biodiversity protection [47]. Member States of the European Union are urged to implement stricter legal protections for Natura 2000 sites, particularly in areas where urban sprawl presents a significant threat. Efforts should focus on minimizing conflicts between biodiversity conservation and transportation development. For instance, connectivity measures can enhance wildlife movement across transport corridors, thereby preserving ecological networks despite infrastructure expansion [48]. Furthermore, advancing methodologies in spatial planning can better align these network objectives. By employing ecological theories and data analysis, planners can more effectively assess the impacts of transit-oriented development on biodiversity. Cross-sector collaboration is crucial, particularly through transnational partnerships that can strengthen conservation initiatives related to Natura 2000. Such cooperation is vital for aligning transportation policies with environmental goals, ultimately enhancing the ecological integrity of protected areas [49,50]. Effective management and strategic planning, supported by strong collaboration among stakeholders, can lead to the successful integration of infrastructure development and biodiversity conservation objectives within the framework of the European Union's environmental policies.

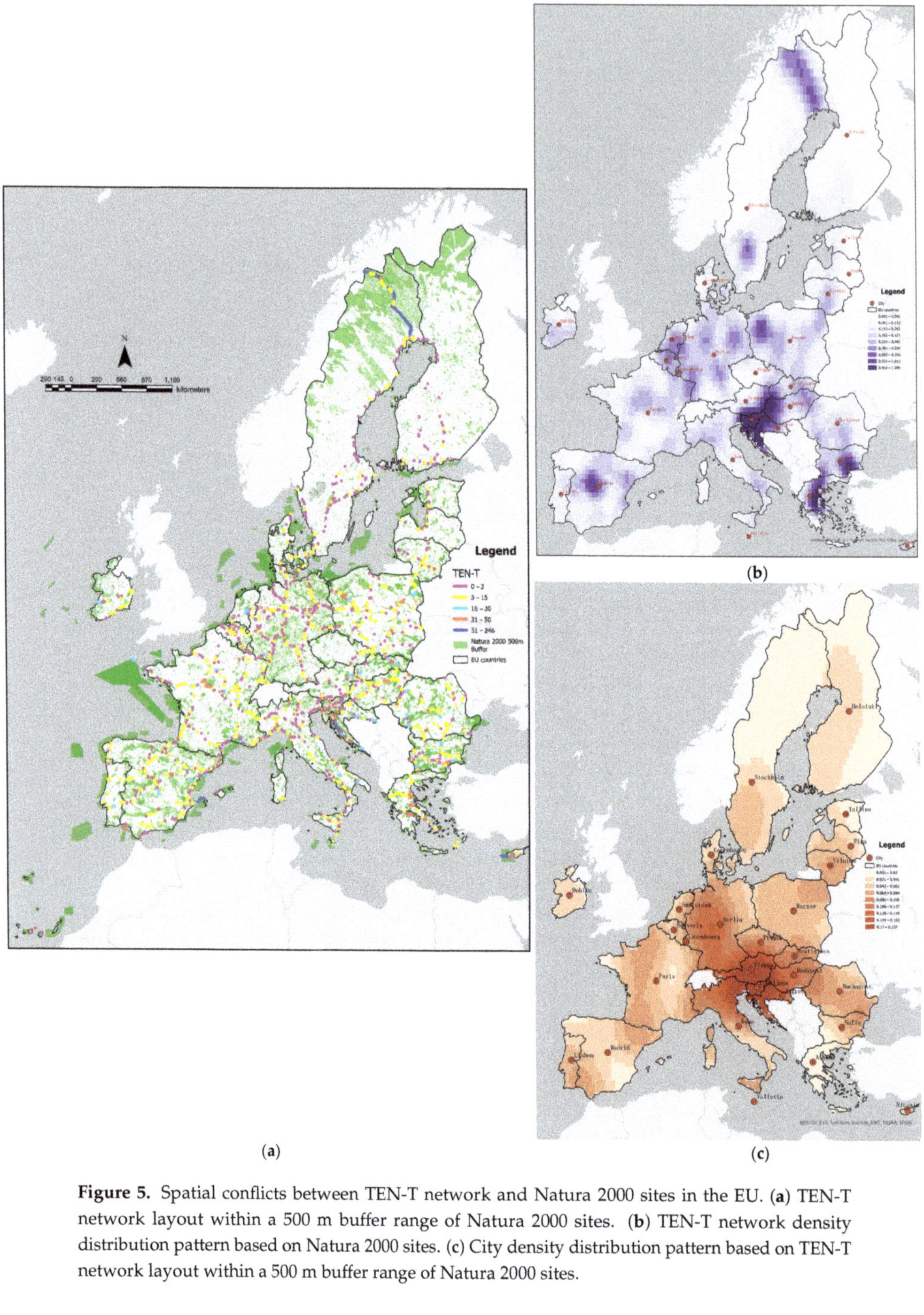

(a) (b) (c)

Figure 5. Spatial conflicts between TEN-T network and Natura 2000 sites in the EU. (**a**) TEN-T network layout within a 500 m buffer range of Natura 2000 sites. (**b**) TEN-T network density distribution pattern based on Natura 2000 sites. (**c**) City density distribution pattern based on TEN-T network layout within a 500 m buffer range of Natura 2000 sites.

Figure 5b shows that the distribution of urban density within the 500 m buffer zone of the TEN-T network in Natura 2000 sites is extremely uneven, with the highest density value reaching 1.991 (such as cities like Bialystok and Galletta), indicating that the urbanization and transportation networks in these areas overlap heavily with the boundaries of protected areas, posing significant potential ecological disturbance risks. Figure 5c shows that the density peak of the TEN-T network in the Natura 2000 core area is only 0.237 (such as Stockholm and Berlin), indicating that the density of transportation infrastructure within the protected area is relatively low. However, the gradient difference between the buffer zone and the core area may still cause fragmentation of ecological corridors.

5.2. Japan's Compact City Model: Lessons for TOD Integration

The concept of compact cities, particularly in relation to Transit-Oriented Development (TOD), is gaining traction in urban planning across various global contexts, including Japan. This approach aims to create high-density, mixed-use environments that optimize access to public transit while promoting sustainable living and reducing the reliance on automobiles. Compact city policies have been implemented to encourage the development of urban areas that are closely connected to public transport systems. This model is characterized by the concentration of residential, commercial, and recreational spaces within walking distance of transit stations. The goal is to facilitate ease of movement and increase public transport usage, thereby enhancing physical activity among residents and contributing to public health objectives [51]. Policies promoting mixed land use and high population densities in proximity to transit hubs are instrumental in achieving sustainable urban development [24]. Furthermore, public transportation services, including buses and rail systems, are designed to integrate with urban planning, thereby enhancing connectivity and accessibility to services and amenities. This integration is crucial for realizing the desired outcomes of compact development [52]. The amalgamation of residential areas with local amenities near transit stops fosters an appealing environment, encouraging individuals to choose public transportation over private vehicles [53]. This paradigm contrasts with the traditional urban sprawl, where residents often rely on cars due to the dispersion of services and amenities. This reliance leads to increased carbon emissions and reduced physical activity levels among the populace [54]. However, compact city policies are not solely focused on optimizing transportation; they also address secondary concerns such as housing affordability and the sustainable use of urban land. While dense development around transit hubs can lead to increased housing prices, strategic planning aims to balance growth with affordability by integrating diverse housing options within these compact areas [55]. This approach enhances social equity by providing accessible living conditions for various socio-economic groups, thereby reinforcing community resilience [56]. By employing techniques such as the node–place model, transportation hubs (nodes) are integrated with community functions (places), emphasizing their dual role as both transportation centers and vibrant spaces for social interaction and community activities [57]. This integration further reflects the holistic approach inherent in Japanese urban planning. Japan's experience with compact city policies significantly shapes Transit-Oriented Development by ensuring that urban environments are designed not only for efficiency but also for livability and accessibility. This multifaceted approach enhances the functionality of public transport systems while promoting environmental sustainability and community engagement, serving as a valuable blueprint for other urban centers grappling with similar challenges.

As shown in Table 4, while the EU relies on legal compliance to mitigate Natura 2000 sites and TEN-T networks conflicts, China's 'ecological veto' empowers administrative agencies to directly block projects, offering a more proactive approach to spatial governance.

Table 4. Comparative analysis of spatial governance tools.

Dimension	China	EU	Japan
Core Tool	EPR	Natura 2000	TOD zoning
Governance Mode	Ecological veto	Legal compliance	Market incentives
Key Challenge	Local–central conflicts	Urban sprawl	Housing affordability

China's 'ecological veto' mechanism offers a scalable model for Global South countries facing rapid urbanization. For instance, Southeast Asian nations could adopt similar institutional constraints to balance Mekong River infrastructure projects with biodiversity conservation.

6. Conclusions and Policy Pathways

6.1. Key Findings

This study demonstrates that China's 'three zones and three lines' territorial spatial planning system, when integrated with a hybrid governance model of top-down ecological constraints and bottom-up technological innovation, can significantly mitigate conflicts between transportation development and ecological conservation. Key empirical results include:

1. Conflict reduction: the implementation of ecological sensitivity veto mechanisms reduced spatial conflicts effectively (e.g., reduced by 18% in the Yangtze River Economic Belt, with rerouting of 32% of railway projects in high-risk zones).
2. Spatial trade-offs: a strong negative correlation (R^2 = 0.75) exists between ecological protection redline (EPR) coverage and transportation network density, necessitating differentiated governance strategies (e.g., stricter veto enforcement in Yunnan vs. connectivity optimization in Shandong).

6.2. Theoretical Contributions

By integrating Douglass North's theory of institutional change with Elinor Ostrom's theory of polycentric governance, this study proposes a three-dimensional analytical framework (Figure 6). This framework introduces ecological veto power as a binding mechanism to disrupt path-dependent transportation planning practices. Unlike North's emphasis on incremental institutional evolution, the 'constraint-coupling-innovation' framework (Figure 7) illustrates how rigid ecological thresholds, such as EPR compliance mandates, can override short-term economic priorities. Additionally, digital tools, including AI conflict detection and GIS simulations, facilitate adaptive governance. This approach addresses the limitations of Ostrom's polycentric theory in hierarchical contexts, offering a novel perspective—spatial institutionalism—for analyzing land use conflicts amid centralized–decentralized tensions.

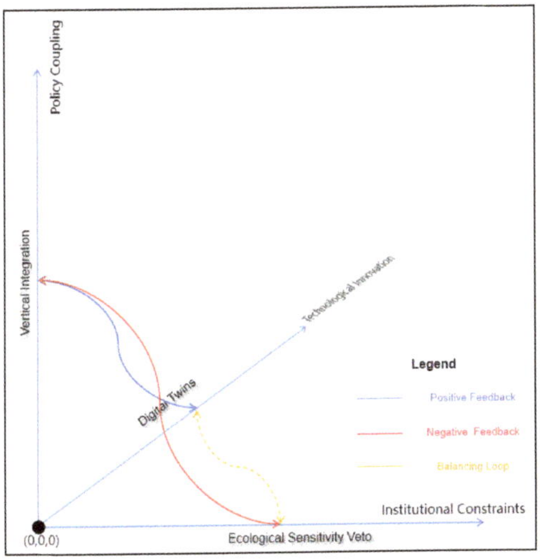

Figure 6. Constraint-coupling-innovation framework: a system dynamics model.

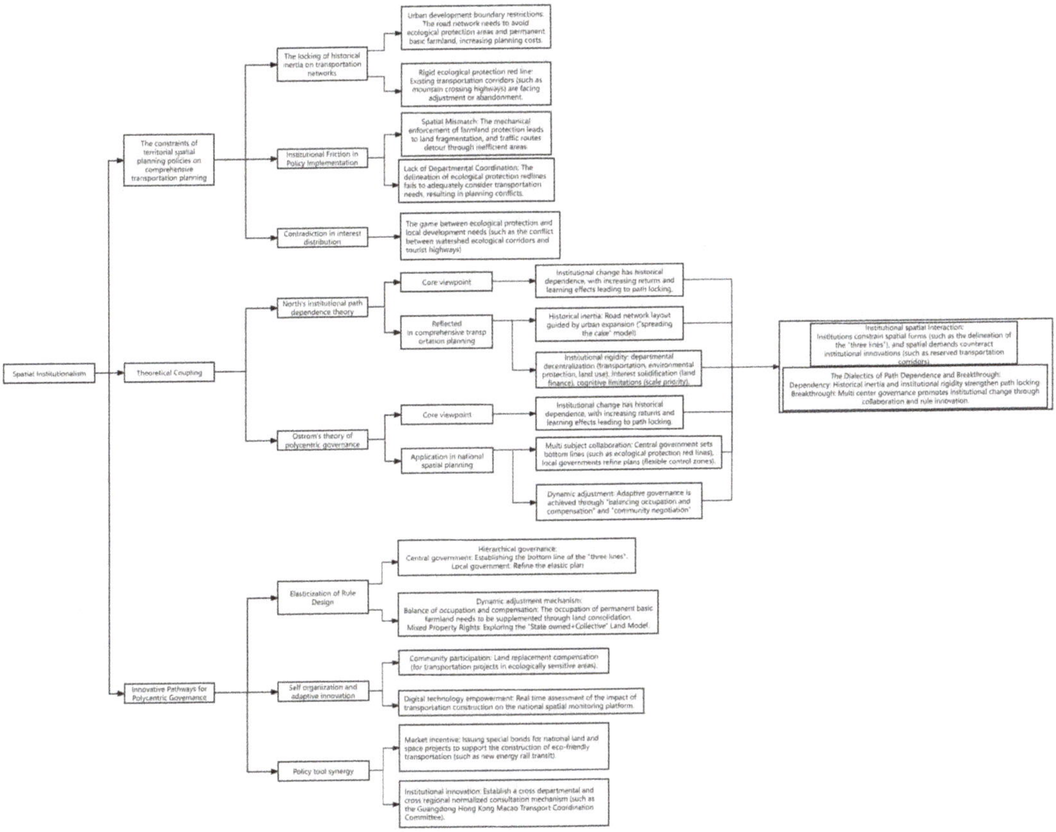

Figure 7. 'Spatial institutionalism' framework.

The 'spatial institutionalism' framework encompasses three pivotal dimensions: (1) Institutional constraints: the ecological sensitivity veto acts as a binding mechanism to mitigate development pressures. (2) Policy coupling: this entails the vertical integration of national objectives, such as the dual carbon initiative, with local experiments, as exemplified by TOER pilots. (3) Technological innovation: the framework utilizes AI-driven conflict detection and digital twin simulations to enhance planning outcomes. By integrating institutional, spatial, and technological dynamics, this framework surpasses the limitations inherent in traditional sectoral analyses.

6.3. Policy Recommendations

First, it is essential to establish the National Transportation Spatial Governance Index (NTSGI) as a valuable framework for evaluating and enhancing the governance and efficacy of the national transportation system. The NTSGI should integrate three sub-indices: (1) Ecological Compliance Rate (ECR), (2) Land Use Efficiency (LUE), (3) Multi-Stakeholder Satisfaction (MSS). Pilot testing in Yunnan Province could prioritize ECR to prevent railway encroachment into protected valleys. Furthermore, the NTSGI encourages interagency collaboration and public participation in the comprehensive transportation planning process, ensuring that the perspectives of all stakeholders are integrated into governance practices.

Second, revising the 'Land and Space Planning Law' to incorporate specific provisions regarding the resilience of integrated transportation networks is imperative. All provincial transport plans should undergo mandatory Environmental Performance Review (EPR) compliance audits, with results made publicly accessible via the Ministry's Geographic Information System (GIS) platform.

Third, establishing market-based trading of development rights (TDR) in ecologically sensitive corridors (e.g., Yangtze River Basin) is crucial for implementing cross-border ecological compensation.

6.4. Future Research

First, research on climate adaptation strategies for key transportation corridors is essential for mitigating the risks that climate change poses to infrastructure. The impacts of climate change, particularly extreme weather events, can significantly affect roads and transportation systems, resulting in increased vulnerability and operational disruptions. Additionally, the rising frequency and intensity of precipitation and storms will influence the design and planning of infrastructure.

Second, it is crucial to enhance the research on the behavioral analysis of stakeholders involved in planning conflicts. This is vital for improving governance and decision-making frameworks, as such analyses can facilitate better conflict resolution and support participatory governance within the planning process. Stakeholders in this context may include local residents, government agencies, non-governmental organizations (NGOs), and private developers. The focus should be on understanding how different stakeholder groups interact, relate, and respond in various planning scenarios and conflicts. This approach not only lends greater legitimacy to proposed solutions but also establishes a foundation for more sustainable and inclusive planning practices.

Author Contributions: Conceptualization, H.Y. and Y.J.; methodology, H.Y. and W.H.; writing—original draft preparation, H.Y. and Y.J.; writing—review and editing, H.Y. and Y.J.; figures, D.Y.; funding acquisition, H.Y. All authors have read and agreed to the published version of the manuscript.

Funding: This research was supported by Fundamental Research Funds for Central-Level Public Welfare Scientific Research Institutes (No. 20256117).

Institutional Review Board Statement: Not applicable.

Informed Consent Statement: Not applicable.

Data Availability Statement: The original contributions presented in this study are included in the article. Further inquiries can be directed to the corresponding author.

Conflicts of Interest: Author Wei Huang was employed by the company CCCC Water Transport Planning and Design Institute Co., Ltd. The remaining authors declare that the research was conducted in the absence of any commercial or financial relationships that could be construed as a potential conflict of interest.

References

1. Oleszczuk, R.; Bajkowski, S.; Urbański, J.; Pawluśkiewicz, B.; Małuszyński, M.J.; Małuszyńska, I.; Jadczyszyn, J.; Hewelke, E. The Impacts of Beaver Dams on Groundwater Regime and Habitat 6510. *Land* **2024**, *13*, 1902. [CrossRef]
2. Hermoso, V.; Morán-Ordóñez, A.; Canessa, S.; Brotons, L. Realising the potential of Natura 2000 to achieve EU conservation goals as 2020 approaches. *Sci. Rep.* **2019**, *9*, 16087. [CrossRef] [PubMed]
3. Concepción, E.D. Urban sprawl into Natura 2000 network over Europe. *Conserv. Biol. J. Soc. Conserv. Biol.* **2021**, *35*, 1063–1072. [CrossRef] [PubMed]
4. Zong, L.; Yang, F.; Pei, X. Implementing Climate Change Adaptation in Territory Spatial Planning Systems: Challenges and Approaches Based on Practices in Guiyang. *Int. J. Environ. Res. Public Health* **2022**, *20*, 490. [CrossRef]
5. Feng, P.; Lu, T.; Yu, Y.; Gao, R.; Ni, T. Effects measurement and spatial differentiation in synergy between pollution and carbon reduction in national urban agglomerations. *PLoS ONE* **2023**, *18*, e0289801. [CrossRef]
6. Xu, Y.; Cui, M. Research on the impact of regional integration policy on carbon emissions—Taking the Yangtze River Economic Belt as an example. *PLoS ONE* **2024**, *19*, e0301968. [CrossRef]
7. Liu, X.M.; Zhang, Y.Q. Digital transformation, green innovation, and carbon emission reduction performance of energy-intensive enterprises. *Sci. Rep.* **2024**, *14*, 3905. [CrossRef]
8. Meng, X.; Shaikh, G.M. Evaluating Environmental, Social, and Governance Criteria and Green Finance Investment Strategies Using Fuzzy AHP and Fuzzy WASPAS. *Sustainability* **2023**, *15*, 6786. [CrossRef]
9. Zeng, S.; Yu, Y.; Li, W. Prediction and evaluation of the energy structure under the green finance development in Chongqing municipality, China. *Heliyon* **2023**, *9*, e22481. [CrossRef]
10. Ostrom, E. A General Framework for Analyzing Sustainability of Social-Ecological Systems. *Science* **2009**, *325*, 419–422. [CrossRef]
11. Keibach, E.; Shayesteh, H. The Implementation of Digital Tools for Climate Adaptation Planning. In Proceedings of the 5th International Conference on Building Energy and Environment, Montréal, QC, Canada, 25–29 July 2022; pp. 2773–2783. [CrossRef]
12. Min, J.; Liu, Z. Science, technology and innovation driving urban land use and urban planning development: Policy and practice. *Appl. Math. Nonlinear Sci.* **2024**, *9*. [CrossRef]
13. Hersperger, A.M.; Thurnheer-Wittenwiler, C.; Tobias, S.; Folvig, S.; Fertner, C. Digitalization in land-use planning: Effects of digital plan data on efficiency, transparency and innovation. *Eur. Plan. Stud.* **2021**, *30*, 2537–2553. [CrossRef]
14. Wei, H.; Zuo, T.; Líu, H.; Yang, Y.-T. Integrating Land Use and Socioeconomic Factors into Scenario-Based Travel Demand and Carbon Emission Impact Study. *Urban Rail Transit* **2017**, *3*, 3–14. [CrossRef] [PubMed]
15. Nault, E.; Jusselme, T.; Aguacil, S.; Andersen, M. Strategic environmental urban planning-A contextual approach for defining performance goals and informing decision-making. *Build. Environ.* **2020**, *168*, 106448. [CrossRef]
16. Zhao, G.; Zheng, X.; Yuan, Z.; Zhang, L. Spatial and Temporal Characteristics of Road Networks and Urban Expansion. *Land* **2017**, *6*, 30. [CrossRef]
17. Griffin, G.P.; Jiao, J. The Geography and Equity of Crowdsourced Public Participation for Active Transportation Planning. *Transp. Res. Rec.* **2019**, *2673*, 460–468. [CrossRef]
18. Lee, R.J.; Sener, I.N. Transportation Planning and Quality of Life: Where Do They Intersect? *Transp. Policy* **2016**, *48*, 146–155. [CrossRef]
19. Logan, J.R. Making a Place for Space: Spatial Thinking in Social Science. *Annu. Rev. Sociol.* **2012**, *38*, 507–524. [CrossRef]
20. Bozorgmehr, K. Rethinking the 'global' in global health: A dialectic approach. *Glob. Health* **2010**, *6*, 19. [CrossRef]
21. Orford, S.; Webb, B. Mapping the interview transcript: Identifying spatial policy areas from daily working practices. *Area* **2018**, *50*, 529–541. [CrossRef]
22. Franklin, R.S.; Delmelle, E.C.; Andris, C.; Cheng, T.; Dodge, S.; Franklin, J.; Heppenstall, A.; Kwan, M.P.; Li, W.; McLafferty, S.; et al. Making Space in Geographical Analysis. *Geogr. Anal.* **2023**, *55*, 325–341. [CrossRef] [PubMed]
23. Tosun, J.; Saad, E.L.; Glückler, J.; Irigoyen Rios, A.; Lehmann, R. Country-Specific Participation Patterns in Transnational Governance Initiatives on Sustainability: Preliminary Insights and Research Agenda. *Glob. Chall.* **2023**, *7*, 2300012. [CrossRef] [PubMed]
24. Das, A.; Roy, S. The new institutional economic approach to land development game: Integrating transaction cost and efficient institutional framework in synergic urban land use-transport development model. *Heliyon* **2023**, *9*, e12793. [CrossRef] [PubMed]

25. Caparros-Midwood, D.; Barr, S.; Dawson, R. Spatial Optimization of Future Urban Development with Regards to Climate Risk and Sustainability Objectives. *Risk Anal. Off. Publ. Soc. Risk Anal.* **2017**, *37*, 2164–2181. [CrossRef]
26. Wang, Z.; Huang, L.; Xu, M.; Wang, S. Bridging the science-practice gaps in nature-based solutions: A riverfront planning in China. *Ambio* **2021**, *50*, 1532–1550. [CrossRef]
27. Hou, Q.; Li, Q.; Yang, Y.; Zhou, J.; Du, Y.; Zhang, Y. Evaluation and optimization of ecological spatial resilience of Yanhe River Basin based on complex network theory. *Sci. Rep.* **2024**, *14*, 1361. [CrossRef]
28. Xu, J.; Xu, D.; Qu, C. Construction of Ecological Security Pattern and Identification of Ecological Restoration Zones in the City of Changchun, China. *Int. J. Environ. Res. Public Health* **2022**, *20*, 289. [CrossRef]
29. Guo, J.; Yue, D.; Li, K.; Hui, C. Biocapacity optimization in regional planning. *Sci. Rep.* **2017**, *7*, 41150. [CrossRef]
30. Ou, M.; Li, J.; Fan, X.; Gong, J. Compound Optimization of Territorial Spatial Structure and Layout at the City Scale from "Production-Living-Ecological" Perspectives. *Int. J. Environ. Res. Public Health* **2022**, *20*, 495. [CrossRef]
31. Qu, Y.; Lu, M. Identifying conservation priorities and management strategies based on ecosystem services to improve urban sustainability in Harbin, China. *PeerJ* **2018**, *6*, e4597. [CrossRef]
32. Gerlak, A.K.; Saguier, M.; Mills-Novoa, M.; Fearnside, P.M.; Albrecht, T.R. Dams, Chinese investments, and EIAs: A race to the bottom in South America? *Ambio* **2020**, *49*, 156–164. [CrossRef] [PubMed]
33. Tampekis, S.; Sakellariou, S.; Samara, F.; Sfougaris, A.; Jaeger, D.; Christopoulou, O. Mapping the optimal forest road network based on the multicriteria evaluation technique: The case study of Mediterranean Island of Thassos in Greece. *Environ. Monit. Assess.* **2015**, *187*, 687. [CrossRef] [PubMed]
34. Li, Z.; Zhu, J. Assessment and spatial partitioning of ecosystem services importance in Giant Panda National Park: To provide targeted ecological protection. *PLoS ONE* **2022**, *17*, e0278877. [CrossRef]
35. Kim, E.S.; Lee, D.K.; Choi, J. Evaluating the effectiveness of mitigation measures in environmental impact assessments: A comprehensive review of development projects in Korea. *Heliyon* **2024**, *10*, e31647. [CrossRef] [PubMed]
36. Han, H.; Li, H.; Zhang, K. Spatial-Temporal Coupling Analysis of the Coordination between Urbanization and Water Ecosystem in the Yangtze River Economic Belt. *Int. J. Environ. Res. Public Health* **2019**, *16*, 3757. [CrossRef]
37. Whitehorn, P.R.; Navarro, L.M.; Schröter, M.; Fernandez, M.; Rotllan-Puig, X.; Marques, A. Mainstreaming biodiversity: A review of national strategies. *Biol. Conserv.* **2019**, *235*, 157–163. [CrossRef]
38. Lu, J.; Wang, K.; Liu, H. Land Development Rights, Spatial Injustice, and the Economic Development in Net-Incremental Reduction Regions of Construction Land: Evidence from Shanghai, China. *Int. J. Environ. Res. Public Health* **2023**, *20*, 2560. [CrossRef]
39. Wang, A.Q.; Chan, E.H.W.; Yeung, S.C.W.; Han, J.B. Urban Fringe Land Use Transitions in Hong Kong: From New Towns to New Development Areas. *Procedia Eng.* **2017**, *198*, 707–719. [CrossRef]
40. Seto, K.; Galland, G.R.; McDonald, A.; Abolhassani, A.; Azmi, K.; Sinan, H.; Timmiss, T.; Bailey, M.; Hanich, Q. Resource allocation in transboundary tuna fisheries: A global analysis. *Ambio* **2021**, *50*, 242–259. [CrossRef]
41. Ye, Y.; Qin, Y.; Yu, R.; Wu, Q. Optimization of Chinese land spatial pattern in the transformation process of resource-based cities: A case study in Tongling City, China. *Sci. Rep.* **2024**, *14*, 6092. [CrossRef]
42. de Souza, A.M.; Maia, G.; Braun, T.; Villas, L.A. An Interest-Based Approach for Reducing Network Contentions in Vehicular Transportation Systems. *Sensors* **2019**, *19*, 2325. [CrossRef] [PubMed]
43. Azami, P.; Jan, T.; Iranmanesh, S.; Ameri Sianaki, O.; Hajiebrahimi, S. Determining the Optimal Restricted Driving Zone Using Genetic Algorithm in a Smart City. *Sensors* **2020**, *20*, 2276. [CrossRef] [PubMed]
44. Ye, X.; Du, J.; Han, Y.; Newman, G.; Retchless, D.; Zou, L.; Cai, Z. Developing Human-Centered Urban Digital Twins for Community Infrastructure Resilience: A Research Agenda. *J. Plan. Lit.* **2022**, *38*, 187–199. [CrossRef] [PubMed]
45. Hasanujjaman, M.; Chowdhury, M.Z.; Jang, Y.M. Sensor Fusion in Autonomous Vehicle with Traffic Surveillance Camera System: Detection, Localization, and AI Networking. *Sensors* **2023**, *23*, 3335. [CrossRef]
46. Kumar, A.; Kannan, K.; Dahiya, M.; Kushwah, V.S.; Siddiqa, A.; Kaur, K.; Rahin, S.A. Advances in Computational Intelligence Techniques-Based Multi-Intersection Querying Theory for Efficient QoS in the Next Generation Internet of Things. *Comput. Intell. Neurosci.* **2023**, *2023*, 1388425. [CrossRef]
47. Fenu, G.; Pau, P.L.; Dessì, D. Functional models and extending strategies for ecological networks. *Appl. Netw. Sci.* **2017**, *2*, 10. [CrossRef]
48. Friedrichs, M.; Hermoso, V.; Bremerich, V.; Langhans, S.D. Evaluation of habitat protection under the European Natura 2000 conservation network—The example for Germany. *PLoS ONE* **2018**, *13*, e0208264. [CrossRef]
49. Nita, A.; Rozylowicz, L.; Manolache, S.; Ciocănea, C.M.; Miu, I.V.; Popescu, V.D. Collaboration Networks in Applied Conservation Projects across Europe. *PLoS ONE* **2016**, *11*, e0164503. [CrossRef]
50. Maes, J.; Barbosa, A.; Baranzelli, C.; Zulian, G.; Batista, E.; Silva, F.; Vandecasteele, I.; Hiederer, R.; Liquete, C.; Paracchini, M.L.; et al. More green infrastructure is required to maintain ecosystem services under current trends in land-use change in Europe. *Landsc. Ecol.* **2015**, *30*, 517–534. [CrossRef]

51. Thrun, E.; Leider, J.; Chriqui, J.F. Exploring the Cross-Sectional Association between Transit-Oriented Development Zoning and Active Travel and Transit Usage in the United States, 2010–2014. *Front. Public Health* **2016**, *4*, 113. [CrossRef]
52. Lu, Y.; Gou, Z.; Xiao, Y.; Sarkar, C.; Zacharias, J. Do Transit-Oriented Developments (TODs) and Established Urban Neighborhoods Have Similar Walking Levels in Hong Kong? *Int. J. Environ. Res. Public Health* **2018**, *15*, 555. [CrossRef] [PubMed]
53. Coxa, L.; Bassi, A.; Kolling, J.; Procter, A.; Flanders, N.; Tanners, N.; Araujo, R. Exploring synergies between transit investment and dense redevelopment: A scenario analysis in a rapidly urbanizing landscape. *Landsc. Urban Plan.* **2017**, *167*, 429–440. [CrossRef] [PubMed]
54. Song, C.; Liu, Q.; Song, J.; Yang, D.; Jiang, Z.; Ma, W.; Niu, F.; Song, J. The Interactive Relationship between Street Centrality and Land Use Intensity-A Case Study of Jinan, China. *Int. J. Environ. Res. Public Health* **2023**, *20*, 5127. [CrossRef] [PubMed]
55. Lu, Y.; Sun, G.; Sarkar, C.; Gou, Z.; Xiao, Y. Commuting Mode Choice in a High-Density City: Do Land-Use Density and Diversity Matter in Hong Kong? *Int. J. Environ. Res. Public Health* **2018**, *15*, 920. [CrossRef]
56. Yan, R.; Yang, B. Interactive effects of the built environment and rail transit on commuting behavior: Evidence from Hefei, China. *Heliyon* **2023**, *9*, e21788. [CrossRef]
57. Hino, K.; Taniguchi, A.; Hanazato, M.; Takagi, D. Modal Shift from Cars and Promotion of Walking by Providing Pedometers in Yokohama City, Japan. *Int. J. Environ. Res. Public Health* **2019**, *16*, 2144. [CrossRef]

land

MDPI

Article

Does New Infrastructure Promote the Development of Rural Industries? A Nonlinear Analysis Based on Provincial Panel Data from China

Lulin Liu [1], Xiaojie Ma [2,*] and Yaolong Li [2]

[1] School of Management, Minzu University of China, Beijing 100081, China; liululin@muc.edu.cn
[2] School of Economics, Minzu University of China, Beijing 100081, China; liyaolong@muc.edu.cn
* Correspondence: maxiaojie@muc.edu.cn

Abstract: Whether and how new infrastructure (NI) can drive the development of rural industries (DRI) is crucial to promoting urban–rural balance and ensuring national food security. Based on panel data from Chinese provincial-level regions (2013–2022), this study constructs a comprehensive DRI evaluation system encompassing production systems, economic benefits, industrial integration, and sustainable development. Using the entropy method to measure NI development levels, we examine its heterogeneous impacts on DRI through multiple analytical dimensions. The results reveal three key findings: First, a robust "inverted U-shaped" relationship exists between NI and DRI. Second, heterogeneity analysis demonstrates that (1) both eastern and western regions show this pattern while central and northeastern regions exhibit infrastructure saturation; (2) intelligent transportation infrastructure critically moderates this relationship—advanced regions achieve greater NI efficiency through digital infrastructure–transportation–industrial synergies, whereas underdeveloped regions face bottlenecks; and (3) provinces with lower population density but higher human capital show enhanced NI absorption capacity. Third, mechanism analysis confirms industrial structure upgrading and market expansion as key transmission channels. These findings suggest that implementing tiered intelligent transportation strategies, differentiated NI policies, human capital investment, and rural market expansion optimize NI's rural development impacts.

Keywords: new infrastructure; development of rural industries; inverted U-shaped; intelligent transportation; impact mechanism

1. Introduction

Within the overarching framework of global economic integration and digital transformation, digital technologies represented by artificial intelligence (AI) have fundamentally transformed the spatial production organization model based on physical space, while the virtual–physical space integration process driven by digital models has reshaped the logical system of rural spatial governance [1]. Against this backdrop, the long-standing impediments to urban–rural factor flow and the urban–rural value divide that have hindered rural development and integrated urban–rural development will gradually be mitigated, with digital rural spatial governance emerging as a critical breakthrough point for advancing the modernization of rural spatial governance and rural industrial development. On one hand, comprehensive urban infrastructure planning, improved accessibility and transportation potential of rural public service facilities [2], the application of Information and Communication Technologies (ICT), and the digital economy [2,3] collectively enhance rural resource efficiency and facilitate urban–rural factor mobility, while rapid transportation infrastructure development and significant progress in information technology promote the rescaling of urban–rural spaces and the flattening of spatial governance [4]. On the other hand, excessive transportation infrastructure and mobility may impose certain pressures on rural activities [5], which compounded by factors such as rural geographical remoteness, sparse population, and technological deficiencies [6], result in inherent weaknesses including overall low competitiveness, limited industrial chain extension, imperfect value chains, and insufficient breadth and depth of agricultural and rural service sector development [7]. These challenges in turn constrain the advancement of information technologies in rural areas, creating a "digital divide" [2] that further exacerbates urban–rural disparities. Consequently, leveraging spatial optimization and industrial upgrading based on information technologies has become a pivotal approach to addressing the issues of inadequate and uneven rural development.

Distinct from traditional infrastructure, new infrastructure (NI) is characterized by technology-driven innovation, information-network foundations, and features of novelty, interconnectivity, and sharing. Its unique capability to transcend physical boundaries may fundamentally transform conventional regional and rural development paradigms [8–10]. Stratan (2024) provides empirical evidence from Moldova showing that both transport and telecommunications infrastructure are critical enablers of business resilience, trade expansion, and regional inclusion [11]. However, Maciulyte and Butkus (2022), through their analysis of data from 28 EU countries, found that not all infrastructure types contribute significantly to economic growth, with only transport, information and communication technologies (ICT), and energy-related infrastructure demonstrating statistically positive impacts [12]. Notably, converged infrastructure (e.g., transportation and utilities) as physical networks and information infrastructure (e.g., the Internet) as relational networks collectively connect regions into an integrated system [13,14], exerting positive influences on regional productivity [15], interregional trade [16], and resource allocation in response to extreme weather events [17]. However, as traditional infrastructure such as roads and railways approaches saturation, diminishing marginal returns on investment have become evident [18,19]. Against this backdrop, a pivotal question emerges: Can NI continue to serve as a catalyst for rural industrial development while overcoming the limitations faced by traditional transportation infrastructure?

Within the context of rural industrial development, there remains ongoing academic debate regarding whether NI continues to drive growth and innovation or, similar to traditional infrastructure, eventually encounters diminishing returns [18,20]? On one hand, NI has the potential to overcome the constraints of traditional infrastructure by enhancing the efficiency and quality of economic operations [21]. Investments in NI such as high-speed rail and postal systems introduce novel products and markets to rural areas, stimulating innovation among agricultural household enterprises and enhancing economic value in rural regions [22]. The development of information infrastructure effectively mitigates rural income inequality [23,24], while enabling farmers to bridge both access and utilization divides [25]. On the other hand, as NI becomes more widespread and mature, the initial burst of growth effects may subside, leading to questions about the sustainability of its impact on rural industries [26,27]. The siting and operation of converged infrastructure projects such as high-speed rail have been shown to exacerbate intra-regional inequalities [28]. Furthermore, NI development alone cannot overcome agricultural specialization constraints imposed by legacy infrastructure systems [29], nor can it serve as an effective short-term stimulus for economic growth [30].

Building upon these considerations, this paper posits that the relationship between NI and the development of rural industries (DRI) may not be a straightforward linear one.

Instead, it may exhibit a nonlinear pattern, where the benefits of NI investment in rural areas could follow an inverted U-shaped curve. The theoretical framework and hypotheses of this study are outlined in the third section, which seeks to unravel this puzzle. The subsequent chapters are arranged as follows: The second section provides a literature review. The third section outlines the theoretical framework and hypotheses of this study. The fourth section provides the model construction and variable explanations, as well as detailing the data sources and providing descriptive statistics. The fifth section presents the empirical analysis, including model regression results, robustness tests, heterogeneity analysis, and mediation effects. The sixth section provides a discussion. The seventh section concludes the study and offers policy implications. By addressing this theoretical conundrum, this paper aims to contribute to the understanding of how NI can be leveraged to promote sustainable DRI, providing insights for policy formulation and rural development strategies both in China and globally.

2. Literature Review

The literature pertinent to this study can be categorized into two main branches: research on the measurement of DRI indices and on the impact effects of NI on rural development.

Regarding the first branch, there is a dearth of research on measuring DRI indices. In 2017, China introduced the Rural Revitalization Strategy, which marked the first formal introduction of the concept of rural industry. The existing literature predominantly features qualitative studies on DRI, with a notable lack of quantitative research. An index for the revitalization of rural industries was developed focusing on four key dimensions: the degree of integration within rural industries, overall production capacity, the progress of specialized industries, and the establishment of interest linkage mechanisms [31,32]. Subsequent research evaluated the influence of this index on the income of those residing in rural areas [32]. It was discovered that the integration of rural industries notably boosts the income of farmers and helps to narrow the disparity between urban and rural regions [33]. Further analysis was conducted to understand how the rural digital economy propels the rejuvenation of rural industries, creating a measurement tool that assesses the level of rural industrial revitalization through three dimensions: the rise in agricultural production, the expansion of agricultural value, and the improvement of farmers' earnings [34]. DRI indices commonly employ the comprehensive index approach, choosing indicators from areas such as rural production, industrial integration, industrial roles, and farmers' incomes. Due to the benefits of objectivity, methodological rigor, and precise weighting, the entropy method or entropy weight method is frequently adopted in specific studies for quantification [35,36].

Regarding the second branch of the literature, existing research posits that NI plays a significant role in promoting agricultural development, increasing farmers' incomes, and enhancing the efficiency of rural governance [37–39]. The primary research conclusions are divided into two types. The first is that NI has a significant positive impact on rural development [30]. NI, incorporating technologies and models like the Internet of Things, big data, and artificial intelligence, has propelled the upgrading and innovation of rural industries [40]. It has facilitated the digital transformation from traditional agriculture to handicrafts [41], enhancing production efficiency [42]. Additionally, NI has expanded the channels and markets for commodity sales, enabling rural industries to enter the global market through e-commerce platforms, attracting more consumers and investors. Furthermore, through services like online education and remote training, NI has improved the knowledge and skills of rural residents, fostering the integration of small-scale farming economies with modern agriculture. This has had a positive impact on food security and the increase in farmers' incomes. The second perspective is the opposite, suggesting that

NI may also have negative impacts on rural industries. Obstacles such as underdeveloped information infrastructure, a shortage of professional digital talent [43], and significant regional development disparities have led to an overall lag in the digital and informational development of rural areas. These factors, however, constrain the penetration and diffusion of the digital economy in rural regions [44], making it difficult for the inclusive effects of NI to benefit farmers widely [28], which could potentially widen the urban–rural digital divide further [45]. Moreover, it is necessary and urgent to investigate this. We need to determine whether NI, like traditional infrastructure, faces the risk of saturation and diminishing marginal returns on investment.

While existing research provides a solid foundation for understanding the impact of NI on DRI, it is not without its limitations. Firstly, the literature has yet to clearly delineate whether the impact of NI on DRI is positive or negative. Second, when studying the influence of NI, existing studies often use the implementation of a specific policy as a proxy variable without fully considering the actual levels and differences in NI across various regions. Third, the mechanisms of action require further validation. In response, this paper constructs an evaluation system for NI and DRI based on provincial panel data and measures both constructs. Utilizing methods like double fixed effects and mediation analysis, the study examines the impact of NI on DRI from the perspectives of heterogeneity and mechanisms of influence. Additionally, the paper employs instrumental variable methods to address endogeneity issues.

The marginal contributions of our findings are as follows: Firstly, this research pioneers in systematically investigating the "inverted U-shaped" relationship between NI and DRI, filling a gap in the existing literature. The empirical validation of this nonlinear relationship offers a novel theoretical lens through which the complex dynamics of NI's impact on DRI can be understood. Secondly, this study identifies and confirms that the upgrading of industrial structures and the expansion of transaction markets serve as pivotal mechanisms mediating the influence of NI on DRI. These findings provide actionable insights for policymakers, underpinning more informed and strategic planning for NI initiatives. Thirdly, this research delves into the heterogeneity of NI's impact on DRI across various regions, considering factors such as geographical location, population density, and human capital. This nuanced analysis enhances the precision and applicability of the study's recommendations for tailored policy interventions.

3. Theoretical Analysis and Hypothesis

3.1. The Impact of NI on DRI

The promotion and application of NI have significantly influenced DRI. The services of NI are continuously expanding into rural areas, with its technology and concepts permeating all aspects of agricultural production [46], rural economic development, and the lives of farmers. This has facilitated the optimization and upgrading of the agricultural industrial structure. It has also enhanced the level of rural public services and rural governance. The construction of NI, such as the internet, has bridged the digital divide between urban and rural areas, reduced the costs associated with information and management for farmers in their production and business activities, innovated agricultural production methods, and markedly enhanced agricultural production efficiency. The widespread availability of these infrastructures has introduced new ideas, knowledge, and technologies to rural areas. Through online platforms, it has promoted the flow of high-quality educational and learning resources to rural communities, which has played a positive role in realizing scientific agricultural production, improving rural human capital, inspiring entrepreneurial spirit among farmers [47], and increasing awareness of ecological conservation. NI, which relies on modern technologies like the internet and cloud computing, helps farmers quickly understand changes in market demand, adjusting production plans and product structures, alleviating the issue of unsold agricultural products, optimizing the allocation of agricultural resources, and achieving multi-level matching between supply and demand. However, research indicates that there is not a simple linear relationship between NI and regional development, a finding verified in the contexts of economic growth, total factor productivity [48], and the efficiency of financial resource allocation [49]. Once regional investment in NI reaches saturation, any further investment may result in resource wastage and diminished operational efficiency, thereby deviating from the optimal level of innovation performance. The digital transformation of agricultural production has not yet eliminated the dependence on traditional production factors. The relative scarcity of these factors determines that the level of agricultural development will not increase without bounds with the improvement of digitalization levels. When the development of NI meets the needs of agricultural growth, it may still face challenges. This is especially true when other supporting systems like policy support, technological research and development, and innovation transformation lag behind. It may lead to information redundancy, intensify intra-industry competition, increase the difficulty of service governance, affect the rational use of digital resources by farmers, and thereby constrain the further improvement of DRI. Thereby, this study posits a tri-phased theoretical framework to explain the "inverted U-shaped" relationship between NI and the DRI. The blue line in Figure 1 illustrates the hypothesized "inverted U-shaped" relationship between NI and DRI. In the initial phase of NI development, the infusion of new technologies and infrastructure, such as the internet and cloud computing, into rural areas will markedly boost DRI by enhancing agricultural production efficiency, lowering information and management costs, and improving access to educational and learning resources. As NI development progresses to the optimal phase, NI will reach a level where its benefits are maximized, and DRI will peak due to the efficient allocation of resources, the alignment of supply and demand across multiple levels, and the adoption of innovative production methods. However, in the saturation and diminishing returns phase, additional NI investment may yield diminishing returns, with the marginal benefits of NI declining because of issues such as resource wastage, operational inefficiencies, information redundancy, heightened competition, and governance challenges, ultimately constraining further advancements in DRI.

Hypothesis 1: *There exists an "inverted U-shaped" relationship between NI and DRI.*

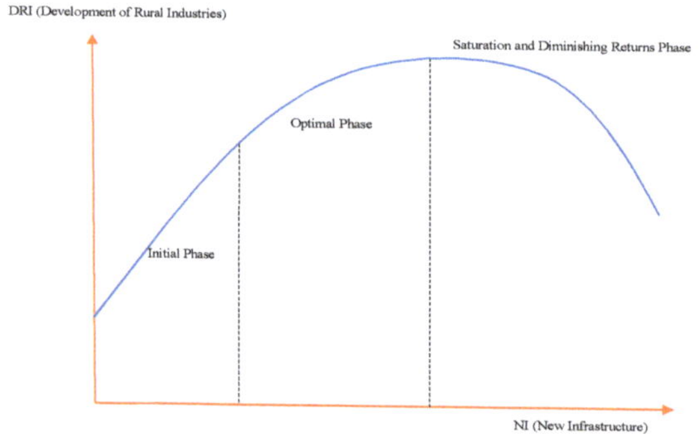

Figure 1. "Inverted U-shaped" relationship between NI and DRI.

The impact of NI on DRI exhibits heterogeneous effects contingent upon regional characteristics including economic base, intelligent transportation infrastructure, population density, and human capital [50]. In areas with higher levels of economic development, the initial investment in NI can quickly integrate with the existing industrial base, generating positive synergistic effects that drive the rapid DRI increase [51]. However, as the infrastructure continues to improve, the marginal benefits gradually diminish [52]. Excessive investment may lead to resource wastage and reduced efficiency, thereby forming the latter half of the "inverted U-shaped" relationship [53]. In contrast, in regions with lower levels of economic development, where infrastructure is underdeveloped due to a late start, the initial investment in NI may not yield the anticipated effects. However, as infrastructure gradually improves, its positive effects will become apparent, though they may not reach the peak seen in regions with a higher economic base. Regions with high-level intelligent transportation infrastructure can enhance the efficiency of element allocation of NI. Firstly, the combination of digital infrastructure and intelligent transportation achieves real-time optimization of agricultural logistics and the expansion of the rural tourism market; secondly, the level of transportation digitization determines the marginal returns of NI, with underdeveloped transportation networks creating a "last mile" bottleneck that constrains the DRI; thirdly, high transportation connectivity expands the service radius of NI by reducing information asymmetry and transaction costs across regions. In densely populated areas, NI can better leverage economies of scale [54], reduce the cost of service provision, and promote the agglomeration and upgrading of rural industries. Nonetheless, an overly high population density can result in the excessive utilization and congestion of infrastructure, which in turn may diminish its effectiveness. In regions with low population density, this might not easily lead to economies of scale due to the lack of demand. As the population density increases moderately, the efficacy of the infrastructure is enhanced over time. In areas abundant in human capital, both farmers and technical experts are able to utilize the NI more efficiently, prompting innovation in agricultural technology and the upgrading of industries [55], thereby speeding up the progress of DRI. However, once human capital reaches a certain level, further investment in infrastructure may fail to be effectively utilized due to a shortage of sufficient high-skilled labor, leading to a decline in benefits. In areas with lower human capital, the potential benefits of NI may not be fully realized due to the insufficient skills of the workforce.

Hypothesis 2: *The NI-DRI relationship is moderated by regional characteristics (economic base, population density, human capital) and intelligent transportation infrastructure levels, with advanced transportation systems enabling higher NI efficiency through digital–logistical synergies.*

3.2. Mechanism of the Impact of NI on DRI

NI, such as the internet, big data, and the Internet of Things, provides new technical support and production methods for rural industries, promoting the optimization and upgrading of the industrial structure [56]. Firstly, by enhancing agricultural production efficiency through the adoption of intelligent and precision agricultural technologies, the aim is to improve the quality and yield of agricultural products [57]. This approach drives the transition from traditional agriculture to modern agriculture by reducing resource wastage and increasing resource utilization efficiency [58]. Secondly, NI facilitates the extension of the industrial chain, contributing to the growth of sectors such as agricultural product processing, logistics, and sales, thereby creating a comprehensive industrial chain and increasing the industry's added value [59]. Thirdly, by relying on NI, rural areas can incubate new industries such as e-commerce, rural tourism, and cultural and creative industries, achieving industrial diversification [60]. Therefore, the upgrading of the industrial structure becomes an important intrinsic mechanism by which NI promotes DRI. The construction and improvement of NI help break through geographical limitations and expand the market space for rural industries [61]. Firstly, the reduction in transaction costs: NI such as high-speed transportation networks and information communication networks can reduce the loss and expenses associated with product transportation and information transmission, thereby improving transaction efficiency. Secondly, the expansion of market scope: the NI allows rural industries to access national and even global markets more conveniently, expanding sales channels and increasing market demand. Finally, the enhancement of market competitiveness: with the NI, rural enterprises can better access market information, enhance product competitiveness, and increase market share. Therefore, the expansion of the transaction market has become an important external mechanism by which NI promotes DRI.

Hypothesis 3a: *NI drives DRI through the upgrading of industrial structures (UISs).*

Hypothesis 3b: *NI promotes DRI by expanding trade markets (ETMs).*

4. Materials and Methods

4.1. Model Specification

This paper employs a two-way fixed effects model and constructs three regression models to test the hypotheses proposed in the previous section. The specific settings are as follows:

$$DRI_{i,t} = \alpha_0 + \alpha_1 NI_{i,t} + \alpha_2 NI^2{}_{i,t} + \omega Control_{i,t} + \mu_i + \varphi_t + \varepsilon_{it} \quad (1)$$

$$Med_{i,t} = \beta_0 + \beta_1 NI_{i,t} + \beta_2 NI^2{}_{i,t} + \lambda Control_{i,t} + \mu_i + \varphi_t + \varepsilon_{it} \quad (2)$$

$$DRI_{i,t} = \gamma_0 + \gamma_1 NI_{i,t} + \gamma_2 NI^2{}_{i,t} + \gamma_3 Med_{i,t} + \pi Control_{i,t} + \mu_i + \varphi_t + \varepsilon_{it} \quad (3)$$

In the aforementioned equations, i and t represent the provincial administrative regions and years, respectively; *DRI* denotes the development of rural industries; *NI* represents new infrastructure; NI^2 denotes the squared term of new infrastructure; Med is the mechanism variable, which includes upgrading industrial structure (UIS) and expanding transaction markets (ETM); Control stands for the control variables; and μ_i, φ_t, and ε_{it} represent the province fixed effects, time fixed effects, and error term, respectively.

In Equation (1), α_1 and α_2 are significantly positive and negative, respectively, and pass the "inverted U-shaped" test, indicating that the "inverted U-shaped" relationship between *NI* and *DRI* is established, thus validating Hypothesis 1. Based on Equation (1), mediation effect models are constructed (Equations (2) and (3)). Firstly, analyze the relationship between *NI* and the mediating variable. If β_1 and β_2 meet the expected signs, proceed to the next step of the test. Secondly, after including the mediating variable, analyze the relationship between *NI* and *DRI*. If γ_3 is significant and the signs of γ_1 and γ_2 remain unchanged, it indicates that the mediating variable plays a mediating role in the process of *NI* promoting *DRI*.

4.2. Variables

4.2.1. Dependent Variable

In order to scientifically evaluate DRI, this study, based on a comprehensive review of relevant research [31,62], fully considers the importance of sustainable development and recognizes that DRI is not only reflected in economic growth but also involves multiple dimensions such as production methods, economic structure, and industrial integration. Therefore, this paper constructs an indicator system to measure the development level of rural industries in China from the following four aspects, to comprehensively and systematically evaluate the status and potential of DRI. The rural industrial production system is key to evaluating the foundation and potential of rural industries, and the perfection of the production system directly affects the industry's sustainable development capacity. The economic benefits of rural industries are the core indicators for measuring the achievements of rural industry development, reflecting the industry's contribution to economic growth and the efficiency of resource allocation. The integration of the three sectors of rural industries represents a new trend in DRI. Through the integration of agriculture with secondary and tertiary industries, it can promote the extension of the industrial chain and industrial upgrading, enhancing the comprehensive competitiveness of rural industries. The sustainable DRI considers the long-term impact of DRI on the environment, society, and economy, and serves as an important basis for evaluating whether rural industries have the capacity for sustainable development.

Based on these four dimensions, this paper selects 11 specific indicators to construct the indicator system. In terms of measurement, the entropy method is used to determine the weight coefficients of each indicator, which can objectively reflect the relative importance of the indicators. Specifically, this study employs MATLAB R2019a software to perform extreme difference standardization on the raw data, thereby eliminating the impact of different indicator dimensions, and calculates the entropy values of each indicator to ultimately ascertain the weight coefficients [63]. The application of this method ensures the scientificity and accuracy of the evaluation results. The detailed calculation results of the weight coefficients are presented in Table 1.

Table 1. Indicators and weights for the level of DRI.

Primary Indicator	Secondary Indicator	Tertiary Indicators	Indicator Type	Weight
Development of rural industries (DRI)	Production system	Cultivated land area grain unit output value	+	0.087
		Agricultural product producer price index	−	0.002
		Agricultural industry structure index	+	0.138
	Economic benefits	Total output value of agriculture, forestry, animal husbandry, and fishery	+	0.079
		Per capita output value of agriculture, forestry, animal husbandry, and fishery	+	0.164
		Disposable income of rural residents	+	0.076
	Integration of three industries	Large-scale agricultural product processing enterprises	+	0.175
		Proportion of agricultural employment in rural population	+	0.067
	Sustainable development	Number of certified green food enterprises	+	0.077
		National agricultural product geographical indications	+	0.121
		Fertilizer use intensity	−	0.015

Note: "+" indicates a positive indicator, "−" indicates a negative indicator.

Considering the differences in units and dimensions among the indicators, it is necessary to standardize the original data into dimensionless indicator evaluation values for comprehensive analysis. The data are standardized using the range standardization method, which scales the indicator values to a range between zero and one, as shown in Formulas (4)–(7). Within the indicator layer, indicators are categorized into positive and negative indicators, indicating the relationship between the selected indicators and the level of DRI. For instance, a higher per capita output value in agriculture, forestry, animal husbandry, and fishery signifies a better level of DRI; similarly, an increased number of certified green food enterprises indicates a higher level of DRI. The method is as follows:

$$\text{Positive indicators}: \; x'_{ij} = \left(x_{ij} - x_{min}\right) / \left(x_{max} - x_{min}\right) \tag{4}$$

$$\text{Negative indicators}: \; x'_{ij} = \left(x_{max} - x_{ij}\right) / \left(x_{max} - x_{min}\right) \tag{5}$$

Considering the need to avoid human interference in the data, this paper opts to use the entropy method to reduce the subjectivity imparted by external factors. The methodology is as follows:

Calculate the weight of the ith subject indicator value under the jth indicator:

$$p_{ij} = x'_{ij} / \sum_{i=1}^{m} x'_{ij} \tag{6}$$

Calculate the entropy of the jth evaluation index:

$$e_j = -1/lnm \sum\nolimits_{i=1}^{m} p_{ij} lnp_{ij}, \tag{7}$$

Calculate the weight of the evaluation index j:

$$w_j = (1 - e_j) / \sum_{j=1}^{n} (1 - e_j) \tag{8}$$

Utilizing the entropy method to measure DRI can assist government departments and policymakers in gaining a more comprehensive and objective understanding of the level of DRI. This, in turn, aids in the formulation of targeted policies and measures, promoting DRI towards a more sustainable and balanced direction.

4.2.2. Core Explanatory Variables

Based on the interpretation of the connotation of NI provided by the National Development and Reform Commission [64], this paper divides the evaluation index system of the NI level into three categories: informational infrastructure, integrated infrastructure, and innovative infrastructure [23]. Ultimately, 17 indicators were selected to construct the index system for the development level of China's NI, as shown in Table 2. The normalization processing consistent with the previous section was adopted, and the entropy method was used to determine the indicator weights. Finally, a comprehensive score was obtained using linear weighting.

4.2.3. Mediating Variables

The upgrading of industrial structure (UIS) is represented by the proxy variable "Advanced Industrial Structure Index (AIS)". The Advanced Industrial Structure (AIS) refers to the process or trend of the industrial structure transitioning from a lower level to a higher level, which is a change in the situation of economic construction and the transformation of economic development patterns. With the integration and development of information technology, intelligent manufacturing, modern agriculture, manufacturing, and services in the industrial chain, this paper selects the upgrade of industrial structure as a mediating variable. According to the characteristics of *AIS*, this study uses the index of advanced industrial structure to measure the upgrade of China's industrial structure, measuring the upgrade of the industrial structure from the proportion of the three types of industries in the total output value. The specific calculation formula is as follows:

$$AIS_{it} = \frac{Y_{1it}}{Y_{it}} \times 1 + \frac{Y_{2it}}{Y_{it}} \times 2 + \frac{Y_{3it}}{Y_{it}} \times 3 \tag{9}$$

Among them, Y_{it} represents the regional GDP of province i in year t, and $\frac{Y_{1it}}{Y_{it}}$, $\frac{Y_{2it}}{Y_{it}}$, and $\frac{Y_{3it}}{Y_{it}}$ represent the ratio of the added value of the primary, secondary, and tertiary industries to the regional GDP of province i in year t, respectively.

The expansion of the trading market (ETM) is represented by the proxy variable "Log of Express Delivery Revenue (lnER)". Express delivery revenue can serve as a basis for the development of e-commerce and the expansion of the trading market in rural areas. By analyzing the relationship between express delivery revenue and the growth of rural industrial sales, as well as the expansion of market coverage, we can examine the impact of trading market expansion on DRI.

Table 2. Indicators and weights of the NI.

<table>
<tr><th>Primary Indicator</th><th>Secondary Indicator</th><th>Tertiary Indicators</th><th>Quaternary Indicators</th><th>Indicator Type</th><th>Weight</th></tr>
<tr><td rowspan="17">New infrastructure development (NI)</td><td rowspan="7">Information infrastructure</td><td rowspan="3">Internet infrastructure</td><td>IPv4 address density</td><td>+</td><td>0.095</td></tr>
<tr><td>Internet domain name density</td><td>+</td><td>0.083</td></tr>
<tr><td>Internet broadband access port density</td><td>+</td><td>0.048</td></tr>
<tr><td rowspan="2">Communications infrastructure</td><td>Fiber optic cable line density</td><td>+</td><td>0.042</td></tr>
<tr><td>Mobile phone base station density</td><td>+</td><td>0.055</td></tr>
<tr><td rowspan="2">New technology infrastructure</td><td>Number of computers used per 100 people</td><td>+</td><td>0.051</td></tr>
<tr><td>Proportion of enterprises with e-commerce transaction activities</td><td>+</td><td>0.036</td></tr>
<tr><td rowspan="7">Converged infrastructure</td><td rowspan="2">Intelligent transportation infrastructure</td><td>Railway operating mileage density</td><td>+</td><td>0.025</td></tr>
<tr><td>Mileage density of high-speed highways</td><td>+</td><td>0.022</td></tr>
<tr><td>Smart weather infrastructure</td><td>Automatic weather station density</td><td>+</td><td>0.035</td></tr>
<tr><td rowspan="2">Smart home infrastructure</td><td>Proportion of cable radio and television subscribers to total households</td><td>+</td><td>0.049</td></tr>
<tr><td>Proportion of digital TV subscribers to total households</td><td>+</td><td>0.044</td></tr>
<tr><td rowspan="2">E-commerce infrastructure</td><td>Number of companies with R&D activities</td><td>+</td><td>0.127</td></tr>
<tr><td>Number of enterprises with R&D institutions</td><td>+</td><td>0.108</td></tr>
<tr><td rowspan="3">Innovative infrastructure</td><td rowspan="2">Scientific and educational infrastructure</td><td>Density of R&D institutions</td><td>+</td><td>0.071</td></tr>
<tr><td>Density of regular institutions of higher learning</td><td>+</td><td>0.026</td></tr>
<tr><td>Innovation infrastructure for high-tech industries</td><td>Density of high-tech enterprises</td><td>+</td><td>0.084</td></tr>
</table>

Note: "+" indicates a positive indicator.

4.2.4. Control Variables

To avoid interference from other variables on the regression results, this study designs four control variables. First is the level of economic development (lngdp), measured by the regional per capita GDP. Second is the degree of population density (Dpd). Population density, to a certain extent, mitigates the negative impact of an aging society, improves the quality structure of the labor force, upgrades consumption supply, and promotes the quality of economic development. It is measured by the natural logarithm of the number of people per square kilometer. Third is the intensity of investment in technological innovation (Iii). Investment in technological innovation can promote social production, enhance economic efficiency, and narrow regional disparities. It is measured by the intensity of

R&D expenditure. Fourth is human capital (Hc), measured by the proportion of college students in the total population.

4.3. Data Sources and Descriptive Statistics

The sample data for this paper consist of panel data from part of provinces, municipalities, and autonomous regions in China from 2013 to 2022, including Beijing, Tianjin, Hebei Province, Shanxi Province, Inner Mongolia Autonomous Region, Liaoning Province, Jilin Province, Heilongjiang Province, Shanghai, Jiangsu Province, Zhejiang Province, Anhui Province, Fujian Province, Jiangxi Province, Shandong Province, Henan Province, Hubei Province, Hunan Province, Guangdong Province, Guangxi Zhuang Autonomous Region, Hainan Province, Chongqing, Sichuan Province, Guizhou Province, Yunnan Province, Shaanxi Province, Gansu Province, Qinghai Province, Ningxia Hui Autonomous Region, and Xinjiang Uyghur Autonomous Region (data of some provincial-level administrative regions of China including Xizang, Hong Kong, Macau, and Taiwan is partially unavailable). The primary data sources include the "China Statistical Yearbook" (2014–2023), the "China Urban and Rural Statistical Yearbook" (2014–2023), the "China Science and Technology Statistical Yearbook" (2014–2023), and the statistical yearbooks of various provinces (regions, cities) (2014–2023) and the EPS data platform at https://www.epsnet.com.cn/index.html#/Index (accessed on 14 December 2024). Missing data are interpolated using the mean substitution method and the regression substitution method. Descriptive statistics for each variable are presented in Table 3.

Table 3. Descriptive statistics.

Variable	Obs	Mean	Std. Dev.	Min	Max
DRI	300	0.154	0.091	0.014	0.503
NI	300	0.092	0.083	0.021	0.532
lngdp	300	9.974	0.868	7.66	11.768
Dpd	300	7.894	0.381	6.965	8.62
Iii	300	1.843	1.165	0.45	6.83
Hc	300	0.024	0.021	0.001	0.246
AIS	300	2.399	0.123	2.177	2.836
lnER	300	13.408	1.495	9.385	17.038

Figure 2 visually depicts the spatial evolutionary trend of China's DRI in 2013, 2016, 2019, and 2022 through maps. The analytical results indicate a significant and steady growth pattern in China's DRI index. In 2013, the lowest recorded DRI index was 0.014, which subsequently increased annually, reaching a minimum value of 0.028 in 2022, signifying a 100% growth over the past decade. This achievement is closely related to China's recent vigorous implementation of the Rural Revitalization Strategy, for example, Guizhou Province has effectively enhanced its local DRI level by developing a characteristic tea industry and expanding sales channels through e-commerce platforms. However, it is noteworthy that despite the overall DRI index in China maintaining a high standard in 2022, there was a declining trend in its maximum value. This phenomenon may foreshadow that some advanced regions are facing pressures such as resource and environmental constraints, market saturation, and the need for economic restructuring, all of which have impacted the further enhancement of the DRI index. Moreover, Figure 3 reveals the dynamic changes in China's NI over the same time series. Observing the data, it is evident that the growth trend of China's NI index is particularly pronounced. Between 2013 and 2022, the minimum NI index value increased from 0.021 to 0.038, while the maximum value surged from 0.199 to 0.532, reflecting a growth of 167.34% over the decade. This significant growth magnitude prominently demonstrates the rapid development of China's NI. For instance, in terms of 5G base station construction, China has built the world's largest 5G network, providing a solid foundation for the digital transformation of various industries. Combined with Figure 2, we can infer that the advancement of new infrastructure has provided new momentum for DRI, for example, the application of 5G technology in the agricultural sector has promoted the development of smart agriculture and enhanced agricultural production efficiency and product quality, thereby promoting the growth of DRI.

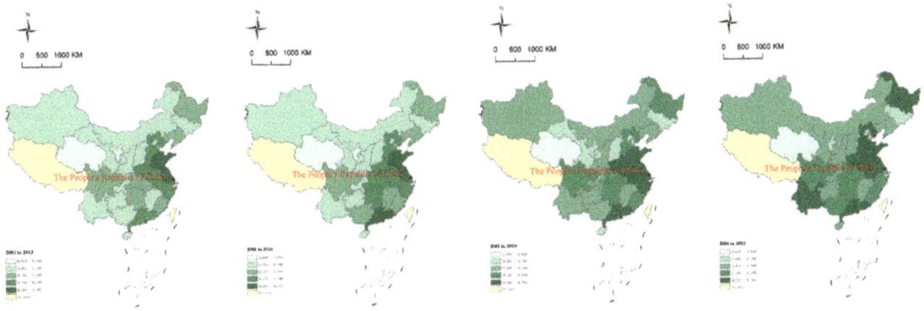

Figure 2. Level of development of rural industries (DRI) of Chinese provinces. Note: This map is produced based on the standard map with approval number GS (2024) 0650, and the base map has not been modified.

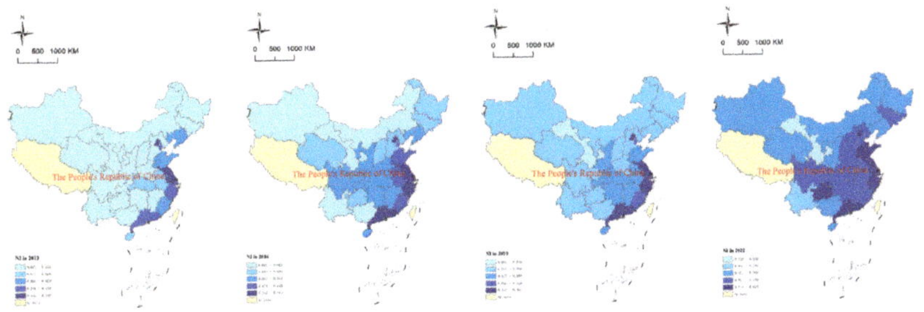

Figure 3. Level of new infrastructure (NI) of Chinese provinces. Note: This map is produced based on the standard map with approval number GS (2024) 0650, and the base map has not been modified.

5. Results

5.1. Baseline Regression Results

The baseline regression results of NI on DRI are shown in Table 4. Models (1) and (2) are OLS regressions of DRI on NI. The first and second terms of NI are significant at the 1% level, verifying the nonlinear impact of NI on DRI. Meanwhile, the F-test rejects the OLS pooled effect, indicating the presence of provincial individual heterogeneity in the model. The Hausman test rejects the random effects model at a 1% significance level. To eliminate provincial individual heterogeneity, this paper establishes a two-way fixed effects model (5). In models (3)–(5), the coefficients of the first and second terms of NI are significant, indicating a clear nonlinear relationship between NI and DRI. The regression results from Model (5) demonstrate that the coefficient for the linear term of NI is 0.264, while the quadratic term shows a coefficient of −0.920 with a 95% confidence interval of [−1.338, −0.502], indicating an "inverted U-shaped" relationship between these variables. When NI = 0, a one-unit increase in NI raises DRI by 0.264 units; when NI reaches the inflection point value of 0.134, the marginal effect declines to zero; and when NI exceeds this threshold, the effect turns negative. This pattern of diminishing returns aligns with the infrastructure productivity threshold hypothesis proposed by Aschauer (1989) [65]. This implies that in the initial stage, a unit increase in NI leads to significant improvements in the annual revenue of rural enterprises, penetration rate of rural e-commerce, and agricultural labor productivity. When NI surpasses the inflection point, the marginal benefit of additional investment becomes negative, typically manifested by near-saturation of digital financial inclusion coverage and relatively high vacancy rates in digital parks in some regions. Furthermore, the "Utest" command in the Stata 16.0 is used to further test the "inverted U-shaped" relationship. The results show that the p-value of the Utest is much less than 0.01, rejecting the null hypothesis of no "inverted U-shaped" relationship at the 1% level. Specifically, the "inverted U-shaped" curve between NI and DRI shows a characteristic of an initially positive average slope (1.784, $p < 0.01$) followed by a negative slope (−2.442, $p < 0.01$), and the inflection point value of NI (0.237) falls within its range of values (0.221, 0.257). This indicates an "inverted U-shaped" relationship between NI and DRI, validating Hypothesis 1.

Table 4. OLS Overall regression results.

	(1)	(2)	(3)	(4)	(5)
NI	1.958 ***	1.225 ***	0.416 *	0.816 ***	0.246 *
	(0.191)	(0.199)	(0.223)	(0.181)	(0.134)
NI^2	−4.137 ***	−2.590 ***	−1.224 ***	−1.652 ***	−0.920 ***
	(0.459)	(0.338)	(0.331)	(0.364)	(0.205)
lngdp		0.070 ***	0.037 ***	0.019 **	0.025 ***
		(0.005)	(0.008)	(0.008)	(0.008)
Dpd		0.023 ***	0.006	0.008	0.003
		(0.009)	(0.011)	(0.012)	(0.012)
Iii		−0.016 **	0.010	0.023 **	0.011
		(0.007)	(0.008)	(0.011)	(0.011)
Hc		−1.132 ***	−0.780 ***	−0.235	−0.551 **
		(0.205)	(0.205)	(0.158)	(0.203)
_cons	0.037 ***	−0.739 ***	−0.299 ***	−0.375 **	−0.158
	(0.010)	(0.073)	(0.107)	(0.151)	(0.131)
id			NO	YES	YES
year			YES	NO	YES
N	300	300	300	300	300
r2_a	0.366	0.686	0.643	0.917	0.582

Standard errors in parentheses. * $p < 0.1$, ** $p < 0.05$, *** $p < 0.01$.

5.2. Robustness Checks

5.2.1. Endogeneity Test

The relationship between NI and DRI is likely to be influenced by political and administrative drivers of investment patterns. This is because, in many countries, especially China, large-scale infrastructure projects are often initiated, funded, and managed by the government. This includes NI projects such as 5G networks, data centers, and intelligent transportation systems. The government's priorities, policy objectives, and administrative decisions play a significant role in determining the location and manner of these investments. Additionally, governments often implement regional development policies aimed at narrowing the urban–rural gap or promoting growth in specific areas. These policies can lead to targeted infrastructure investments in certain rural areas, thereby influencing the observed relationship between NI and DRI. Therefore, one can attempt to use instrumental variables (IVs) that are related to NI but not directly influenced by political or administrative decisions, which can help identify the causal effect of NI on DRI. Meanwhile, the instrumental variable method can ensure that estimates are not biased by omitted variable problems or two-way causality.

The Hausman test's p-value of 0.007 indicates that the baseline regression model has endogeneity issues. Considering the fact that digital infrastructure is importantly carried by information networks, we selected the number of post offices per million people in 1984 as the instrumental variable for NI. The rationale behind this choice is as follows: On one hand, as an important part of traditional communication, post offices to some extent foreshadow a region's communication capabilities and information dissemination efficiency, reflecting the level of early information transmission infrastructure. Similarly, the development of NI cannot be separated from the evolution of traditional communication technologies. Areas with a high number of post offices in history are likely to also have better-developed telecommunications infrastructure. On the other hand, the impact of historical post office numbers on the economies of urban and rural areas is gradually diminishing and so currently struggles to affect the efficiency of DRI, which satisfies the relevance and exogeneity principles of instrumental variable selection. Logically, the current DRI will not affect the number of local post offices. Therefore, there is no reverse causality. The instrumental variable method can be used to alleviate potential omitted variable bias and reverse causality issues between NI and DRI. Additionally, because the sample in this study is made up of balanced panel data, using only the number of post offices per million people in 1984 at the city level as an instrumental variable would pose measurement challenges due to the application of the fixed effects model. To address this, we drew on the setup method by Nunn and Qian [66] and constructed an interaction term between the number of post offices per million people in 1984 (cross-sectional data) and the national IT service revenue of the previous year (time-series data) as the IV for NI. Then, we employed the Two-Stage Least Squares (2SLS) method to identify the causal relationship between NI and DRI. The results, as shown in Table 5 Column (1), indicate that the regression coefficient of the instrumental variable is positive and passes the 1% significance test, suggesting that NI still has a significant "inverted U-shaped" relationship with DRI, validating the stability of the baseline results. The chosen instrumental variable passed the weak instrument test (F-statistic 22.68) and the under-identification test (Anderson LM statistic 12.666), rejecting the hypotheses of weak instruments and non-identification, indicating that the instrumental variable selection is reasonable and effective. Therefore, the instrumental variable regression results also confirm that the impact of NI on DRI is robust.

5.2.2. Replacing the Core Explanatory Variable

Mobile base stations are an important component of NI, particularly in terms of communication infrastructure. Therefore, the distribution and development of mobile base stations can reflect the penetration and progress of NI to a certain extent. To avoid research errors caused by variable setting bias, we use the number of mobile phone base stations (NI_p) to replace the original core explanatory variable and re-estimate the baseline model. The results are shown in Table 5, Column (2). It is not difficult to see that there have been no significant changes in the significance and direction of the coefficient of the core explanatory variable, demonstrating that the estimation results of the baseline model are robust.

5.2.3. Winsorization Treatment

To avoid the potential bias that extreme values may introduce into the research results, the relevant data for the variables used in the article were trimmed at the top and bottom 1%, and the regression analysis was conducted again. The results are shown in Column (3) of Table 5. Observing the data, there is no significant difference in the sign and significance of the regression coefficients for NI compared to the baseline regression results, indicating that the baseline regression results are reliable.

Table 5. Endogeneity and robustness tests.

	(1)	(2)	(3)	(4)
NI	0.158		0.315 **	
	(1.054)		(0.140)	
NI^2	−1.689 *		−1.158 ***	
	(0.990)		(0.238)	
NI_p		0.003 ***		
		(0.001)		
NI_p^2		−0.000 **		
		(0.000)		
NI_lag				0.149
				(0.314)
NI_lag^2				−0.883 ***
				(0.315)
lngdp	0.0316 **	0.021 *	0.028 ***	0.023 ***
	(0.0135)	(0.010)	(0.008)	(0.008)
Dpd	0.0106	0.014	0.007	0.004
	(0.0120)	(0.012)	(0.011)	(0.012)
Iii	0.0324	0.012	0.008	0.013
	(0.0200)	(0.015)	(0.010)	(0.011)
Hc	−0.805 *	−0.421 *	−0.887 ***	−0.486 **
	(0.454)	(0.205)	(0.266)	(0.182)
_cons	−0.302 *	−0.209	−0.211	−0.135
	(0.168)	(0.142)	(0.127)	(0.134)
id	YES	YES	YES	YES
year	YES	YES	YES	YES
N	300	300	300	270
r2_a	0.442	0.330	0.391	0.311

Standard errors in parentheses. * $p < 0.1$, ** $p < 0.05$, *** $p < 0.01$.

5.2.4. Reverse Causality Test

To further test for reverse causality, we also use the lagged values of the independent variable to regress the current values of the dependent variable, that is, we used the one-period lag of the explanatory variable NI to regress DRI. The results, shown in Column (4) of Table 5, indicate a significant nonlinear relationship between NI and DRI, indicating that the baseline regression results are reliable.

5.3. Heterogeneity Analysis

5.3.1. Regional Heterogeneity

Due to China's vast territory, there are significant gaps among various regions in terms of the foundation for the development of the digital industry, digital talent, and investment in capital elements. This leads to the possibility of differences in the impact of NI on DRI. Therefore, according to the classification standards of the National Bureau of Statistics, the total sample of the study is divided into four regions: east, central, west, and northeast. We conducted an analysis on the heterogeneity of the impact of NI on DRI in different regions, with the regression results shown in Table 6, Columns (1)–(4). The regression results of Model (1) demonstrate a statistically significant inverted U-shaped relationship between NI and DRI in eastern regions, with the linear term coefficient of NI being 0.353 ($p < 0.01$) and the quadratic term coefficient being −0.605 (95% CI [−1.047, −0.163]). Calculations show that when NI levels are below the inflection point of 0.292, each additional unit of NI investment increases DRI by 0.353 units, whereas when NI exceeds this threshold, the marginal returns of additional investment approach zero. The estimation results of Model (3) reveal that although western regions exhibit a similar inverted U-shaped pattern (linear term coefficient 2.044, $p < 0.05$; quadratic term coefficient −5.662, 95% CI [−11.353, 0.029]), the quadratic term's confidence interval including zero indicates weaker statistical significance of the nonlinear relationship. Notably, the inflection point value in western regions (0.181) is significantly lower than that in eastern regions (0.292). Compared to the western region, the inflection point in the eastern region is further to the right, which may be because the level of economic development is higher, the infrastructure is more comprehensive, and the industrial base is stronger, allowing for more effective utilization of the advantages brought by NI. Therefore, it can withstand a higher level of NI investment before negative effects appear at a higher investment level. Although the western region has a relatively lower level of economic development, the marginal benefit of NI may be higher because improvements in infrastructure can bring about greater development space. The industrial structures in the central and northeast regions of China are currently in a period of transition and may be approaching the point of infrastructure saturation, with NI failing to effectively translate into a driving force for DRI. This is primarily attributed to the distinctive nature of their economic structures, insufficient innovation-driven capabilities, deficiencies in policy guidance and financial investment, as well as limitations in the level of marketization and openness. Specifically, the heavy industry and agriculture-oriented economic structures in these two regions result in a differentiated dependency and demand for NI compared to other areas, and the integration and adaptation of NI require time. Moreover, the relatively weak innovation-driven capacity and the lagging development of high-tech industries constrain the technological integration and application of NI, thereby affecting its promotional role in DRI. In addition, insufficient policy support, an imbalanced investment structure, and the inadequate role of market mechanisms in resource allocation collectively contribute to the failure of NI resources to effectively transform into a driving force for DRI.

Table 6. Regional heterogeneity analysis.

	(1)	(2)	(3)	(4)
NI	0.353 **	−0.562	2.044 *	3.267
	(0.176)	(2.424)	(1.194)	(13.238)
NI^2	−0.605 ***	−3.300	−5.662 *	−19.540
	(0.225)	(9.572)	(2.903)	(105.728)
_cons	0.627 **	−0.479	0.000	0.028
	(0.266)	(0.726)	(0.000)	(1.103)
id	YES	YES	YES	YES
year	YES	YES	YES	YES
N	100	60	110	30
r2_a	0.956	0.926	0.878	0.841

Standard errors in parentheses. * $p < 0.1$, ** $p < 0.05$, *** $p < 0.01$.

5.3.2. Heterogeneity of Intelligent Transportation Infrastructure

Intelligent transportation infrastructure serves as a core carrier of NI, and its development directly impacts the penetration efficiency of technologies such as 5G and the Internet of Things in rural areas. Some studies suggest that NI exhibits inclusive characteristics [67], while others point out that it may exacerbate regional imbalances [22]. Intelligent transportation infrastructure may act as a key variable in modulating the inflection point of the "inverted U-shaped" relationship between NI and DRI by altering technology application scenarios, logistics efficiency, or market accessibility. To test the heterogeneity of the impact of NI on DRI under varying levels of intelligent transportation infrastructure, this study uses "Railway operating mileage density" and "Mileage density of high-speed highways" as indicators to measure the level of regional intelligent transportation infrastructure. The entropy method is employed to assign weights and calculate the level of intelligent transportation infrastructure for each province. The median value of 32.723 is selected as the threshold, with provinces below this median classified as having a low level of intelligent transportation infrastructure and those at or above the median as having a high level. The regression results are presented in Table 7, Columns (1) and (2). The results indicate that in provinces with high levels of intelligent transportation infrastructure, there exists an "inverted U-shaped" relationship between NI and DRI, with a linear term coefficient of 0.262 and a quadratic term coefficient of −0.864 (95% CI [−1.047, −0.163]). This implies that when intelligent transportation levels are high, each unit increase in NI investment raises DRI by 0.262 units during the phase when NI is below 0.152, while the marginal effect turns negative when NI exceeds 0.152. In provinces with low levels of intelligent transportation infrastructure, the regression coefficients are statistically insignificant. This may occur because when intelligent transportation infrastructure levels are low, NI investments fail to form complete application scenarios, face constrained service radii, and quickly reach investment saturation, leading to an earlier emergence of the inflection point (0.085). In regions with higher levels of intelligent transportation infrastructure, the factor allocation efficiency of NI is significantly enhanced, industrial integration space expands, and the inflection point appears later (0.152). These findings align with the conclusions of Banerjee et al. (2020) regarding the threshold effects of transportation infrastructure [68]. It follows that intelligent transportation infrastructure serves as a key moderating variable determining the inflection point position in the "inverted U-shaped" relationship between NI and DRI.

Table 7. Analysis of heterogeneity in intelligent transportation infrastructure, population density, and human capital.

	(1)	(2)	(3)	(4)	(5)	(6)
NI	0.469	0.262	1.195 **	0.246	1.053 **	1.312 **
	(1.006)	(0.185)	(0.466)	(0.240)	(0.460)	(0.648)
NI^2	−2.755	−0.864 ***	−2.763 ***	−0.871 ***	−2.699 ***	−2.158 ***
	(1.917)	(0.249)	(0.994)	(0.334)	(1.032)	(0.791)
_cons	−0.178	−0.198	0.275	−0.394 *	0.000	−0.546 *
	(0.228)	(0.207)	(0.176)	(0.211)	(.0.000)	(0.287)
id	YES	YES	YES	YES	YES	YES
year	YES	YES	YES	YES	YES	YES
N	150.000	150.000	147	153	214	86
r2_a	0.906	0.941	0.967	0.920	0.948	0.938

Standard errors in parentheses. * $p < 0.1$, ** $p < 0.05$, *** $p < 0.01$.

5.3.3. Heterogeneity of Population Density

To test the heterogeneity of the impact of NI on DRI in provinces with different population densities, the logarithm of the number of people per square kilometer is used as an indicator to measure regional population density. The median population density of 7.894 is selected, with provinces below the median being classified as low-population-density provinces and those equal to or above the median as high-population-density provinces. The regression results are shown in Table 7, Columns (3) and (4). The empirical results demonstrate significant population density heterogeneity in the relationship between NI and DRI. The low population density subsample exhibits a statistically significant inverted U-shaped relationship, with a quadratic term coefficient of −2.763 (95% CI [−4.711, −0.814]) and an inflection point at 0.216. Similarly, the high population density subsample shows an inverted U-shaped pattern, but with a smaller absolute value of the quadratic term coefficient (−0.871, 95% CI [−1.526, −0.215]) and an earlier inflection point at 0.141. These findings robustly support Hypothesis 2. The regional differences in inflection points primarily stem from the following mechanisms. First, there are regional differences in the marginal diminishing pattern of resource utilization efficiency. In provinces with low population density, although the initial construction of NI faces higher unit costs, the subsequent marginal utility diminishes at a slower rate due to the relatively higher per capita infrastructure capital stock. In contrast, in densely populated regions, while digital infrastructure initially yields higher returns, the marginal benefits turn negative more rapidly due to faster saturation of service demand. Second, the moderating effect of economic development stages is significant. Less developed regions (mostly provinces with low population density) exhibit higher "demand elasticity" for digital infrastructure, allowing the positive effects of NI to persist for a longer duration. Finally, there are distinct heterogeneous responses across industrial structures. Provinces with low population density are predominantly characterized by agriculture-dominated industries, where digital technologies have greater transformative potential for traditional agriculture. Conversely, in provinces with high population density, where the service sector accounts for over 45% of the economy, the competitive saturation effect in digital services leads to a more rapid decline in the marginal returns of NI.

5.3.4. Heterogeneity of Human Capital

Considering that human capital is also an important factor influencing the role of NI in DRI, this paper uses the proportion of college students in the total population as an indicator to measure regional human capital. The median level of human capital is set at 0.236. Provinces with a proportion below the median are classified as having low human capital, while those at or above the median are classified as having high human capital. The regression results are shown in Table 7, Columns (5) and (6). The results indicate an "inverted U-shaped" relationship between NI and DRI, further validating Hypothesis 1. In the coefficient of the impact of digital NI on DRI, the inflection point for provinces with high human capital (0.304) occurs later than that for provinces with low human capital (0.195). This is mainly because provinces with high human capital have a larger labor force with professional skills and innovative capabilities, who are more adept at learning new knowledge and skills, thus being able to maintain the marginal benefits of NI for a longer period of time. Additionally, provinces with high human capital may have a larger market demand and potential, more reasonable industrial structures, and better resource allocation and investment efficiency, allowing these regions to maintain the marginal benefits of NI for a longer time. This part of the empirical results validates Hypothesis 2.

5.4. Mechanism Analysis

Table 8 shows the analysis of the transmission pathways of NI on DRI. The text previously discussed that NI may affect DRI through channels such as industrial structure upgrading and the expansion of transaction markets. To test whether these transmission pathways exist, the study explores the ways in which NI impacts DRI based on Equations (1)–(4). The regression results are presented in Table 8. Table 8 Column (1) shows that the coefficient of NI is significantly positive, and the coefficient of the quadratic term is negative, indicating that NI can promote the upgrading of the industrial structure. There is an "inverted U-shaped" relationship between NI and industrial structure upgrading. In Column (2), both the quadratic term of NI and the coefficient of the industrial structure upgrading index are significant, but compared to Column (4) in Table 1, the coefficient of NI has decreased. This may be because industrial structure upgrading absorbs part of the positive impact of NI on DRI, thereby refining and adjusting the impact of NI. This suggests that industrial structure upgrading can explain part of the impact of NI, making its effect on DRI more moderate. Additionally, the study uses Sobel tests and Bootstrap tests to demonstrate the robustness of the above mechanism tests. After 1000

Bootstrap samples, the mediating effect of industrial structure upgrading is confirmed with a p-value less than 0.01. It is noteworthy that the indirect effect is negative, indicating a substitution effect between NI and industrial structure upgrading. This could be due to the construction of NI attracting resources away from rural areas to cities or more favorable regions, leading to hindered industrial structure upgrading or resource scarcity in rural areas, and consequently negatively affecting DRI. Columns (3)–(4) of Table 8 test the mechanism of transaction market expansion. The results indicate that NI can promote the expansion of transaction markets, and there is an "inverted U-shaped" relationship between NI and transaction market expansion. In Column (2), both the quadratic term of NI and the coefficient of the transaction market expansion index are significant, but compared to Column (5) in Table 1, the coefficient of NI has decreased, suggesting that transaction market expansion plays a partial mediating role between NI and DRI. This part of the test validates Hypotheses 3a and 3b.

Table 8. Mediation analysis.

	(1)	(2)	(3)	(4)
	AIS	DRI	lnER	DRI
NI	1.627 **	0.048	4.581 ***	0.110
	(0.652)	(0.158)	(0.704)	(0.162)
NI^2	−2.096 **	−0.665 ***	−10.878 ***	−0.597 **
	(0.766)	(0.191)	(1.093)	(0.261)
AIS		0.122 **		
		(0.053)		
lnER				0.030 *
				(0.017)
_cons	2.174 ***	−0.422 **	11.488 ***	−0.499 **
	(0.256)	(0.169)	(0.368)	(0.239)
id	YES	YES	YES	YES
year	YES	YES	YES	YES
N	300	300	300	300
r2_a	0.642	0.394	0.980	0.381

Standard errors in parentheses. * $p < 0.1$, ** $p < 0.05$, *** $p < 0.01$.

6. Discussion

The findings of this study contribute significantly to the ongoing discourse on the relationship between NI and DRI. The discussion below aims to elaborate on the theoretical insights, policy implications, and the limitations of the study, while also suggesting directions for future research.

6.1. Discussion on Theoretical Implications

The discovery of an "inverted U-shaped" relationship between NI and DRI is a particularly intriguing theoretical contribution. This finding challenges the monolithic view that NI is either entirely beneficial or detrimental to rural revitalization [69,70]. Instead, it suggests a nuanced perspective that acknowledges the complexity of NI's impact. The identification of an inflection point where NI's effects shift from positive to negative is a critical insight that can guide future theoretical development. This revelation suggests that the relationship between NI and the DRI is not static but dynamic, and it depends on specific contextual factors, a perspective that resonates with the multifaceted and complex nature of rural development. In this sense, the current study builds upon the work of previous scholars who have emphasized the environmental variability of infrastructure impacts [71] while also providing a more precise empirical basis for the concept of optimal levels of NI investment. Exploring the mechanisms through which NI affects non-agricultural industries, particularly via industrial structure upgrading and transaction market expansion, fills a significant gap in the literature. While prior research has touched upon the role of infrastructure in promoting economic growth [72], this study delves deeper into the specific channels through which NI can facilitate rural revitalization. The empirical validation of these mechanisms not only offers a clearer understanding of how NI influences rural economies but also provides a more granular analysis that can guide more targeted and effective policy interventions.

6.2. Discussion on Heterogeneity

The heterogeneity of geographical location is a particularly interesting aspect of this study. The variation in the "inverted U-shaped" relationship between NI and DRI across different regions highlights the importance of context in infrastructure development. Both the eastern and western regions exhibit this relationship, but with different inflection points, indicating that the optimal level of NI investment varies by region. Compared to the western region's inflection point of 0.181, the higher inflection point in the eastern region (0.292) suggests that more developed areas may have a higher capacity to absorb and utilize NI before reaching saturation. In contrast, the central and northeastern regions are closer to infrastructure saturation, emphasizing the need for a more nuanced approach to NI investment in these areas where the traditional infrastructure-led growth model may no longer be as effective.

This study reveals the critical moderating role of intelligent transportation infrastructure in the relationship between NI and DRI. In regions with advanced intelligent transportation systems, NI establishes a virtuous cycle of "digital infrastructure—intelligent transportation—industrial upgrading" by enhancing factor allocation efficiency and industrial synergy, demonstrating a significant "inverted U-shaped relationship" with DRI. Conversely, in areas with underdeveloped intelligent transportation, new NI investments fail to overcome developmental thresholds due to the "bucket effect" of infrastructure deficiencies and factor outflows, resulting in statistically insignificant policy effects. These findings not only validate the applicability of Krugman's core–periphery theory in the digital economy era but also provide a novel explanatory perspective for reconciling the academic debate between the "inclusive nature" and "regional disparities" of NI and the developmental level of intelligent transportation essentially constitutes a threshold condition for the effective implementation of NI policies.

The study's examination of the heterogeneity of population density and human capital provides further theoretical depth. Provinces with low population density and high human capital can utilize NI more effectively, indicating certain types of NI with a high concentration of human capital can be better utilized in areas with a skilled labor force [73]. This finding supports the theory that the interaction between infrastructure and human capital is a key determinant of development outcomes.

6.3. Discussion on Mechanisms

The upgrading of industrial structure serves as a significant pathway through which NI promotes rural economic growth, involving multiple dimensions such as technological advancement, extension of industrial chains, and increased added value. Specifically, NI facilitates the transformation of agriculture towards high-value-added industries by providing advanced information technology and logistics support, thereby enhancing the economic benefits of rural areas [60]. Concurrently, the expansion of transaction markets creates favorable conditions for the development of non-agricultural industries in rural areas by improving market access and reducing transaction costs [61]. These two mechanisms are not isolated but rather mutually reinforcing and interdependent. For instance, market expansion may further stimulate the demand for industrial structure upgrading, while the optimization of the industrial structure provides more quality products and services for market expansion. These mechanisms interact with policy support, social capital, and the natural environment in rural settings, collectively shaping the complex landscape of rural development. Policy support, as a crucial external force driving industrial structure upgrading, encourages technological innovation and industrial transformation through measures such as tax incentives, fiscal subsidies, and credit support [74,75]. Investments in areas like smart agriculture promote the upgrading of the industrial structure towards higher-value-added service industries and high-tech industries, providing a dual guarantee of funds and markets for this upgrading. Social capital, including trust, norms, and networks accumulated in rural areas, facilitates information sharing, resource integration, and technology dissemination. Social networks like farmers' cooperatives and industry associations promote the application of new technologies, accelerating the upgrading of the industrial structure [76]. The condition of the natural environment directly affects the choice and upgrading path of the industrial structure. NI investments in ecologically vulnerable areas focus on ecological protection, promoting the development of green industries, and a favorable natural environment is also the basis for developing specialized industries such as tourism. In terms of transaction market expansion, policy support reduces transaction costs and optimizes the market environment, such as the construction of agricultural product trading platforms and the improvement of logistics distribution systems. Social capital reduces transaction uncertainties, enhances market trust, and forms stable trading partner relationships. The natural environment influences the expansion of transaction markets, where transportation convenience and resource abundance determine the types and quantities of products, affecting market size and structure. In summary, these interacting elements constitute a dynamic system, and understanding them is essential for formulating effective rural revitalization strategies.

6.4. Discussion on Limitations and Future Research Prospects

This paper, from the perspective of NI and DRI, empirically tests the relationship between NI and DRI and its mechanism of action. There is a certain degree of innovation in the research perspective, but there are still some shortcomings. Firstly, the statistical data of rural areas in China are yet to be enriched, so the selected indicator dimensions in the process of indicator construction are relatively few, which can only capture the characteristics of DRI as accurately as possible, but it is difficult to achieve absolute accuracy. Second, the data used in this study are provincial-level panel data in China, without considering the urban–rural differences and NI differences between and within provinces. Meanwhile, since the concept of NI has not been proposed for a long time, the amount of time-series data on this is small. Although the conclusions of this study have been verified by many empirical processes, there is still room for further exploration of the data and methods. Therefore, future research can proceed from three angles: (1) Enrich research data. Collect more comprehensive and detailed statistical data on rural areas, refine the dimensions of DRI. Continuously collect relevant time-series data to facilitate long-term trend analysis. (2) Consider regional differences and hierarchy. Focus on the urban–rural differences between and within provinces in terms of NI, as well as regional differences. Collect as much diverse data as possible, such as city and county or even township-level data, to more accurately analyze the impact of NI on DRI. (3) Diverse methodological research. With the improvement of data availability in rural areas in the future and the increasingly mature mechanism testing techniques, follow-up studies can further diversify the dimensions of indicator construction and can also use machine learning algorithms for preliminary dimension screening to improve the accuracy of index construction and enrich the connotation of indicators. Use spatial econometric methods to analyze the spatial distribution of NI and its impact on DRI.

7. Conclusions and Policy Recommendations

7.1. Conclusions

This study examines the relationship between NI and DRI using panel data from Chinese provincial-level regions (2013–2022). The principal findings demonstrate three key insights. First, there is a robust "inverted U-shaped" relationship between NI and DRI, with an inflection point at 0.134, consistent across instrumental variable analysis, alternative variable specifications, and winsorization procedures. Second, heterogeneity analysis reveals significant variations: regional heterogeneity analysis shows eastern and western regions exhibit the "inverted U-shaped" relationship with rightward-shifted inflection points at 0.292 and 0.181, respectively, while central and northeastern regions display infrastructure saturation; intelligent transportation infrastructure analysis indicates advanced regions achieve greater NI efficiency with a 0.152 inflection point through digital infrastructure-intelligent transportation–industrial upgrading synergies, whereas underdeveloped regions face infrastructure bottlenecks with a 0.085 inflection point; population density and human capital analysis reveals provinces combining lower population density with higher human capital demonstrate superior NI absorption capacity. Third, mechanism analysis establishes industrial structure upgrading and market expansion as critical channels through which NI influences DRI development.

7.2. Policy Recommendations

Our findings offer several valuable insights for policymakers:

(1) Differentiated policy support and optimal resource allocation. Given the varying "inverted U-shaped" relationships between NI and DRI across different regions, policymakers should consider regional heterogeneity when formulating policies. Specifically, the eastern and western regions should pay attention to the inflection points of infrastructure construction. Considering the eastern region, it is recommended to establish an investment early warning mechanism triggered by an assessment when NI approaches saturation levels, and to set up an industrial upgrading fund, allocating a certain proportion of NI investment to support the transformation of traditional industries towards digitalization and intelligentization. Considering the western region's larger space for infrastructure development, it is proposed to implement a "NI+" bundled investment strategy, integrating transportation infrastructure construction with digital infrastructure development, and advancing infrastructure upgrades in stages, from basic communication network coverage to industrial digital applications, and finally to intelligent agricultural production. In the central and northeastern regions, the focus should be on improving the utilization efficiency of existing infrastructure, conducting efficiency audits of existing facilities, and upgrading facilities with low utilization rates. Implement a "Digital Empowerment" initiative, supporting a certain number of enterprises in digital transformation annually, and innovate policy instruments, such as issuing special bonds and establishing regional talent centers, to promote industrial upgrading and innovative development.

(2) Implementation of a tiered development strategy for intelligent transportation infrastructure. First, establish an evaluation system for intelligent transportation infrastructure, including core indicators such as road network digitalization rate and vehicle networking penetration rate, and classify regions into different levels. For regions with a high level of intelligent transportation infrastructure, a leading development strategy should be formulated; for regions with a low level, a strategy of addressing weaknesses and strengthening foundations should be adopted. Second, for regions with a high level of

intelligent transportation infrastructure, focus on promoting the construction of 5G-V2X deep integration infrastructure, including real-time high-precision map updating systems and intelligent logistics upgrade projects, thereby improving the efficiency of agricultural product cold-chain logistics and providing technical support for the intelligent management of rural tourism. Third, for regions with a low level of intelligent transportation infrastructure, focus on implementing basic digital transformations such as intelligent upgrades of rural roads, digital upgrades of village-level logistics outlets, and popularization of Beidou navigation terminals, driving the development of rural e-commerce clusters, and supporting collaborative layouts of e-commerce warehousing and low-cost IoT solutions. Fourth, all construction projects must be equipped with dynamic monitoring systems to achieve benefit evaluation and plan adjustment, ensuring maximized investment returns and realizing the coordinated DRI and NI.

(3) Implementation of a differential human capital development strategy. In view of the important role of human capital in promoting rural industry development through NI, precise policies should be implemented for regions with different levels of human capital. For regions with a high level of human capital, policies should focus on improving the investment benefits of NI. Specific measures include establishing university–enterprise joint R&D centers, focusing on applied research of NI in rural development; and setting up a special talent introduction plan for NI. For regions with a weaker human capital base, promote the "Internet + Education" model, increase investment in rural education, implement rural teacher training and improvement programs, improve the infrastructure of rural schools, and ensure the quality of education. At the same time, carry out vocational skills training programs, cultivate professional and technical personnel according to the needs of NI, such as professionals in information technology, new energy, and intelligent manufacturing. Finally, establish and improve the rural talent introduction and incentive mechanism, attract high-level talents to return to their hometowns for employment and entrepreneurship, and enhance the overall level of human capital in rural areas.

(4) Expanding transaction markets and strengthening the marginal benefits of NI. Firstly, efforts should be made to promote the establishment and widespread use of e-commerce platforms in rural areas, optimizing the logistics and distribution network to ensure the rapid circulation of agricultural products. Secondly, information infrastructure should be strengthened to improve network coverage and support rural e-commerce. In provinces with low population density, focus should be on investing in transportation and information infrastructure to enhance positive marginal benefits, promote effective resource flow, and integrate regional economies. Thirdly, measures should be taken to avoid overinvestment in NI. This includes establishing a pre-investment evaluation system to ensure investment efficiency, building logistics hubs in key areas to reduce the distance between urban and rural logistics, and promoting intelligent agricultural technologies to improve agricultural production efficiency, thereby maximizing the economic benefits of NI.

Author Contributions: L.L.: Formal analysis, writing—original draft, reviewed, and edited the manuscript, managed the project, and supervised the research process. X.M.: conception of the study, prepared the original draft of the manuscript, and was responsible for data curation. Y.L.: software and data processing, formal analysis, and contributed to writing, reviewing, and editing the manuscript. All authors have read and agreed to the published version of the manuscript.

Funding: This work has been partially supported by the National Social Science Fund of China (No. 24BMZ025).

Data Availability Statement: The data used in this study are available to the public for download and verification. The original data presented in the study are openly available on the EPS data platform at https://www.epsnet.com.cn/index.html#/Index (accessed on 20 October 2024). Interested parties can access the datasets by visiting the provided link and navigating to the appropriate data sections.

Conflicts of Interest: The authors declare no conflicts of interest.

References

1. Tong, L.; Ji, H.; Zhang, H.; Chen, L.; Cheng, X. Can information and communication technology drive effective spatial governance in rural areas? Emergency of e-participation ecosystem from rural China. *Inf. Syst. J.* **2024**, *35*, 674–709. [CrossRef]
2. Zhang, Y.; Love, D.J.; Krogmeier, J.V.; Anderson, C.R.; Heath, R.W.; Buckmaster, D.R. Challenges and Opportunities of Future Rural Wireless Communications. *IEEE Commun. Mag.* **2021**, *59*, 16–22. [CrossRef]
3. Wang, C.; Zhou, T.; Ren, M. Driving spatial network connections in rural settlements: The role of e-commerce. *Appl. Geogr.* **2023**, *159*, 103067. [CrossRef]
4. Sun, S.L.; Jia, P.; Qi, J.; Guo, C. Shaping opportunity structures: How transport infrastructure fuels social ventures and reduces income inequality. *Entrep. Reg. Dev.* **2025**, *37*, 618–641. [CrossRef]
5. Hamim, O.F.; Hossain, M.S.; Hadiuzzaman, M. Developing empirical model with graphical tool to estimate and predict capacity of rural highway roundabouts. *Int. J. Transp. Sci. Technol.* **2022**, *11*, 726–737. [CrossRef]
6. Perumal, S.; Raman, V.; Samy, G.N.; Shanmugam, B.; Kisenasamy, K.; Ponnan, S. Enhanced disruption tolerant network (DTN) framework for improving network efficiency in rural areas. *Int. J. Syst. Assur. Eng. Manag.* **2022**, *13*, 710–717. [CrossRef]
7. Wang, R.; Shi, J.; Hao, D.; Liu, W. Spatial–Temporal Characteristics and Driving Mechanisms of Rural Industrial Integration in China. *Agriculture* **2023**, *13*, 747. [CrossRef]
8. Philip, L.; Cottrill, C.; Farrington, J.; Williams, F.; Ashmore, F. The digital divide: Patterns, policy and scenarios for connecting the 'final few' in rural communities across Great Britain. *J. Rural. Stud.* **2017**, *54*, 386–398. [CrossRef]
9. Ding, W.; Wu, Q.; Xu, X. Digital Infrastructure Construction and Improvement of Non-Farm Employment Quality of Rural Labor Force—From the Perspective of Informal Employment. *Sustainability* **2024**, *16*, 5345. [CrossRef]
10. Xie, Y.-Q.; Liu, H. The welfare effects of digital infrastructure: Micro-level evidence from China. *Ekon. A Manag. E+M* **2024**, *27*, 1–21. [CrossRef]
11. Stratan, A. Infrastructure as a Determinant of the National Competitiveness. *Across* **2024**, *8*, 31–38.
12. Maciulyte-Sniukiene, A.; Butkus, M. Does Infrastructure Development Contribute to EU Countries' Economic Growth? *Sustainability* **2022**, *14*, 5610. [CrossRef]
13. Wan, J.; Xie, Q.; Fan, X. The impact of transportation and information infrastructure on urban productivity: Evidence from 256 cities in China. *Struct. Change Econ. Dyn.* **2024**, *68*, 384–392. [CrossRef]
14. National Internet Information Office. National Internet Information Office Releases Digital China Development Report. Available online: https://www.cac.gov.cn/2023-05/22/c_1686402318492248.htm (accessed on 20 December 2024).
15. Straub, S. Infrastructure and Development: A Critical Appraisal of the Macro-level Literature. *J. Dev. Stud.* **2011**, *47*, 683–708. [CrossRef]
16. Donaldson, D. Railroads of the Raj: Estimating the Impact of Transportation Infrastructure. *Am. Econ. Rev.* **2018**, *108*, 899–934. [CrossRef]
17. Jia, Z.; Donaldson, D.L.; Ferranti, E. Weather-related fragility modelling of critical infrastructure: A power and railway case study. *Proc. Inst. Civ. Eng.-Civ. Eng.* **2023**, *177*, 50–58. [CrossRef]
18. Yin, F.; Qian, Y.; Zeng, J.; Wei, X. The Spatial Spillover Effects of Transportation Infrastructure on Regional Economic Growth—An Empirical Study at the Provincial Level in China. *Sustainability* **2024**, *16*, 8689. [CrossRef]
19. Loder, A.; Ambühl, L.; Menendez, M.; Axhausen, K.W. Understanding traffic capacity of urban networks. *Sci. Rep.* **2019**, *9*, 16283. [CrossRef]
20. Wu, Z.; Long, H.; Song, H. The impact of infrastructure investment on multidimensional poverty. Evidence from Chinese rural migrant workers. *Econ. Syst.* **2024**, *48*, 101239. [CrossRef]
21. Wu, X.; Mao, C. Evaluation on the New Infrastructure Drive in the Digital Economy Era Based on the Internet of Things. *Inf. Syst. Econ.* **2024**, *5*, 148–159. [CrossRef]
22. Wang, F.; Wang, M.; Yuan, S. Spatial Diffusion of E-Commerce in China's Counties: Based on the Perspective of Regional Inequality. *Land* **2021**, *10*, 1141. [CrossRef]
23. Zhang, J.; Li, M. Digital technology access, labor market behavior, and income inequality in rural China. *Heliyon* **2024**, *10*, e33528. [CrossRef]
24. de Moraes, C.O.; Roquete, R.M.; Gawryszewski, G. Who needs cash? Digital finance and income inequality. *Q. Rev. Econ. Financ.* **2023**, *91*, 84–93. [CrossRef]
25. Meng, X.; Wang, X.; Nisar, U.; Sun, S.; Ding, X. Mechanisms and heterogeneity in the construction of network infrastructure to help rural households bridge the "digital divide". *Sci. Rep.* **2023**, *13*, 19283. [CrossRef] [PubMed]

26. Kwilinski, A. Understanding the nonlinear effect of digital technology development on CO_2 reduction. *Sustain. Dev.* **2024**, *32*, 5797–5811. [CrossRef]
27. Cao, L.; Niu, H.; Wang, Y. Utility analysis of digital villages to empower balanced urban-rural development based on the three-stage DEA-Malmquist model. *PLoS ONE* **2022**, *17*, e0270952. [CrossRef] [PubMed]
28. Xu, Y.; Zhu, S. Transport Infrastructure, Intra-Regional Inequality and Urban-Rural Divide: Evidence From China's High-Speed Rail Construction. *Int. Reg. Sci. Rev.* **2024**, *47*, 378–406. [CrossRef]
29. Han, H.; Yuan, Z.; Zou, K. Does new infrastructure outperform old infrastructure in China's Agricultural Specialization: An empirical examination based on "One village One product". *Int. Food Agribus. Manag. Rev.* **2023**, *26*, 1–19. [CrossRef]
30. Meng, J.; Zhu, Y.; Han, Y. Can 'new' infrastructure become an engine of growth for the Chinese economy? *J. Chin. Econ. Bus. Stud.* **2023**, *21*, 341–362. [CrossRef]
31. Xu, Z.; Yang, J. Impact of digital finance on rural industry revitalization. *Int. Rev. Econ. Financ.* **2025**, *97*, 103820. [CrossRef]
32. Peng, H.; Yang, F.; Yue, O. How Rural Industry Revitalization Affects Farmers' Incomes in China. *Sustainability* **2024**, *16*, 9182. [CrossRef]
33. Qi, L. Can Rural Industry Integration Improve Rural Household Income? Empirical Analysis Based on CFPS Data. *Acad. J. Bus. Manag.* **2024**, *6*, 1–10. [CrossRef]
34. Tian, Y.; Liu, Q.; Ye, Y.; Zhang, Z.; Khanal, R. How the Rural Digital Economy Drives Rural Industrial Revitalization-Case Study of China's 30 Provinces. *Sustainability* **2023**, *15*, 6923. [CrossRef]
35. Li, Y.; Ma, X.; Liu, Y.; Zhong, F. Can China's New Infrastructure Promote Urban–Rural Integrated Development? Evidence from 31 Chinese Provinces. *Buildings* **2024**, *14*, 3978. [CrossRef]
36. Qian, G. Research on the integration management innovation path of tea culture and rural tourism industry under the background of information technology. *Appl. Math. Nonlinear Sci.* **2024**, *9*, 2444–8656. [CrossRef]
37. Karaivanov, A. Financial constraints and occupational choice in Thai villages. *J. Dev. Econ.* **2012**, *97*, 201–220. [CrossRef]
38. Demir, A.; Pesqué-Cela, V.; Altunbas, Y.; Murinde, V. Fintech, financial inclusion and income inequality: A quantile regression approach. *Eur. J. Financ.* **2022**, *28*, 86–107. [CrossRef]
39. Zhao, Y. Research and Practice of Digital Economy Helping Rural Industry Revitalization in the Information Age. *Appl. Math. Nonlinear Sci.* **2024**, *9*, 1–15. [CrossRef]
40. Wu, W.; Ji, Z.; Liang, H. The Impact of New infrastructure Construction on Optimization and Upgrading of Industrial Structure. *Buildings* **2023**, *13*, 2580. [CrossRef]
41. Costa, F.; Frecassetti, S.; Rossini, M.; Portioli-Staudacher, A. Industry 4.0 digital technologies enhancing sustainability: Applications and barriers from the agricultural industry in an emerging economy. *J. Clean. Prod.* **2023**, *408*, 137208. [CrossRef]
42. Kumar, T. The Impact of Regional Infrastructure Investment in India. *Reg. Stud.* **2002**, *36*, 194–200. [CrossRef]
43. Qiu, Z.; Wang, Y.; Bao, L.; Yun, B.; Lu, J. Sustainability of Chinese Village Development in a New Perspective: Planning Principle of Rural Public Service Facilities Based on "Function-Space" Synergistic Mechanism. *Sustainability* **2022**, *14*, 8544. [CrossRef]
44. Peng, Z.; Dan, T. Digital dividend or digital divide? Digital economy and urban-rural income inequality in China. *Telecommun. Policy* **2023**, *47*, 102616. [CrossRef]
45. Lythreatis, S.; Singh, S.; El-Kassar, A. The digital divide: A review and future research agenda. *Technol. Forecast. Soc. Change* **2022**, *175*, 121359. [CrossRef]
46. Mushtaq, R.; Bruneau, C. Microfinance, financial inclusion and ICT: Implications for poverty and inequality. *Technol. Soc.* **2019**, *59*, 101154. [CrossRef]
47. Tian, Y.; Du, W.; Hua, Y.; Fan, L. Will infrastructure construction drive Chinese rural families to develop their own enterprises? *Appl. Econ. Lett.* **2022**, *29*, 1665–1669. [CrossRef]
48. Liu, Y.; Wang, Y.; Yan, B.; Zhang, H. The impact of new infrastructure development on urban total factor carbon productivity-empirical evidence based on Chinese cities. *Front. Environ. Sci.* **2024**, *12*, 1432534. [CrossRef]
49. Liu, X. New Digital Infrastructure, Financial Resource Allocation and High Quality Economic Development in the Internet Era. *Appl. Math. Nonlinear Sci.* **2023**, *9*, 1–17. [CrossRef]
50. Shi, Y.; Du, B.; Yang, B. How does digital inclusive finance affect common prosperity: Empirical evidence from China's underdeveloped and developed regions. *Cities* **2025**, *158*, 105640. [CrossRef]
51. Zhao, W.; Liang, Z.; Li, B. Realizing a Rural Sustainable Development through a Digital Village Construction: Experiences from China. *Sustainability* **2022**, *14*, 14199. [CrossRef]
52. Kong, F.; Wang, Q.; Liu, X.; Wen, C.; He, Q. Green development effect assessment and driving force analysis of the new infrastructure construction in China. *Environ. Dev. Sustain.* 2024; *in press*. [CrossRef]
53. Mao, F.; Wei, Y.; Wang, Y. Impact of computing infrastructure on carbon emissions in China. *Sci. Rep.* **2024**, *14*, 29814. [CrossRef] [PubMed]

54. Chen, J.; Li, H.; Luo, S.; Su, D.; Zang, T.; Kinoshita, T.; Yang, L. How do economic levels, urbanization, and infrastructure investments influence inequality in urban green space exposure? Insights from Japanese municipalities. *Urban For. Urban Green.* **2025**, *104*, 128649. [CrossRef]
55. Min, S.; Liu, M.; Huang, J. Does the application of ICTs facilitate rural economic transformation in China? Empirical evidence from the use of smartphones among farmers. *J. Asian Econ.* **2020**, *70*, 101219. [CrossRef]
56. Su, J.; Su, K.; Wang, S. Does the Digital Economy Promote Industrial Structural Upgrading?—A Test of Mediating Effects Based on Heterogeneous Technological Innovation. *Sustainability* **2021**, *13*, 10105. [CrossRef]
57. Shepherd, M.; Turner, J.; Small, B.; Wheeler, D. Priorities for science to overcome hurdles thwarting the full promise of the 'digital agriculture' revolution. *J. Sci. Food Agric.* **2020**, *100*, 5083–5092. [CrossRef]
58. Zhao, Q.; Pan, Y.; Xia, X. Internet can do help in the reduction of pesticide use by farmers: Evidence from rural China. *Environ. Sci. Pollut. Res.* **2021**, *28*, 2063–2073. [CrossRef]
59. Tiwasing, P.; Clark, B.; Gkartzios, M. How can rural businesses thrive in the digital economy? A UK perspective. *Heliyon* **2022**, *8*, e10745. [CrossRef]
60. Leng, C.; Ma, W.; Tang, J.; Zhu, Z. ICT adoption and income diversification among rural households in China. *Appl. Econ.* **2020**, *52*, 3614–3628. [CrossRef]
61. Rijswijk, K.; Klerkx, L.; Bacco, M.; Bartolini, F.; Bulten, E.; Debruyne, L.; Dessein, J.; Scotti, I.; Brunori, G. Digital transformation of agriculture and rural areas: A socio-cyber-physical system framework to support responsibilisation. *J. Rural. Stud.* **2021**, *85*, 79–90. [CrossRef]
62. Liu, X. Impact of digital inclusive finance development on rural industry revitalization—Observations of rural China. *Financ. Eng. Risk Manag.* **2024**, *7*, 130–143. [CrossRef]
63. Rao, C.; Gao, Y. Evaluation Mechanism Design for the Development Level of Urban-Rural Integration Based on an Improved TOPSIS Method. *Mathematics* **2022**, *10*, 380. [CrossRef]
64. Xu, W. The National Development and Reform Commission Clarifies the Scope of "New Infrastructure" and Will Focus on Four Key Areas of Work. Available online: http://finance.people.com.cn/n1/2020/0420/c1004-31680443.html (accessed on 12 December 2024).
65. Aschauer, D.A. Is public expenditure productive? *J. Monet. Econ.* **1989**, *23*, 177–200. [CrossRef]
66. Nunn, N.; Qian, N. US Food Aid and Civil Conflict. *Am. Econ. Rev.* **2014**, *104*, 1630–1666. [CrossRef]
67. Meng, S.; Xu, X. Can digital infrastructure improve corporate productivity? Evidence from a quasi-natural experiment in China. *Econ. Anal. Policy* **2025**, *85*, 1867–1884. [CrossRef]
68. Banerjee, A.; Duflo, E.; Qian, N. On the road: Access to transportation infrastructure and economic growth in China. *J. Dev. Econ.* **2020**, *145*, 102442. [CrossRef]
69. Yang, Y. Research on the Strategy of Sustainable Development of Leisure Sports and Rural Tourism Industry with the Help of Artificial Intelligence. *Appl. Math. Nonlinear Sci.* **2024**, *9*, 1–17. [CrossRef]
70. Chen, X.; Wang, B.; Zhang, G. Research on the Key Path of Digital Economy Empowering Rural Revitalization. *Popul. Resour. Environ. Econ.* **2023**, *4*, 125–132. [CrossRef]
71. Zheng, S.; Ding, M.; Li, Q.; Sohail, S.; Ullah, S. How does environmental policy stringency affect clean energy justice? The role of digital infrastructure. *Energy* **2025**, *318*, 134800. [CrossRef]
72. Du, X.; Zhang, H.; Han, Y. How Does New Infrastructure Investment Affect Economic Growth Quality? Empirical Evidence from China. *Sustainability* **2022**, *14*, 3511. [CrossRef]
73. Chen, Y.; Jiang, C.; Peng, L.; Zhao, S.; Chen, C. Digital infrastructure construction and urban industrial chain resilience: Evidence from the "Broadband China" strategy. *Sustain. Cities Soc.* **2025**, *121*, 106228. [CrossRef]
74. Zhu, F.; Zhang, F.; Ke, X. Rural industrial restructuring in China's metropolitan suburbs: Evidence from the land use transition of rural enterprises in suburban Beijing. *Land Use Policy* **2018**, *74*, 121–129. [CrossRef]
75. Wu, W.; Li, Y.; Liu, Y. What constrains impoverished rural regions: A case study of Henan Province in central China. *Habitat Int.* **2022**, *119*, 102477. [CrossRef]
76. Rahe, M.L.; Van Leuven, A.J.; Malone, T. Leveraging social ties to financial gains: Exploring the impact of social capital in rural development. *J. Rural. Stud.* **2025**, *114*, 103539. [CrossRef]

land

MDPI

Article

Does Intercity Transportation Accessibility Matter? Its Effects on Regional Network Centrality in South Korea

Sangwan Lee , Jeongbae Jeon *, Kuk Cho and Junhyuck Im

LX Spatial Information Research Institute, Korea Land and Geospatial Informatix Corporation, Jeonju 54870, Republic of Korea; esangwan@lx.or.kr (S.L.); kcho@lx.or.kr (K.C.)
* Correspondence: jbjeon@lx.or.kr

Abstract: This study investigates the relationship between intercity transportation accessibility and network centrality across South Korea by integrating Global Positioning System (GPS)-based mobility data with graph-theoretic centrality measures, including degree, PageRank, local clustering coefficient, harmonic, Katz, and information centrality. Employing both statistical modeling and machine learning techniques, this analysis uncovers key structural patterns and interaction effects within the national mobility network. The findings yield several important insights. First, the Seoul Metropolitan Area emerges as the dominant mobility hub, with Busan, Daegu, and Daejeon functioning as secondary centers, reflecting a polycentric urban configuration. Second, intermediary transfer hubs—despite having lower direct connectivity—substantially enhance overall network efficiency and interregional mobility. Third, transportation accessibility, particularly in relation to regional transit and highway infrastructure, exhibits a significant association with centrality measures and strong feature importance, identifying these modes as primary determinants of spatial connectivity. Fourth, the impact of accessibility on centrality is characterized by nonlinear relationships and threshold effects. By elucidating the complex interplay between mobility infrastructure and spatial network dynamics, this study contributes to a more comprehensive understanding of regional connectivity and network centrality and offers policy-relevant insights for future transportation planning.

Keywords: transportation and land use interaction; network centrality; transportation accessibility; GPS-based mobility data; interpretable machine learning

1. Introduction

Cities operate as complex systems comprising numerous interdependent components, including transportation networks, streets, buildings, and public spaces [1]. The dynamic interactions among these elements continuously shape urban form and function, influencing spatial organization and the distribution of human activities [2–4]. Among these, the spatial configuration of transportation infrastructure plays a particularly critical role in shaping patterns of accessibility, mobility, and land use [5]. Well-planned and strategically located transportation infrastructure plays a critical role in enhancing the efficient movement of people and goods, stimulating economic activity, mitigating spatial fragmentation, and guiding patterns of urban expansion by strengthening regional connectivity and spatial integration [6,7]. As such, transportation accessibility, determined by the location of transportation infrastructures, is fundamental to regional development, as it facilitates economic integration, population movement, and spatial connectivity [8,9].

Beyond their operational role, transportation systems also structure the spatial distribution of opportunity, directly shaping regional connectivity and influencing the geographic allocation of resources and populations [10]. Empirical studies have consistently shown that improved accessibility is associated with economic growth, productivity gains, and increased urban competitiveness [11]. However, the effects of enhanced transportation accessibility are not uniformly positive across all regions [12]. While it can support regional integration, it can also exacerbate spatial inequalities by disproportionately benefiting major metropolitan areas while leaving peripheral or rural regions behind [13,14].

Indeed, large urban centers often capture the bulk of the benefits generated by new or improved transportation infrastructure, while less connected regions may experience stagnation or even decline [15,16]. This uneven distribution of accessibility benefits has been documented in the context of multimodal transportation hubs, which tend to reinforce existing regional advantages and deepen disparities [16,17]. For instance, research in Greece has shown that improvements in transportation infrastructure can exacerbate socio-spatial inequalities by reinforcing hierarchical urban systems rather than fostering balanced development [18]. As regional inequalities deepen—especially between metropolitan cores and declining rural areas [19,20]—it becomes increasingly important to understand how transportation accessibility shapes the spatial structure of regional systems in order to promote equitable and sustainable development.

Parallel to discussions on accessibility, the concept of network centrality has gained increasing prominence in transportation and spatial network analysis. Centrality metrics are used to evaluate the structural importance of nodes (e.g., cities or regions) within a broader network based on their position and connectivity [21–23]. The degree to which a location is centrally positioned within a transportation network can influence its economic potential, accessibility to other regions, and vulnerability to spatial inequality [24]. Empirical research has demonstrated that regions with higher network centrality are often characterized by greater economic activity, denser urban development, and more substantial infrastructure investment [25–27]. In this context, centrality measures such as degree, betweenness, and PageRank have been applied to analyze the efficiency and functional hierarchy of urban and regional networks [28,29].

While accessibility describes the ease with which people and goods move between locations [30], centrality captures the strategic significance of a city or region within the transportation system [31]. The former is typically examined in relation to outcomes such as economic growth, land use, travel behavior, and housing markets [32–34], while the latter is employed to assess the structural integration and spatial embeddedness of cities [35,36]. Despite the conceptual overlap and practical interdependence of these two constructs, limited research has explicitly examined their interaction. Few studies have investigated how transportation accessibility influences the regional centrality within the network, and even fewer have attempted this using high-resolution spatial data in the context of rapidly evolving national urban systems.

To address these gaps, this study examines the relationship between intercity transportation accessibility—specifically, access to regional transit, intercity buses, and highways—and regional network centrality in South Korea. The study quantifies six network centrality indices using Global Positioning System (GPS)-based mobility data and evaluates them within three conceptual frameworks: (1) node importance and influence (degree and PageRank centrality), (2) network embeddedness and structural cohesion (local clustering coefficient and harmonic centrality), and (3) destination reachability and network efficiency (Katz and information centrality). This analysis employs correlation analysis (COR), ordinary least squares (OLS) regression, and spatial lag models (SLMs) to estimate the effect size and statistical significance of intercity transportation accessibility on these network centrality measures. Additionally, advanced machine learning (ML) techniques—including extreme gradient boosting (XGB), permutation-based feature importance (PBFI), and partial dependence plots (PDP)—are incorporated to identify feature importance and nonlinear and threshold relationships.

This study makes three contributions. First, it advances network theory in urban and transportation planning by integrating transportation accessibility and network centrality within a unified analytical framework. Second, it provides empirical insights into how intercity transit, buses, and highways collectively shape regional connectivity and spatial organization. Third, it offers actionable policy insights to support more balanced infrastructure development, mitigate regional disparities, and enhance the functional efficiency of transportation systems.

2. Materials and Methods

2.1. Study Area

This study focuses on South Korea, which is characterized by a highly developed yet spatially imbalanced transportation system [37,38]. It offers a valuable empirical setting for investigating the interaction between intercity transportation accessibility and network centrality for several reasons. First, its well-developed but unevenly utilized transportation infrastructure allows for a comparative analysis of regions with varying levels of accessibility [39]. Second, despite significant investments in high-speed rail (Korean Train eXpress, KTX), intercity bus services, and highways [40,41], regional disparities, shown in Figure 1, remain a persistent challenge [42,43], making South Korea an ideal case for examining the relationship between intercity transportation accessibility and regional network centrality.

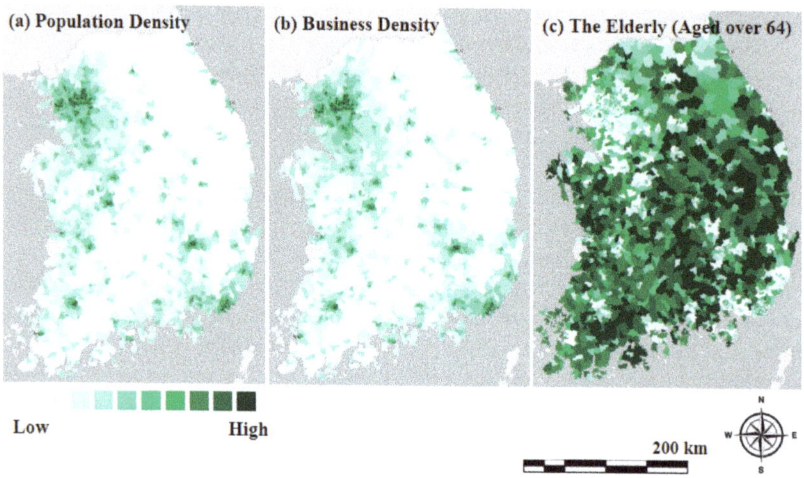

Figure 1. Characteristics of study area (South Korea).

2.2. Variables

2.2.1. Dependent Variables

This study examines the relationship between transportation accessibility and regional network centrality by employing six network centrality indices as dependent variables (see Table 1). Network centrality serves as a fundamental theoretical framework for understanding regional connectivity by quantifying the structural and functional significance of individual locations within a transportation or mobility network [29]. Among the various centrality indices, measures such as degree and PageRank have been widely adopted to evaluate the extent of a node's connectedness and its influence within a networked system [44,45]. These indicators are instrumental in identifying the relative importance of locations based on both direct and indirect connections [46].

Moreover, recent advances in network science suggest that relying on a limited set of centrality metrics may oversimplify the complex dynamics of spatial systems. Accordingly, emerging measures—such as harmonic, load, or percolation centrality—offer additional insights into dimensions such as network robustness, spatial embeddedness, and system-wide efficiency [47,48]. To comprehensively capture the multifaceted nature of regional connectivity, this study employs six key centrality indices, organized into three conceptual dimensions: positional accessibility, structural importance, and spatial cohesion (see Table 1).

- Node Importance and Influence: Degree centrality and PageRank centrality measure the prominence of a node within a network, indicating its connectivity and influence in facilitating mobility flows.
- Network Embeddedness and Structural Cohesion: Local clustering coefficient and harmonic centrality assess the extent to which a node contributes to regional cohesion and the overall network structure.
- Destination Reachability and Network Efficiency: Katz centrality and information centrality evaluate how effectively a node enables movement beyond its immediate connections, reflecting its role in ensuring network-wide accessibility and efficiency.

Table 1. Conceptual frameworks and descriptions of the six network centrality indices.

Category	Index	Conceptualization	Mean	S. D.
Centrality (Influence and Importance of Node)	Degree Centrality	This quantifies the number of direct connections a node has, providing a fundamental measure of network prominence. In transportation networks, high-degree nodes correspond to major transit hubs or well-connected urban centers that serve as primary points of access within the system [49].	3.62	0.61
	PageRank Centrality	This extends degree centrality by considering not just the number of connections but also the influence of those connections [50].	2.10	0.56
Proximity (Network Embeddedness and Structure)	Local Clustering Coefficient	This measures the tendency of nodes to form tightly knit clusters, indicating regional integration and resilience. A high clustering coefficient suggests that a node's neighbors are well connected, reinforcing spatial cohesion and localized accessibility [51].	2.31	0.71
	Harmonic Centrality	This accounts for the inverse of the shortest path distances from a node to all other nodes, emphasizing the efficiency of information or mobility flow within the network. Unlike closeness centrality, harmonic centrality remains well defined for disconnected networks, making it particularly relevant for assessing transportation accessibility in spatially fragmented regions [52].	4.40	0.14

Table 1. *Cont.*

Category	Index	Conceptualization	Mean	S. D.
Accessibility (Destination Reachability and Network Efficiency)	Katz Centrality	This extends degree centrality by incorporating indirect connections, assigning greater importance to nodes that are connected to other influential nodes. This metric is particularly relevant in transportation networks as it captures the long-range accessibility of regions beyond their immediate connections [53].	0.65	0.49
	Information Centrality	This evaluates the efficiency of information flow by considering all possible paths within the network rather than just shortest paths. In transportation systems, information centrality has been used to assess redundancy and resilience, ensuring that mobility networks remain functional despite disruptions [54]	3.91	0.45

2.2.2. Independent Variables

The independent variables selected for this study—network distances to the nearest transit station, intercity bus station, and highway interchange—reflect key components of intercity transportation accessibility (see Table 2). These variables were chosen based on their critical roles in shaping regional mobility patterns, influencing accessibility, and determining network centrality within urban and rural contexts [55,56]. Prior research has shown that proximity to transportation nodes such as rail and bus stations significantly affects spatial inequality, transit usage, and accessibility-related development outcomes [57,58]. By incorporating these three modes of intercity transportation—regional transit, intercity buses, and highways—this study seeks to provide a comprehensive assessment of how varying forms of intercity accessibility shape regional centrality patterns within the national spatial network.

Table 2. Descriptive statistics of independent variables.

Name	Description	Mean	S. D.
Transit	Log-transformed network distance, in meters, from the centroid of each EMD to the nearest regional transit station, including a high-speed rail (KTX) station	9.20	1.00
Bus	Log-transformed network distance, in meters, from the centroid of each EMD to the nearest intercity bus station	5.72	1.11
Highway	Log-transformed network distance, in meters, from the centroid of each EMD to the nearest highway interchange	9.01	0.77

2.3. Method

2.3.1. Method 1: Network Centrality and Accessibility Estimation

We first construct an empirical mobility network using GPS-based origin–destination (OD) data provided by Korea Telecom (KT), covering the period from 13 June to 19 June 2022, at the Eup/Myeon/Dong (EMD) level. The dataset captures the actual movement patterns of individuals across South Korea. By leveraging these GPS-derived

OD matrices, we calculate a suite of centrality indices (see Table 3)—including betweenness, closeness, and PageRank—that reflect not only the structural position of each node (i.e., city or region) but also its functional role in facilitating the circulation of people [59]. These indices were estimated using Python packages such as NetworkX, scaled to a 0–100 range for comparability, and subsequently log-transformed to address skewed distributions prior to statistical modeling and machine learning.

Table 3. Equations of the six network centrality indices.

Indices	Equation
Degree Centrality	$s_{total}(i) = s_{in}(i) + s_{out}(i) = \sum_j w_{ji} + \sum_j w_{ij}$ where $s_{\text{total}}(i)$ represents the weighted total degree of node i, $s_{\text{in}}(i)$ represents the weighted in-degree of node i (the sum of weights of incoming edges), $s_{\text{out}}(i)$ represents the weighted out-degree of node i (the sum of weights of outgoing edges), w_{ji} is the edge weight from node j to node i, and w_{ij} is the edge weight from node i to node j.
PageRank Centrality	$PR(i) = \frac{1-d}{N} + d \sum_{j \in M(i)} PR(j) \frac{w_{ji}}{\sum_k w_{jk}}$ where $PR(i)$ represents the PageRank value of node i, d is the damping factor (typically 0.85), N is the total number of nodes in the graph, $M(i)$ is the set of nodes that have links (edges) coming into node i, w_{ji} is the edge weight from node j to node i, and $\sum_k w_{jk}$ is the sum of weights of all edges going out from node j.
Local Clustering Coefficient	$C_w(i) = \frac{1}{s_i(k_i-1)} \sum_{j,h} \frac{w_{ij}+w_{ih}}{2} a_{ij}a_{ih}a_{jh}$ where $C_w(i)$ represents the weighted local clustering coefficient of node i, s_i is the strength of node i (the sum of all connected edge weights), k_i is the degree of node i, a_{ij} is an adjacency matrix element (1 if there is a connection between nodes i and j, 0 if not), w_{ij} is the edge weight between nodes i and j, and $a_{ij}a_{ih}a_{jh}$ indicates whether nodes i, j, and h form a triangle.
Harmonic Centrality	$H(i) = \sum_{j \neq i} \frac{1}{d_w(i,j)}$ where $H(i)$ represents the harmonic centrality value of node i, $d(i,j)$ is the shortest path distance from node i to node j, and the sum is calculated for all $j$nodes, excluding node i. If there is no path from node i to node j, $\frac{1}{d(i,j)}$ is treated as 0.
Katz Centrality	$I(i) = \left(\sum_{j=1}^{n} \frac{1}{D_{ij}}\right)^{-1}$ where $I(i)$ represents the information centrality value of node i, D_{ij} is the information distance between nodes i and j (calculated as $D_{ij} = (R_{ii} + R_{jj} - 2R_{ij})$), and R is the resistance matrix of the network.
Information Centrality	$K(i) = \sum_{k=1}^{\infty} \sum_j \alpha^k (A^k)_{ji}$ where $K(i)$ represents the Katz centrality value of node i, α is the attenuation factor (typically $\alpha < 1/\lambda_{max}$), λ_{max} is the largest eigenvalue of the adjacency matrix A, A is the adjacency matrix of the network, $(A^k)_{ji}$ is the (j,i)element of the $k-$th power of matrix A, k is the path length, and the sum is calculated for all $k \geq 1$ up to infinity.

Moreover, we compute the accessibility using network analysis techniques in ArcGIS Pro 3.3, incorporating national-level road network and transit infrastructure datasets to reflect actual travel paths. This measure is quantified by calculating the shortest network distance, in meters, from the centroid of each EMD administrative unit to the nearest node of each respective transportation mode. Also, to account for the highly skewed nature of raw distance values and to better capture the nonlinear relationship between accessibility and regional connectivity, all distance measures were log-transformed. The resulting maps in Figure 2 illustrate clear spatial disparities, with higher accessibility (i.e., shorter distances) concentrated around major metropolitan regions and limited accessibility observed in peripheral or rural areas.

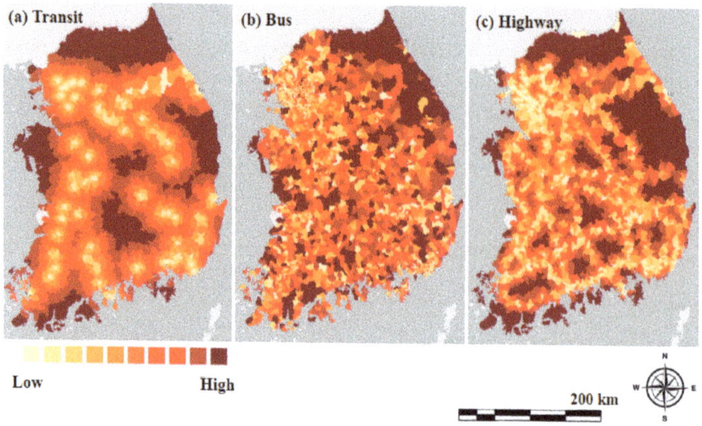

Figure 2. Spatial distributions of the transportation accessibility of South Korea.

2.3.2. Method 2: Econometrics Models

To explore the effect size and statistical significance of transportation accessibility on network centrality indices, this study first conducts a correlation analysis by using packages in R Studio 4.4.1, such as spdep and spatialreg. The correlation analysis provides an initial understanding of the linear relationships between transportation accessibility (measured by KTX, intercity buses, and highways) and the six network centrality indices [60]. Following this, an OLS regression model is applied to estimate the direct effects of transportation accessibility on each network centrality index, while controlling for potential confounding variables [61]. The OLS model helps quantify the impact of accessibility variables on the centrality of regions within the transportation network, with separate models being built for each centrality index to capture the unique dynamics associated with different dimensions of network centrality. Furthermore, to account for potential spatial dependencies—where the centrality of one region may be influenced by neighboring regions—this study employs an SLM. The SLM incorporates spatial interactions between regions, providing a more nuanced understanding of how transportation accessibility in one area may influence the centrality of surrounding regions [58,62].

2.4. Method 3: Machine Learning

This study also uses ML techniques, specifically XGB, an ensemble learning method that builds a series of decision trees, to predict complex, nonlinear relationships by using Python 3.13 packages, including Scikit-learn and XGBoost [63,64]. The decision to employ ML in this study, rather than conventional econometric approaches such as OLS, stems from the complex and potentially nonlinear nature of the relationships between network centrality metrics, transportation accessibility, and floating population patterns [65,66]. Traditional econometric methods, while statistically robust and interpretable, generally assume linearity and rely on predefined functional forms [67]. Although spatial regression models can account for spatial dependence [62], they still face limitations in capturing intricate nonlinear interactions or variable thresholds unless explicitly modeled [68]. In contrast, ML techniques, particularly ensemble methods like random forests and gradient boosting, offer greater flexibility in modeling high-dimensional and nonlinear patterns without requiring strict assumptions about data distribution or variable interactions [69,70].

To find the most optimal algorithm for this study, 10-fold cross-validation is employed to evaluate and compare the performance of three ML algorithms [71]: Random Forest, Gradient Boosting, and XGB. The results of the cross-validation, summarized in Table 4, demonstrate that XGB outperforms the other algorithms in terms of predictive accuracy and robustness, making it the optimal choice for this study. To further enhance model performance, hyper-parameter tuning is conducted through grid search, a method that systematically tests a predefined set of hyper-parameters to identify the configuration that maximizes the model performance [72,73]. We selected four key hyperparameters for XGB—Number of Estimators (NE), Maximum Depth (MD), Learning Rate (LR), and Subsample (SS)—based on their commonly recognized influence on model performance and generalization. The optimal hyperparameter values identified are presented in Table 5.

Table 4. Model performance comparisons using R-Squared.

Models	Degree	PageRank	Local	Harmonic	Information	Katz
RF	0.255	0.353	0.289	0.118	0.250	0.193
GB	0.282	0.480	0.295	0.177	0.266	0.207
XGB	0.292	0.485	0.304	0.209	0.268	0.215

Table 5. Hyper-parameters of optimal XGB algorithms.

Parameters	Degree	PageRank	Local	Harmonic	Information	Katz
LR	0.1	0.01	0.1	0.01	0.1	0.1
MD	3	5	3	5	3	3
NE	50	200	50	100	50	50
SS	0.8	0.8	0.8	0.8	0.8	0.9

Hyper-parameters: number of estimators (NE), maximum depth (MD), learning rate (LR), and sub-sample (SS).

Once the XGB model is optimized, PBFI and PDP are used to explore feature importance and nonlinear relationships between transportation accessibility and network centrality indices. First, to evaluate the relative importance of each feature—specifically, transportation accessibility measures and control variables—this study employs PBFI. This technique assesses the impact of each feature by randomly permuting its values and measuring the corresponding changes in model performance, thereby revealing the significance of individual variables [74]. Additionally, PDPs are utilized to explore the nonlinear relationships between transportation accessibility and network centrality indices. PDPs visualize the marginal effect of one or more features on the predicted outcome, providing insights into how variations in transportation accessibility influence network centrality while controlling for other factors [68].

3. Results

3.1. Network Centrality Structure of South Korea

Figure 3 presents the spatial distributions of six network centrality indices estimated from GPS-based mobility data.

The key findings are as follows. Degree centrality exhibits high values in the Seoul Metropolitan Area and other major urban centers, indicating that these locations serve as primary hubs with a dense network of direct connections. PageRank centrality, while generally aligned with degree centrality, also highlights regions that possess fewer direct connections but maintain high importance due to their linkages with critical nodes. The local clustering coefficient is notably high in metropolitan areas and some regional centers, indicating strong intra-regional connectivity. Harmonic centrality exhibits its highest values in densely connected yet non-peripheral urban areas, particularly in areas with extensive transit and road networks. Katz centrality closely mirrors the spatial patterns of PageRank centrality, emphasizing the significance of both direct and indirect connectivity to major hubs. Information centrality is concentrated in large metropolitan areas and along major transportation corridors, particularly in and around Seoul.

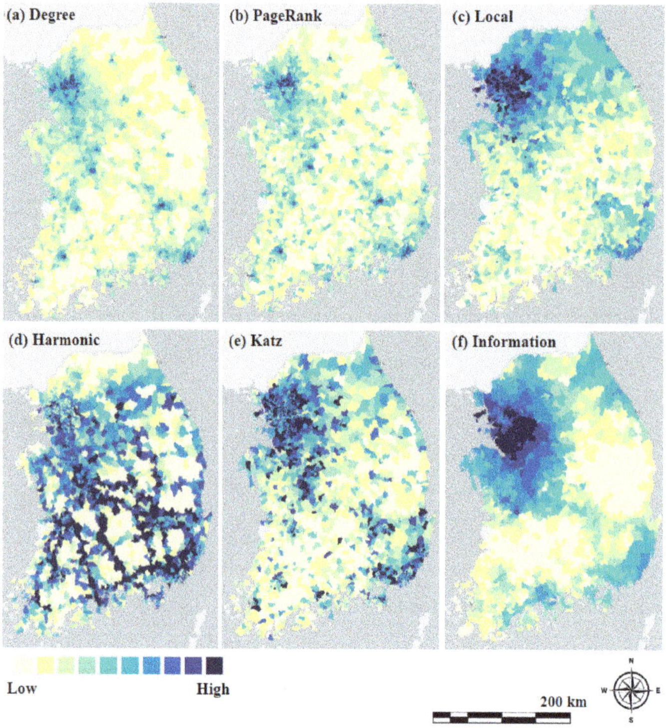

Figure 3. Spatial distributions of the network centrality indices of South Korea.

3.2. Effect Size and Significance of Transportation Accessibility on Network Centrality

Table 6 presents the associations between transportation accessibility—measured as network distance to regional transit, intercity bus, and highways—and six network centrality indices using COR, OLS, and SLM. The results highlight several notable patterns. First, the negative coefficients across most models indicate that shorter distances to transit, bus, and highway infrastructure are generally associated with higher network centrality. This suggests that areas with better transportation accessibility tend to function as key nodes in the mobility network, facilitating movement and interaction. Second, Katz centrality, which captures both direct and indirect connections, is significantly influenced by transit, bus, and highway accessibility, with shorter distances corresponding to higher connectivity (coefficient of −0.018, −0.036, and −0.063 in SLM).

Table 6. Results of correlation, ordinary least squares regression, and spatial lag models.

Models	Degree	PageRank	Local	Harmonic	Katz	Information
			Effect of Regional Transit Accessibility			
COR	−0.542 ***	−0.213 ***	−0.541 ***	−1.922 **	−0.556 ***	−0.610 ***
OLS	−0.118 ***	−0.042 ***	−0.176 ***	−0.029 ***	−0.064 ***	−0.056 ***
SLM	−0.033 ***	−0.023 **	−0.025 ***	−0.016 ***	−0.018 ***	−0.003 *

Table 6. *Cont.*

Models	Degree	PageRank	Local	Harmonic	Katz	Information
			Effect of Intercity Bus Accessibility			
COR	−0.470 ***	−0.184 **	−0.570 **	−1.438 **	−0.673 **	−0.693 **
OLS	−0.055 ***	−0.020 ***	−0.144 ***	−0.010 ***	−0.076 ***	−0.059 ***
SLM	−0.017 ***	−0.018 **	−0.033 ***	−0.005 **	−0.036 ***	−0.005 *
			Effect of Highway Accessibility			
COR	−0.461 **	−0.153 **	−0.286 **	0.983 **	−0.492 **	−0.606 **
OLS	−0.216 ***	−0.055 ***	−0.110 ***	−0.018 ***	−0.140 ***	−0.157 ***
SLM	−0.092 ***	−0.045 ***	−0.004 *	−0.006 **	−0.063 ***	−0.025 ***

Models: correlation (COR), ordinary least squares regression (OLS) model, and spatial lag model (SLM), Significance level: * $p < 0.1$; ** $p < 0.05$; *** $p < 0.01$.

Third, harmonic centrality, which emphasizes global efficiency in shortest-path distances, displays the strongest response to transit accessibility, with a coefficient of −1.922 in the COR model and significant effects in the OLS and SLM. This suggests that proximity to regional transit hubs enhances the overall efficiency of the mobility network by reducing travel distances across the system. Fourth, a key insight from the findings is that the local clustering coefficient, which measures neighborhood-level connectivity, is negatively associated with transportation accessibility across all modes. This suggests that areas with strong transit and highway accessibility tend to be more integrated into the broader network rather than functioning as self-contained clusters.

3.3. Feature Importance of Accessibility Factors in Predicting Network Centrality

Table 7 presents the PBFI results from the XGB, assessing the relative contributions of the accessibility in predicting network centrality indices. Some of the major findings are as follows. First, highway accessibility emerges as the most influential factor, exhibiting the highest feature importance values for degree (0.555), PageRank (0.556), harmonic (0.562), information (0.501), and Katz (0.526) centralities. These findings suggest that proximity to highways is a primary determinant of network centrality, likely due to the extensive spatial coverage and direct connectivity that highways provide.

Table 7. Permutation-based feature importance results.

Variables	Degree	PageRank	Local	Harmonic	Katz	Information
Transit	0.410	0.343	0.533	0.409	0.229	0.349
Bus	0.035	0.101	0.272	0.029	0.245	0.150
Highway	0.555	0.556	0.195	0.562	0.526	0.501

Second, transit accessibility exhibits notable significance, particularly in the local clustering coefficient (0.533), reinforcing the notion that transit infrastructure fosters localized network cohesion. This finding suggests that areas with higher transit accessibility are more likely to exhibit strong intra-regional connectivity, likely driven by dense transit-oriented development patterns and frequent service intervals. Additionally, transit accessibility contributes moderately to degree (0.410) and harmonic (0.409) centrality, indicating its role in both direct network connectivity and overall accessibility efficiency.

Third, intercity bus accessibility demonstrates the lowest influence, a marginal one, among the three transportation modes. The feature importance values for bus accessibility remain relatively low across all centrality measures, with the highest being 0.272 for the local clustering coefficient. This suggests that intercity bus services, while relevant for localized connectivity, do not exert a dominant influence on the broader mobility network compared to highways and transit systems.

3.4. Nonlinear Associations Between Transportation Accessibility and Network Centrality

The following subsections present the results of PDPs, illustrating the nonlinear associations between network distance to transit, bus, and highway infrastructures and six network centrality indices. The PDP results in Figure 4 indicate that degree centrality initially increases slightly with transit accessibility but starts to decline beyond a certain threshold. This pattern suggests that while proximity to transit infrastructure enhances local connectivity, excessive distance weakens it, reinforcing the presence of a threshold effect. Similarly, degree centrality declines as distance from highways increases, with a steep drop at a specific point. In contrast, the bus network exhibits a relatively stable effect on degree centrality, implying that bus accessibility has a less pronounced impact on regional connectivity.

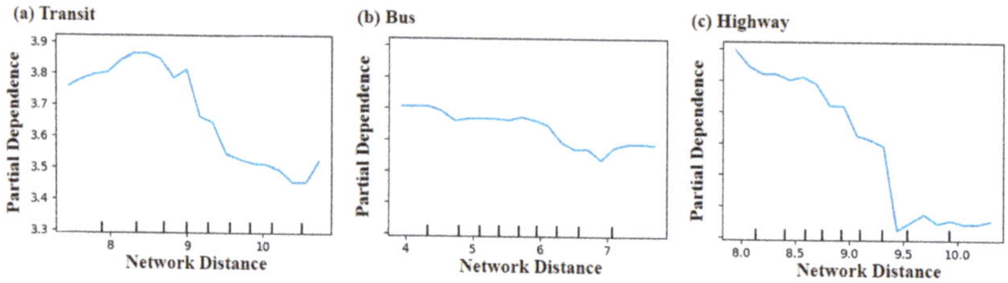

Figure 4. Partial dependence plots on degree centrality Index.

Figure 5 presents the PDP results for PageRank centrality, revealing a general declining trend for transit accessibility. As distance from transit increases, the relative influence of nodes within the network diminishes, suggesting that accessibility plays a key role in maintaining central nodes. Additionally, the highway PDP demonstrates a sharp decline, reinforcing the idea that highway proximity contributes significantly to maintaining network importance.

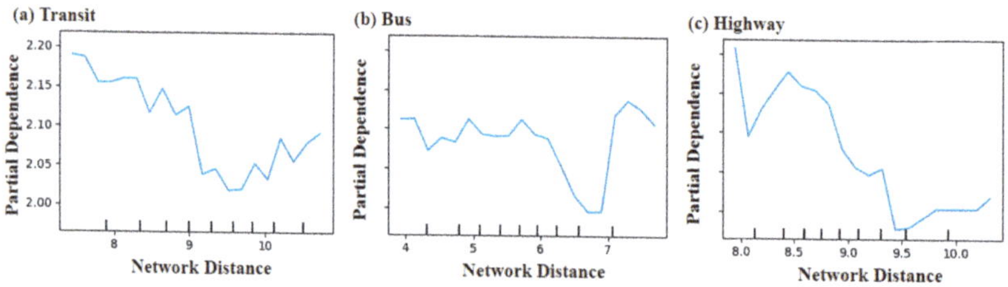

Figure 5. Partial dependence plots on PageRank centrality index.

Figure 6 illustrates the PDPs for the local clustering coefficient, a measure of network embeddedness and cohesion. Transit accessibility exhibits a gradual decline beyond a certain threshold, indicating that increased distance from transit reduces the likelihood of nodes forming tightly clustered sub-networks. Similarly, bus accessibility follows a consistent downward trend. Highway accessibility demonstrates a clear decreasing trend beyond a certain threshold, highlighting that nodes near highways tend to be more locally embedded, whereas those farther away exhibit reduced cohesion.

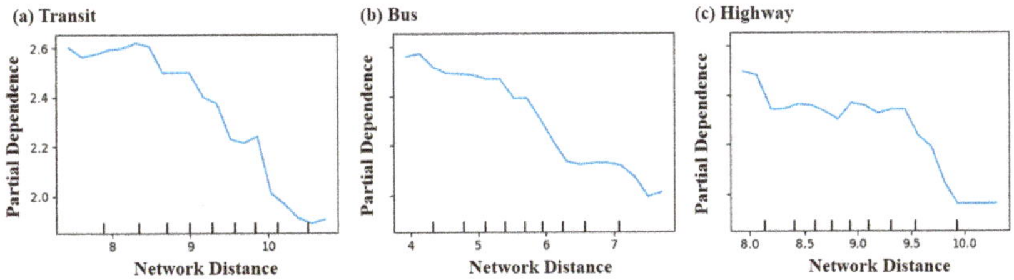

Figure 6. Partial dependence plots on local clustering coefficient centrality index.

PDPs in Figure 7 reveal that harmonic centrality, which captures the efficiency of a node in reaching other nodes, declines sharply with increasing distance from transit infrastructure and highways. For transit, this decline begins beyond a critical threshold, indicating a significant loss of efficiency as accessibility decreases. In contrast, the bus network exhibits a relatively stable effect, with minor variations suggesting a weaker relationship between bus accessibility and harmonic centrality.

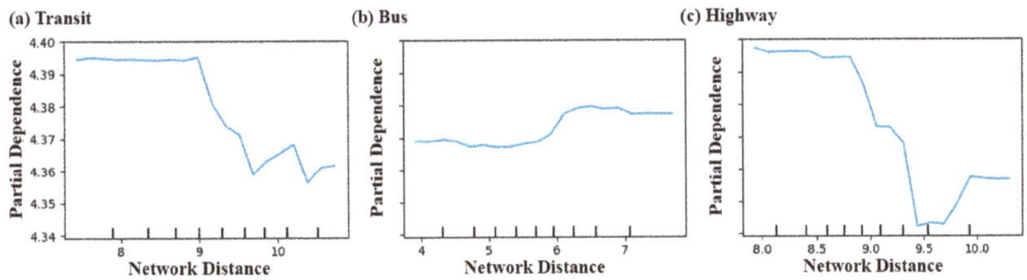

Figure 7. Partial dependence plots on harmonic centrality index.

Figure 8 presents the PDP results for Katz centrality. Transit accessibility initially increases Katz centrality, peaking at a specific threshold before declining significantly. This suggests that moderate proximity to transit infrastructure enhances network influence, while excessive distance weakens it. The bus network exhibits a steady decline. Highways display a pronounced threshold effect, beyond which Katz centrality sharply drops, reinforcing the importance of highway proximity for regional influence.

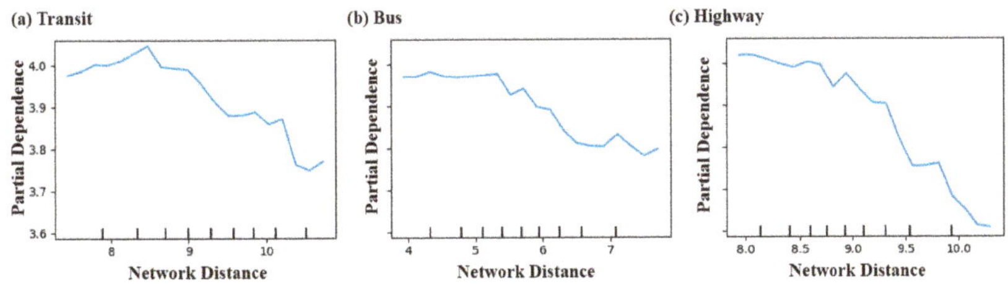

Figure 8. Partial dependence plots on Katz centrality index.

Figure 9 illustrates the PDP results for information centrality, which evaluates the role of a node in efficient information transfer within the network. Transit accessibility exhibits a declining trend, with a significant drop beyond a threshold distance. Similarly, bus accessibility reveals a gradual decline with a steeper drop beyond a certain threshold, indicating a critical threshold beyond which network efficiency deteriorates. Highways demonstrate the strongest threshold effect, with a sharp decline beyond a certain distance, suggesting that highway proximity is essential for maintaining network-wide communication efficiency.

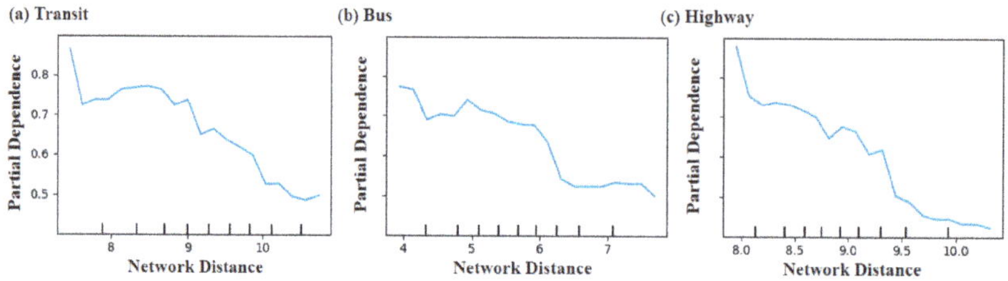

Figure 9. Partial dependence plots on information centrality index.

4. Discussions

4.1. Network Hierarchy and Structure of South Korea

Consistent with findings from previous studies [37,75], the pronounced concentration of high centrality values in the Seoul Metropolitan Area reaffirms its dominant position as the principal hub for both regional and national mobility. In addition, the emergence of secondary hubs in cities such as Busan, Daegu, and Daejeon reflects a polycentric urban structure that promotes inter-regional connectivity and reduces overdependence on a single metropolitan core.

Beyond reinforcing the findings from the existing literature, this study also reveals the strategic significance of intermediary cities that exhibit high centrality despite having relatively few direct connections. In particular, the results derived from Katz centrality emphasize the critical role of indirect linkages, suggesting that network-wide accessibility is sustained not only by direct connections but also through cumulative, long-range interactions. Such findings highlight the necessity of incorporating measures of indirect access when assessing the functional importance of nodes within the national transportation system.

Furthermore, the spatial heterogeneity observed in harmonic and information centrality underscores the multifaceted nature of an efficient and resilient transportation network. While certain regions enhance network performance by contributing to efficient shortest-path routing, others strengthen overall robustness by providing redundant pathways that can mitigate the impacts of localized disruptions. This balance between efficiency and resilience is vital for ensuring stable and equitable connectivity across regions.

4.2. The Role of Intermediary Transfer Hubs in Spatial Hierarchy

Departing from conventional accessibility research that primarily focuses on metropolitan cores such as Seoul [76], this study underscores the strategic role of intermediary transfer hubs in enhancing regional connectivity and mitigating spatial inequality. Although these hubs may not exhibit the highest levels of direct connectivity, they play a pivotal role in strengthening network centrality by facilitating efficient interregional linkages. By serving as critical junctions within the national transportation system, these intermediary nodes help redistribute mobility flows and reduce the excessive concentration of accessibility benefits in core metropolitan areas. This decentralizing effect not only enhances the structural cohesion of the network but also supports more balanced regional development.

These findings align with the existing literature on urban mobility systems [77,78], which highlights the significance of well-integrated transfer hubs in improving overall network performance. In the context of South Korea—where long-standing disparities between urban cores and peripheral regions persist [41,79]—strengthening such intermediary hubs offers a practical pathway toward more equitable spatial outcomes. Strategic investments in multimodal transfer infrastructure, such as express bus terminals and high-speed rail stations, can enhance the connectivity and functional importance of semi-core regions, thereby elevating their centrality within the broader transportation network.

Looking forward, the emergence of advanced mobility solutions, including urban air mobility (UAM), may present opportunities to further reinforce the role of intermediary cities. Recent studies on vertiport location optimization and UAM-integrated urban design [80–82] suggest that these cities could become increasingly important nodes in future hybrid mobility systems. By positioning intermediary hubs as central components of both current and next-generation transport networks, policymakers can advance a more resilient, connected, and spatially inclusive national development strategy.

4.3. Nonlinear Dynamics and Multifacted Impacts of Accessibility on Network Centrality

This analysis reveals that the relationship between transportation accessibility and network centrality is characterized by distinct nonlinear patterns, suggesting the presence of threshold effects that have been largely underexplored in previous studies. In particular, degree centrality exhibits a sharp increase with improved transit accessibility up to a certain threshold, beyond which the marginal gains in centrality diminish. This pattern reflects the saturation point at which additional transit proximity yields limited improvements in nodal influence within the network. Similarly, PageRank centrality shows a steep decline as the distance from highway infrastructure increases, reaffirming the foundational role that expressways play as structural backbones in the national mobility system. In contrast, intercity bus accessibility demonstrates more stable and moderate effects across various centrality indices, implying a dispersed yet consistently supportive role in enhancing connectivity—particularly for regions that may lack access to high-capacity transit infrastructure.

Beyond reaffirming the established association between transportation accessibility and network centrality [16,57,83], this study contributes to the literature by disentangling the heterogeneous effects of different transportation modes across a range of centrality measures. Notably, transit accessibility emerges as the most influential determinant of the local clustering coefficient, highlighting its key role in strengthening intra-regional cohesion and connectivity between proximate nodes.

Intercity bus accessibility, while exhibiting a comparatively weaker influence, nonetheless enhances the continuity of the national network by facilitating linkages in underserved or peripheral areas. Highway accessibility, on the other hand, exerts the most substantial influence across multiple centrality metrics—including degree, PageRank, and harmonic centrality—underscoring its central role in maintaining large-scale spatial integration and supporting the hierarchical structure of the transportation network.

4.4. Policy Implications for Transportation Infrastructure and Spatial Inequality

The findings provide several policy implications. First, enhancing regional transportation networks—especially high-frequency regional rail and intercity bus systems—can significantly improve national network performance by elevating centrality measures such as harmonic and Katz centrality, which reflect systemic resilience and indirect accessibility. These investments have the potential to redistribute connectivity more equitably across regions, especially benefiting peripheral or structurally disadvantaged areas that have experienced sustained population decline and economic stagnation.

Second, while the positive association between transportation accessibility and network centrality underscores the developmental benefits of proximity to major transport infrastructure, this relationship also poses risks. Infrastructure investments that disproportionately concentrate in already-centralized urban cores can exacerbate spatial polarization, reinforcing the dominance of major metropolitan regions such as Seoul and accelerating the marginalization of rural and shrinking cities. This pattern aligns with the core–periphery dynamic often observed in the theory of spatial polarization. As such, infrastructure planning needs to be guided by principles of spatial equity, ensuring that underdeveloped regions are not further excluded from the benefits of regional connectivity.

Third, large-scale infrastructure projects can produce dual outcomes in lagging regions. On one hand, improved accessibility can catalyze economic revitalization; on the other, such projects may bypass smaller towns or reinforce the concentration of resources and services in dominant hubs. To mitigate these unintended consequences, complementary policies—such as the development of local feeder networks, regional multimodal transfer hubs, and last-mile connectivity—should be implemented to ensure that smaller municipalities are meaningfully integrated into the broader mobility system.

Fourth, the expansion of highway networks offers targeted potential to reconnect shrinking cities and rural areas, particularly in regions that lack robust transit options. However, these investments must be accompanied by land use coordination and environmental safeguards to prevent negative externalities such as sprawl, ecological degradation, and infrastructure redundancy. A strategic integration of transportation planning with broader regional development goals is critical to promote long-term territorial cohesion.

Finally, future mobility strategies must be adaptive and forward-looking, particularly as emerging technologies reshape intercity connectivity. The development of multimodal mobility corridors that integrate traditional modes with novel services—such as UAM—could enhance accessibility and resilience for intermediary and declining regions. Planning frameworks should prioritize inclusivity and balanced development, ensuring that innovation in mobility infrastructure does not amplify existing inequalities but rather contributes to a more spatially equitable and sustainable national development strategy.

4.5. Limitations of This Study

Despite its contributions, this study has several limitations. First, the analysis is based on a cross-sectional dataset, limiting the ability to infer causal relationships over time. A longitudinal approach incorporating temporal dynamics would provide deeper insights into how transportation accessibility influences network centrality across different periods. Second, while this study controls for spatial dependencies, it does not fully account for other socio-economic factors—such as land use patterns, economic activity, and demographic shifts—that may also shape network centrality. Incorporating these factors in future research could provide a more comprehensive understanding of regional mobility patterns. Lastly, although this study centers on South Korea, its insights hold broader relevance. While the core mechanism—where enhanced accessibility increases regional network centrality—may be widely applicable, context-specific factors such as governance capacity, population distribution, and planning regimes should be accounted for.

4.6. Future Research Direcitons

Future research can build upon this study in several key directions. First, incorporating longitudinal data would facilitate dynamic analyses of how transportation investments influence regional connectivity over time. Second, integrating additional socio-economic and land use variables could provide a more comprehensive understanding of the multifaceted drivers of network centrality. Third, combining network analysis with agent-based modeling or simulation techniques may offer deeper insights into the interplay between transportation accessibility, mobility behavior, and regional development trajectories. Fourth, future studies should explore the implications of emerging mobility technologies—particularly UAM vertiport—on regional network structures. Lastly, comparative cross-national research could offer valuable contributions to the evolving discourse on transportation infrastructure and regional equity.

5. Conclusions

This study examines the relationship between intercity transportation accessibility and regional network centrality in South Korea. Specifically, it evaluates how access to regional transit, intercity buses, and highways influences the structural positioning of regions within transportation networks. We employ both econometric models—such as COR, OLS, and SLM—and ML techniques, including XGB with PBFI and PDP. This study contributes by (1) advancing network theory in urban and transportation planning through the introduction of a novel analytical framework, (2) offering empirical evidence on the influence of transit, buses, and highways in shaping regional connectivity and spatial structure, and (3) providing policy recommendations for optimizing transportation networks, mitigating regional inequalities, and enhancing infrastructure planning.

Author Contributions: Conceptualization, S.L., J.J., K.C. and J.I.; methodology, S.L., J.J. and K.C.; software, S.L., J.J. and J.I.; validation, K.C. and J.I.; formal analysis, S.L. and J.J.; investigation, S.L. and J.J.; data curation, J.J. and J.I.; writing—original draft preparation, S.L. and J.J.; writing—review and editing, K.C. and J.I.; visualization, S.L.; supervision, J.J. and J.I.; project administration, K.C.; funding acquisition, K.C. All authors have read and agreed to the published version of the manuscript.

Funding: This research was supported by the Ministry of Trade, Industry, and Energy in the Republic of Korea (grant number: P0020670, Research Title: Establishing a Demonstration Infrastructure of Autonomous Cargo Transportation Service for Commercial Vehicles in Saemangeum).

Data Availability Statement: The data presented in this study are available on request from the corresponding author.

Conflicts of Interest: Authors Sangwan Lee, Jeongbae Jeon, Kuk Cho and Junhyuck Im were employed by the company Korea Land and Geospatial Informatix Corporation. The remaining authors declare that the research was conducted in the absence of any commercial or financial relationships that could be construed as a potential conflict of interest.

References

1. Bettencourt, L.M.A. *Introduction to Urban Science: Evidence and Theory of Cities as Complex Systems*; MIT Press: Cambridge, MA, USA, 2021; ISBN 978-0-262-04600-8.
2. Anas, A.; Arnott, R.; Small, K.A. Urban Spatial Structure. *J. Econ. Lit.* **1998**, *36*, 1426–1464.
3. Hou, Q.; Li, S.-M. Transport Infrastructure Development and Changing Spatial Accessibility in the Greater Pearl River Delta, China, 1990–2020. *J. Transp. Geogr.* **2011**, *19*, 1350–1360. [CrossRef]
4. Bardaka, E.; Delgado, M.S.; Florax, R.J.G.M. A Spatial Multiple Treatment/Multiple Outcome Difference-in-Differences Model with an Application to Urban Rail Infrastructure and Gentrification. *Transp. Res. Part A Policy Pract.* **2019**, *121*, 325–345. [CrossRef]
5. Rode, P.; Floater, G.; Thomopoulos, N.; Docherty, J.; Schwinger, P.; Mahendra, A.; Fang, W. Accessibility in Cities: Transport and Urban Form. In *Disrupting Mobility: Impacts of Sharing Economy and Innovative Transportation on Cities*; Meyer, G., Shaheen, S., Eds.; Springer International Publishing: Cham, Switzerland, 2017; pp. 239–273. ISBN 978-3-319-51602-8.
6. Zhou, H.; Gao, H. The Impact of Urban Morphology on Urban Transportation Mode: A Case Study of Tokyo. *Case Stud. Transp. Policy* **2020**, *8*, 197–205. [CrossRef]
7. Liu, X.; Derudder, B.; Wu, K. Measuring Polycentric Urban Development in China: An Intercity Transportation Network Perspective. *Reg. Stud.* **2016**, *50*, 1302–1315. [CrossRef]
8. Levine, J.; Grengs, J.; Merlin, L.A. *From Mobility to Accessibility: Transforming Urban Transportation and Land-Use Planning*; Cornell University Press: Ithaca, NY, USA, 2019; ISBN 978-1-5017-1610-2.
9. Dadhich, P.N.; Hanaoka, S. Spatial Investigation of the Temporal Urban Form to Assess Impact on Transit Services and Public Transportation Access. *Geo-Spat. Inf. Sci.* **2012**, *15*, 187–197. [CrossRef]
10. Gil, J. Analyzing the Configuration of Multimodal Urban Networks. *Geogr. Anal.* **2014**, *46*, 368–391. [CrossRef]
11. Vickerman, R.; Spiekermann, K.; Wegener, M. Accessibility and Economic Development in Europe. *Reg. Stud.* **1999**, *33*, 1–15. [CrossRef]
12. Yang, J.; Lee, S.; Im, J.; Cho, K.; Park, H. Perceptual Disparities in Transportation Accessibility, Neighborhood Quality, and Satisfaction across Metro, Urban, and Rural Areas. *Sens. Mater.* **2024**, *36*, 3899. [CrossRef]

13. Câmara, P.; and Banister, D. Spatial Inequalities in the Provision of Public Transport in Latin American Cities. *Transp. Rev.* **1993**, *13*, 351–373. [CrossRef]
14. Chang, Z.; Chen, J.; Li, W.; Li, X. Public Transportation and the Spatial Inequality of Urban Park Accessibility: New Evidence from Hong Kong. *Transp. Res. Part D Transp. Environ.* **2019**, *76*, 111–122. [CrossRef]
15. Israel, E.; Frenkel, A. Social Justice and Spatial Inequality: Toward a Conceptual Framework. *Prog. Hum. Geogr.* **2017**, *42*, 030913251770296. [CrossRef]
16. Lee, S. Literature Review on High-Speed Rail and Regional Uneven Development. *J. Korea Plan. Assoc.* **2023**, *58*, 5–18. [CrossRef]
17. Quiroz Villanueva, M.J.; Guglielmi, F.; De Fabiis, F.; Coppola, P. Social and Economic Impacts of Transportation Multi-Modal and Multi-Service Hub Development. *Sustainability* **2025**, *17*, 1767. [CrossRef]
18. Panagiotopoulos, G.; Kaliampakos, D. Accessibility and Spatial Inequalities in Greece. *Appl. Spat. Anal.* **2019**, *12*, 567–586. [CrossRef]
19. Modai-Snir, T.; van Ham, M. Neighbourhood Change and Spatial Polarization: The Roles of Increasing Inequality and Divergent Urban Development. *Cities* **2018**, *82*, 108–118. [CrossRef]
20. Wei, Y.D.; Wu, Y.; Liao, F.H.; Zhang, L. Regional Inequality, Spatial Polarization and Place Mobility in Provincial China: A Case Study of Jiangsu Province. *Appl. Geogr.* **2020**, *124*, 102296. [CrossRef]
21. Jang, S.; An, Y. Influence of Centrality Indices of Urban Railway Stations: Social Network Analysis of Transit Ridership and Travel Distance. *J. Reg. City Plan.* **2023**, *33*, 323–344. [CrossRef]
22. Piraveenan, M.; Saripada, N.B. Transportation Centrality: Quantifying the Relative Importance of Nodes in Transportation Networks Based on Traffic Modeling. *IEEE Access* **2023**, *11*, 142214–142234. [CrossRef]
23. Pflieger, G.; Rozenblat, C. Introduction. Urban Networks and Network Theory: The City as the Connector of Multiple Networks. *Urban Stud.* **2010**, *47*, 2723–2735. [CrossRef]
24. Fleming, D.K.; Hayuth, Y. Spatial Characteristics of Transportation Hubs: Centrality and Intermediacy. *J. Transp. Geogr.* **1994**, *2*, 3–18. [CrossRef]
25. Joseph, A.; Chen, G. Network Centrality and Key Economic Indicators: A Case Study. In *Network Models in Economics and Finance*; Kalyagin, V.A., Pardalos, P.M., Rassias, T.M., Eds.; Springer International Publishing: Cham, Switzerland, 2014; pp. 159–180. ISBN 978-3-319-09683-4.
26. Wang, H.; Zhao, Y.; Dang, B.; Han, P.; Shi, X. Network Centrality and Innovation Performance: The Role of Formal and Informal Institutions in Emerging Economies. *J. Bus. Ind. Mark.* **2019**, *34*, 1388–1400. [CrossRef]
27. Burger, M.J.; Meijers, E.J. Agglomerations and the Rise of Urban Network Externalities. *Pap. Reg. Sci.* **2016**, *95*, 5–16. [CrossRef]
28. Tsiotas, D.; Polyzos, S. Introducing a New Centrality Measure from the Transportation Network Analysis in Greece. *Ann. Oper. Res.* **2015**, *227*, 93–117. [CrossRef]
29. Rodrigues, F.A. Network Centrality: An Introduction. In *A Mathematical Modeling Approach from Nonlinear Dynamics to Complex Systems*; Macau, E.E.N., Ed.; Springer International Publishing: Cham, Switzerland, 2019; pp. 177–196. ISBN 978-3-319-78512-7.
30. Ana, V. Accessibility Concepts and Indicators in Transportation Strategic Planning Issues: Theoretical Framework and Literature Review. *Logist. Sustain. Transp.* **2016**, *7*, 58–67.
31. Chakrabarti, S.; Kushari, T.; Mazumder, T. Does Transportation Network Centrality Determine Housing Price? *J. Transp. Geogr.* **2022**, *103*, 103397. [CrossRef]
32. Saif, M.A.; Zefreh, M.M.; Torok, A. Public Transport Accessibility: A Literature Review. *Period. Polytech. Transp. Eng.* **2019**, *47*, 36–43. [CrossRef]
33. Lee, S. Measuring the Influence of Transit Accessibility on Housing and Transportation Expenditures in the U.S.: A Propensity Score Matching Approach. *Case Stud. Transp. Policy* **2023**, *14*, 101092. [CrossRef]
34. He, S.Y. Regional Impact of Rail Network Accessibility on Residential Property Price: Modelling Spatial Heterogeneous Capitalisation Effects in Hong Kong. *Transp. Res. Part A Policy Pract.* **2020**, *135*, 244–263. [CrossRef]
35. Zhong, C.; Arisona, S.M.; Huang, X.; Batty, M.; Schmitt, G. Detecting the Dynamics of Urban Structure through Spatial Network Analysis. *Int. J. Geogr. Inf. Sci.* **2014**, *28*, 2178–2199. [CrossRef]
36. Ding, R.; Ujang, N.; Hamid, H.B.; Manan, M.S.A.; Li, R.; Albadareen, S.S.M.; Nochian, A.; Wu, J. Application of Complex Networks Theory in Urban Traffic Network Researches. *Netw. Spat. Econ.* **2019**, *19*, 1281–1317. [CrossRef]
37. Lee, S.; Im, J.; Cho, K. Understanding Spatial Inequalities and Stratification in Transportation Accessibility to Social Infrastructures in South Korea: Multi-Dimensional Planning Insights. *Sci. Rep.* **2024**, *14*, 18445. [CrossRef] [PubMed]
38. Lee, S. Spatial and Socioeconomic Inequalities in Accessibility to Healthcare Services in South Korea. *Healthcare* **2022**, *10*, 2049. [CrossRef]
39. Kim, H.; Sultana, S. The Impacts of High-Speed Rail Extensions on Accessibility and Spatial Equity Changes in South Korea from 2004 to 2018. *J. Transp. Geogr.* **2015**, *45*, 48–61. [CrossRef]
40. Kim, H.; Lee, K.; Park, J.S.; Song, Y. Transit Network Expansion and Accessibility Implications: A Case Study of Gwangju Metropolitan Area, South Korea. *Res. Transp. Econ.* **2018**, *69*, 544–553. [CrossRef]

41. Lee, S.; Ock, Y.; Kim, M.; Schrock, G. The Regional Uneven Development and the State Intervention to Reshape the Spatial Hierarchy. *Int. J. Reg. Dev.* **2021**, *8*, 1. [CrossRef]
42. van Geenhuizen, M. Interconnectivity of Transport Networks: A Conceptual and Empirical Exploration. *Transp. Plan. Technol.* **2000**, *23*, 199–213. [CrossRef]
43. Alstadt, B.; Weisbrod, G.; Cutler, D. Relationship of Transportation Access and Connectivity to Local Economic Outcomes: Statistical Analysis. *Transp. Res. Rec.* **2012**, *2297*, 154–162. [CrossRef]
44. Bergamini, E.; Crescenzi, P.; D'angelo, G.; Meyerhenke, H.; Severini, L.; Velaj, Y. Improving the Betweenness Centrality of a Node by Adding Links. *ACM J. Exp. Algorithmics* **2018**, *23*, 1.5:1–1.5:32. [CrossRef]
45. Evans, T.S.; Chen, B. Linking the Network Centrality Measures Closeness and Degree. *Commun. Phys.* **2022**, *5*, 172. [CrossRef]
46. van der Grinten, A.; Angriman, E.; Meyerhenke, H. Scaling up Network Centrality Computations—A Brief Overview. *It-Inf. Technol.* **2020**, *62*, 189–204. [CrossRef]
47. Lin, H.; Xia, Y.; Li, X.; Gao, X. A Routing Strategy for Spatial Networks Based on Harmonic Centrality. *Front. Phys.* **2023**, *11*, 1203665. [CrossRef]
48. Zhan, J.; Gurung, S.; Parsa, S.P.K. Identification of Top-K Nodes in Large Networks Using Katz Centrality. *J. Big Data* **2017**, *4*, 16. [CrossRef]
49. Zhang, J.; Luo, Y. Degree Centrality, Betweenness Centrality, and Closeness Centrality in Social Network. In Proceedings of the 2017 2nd International Conference on Modelling, Simulation and Applied Mathematics (MSAM2017), Bangkok, Thailand, 26–27 March 2017; Atlantis Press: Bangkok, Thailand, 2017.
50. Zhang, P.; Wang, T.; Yan, J. PageRank Centrality and Algorithms for Weighted, Directed Networks. *Phys. A Stat. Mech. Its Appl.* **2022**, *586*, 126438. [CrossRef]
51. Zhao, X.; Liu, F.; Wang, J.; Li, T. Evaluating Influential Nodes in Social Networks by Local Centrality with a Coefficient. *ISPRS Int. J. Geo-Inf.* **2017**, *6*, 35. [CrossRef]
52. Rochat, Y. *Closeness Centrality Extended To Unconnected Graphs: The Harmonic Centrality Index*; University of Lausanne: Lausanne, Switzerland, 2009.
53. Sharkey, K.J. A Control Analysis Perspective on Katz Centrality. *Sci. Rep.* **2017**, *7*, 17247. [CrossRef]
54. Fortunato, S.; Latora, V.; Marchiori, M. Method to Find Community Structures Based on Information Centrality. *Phys. Rev. E* **2004**, *70*, 056104. [CrossRef]
55. Beaudoin, J.; Lin Lawell, C.-Y.C. The Effects of Public Transit Supply on the Demand for Automobile Travel. *J. Environ. Econ. Manag.* **2018**, *88*, 447–467. [CrossRef]
56. Cervero, R.; Kang, C.D. Bus Rapid Transit Impacts on Land Uses and Land Values in Seoul, Korea. *Transp. Policy* **2011**, *18*, 102–116. [CrossRef]
57. Liu, C.; Duan, D. Spatial Inequality of Bus Transit Dependence on Urban Streets and Its Relationships with Socioeconomic Intensities: A Tale of Two Megacities in China. *J. Transp. Geogr.* **2020**, *86*, 102768. [CrossRef]
58. Wang, C.; Park, J. A Multi-Level Analysis of Bus Ridership in Buffalo, New York. *ISPRS Int. J. Geo-Inf.* **2024**, *13*, 443. [CrossRef]
59. Wang, M.; Wang, H.; Zheng, H. A Mini Review of Node Centrality Metrics in Biological Networks. *Int. J. Netw. Dyn. Intell.* **2022**, *1*, 99–110. [CrossRef]
60. Asuero, A.G.; Sayago, A.; and González, A.G. The Correlation Coefficient: An Overview. *Crit. Rev. Anal. Chem.* **2006**, *36*, 41–59. [CrossRef]
61. Leng, L.; Zhang, T.; Kleinman, L.; Zhu, W. Ordinary Least Square Regression, Orthogonal Regression, Geometric Mean Regression and Their Applications in Aerosol Science. *J. Phys. Conf. Ser.* **2007**, *78*, 012084. [CrossRef]
62. Anselin, L. Spatial Econometrics. In *A Companion to Theoretical Econometrics*; Baltagi, B.H., Ed.; Blackwell Publishing Ltd.: Malden, MA, USA, 2003; pp. 310–330. ISBN 978-0-470-99624-9.
63. Chang, Y.-C.; Chang, K.-H.; Wu, G.-J. Application of eXtreme Gradient Boosting Trees in the Construction of Credit Risk Assessment Models for Financial Institutions. *Appl. Soft Comput.* **2018**, *73*, 914–920. [CrossRef]
64. Lee, S.; Yang, J.; Cho, K.; Cho, D. The Influence of Transportation Accessibility on Traffic Volumes in South Korea: An Extreme Gradient Boosting Approach. *Urban Sci.* **2023**, *7*, 91. [CrossRef]
65. Ding, C.; Cao, X.; Næss, P. Applying Gradient Boosting Decision Trees to Examine Non-Linear Effects of the Built Environment on Driving Distance in Oslo. *Transp. Res. Part A Policy Pract.* **2018**, *110*, 107–117. [CrossRef]
66. Dong, W.; Wang, N.; Dong, Y.; Cao, J. Examining the Nonlinear and Interactive Effects of Built Environment Characteristics on Travel Satisfaction. *J. Transp. Geogr.* **2025**, *123*, 104111. [CrossRef]
67. Murdoch, W.J.; Singh, C.; Kumbier, K.; Abbasi-Asl, R.; Yu, B. Definitions, Methods, and Applications in Interpretable Machine Learning. *Proc. Natl. Acad. Sci. USA* **2019**, *116*, 22071–22080. [CrossRef]
68. Lee, S. Transportation Mode Choice Behavior in the Era of Autonomous Vehicles: The Application of Discrete Choice Modeling and Machine Learning. Ph.D. Thesis, Portland State University, Portland, OR, USA, 2022.

69. Bhavsar, P.; Safro, I.; Bouaynaya, N.; Polikar, R.; Dera, D. Machine Learning in Transportation Data Analytics. In *Data Analytics for Intelligent Transportation Systems*; Elsevier: Amsterdam, The Netherlands, 2017; pp. 283–307. ISBN 978-0-12-809715-1.
70. Bi, Q.; Goodman, K.E.; Kaminsky, J.; Lessler, J. What Is Machine Learning? A Primer for the Epidemiologist. *Am. J. Epidemiol.* **2019**, *188*, 2222–2239. [CrossRef]
71. Malakouti, S.M.; Menhaj, M.B.; Suratgar, A.A. The Usage of 10-Fold Cross-Validation and Grid Search to Enhance ML Methods Performance in Solar Farm Power Generation Prediction. *Clean. Eng. Technol.* **2023**, *15*, 100664. [CrossRef]
72. Shams, M.Y.; Elshewey, A.M.; El-kenawy, E.-S.M.; Ibrahim, A.; Talaat, F.M.; Tarek, Z. Water Quality Prediction Using Machine Learning Models Based on Grid Search Method. *Multimed. Tools Appl.* **2024**, *83*, 35307–35334. [CrossRef]
73. Lee, S. Relationships between Transportation Expenditures and Built Environment in the United States: Insights from Interpretable Machine-Learning Approach. *J. Plan. Educ. Res.* **2024**. [CrossRef]
74. Huang, N.; Lu, G.; Xu, D. A Permutation Importance-Based Feature Selection Method for Short-Term Electricity Load Forecasting Using Random Forest. *Energies* **2016**, *9*, 767. [CrossRef]
75. Lee, Y.; Seo, D. Identifying Relationship between Regional Centrality and POI Facilities: A Case Study of Seoul Metropolitan Area. *ISPRS Int. J. Geo-Inf.* **2024**, *13*, 12. [CrossRef]
76. Kang, C.-D. The Effects of Spatial Accessibility and Centrality to Land Use on Walking in Seoul, Korea. *Cities* **2015**, *46*, 94–103. [CrossRef]
77. O'Connor, K. Global City Regions and the Location of Logistics Activity. *J. Transp. Geogr.* **2010**, *18*, 354–362. [CrossRef]
78. Yuan, G.; Sun, L.; Kong, D.; Bai, Z.; Shao, J. Supernetwork Perspective on Studying the Method of Identifying the Important Urban Transport Hub. *J. Highw. Transp. Res. Dev. (Engl. Ed.)* **2023**, *17*, 82–91. [CrossRef]
79. Chun, Y.; Kim, K. Temporal Changes in the Urban System in South Korea. *Front. Sustain. Cities* **2022**, *4*, 1013465. [CrossRef]
80. Zhao, Y.; Feng, T. Commuter Choice of UAM-Friendly Neighborhoods. *Transp. Res. Part A Policy Pract.* **2025**, *192*, 104338. [CrossRef]
81. Coppola, P.; De Fabiis, F.; Silvestri, F. Urban Air Mobility Demand Forecasting: Modeling Evidence from the Case Study of Milan (Italy). *Eur. Transp. Res. Rev.* **2025**, *17*, 2. [CrossRef]
82. Jiang, Y.; Li, Z.; Wang, Y.; Xue, Q. Vertiport Location for eVTOL Considering Multidimensional Demand of Urban Air Mobility: An Application in Beijing. *Transp. Res. Part A Policy Pract.* **2025**, *192*, 104353. [CrossRef]
83. Liu, T.-Y. Spatial Structure Convergence of China's Transportation System. *Res. Transp. Econ.* **2019**, *78*, 100768. [CrossRef]

land

Article

Assessing Users' Satisfaction with the Urban Central Metro Station Area in Chengdu: An SEM-IPA Approach

Jiexi Ma, Zhongwei Shen *, Pengpeng Liang *, Yu Zhao and Wen Song

School of Architecture, Southwest Jiaotong University, 999 Xi'an Road, Chengdu 610097, China; mjx@my.swjtu.edu.cn (J.M.)
* Correspondence: szwpaper@163.com (Z.S.); liang_pengpeng@my.swjtu.edu.cn (P.L.)

Abstract: An urban central metro station area is a core hub within the high-quality Transit-Oriented Development (TOD) model. This study explores users' perceptions of built environments around urban central metro stations to investigate the critical determinants of user satisfaction and proposes strategies to enhance the quality of these environments. First, a comprehensive perception system, including location situation, field environment, and urban aesthetics, was developed through literature reviews and expert consultation. Secondly, three typical central metro station areas in Chengdu were selected as study cases, and 425 questionnaires were collected from August to October 2024. The data were analyzed using a structural equation model (SEM) to reveal the impact of built environment perception on overall satisfaction. The results indicate that the field environment has the strongest direct influence on satisfaction. Urban aesthetics impacts satisfaction both directly and indirectly, making its overall effect the most significant. While the location situation does not directly affect satisfaction, it indirectly influences satisfaction through its impact on the field environment and urban aesthetics. Subsequently, based on the satisfaction performance and SEM outcomes, an importance–performance analysis (IPA) was conducted to identify specific areas needing enhancement. Finally, we integrated environmental assessments with the above findings and put forth strategic recommendations to enhance the quality of the built environment.

Keywords: TOD; central metro station area; built environment; public transport satisfaction; travel behavior; SEM-IPA

1. Introduction

A central metro station area (CMSA), as a practical implementation of the Transit-Oriented Development (TOD) model, plays a pivotal role in urban centers and sub-centers. It integrates crucial functions such as transportation, employment, and daily living, positioning itself as a critical public space within the city [1]. TOD has been widely recognized as a sustainable growth strategy, guiding cities towards a more polycentric structure and fostering territorial cohesion [2–4]. CMSAs are designed to encourage "metro + walking" travel patterns, where users can complete their remaining activities through short walking or cycling distances after using public transport. This concept aligns with Carlos Moreno's "15 min city" model, which, when adopted in long-term planning, would improve quality of life by reducing time wasted in traffic and promoting sustainable mobility [5,6].

In practical design, however, due to the complex environment of urban centers and the competition among various elements, the built environment design of a CMSA faces numerous conflicts between economic benefits and user experience [7]. The latest research

 https://doi.org/10.3390/land14051023

on public perception of urban space shows that users are more likely to show negative emotions in urban transportation nodes than at other places [8]. These metro station areas located in the center or sub-center of the city bear more social responsibility than other ordinary TOD areas [9], necessitating design considerations that extend beyond mere transportation and economic considerations to encompass public sentiment and perception [10]. The primary way to enhance public satisfaction is to create a people-oriented, efficient, and comfortable pedestrian environment [11]. Consequently, a comprehensive understanding of public perceptions of central metro station areas is imperative for formulating socially oriented design strategies and policies to enhance overall urban livability [7].

The station catchment area represents an intersection of transportation planning and urban design. Currently, traffic researchers focus primarily on the impact of macro-level indicators on passenger travel and the complex interrelationships among these variables [12,13]. In contrast, urban design researchers concentrate on universal methodologies to enhance the satisfaction of urban public spaces [14,15]. However, there exists a significant gap in the literature regarding the in-depth exploration of urban spatial perception within the context of traffic characteristics. The primary reason for this gap lies in the fact that traffic research typically focuses on macro-level indicators such as traffic flow and travel patterns, while urban design research tends to emphasize spatial form and aesthetic experience. The lack of an integrated interdisciplinary perspective has resulted in insufficient exploration of how traffic characteristics influence urban spatial perception. Thus, establishing a user satisfaction research framework that combines the dual attributes of urban public spaces and traffic node functions in central metro station areas is essential. Specifically, the evaluation of transit catchment areas should include the integrated development characteristics of TOD, such as the 5D principles and node place attributes, while also incorporating aspects like urban aesthetics and service experience. This is necessary for shaping dual benefits that balance both urban economic development and a human-centered society [10,16].

This study delves into the determinants of user satisfaction with the built environment of a CMSA through the application of a combined approach involving the structural equation modeling (SEM) and importance–performance analysis (IPA) methodologies [17]. By assessing the comparative impact intensity and satisfaction levels of these factors, this study aims to pinpoint areas requiring enhancement. First, we devised a conceptual framework for the CMSA that includes location situation, field environment, and urban aesthetics. Subsequently, a selection of observed variables was curated through word frequency analysis and expert consultation, culminating in the construction of a second-order structural equation model. Drawing upon the SEM analysis of 425 questionnaires collected from three typical CMSAs in Chengdu, our investigation elucidates the intricate influence mechanisms of location situation, density, diversity, distance to transit, destination accessibility, road networks, connection design, street characteristics, and order management on user satisfaction. Then, we assessed the alignment between the significance and performance of objective environmental elements by IPA. Finally, tailored strategic recommendations were formulated for the CMSA by synthesizing these findings with on-site investigations. This study offers a dual contribution: (1) It develops a perception assessment framework for the CMSA, which aids in evaluating user satisfaction and analyzing potential correlations between the perception of individual environmental elements and public sentiment. (2) The findings provide various policy recommendations for urban designers and TOD developers to enhance CMSA design, ultimately improving user satisfaction and promoting human-centered development.

The contents are organized as follows: The second section describes the theoretical framework, hypothesis, and CMSA perception index system. The third section details the research method, encompassing the methodology of SEM-IPA, study areas, data sources, and research process. The fourth section presents the empirical study results, focusing on survey data characteristics, model testing and fitness, path impact outcomes, and the alignment identified by importance–performance. In the last section, some design lessons and key points are summarized.

2. Theoretical Framework and Index

2.1. Theoretical Framework and Hypothesis

A conceptual framework is constructed based on the interpretation of the content of the catchment area of a metro station by existing academic theory. The 3D theory (and its extensions) and node–place model are the academic theories commonly used to characterize the built environment of a catchment area. The mesoscale factors of the built environment are gauged within an area and include land use diversity, street designs, and densities. These factors are designated as the 3Ds; later, these 3Ds were expanded to 5Ds by the addition of two more factors: distance to transit and destination accessibility [18]. Banerjee [19] proposed the concept of 'railscape', emphasizing the role of streetscape aesthetics, historical imagery, and user perception in the built environment surrounding railways. The node–place model introduced by Bertolini [20] underscores the necessity to characterize the catchment area by considering both macroscopic traffic performance and the mesoscale place design. The descriptions of the built environment by the NP model and "D" theory overlap. Cao et al. [21] took the 5Ds as the second-level index of the NP model. Deng et al. [22] combined the 5Ds and NP model to evaluate the catchment area. They identified density, diversity, and design as indicators of place design and added a 'linkage' dimension to represent the distance to transit and destination accessibility.

By the explanation of the built environment of the catchment area in different scales, we intended to deconstruct the theoretical connotation of the built environment into the location situation, field environment, and urban aesthetics. Figure 1 is a diagrammatic representation of built environment quality attributes for user satisfaction as a theoretical framework for this study, which is also the basis for the structural equation model. There are three first-order latent variables and five second-order latent variables.

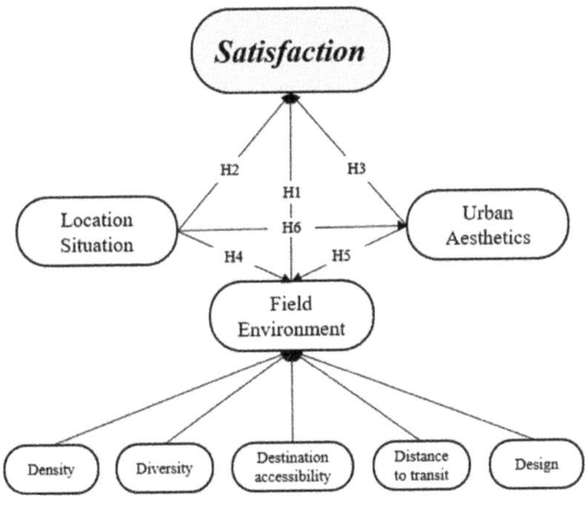

Figure 1. Theoretical framework and hypothesis.

Hypotheses regarding the direct and indirect effects among latent constructs may be verified after an SEM analysis has been conducted. Table 1 shows six hypotheses between latent variables and satisfaction and between latent variables.

Table 1. Theoretical hypothesis.

Hypothesis	Description	Reference
H1	Field environment directly affects satisfaction	[14,23,24]
H2	Location situation directly affects satisfaction	[25,26]
H3	Urban aesthetics directly affects satisfaction	[14,27,28]
H4	Location situation affects field environment	[29]
H5	Location situation affects urban aesthetics	[30,31]
H6	Urban aesthetics affects field environment	[32,33]

2.2. Index System

Based on the conceptual framework, the corresponding observed variables were selected through word frequency analysis and expert consultation to form a measurable model. We selected the indicators through word frequency analysis and expert consultation. As one of the bibliometric methods, the word frequency analysis is a research method that uses quantitative statistical analysis to describe the external characteristics of publications and evaluate the critical spots in a particular research field [34]. Expert consultation is seen as a consensus-oriented technique where experts orient themselves towards a greater or lesser degree of consensus around the critical projection or visions of the future within a specific field [35].

We searched for relevant studies that met two conditions: first, a title containing the words "Station Area", "Transit Oriented", "TOD", "Transit Station", and so on; second, an abstract or keywords containing words such as "performance", "influence", "impact", "effect", "quality", "evaluate", "assess", or "measure". In total, 158 studies from 2019 to 2023 were collected, and 94 indicators were selected. According to the frequency of these words, we divided them into high, mid, and low frequencies. All the high-frequency indicators were adopted, most mid-frequency indicators were retained, the indicators that did not meet the research theme were deleted, and only a few of the low-frequency indicators were retained. Finally, 58 primary indicators were obtained.

We consulted 13 experts on these 58 indicators and asked them to rate the impact of each indicator on user perception on a scale of 1–5. The experts were from university scholars, designers, TOD developers, and metro companies, and 12 experts responded—a positive degree of 93.3%. The index was further selected by calculating and comparing the average value and coefficient of variation of each indicator. The average value reflects the influence of the indicator on the perception, and the coefficient of variation reflects the degree of expert consensus. The calculation results are shown in Figure 2. Indicators with low average values and a high coefficient of variation were deleted. Table 2 shows the index system after preliminary selection. We designed the questionnaire based on the indicator system. Specifically, for each indicator in the table, respondents were asked to provide their perceptions, which were quantified using a Likert scale. A score of 5 indicated "very satisfied", 4 indicated "somewhat satisfied", 3 indicated "neutral", 2 indicated "somewhat dissatisfied", and 1 indicated "very dissatisfied".

Table 2. Perception index of the built environment of the CMSA.

Dimension	1st-Order Index	2nd-Order Index
Location Situation	/	L1. Location in the city L2. The rationality of subway line planning L3. Number of subway lines L4. Accessibility to other centers

Table 2. *Cont.*

Dimension	1st-Order Index	2nd-Order Index
Field Environment	Density	A1. Development intensity A2. Building volume A3. Underground space development intensity A4. Adequacy of open space A5. Crowdedness A6. Traffic jam A7. The interference between people and vehicles
	Diversity	B1. Satisfaction of basic needs B2. Adequacy of function selection B3. The rationality of function ratio
	Destination accessibility	C1. Accessibility of commercial facilities C2. Accessibility of official facilities C3. Accessibility of residential facilities C4. Accessibility of public service facilities
	Distance to transit	D1. The convenience of the bus transfer D2. The convenience of the bicycle transfer
	Design	E1. Road connectivity E2. Road directivity E3. The convenience of getting to the destination E4. Entrance quantity E5. Entrance location E6. Degree of difficulty of finding entrance E7. Form of entrance E8. Entrance connection with building
Urban Aesthetics	/	U1. Proportion and scale of the street U2. Color and style of the street U3. Street furniture U4. Culture of the street U5. Street hygiene U6. Landscape maintenance U7. Perceived safety

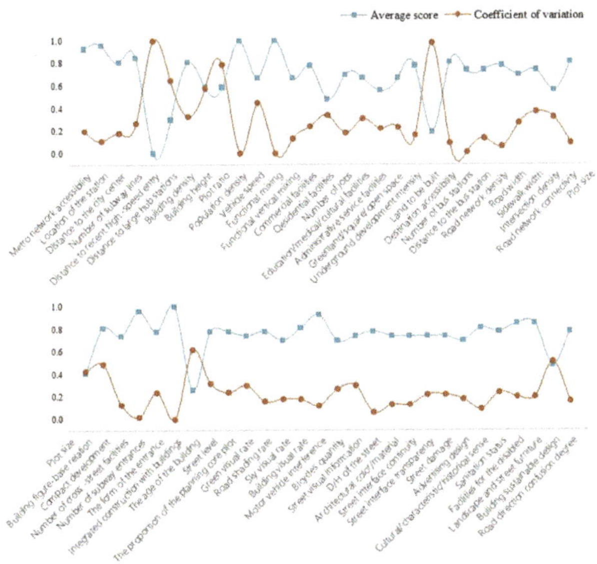

Figure 2. Index average score and coefficient of variation of expert consultation.

3. Method and Data

3.1. SEM-IPA Model

3.1.1. Structural Equation Model (SEM)

Structural equation modeling (SEM) is a statistical method for analyzing variables based on their covariance matrix to find their inherent structural relationships. Users provided judgments for each aspect, allowing them to be evaluated as observed endogenous latent variables in the model [36]. A complete SEM includes structural and measurement models. Of these, the former reflects the relationship between latent variables, which is expressed as

$$\eta = \beta\eta + \Gamma\xi + \xi, \tag{1}$$

where η denotes the endogenous latent variable, β is the influence relationship between the endogenous latent variables, Γ is the influence relationship between the exogenous latent variable and the endogenous latent variable, and ξ is the residual term of the endogenous latent variable.

The measurement model reflects the relationship between latent variables and explicit variables, which can be expressed as

$$\begin{aligned} X &= \Lambda x \times \xi + \delta, \\ Y &= \Lambda y \times \eta + \varepsilon, \end{aligned} \tag{2}$$

where X is the exogenous explicit variable, Y is the endogenous explicit variable, Λx is the relationship matrix between the exogenous explicit variable X and the exogenous latent variable ξ, Λy is the relationship between the endogenous explicit variable Y and the endogenous latent variable η, and δ and ε are X and Y error terms.

3.1.2. Importance–Performance Analysis (IPA)

Importance–performance analysis (IPA) is a method for identifying the alignment of importance and performance and assumes a distribution of a set of attributes in four sets, according to the four possible combinations, with respect to the performance and importance of attributes [37,38]. Quadrant I is the area of "high importance and high performance". Quadrant II is the area of "high importance and low performance". Quadrant III is the area of "low importance and low performance". Quadrant IV is the area of "low importance and high performance". The recommendable actions given by the IPA theory for the four areas of appeal are "Keep up the good work", "Concentrate here", "Low priority", and "Possible overkill" [39].

3.1.3. SEM-IPA Method

The SEM-IPA method is also presented as an importance–performance matrix (Figure 3), and the quantification of importance is based on the results of SEM. We calculate the importance of these variables based on their total contribution to satisfaction, as the greater the contribution of a factor to satisfaction, the higher its importance. For both first-order and second-order latent variables, their contribution to satisfaction is the sum of their direct and indirect effects, i.e., the total effect on satisfaction. For observed variables, their importance is determined by the product of their factor loadings and total effects. Similarly, the performance value is defined by the satisfaction score; the higher the satisfaction, the better the perceived performance.

In the IPA results, the alignment of importance and performance is essential for the evaluation of user perception. Specifically, a higher perception of elements should correspond to increased satisfaction, indicating that users have had a positive experience with these environmental factors. Based on the assessment of this alignment, we categorized the quality attributes in the four quadrants into two groups: matched (Quadrants I and III) and unmatched (Quadrants II and IV). When importance and performance align, they can be either high or low; in both scenarios, the strategy is to maintain the current state. Conversely, when there is a mismatch, two situations arise: in Quadrant II, the importance of factors significantly exceeds their performance, necessitating additional investment for improvement. In Quadrant IV, the importance of elements is considerably lower than their performance, indicating a need to reduce input.

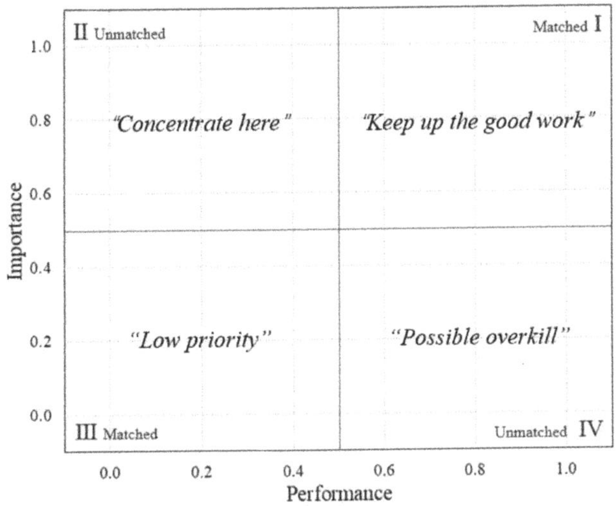

Figure 3. Importance–performance analysis [40].

3.2. Study Area and Data

Based on the "Guidelines for Integrated Urban Design of Chengdu Metro Stations", the differences among several city-level central metro stations are primarily reflected in two aspects: location and dominant function. Location differences refer to whether the stations are situated in the new city or the old city. Functional differences are seen in whether the stations are dominated by historical landmarks, comprehensive commercial, business, and administration areas, or large transit hubs. Due to their unique characteristics in terms of function, form, and clientele, stations dominated by large transit hubs have been excluded from this study. The selection of the study samples was based on three principles: first, including the representation of different locations; second, covering stations with different dominant functional types; and third, including stations with a relatively high level of development. Three stations were chosen: Tianfu Square, 3rd Tianfu Street, and Chunxi Road. Tianfu Square, located in the old city center, primarily serves administrative and historical functions; Chunxi Road is a commercial hub; and 3rd Tianfu Street is predominantly a business and office district. These three stations are well-established urban centers, each with distinct functional characteristics (Figure 4).

The scale range of the catchment area is defined by the effective perception range of the user in actual use. We simulated the 5 min walk at the outermost entrance and exit of the site through Mapbox, combined with the 5 min range and the shape of the plot, and manually outlined the effective perception range. The overall size is about 400–800 m, which is similar to previous research [41,42].

Based on the theoretical framework, this study utilized a questionnaire consisting of self-formulated questions of the index system. The questionnaire comprised two sections: personal information and a satisfaction survey. The personal information section gathered data on participants' gender, age, educational background, occupation, purpose of visit, and frequency of visits. The satisfaction survey required participants to evaluate various components of the catchment area using a standardized rating scale ranging from 1 (dissatisfaction) to 5 (satisfaction). Questionnaires were distributed near different entrances of the three stations. This approach ensured a diverse sampling of individuals using these stations, thereby enhancing the representativeness and reliability of the collected information.

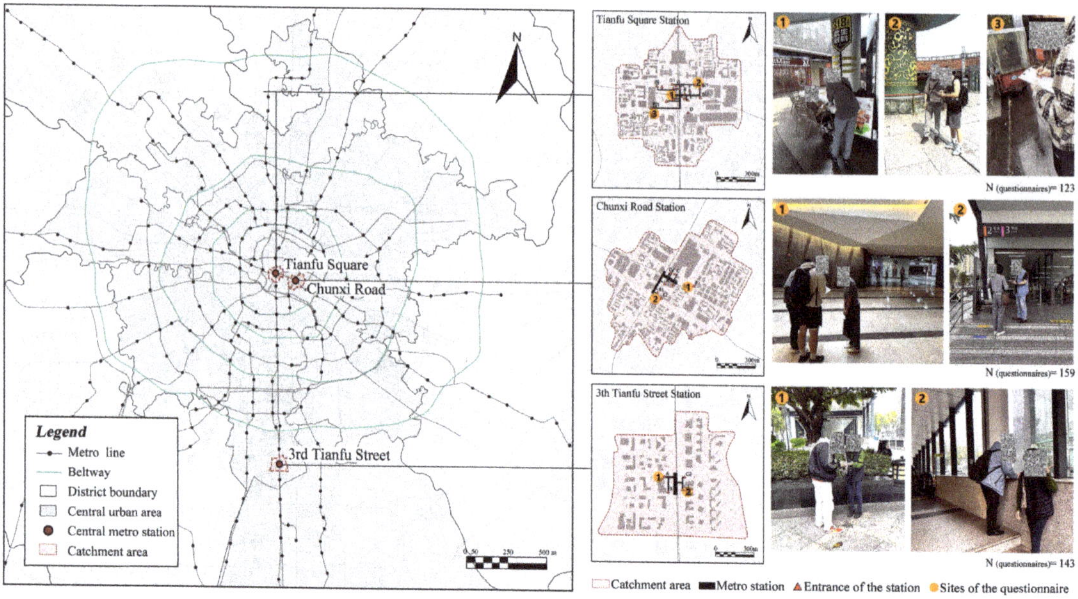

Figure 4. Study areas and questionnaire survey.

3.3. Research Process

We developed the following framework (Figure 5): The initial phase established the research foundation by conducting a literature review to construct a theoretical framework. This phase also involved administering a satisfaction questionnaire to gather fundamental study data. The theoretical framework was developed based on the built environment of the transport area theory, with formulated study hypotheses. The framework was refined through word frequency analysis and expert consultation into a specific and measurable perceptual evaluation index system supported by data from a questionnaire survey.

In the subsequent phase, SEM was employed to depict the interplay among different dimensions of the perception of the catchment area. Concurrently, the IPA method was utilized to evaluate the alignment between the importance and performance of individual spatial elements. A two-order SEM model was constructed based on the theoretical framework and questionnaire data, adjusted iteratively to test the validity of the proposed hypotheses. The SEM results unveiled each factor's influence mechanisms and intensities in relation to satisfaction, providing an importance value for the IPA analysis. IPA analysis categorized indicators based on perceived importance and performance, distinguishing matched and unmatched types.

In the final phase, the research findings were carefully examined, culminating in the development of strategic recommendations. Drawing from the SEM-IPA analysis, additional insights into the perceptual attributes of elements are presented. These insights, combined with on-site structural assessments, led to the identification of deficient built environment elements and the provision of targeted recommendations. Furthermore, some critical environmental features conducive to enhancing satisfaction in the central catchment area were summarized.

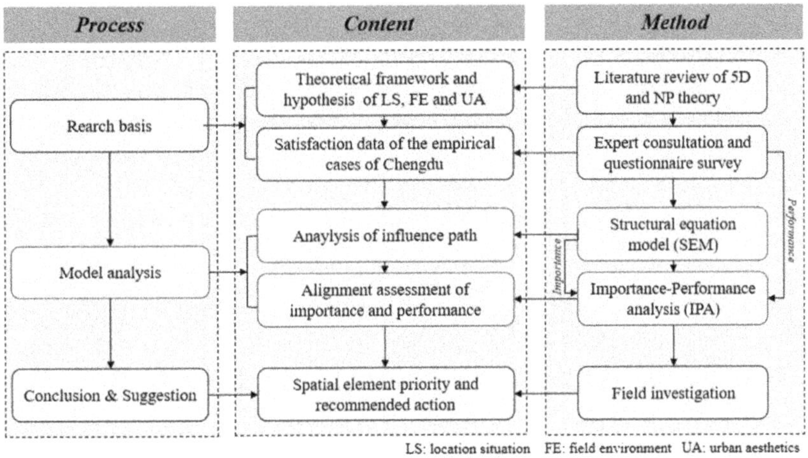

Figure 5. Research process.

4. Results

4.1. Data Description

In total, 425 valid questionnaires were collected: 167 at Tianfu Square Station, 138 at Chunxi Road Station, and 120 at 3rd Tianfu Street Station. Table 3 presents a statistical overview of the respondents' information, revealing the following key characteristics: (1) The gender distribution of the respondents was nearly balanced, with an average age of 28.6 years. The education level was primarily dominated by bachelor's degrees. (2) Most respondents have low knowledge of TOD and urban design. (3) The primary purposes for visiting the stations were work, official business, shopping, and entertainment, with a low frequency.

Table 3. Respondents' characteristics.

Personal Particulars	Item	Percent (%)	Personal Particulars	Item	Percent (%)
Gender	Male	47.29	Purpose of visit	Transfer	1.65
	Female	52.71		Shopping	15.06
Age	<30	50.82		Work	33.65
	≥30	49.18		Residence	7.06
Education	Sub-bachelor's	56.94		Affairs	32.94
	Bachelor's degree or above	43.06		Travel	5.42
Professional relevance	Unrelated	74.82		Others	4.22
	Relevant	25.18			
Visiting frequency	<2 times/month	43.53			
	3–4 times/month	33.88			
	Almost every day	22.59			

4.2. Reliability and Validity

Using SPSS 25.0, we evaluated the reliability and validity of the questionnaire data to assess their rationality. Reliability is the evaluation of the stability and consistency of the results. Validity is used to assess the consistency between predicted and actual measured data results.

We adopted Cronbach's alpha (α) to evaluate the reliability of the scale. It is generally believed that alpha should be greater than 0.6 to indicate that the internal consistency of the questionnaire is accepted [43]. The α for the destination accessibility dimension is 0.596. Upon examining the α coefficient after item deletion, it was found that the α coefficient for C3 (accessibility of residential facilities) increased to 0.720. Consequently, the C3 item was removed. The overall α coefficient is 0.871, with the coefficients for each dimension ranging from 0.720 to 0.875 (Table 4), indicating good internal consistency. The results showed that KMO was greater than 0.8 [44] (Table 5), indicating the data were suitable for factor analysis to test the validity and the variables had good independence.

Table 4. Reliability analysis of latent variables.

Latent Variable	Cronbach's Alpha	Latent Variable	Cronbach's Alpha
Location situation	0.825	Distance to transit	0.835
Density	0.872	Design	0.763
Diversity	0.875	Urban aesthetics	0.801
Destination accessibility	0.720	Satisfaction	0.858

Table 5. Results of KMO test and the Bartlett's test of sphericity for the questionnaires.

KMO Measure of Sampling Adequacy		0.847
Bartlett Test of Sphericity	Approximate Chi-Square	6977.882
	df	561
	Sig.	0.000

Exploratory factor analysis strives to uncover the coherence between the variables and the primary component factors extracted from empirical data. It can also conduct mathematical tests on the previously constructed observed variables. With the principal component extraction method and the maximum variance rotation, nine factors with eigenvalues greater than 1.0 were extracted, and the cumulative explanatory variance was 69.014%, which is above 50% and is acceptable [45] (Table 6). Therefore, we adjusted the original index system: The first-order latent variable named design dimension is refined into road network design and connection design, and the first-order latent variable named urban aesthetics is refined into street characteristics and order management (Table 7). After principal component analysis, all factors loading above 0.4 can be retained [46]. Among the observed variables, the standard factor load of E6 (difficulty of finding entrance) is lower than 0.4, so the item is deleted.

Table 6. Total variance explained based on the field questionnaire.

Factors	Total	% of Variance	Cumulative %	Total	% of Variance	Cumulative %	Total	% of Variance	Cumulative %
1	7.435	21.867	21.867	7.435	21.867	21.867	4.207	12.372	12.372
2	3.586	10.547	32.414	3.586	10.547	32.414	3.056	8.987	21.359
3	2.503	7.362	39.776	2.503	7.362	39.776	2.983	8.773	30.133
4	2.145	6.31	46.086	2.145	6.31	46.086	2.679	7.88	38.013
5	1.826	5.371	51.457	1.826	5.371	51.457	2.447	7.197	45.21
6	1.753	5.157	56.614	1.753	5.157	56.614	2.259	6.644	51.853
7	1.612	4.741	61.355	1.612	4.741	61.355	2.069	6.086	57.94
8	1.338	3.937	65.291	1.338	3.937	65.291	1.965	5.778	63.718
9	1.266	3.723	69.014	1.266	3.723	69.014	1.801	5.297	69.014

Table 7. Factor analysis.

	Component	Factor Loading			
1	Density	A1/0.576 A5/0.761	A2/0.715 A6/0.806	A3/0.710 A7/0.728	A4/0.669
2	Connection design	E4/0.797	E5/0.806	E7/0.857	E8/0.830
3	Location situation	L1/0.783	L2/0.796	L3/0.830	L4/0.795
4	Urban characteristics	U1/0.823	U2/0.859	U3/0.837	U4/0.842
5	Diversity	B1/0.762	B2/0.801	B3/0.814	
6	Order management	U5/0.845	U6/0.830	U7/0.737	
7	Road network design	E1/0.859	E2/0.879	E3/0.838	
8	Destination accessibility	C1/0.804	C2/0.732	C4/0.781	
9	Distance to transit	D1/0.871	D2/0.859		

4.3. Influence Path

We used Amos 28 software to build and compute the structural equation model, establishing a model that includes four first-order latent variables and eight second-order latent variables, and imported the questionnaire data into the model. The model (Figure 6) demonstrated that location situation (H2, p-value > 0.05, C.R. < 1.96) had no significant effect on satisfaction; thus, these exogenous latent variables were removed. This was represented by the broken lines. The remaining five hypothetical paths are all supported (Table 8). In addition, items with values > 0.50 were deemed significant [47]. In the current model, all items exceed this threshold. The SEM was rerun after the removal of the exogenous latent variable and the consideration of the parameters to create the final model in evaluating the effect on satisfaction.

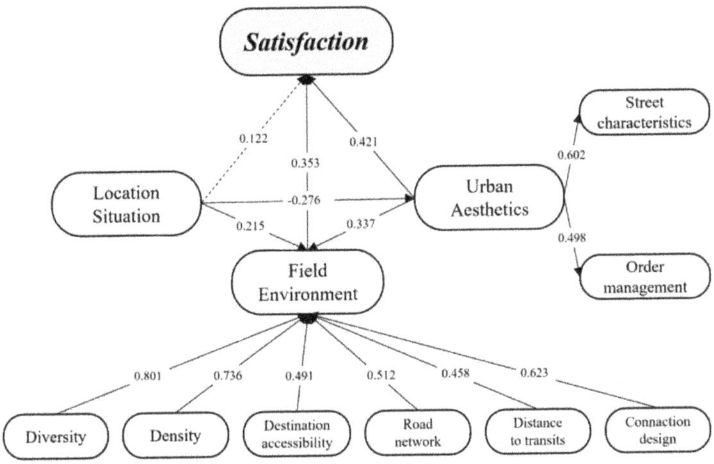

Figure 6. The SEM for the perception of the CMSA.

Table 8. Model path coefficients and hypothesis results.

Hypothesis	Path	C.R.	Result
H1	Field environment—>Satisfaction	4.789 ***	True
H2	Location situation—>Satisfaction	0.069	False
H3	Urban aesthetics—>Satisfaction	3.242 **	True
H4	Location situation—>Field environment	3.006 **	True
H5	Location situation—>Urban aesthetics	−3.151 **	True
H6	Urban aesthetics—>Field environment	2.932 **	True

* denotes a p-value less than 0.05, ** denotes a p-value less than 0.01, and *** denotes a p-value less than 0.001.

Table 9 demonstrates the IFI, CFI, GFI, AGFI, and RMSEA of the adjusted model. By the standard [48,49], values higher than 0.800 are cut off. The value of x^2/df ranges from 1 to 3, and RMSEA < 0.07 is acceptable, which provides the reliability of a good representation of the data observed.

Table 9. Fitness of the SEM.

Goodness of Fit Measures	IFI	CFI	GFI	AGFI	x^2/df	RMSEA
Parameter estimates	0.949	0.948	0.890	0.874	1.654	0.039
Minimum cut-off	>0.80	>0.80	>0.80	>0.80	1~3	<0.07

Table 10 presents the AVE and CR values of the adjusted model. AVE reflects the explanatory power of latent variables on their respective indicators, while CR is used to assess the reliability of latent variables. The majority of the latent variables exhibit an AVE greater than 0.5. According to Hair et al. [50], factor loadings greater than 0.5 and CR values above 0.70 are typically considered ideal. In this study, a few dimensions have AVE values slightly below 0.5, which can be considered acceptable as they demonstrate moderate explanatory power.

The final model (Figure 7 and Table 11) shows the impact path results as follows: (1) Field environment perception has the most significant direct perceptual effect on station satisfaction, and it is also affected by urban aesthetics and location perception. (2) Urban aesthetic perception exerts a significant direct influence on satisfaction and also serves as a mediator by positively impacting the field environment. Consequently, urban aesthetic perception demonstrates the most substantial total effect on satisfaction. (3) The direct influence of location perception on satisfaction is relatively modest. However, it impacts the field environment and urban aesthetic perception, indirectly affecting satisfaction. The positive correlation between location conditions and the field environment suggests that superior location conditions enhance users' inclusiveness towards the field environment. Conversely, the negative relationship between location conditions and urban aesthetics implies that a more centralized site elevates public expectations regarding its portal effect.

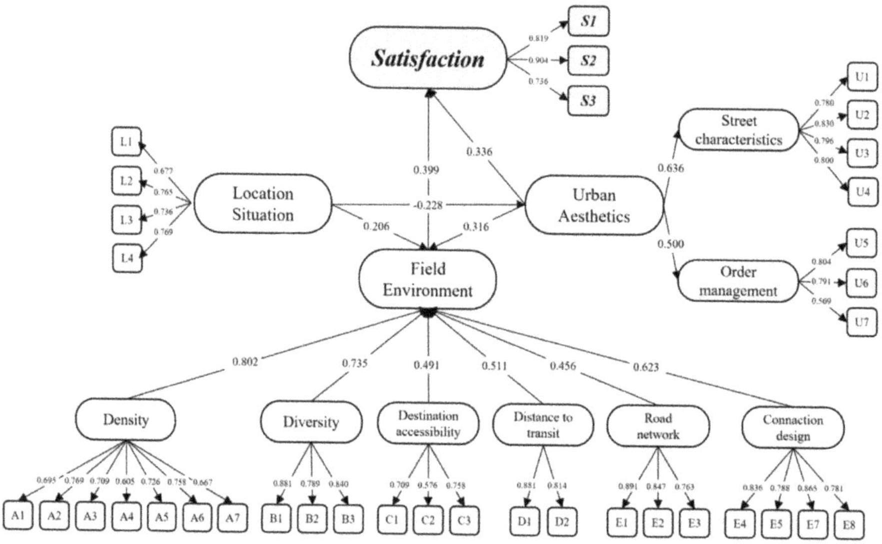

Figure 7. The adjusted final SEM for the perception of the CMSA.

Table 10. AVE and CR scores.

Observed Variable	Latent Variable	Estimate	AVE	CR	Observed Variable	Latent Variable	Estimate	AVE	CR
L4		0.769			D2	Distance to transit	0.814	0.719	0.837
L3	Location Situation	0.736	0.542	0.827	D1		0.881		
L2		0.765			E3		0.763		
L1		0.677			E2	Road network design	0.847	0.698	0.873
A7		0.667			E1		0.891		
A6		0.758			E7		0.781		
A5		0.726			E6	Connection design	0.865	0.670	0.89
A4	Density	0.605	0.485	0.874	E5		0.788		
A3		0.709			E4		0.836		
A2		0.769			U1		0.78		
A1		0.695			U2	Urban characteristics	0.83	0.643	0.878
B3		0.84			U3		0.796		
B2	Diversity	0.789	0.701	0.876	U4		0.8		
B1		0.881			U5		0.804		
C3		0.758			U6	Order management	0.791	0.532	0.769
C2	Destination accessibility	0.576	0.470	0.724	U7		0.569		
C1		0.709							

Table 11. Influence path analysis.

Independent Variable	Dependent Variable	Direct Effect	Indirect Effect	Total Effect
Field environment	Satisfaction	0.399	/	0.399
Location situation		/	−0.023	−0.023
Urban aesthetics		0.336	0.126	0.462
Location situation	Field environment	0.206	−0.072	0.134
Location situation	Urban aesthetics	−0.228	/	0.228
Urban aesthetics	Field environment	0.316	/	0.316

4.4. Importance–Performance Alignment

Figure 8 and Table 12 displays the IPA results. The values of the contribution of each element to perception and the satisfaction scores represent the importance and performance, respectively. Using the average values of all elements' importance and performance as the thresholds, two dividing lines are drawn vertically and horizontally on the coordinates, categorizing the elements into four quadrants.

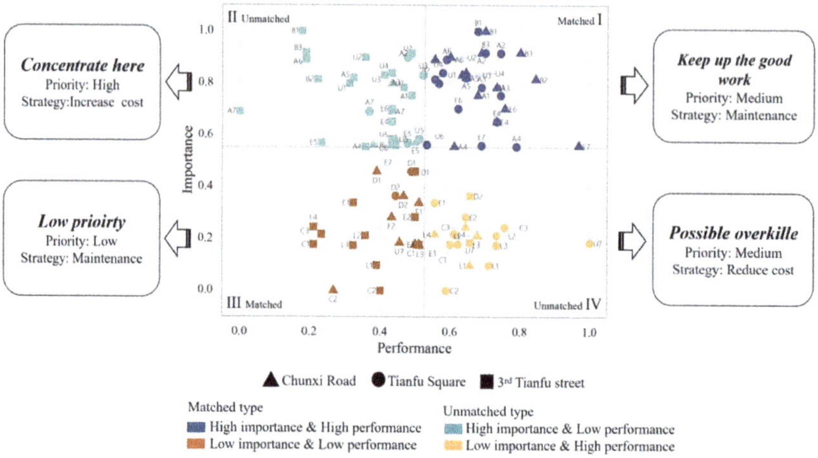

Figure 8. The matching relationship between the importance and performance of the environment element.

Table 12. The distribution of elements in the IPA quadrants.

Station	First Quadrant	Second Quadrant	Third Quadrant	Fourth Quadrant
Tianfu Square	A1, A2, A3, A4, A5, A6, B1, B2, B3, E4, E6, E7, U1, U4, U6	A7, E5, U2, U3, U5	D1, D2	L1, L2, L3, L4, C1, C2, C3, E1, E2, E3, U7
Chunxi Road	A1, A2, A3, A4, A5, A6, B1, B2, B3, E4, E6, E7, U2, U3, U4	A7, E5, U1, U5, U6	L3, C1, C2, D1, D2, E1, E2, E3	L1, L2, L4, C3
3rd Tianfu Street	/	A1, A2, A3, A4, A5, A6, A7, B1, B2, B3, E4, E5, E6, E7, U1, U2, U3, U4, U5, U6	L1, L2, L3, L4, C1, C2, C3, D1, E1, E2, E3	D2, D7

Quadrant I: The elements located in this area exhibit both high importance and high performance, indicating a well-aligned relationship between their perception influence and satisfaction situation. These elements represent significant strengths and are the primary contributors to users' positive experiences within the catchment area. Notably, elements in this quadrant are predominantly found in Chunxi Road and Tianfu Square, accounting for 48.3% and 51.7%, respectively.

Quadrant II: The elements in quadrant II exhibit a stark mismatch between their high importance and their performance, which is notably low. These elements significantly influence user satisfaction but fail to receive the corresponding user praise, leading to a diminished user experience. Elements in this quadrant are predominantly found in 3rd Tianfu Street, accounting for 66.7%. Tianfu Square and Chunxi Road both accounted for 16.7%.

Quadrant III: The elements located in this area exhibit both low importance and low performance, indicating a matching relationship between their perceived influence and satisfaction levels. These elements have a minimal impact on user satisfaction, and their evaluations by users are correspondingly low. Elements in this quadrant are present across all three stations, with proportions of 50.0% at 3rd Tianfu Street, 40.1% at Chunxi Road, and 10.0% at Tianfu Square.

Quadrant IV: The elements located in this quadrant exhibit a mismatch between their importance and performance, with performance significantly exceeding importance. These elements have a minimal impact on user satisfaction, yet they receive disproportionately positive feedback from users. Elements in this quadrant are present across all three stations, with proportions of 64.7% at Tianfu Square, 23.5% at Chunxi Road, and 11.8% at 3rd Tianfu Street.

Figure 9 utilizes radar charts to illustrate the standardized comparison of the importance and performance of different level variables in each station area. These charts facilitate the easy identification of defect elements. An image is formed by connecting the importance and performance values of each area. The higher the resemblance in shape and area between the importance and performance representations, the greater the matching degree of the catchment area. Conversely, notable disparities in shape indicate potential design flaws within the catchment area. A performance graph area surpassing that of importance signifies inadequacy in the station area, prompting a need to "Concentrate here", while an importance graph area exceeding that of performance suggests a potential "Possible overkill" in the area. Below, we elaborate the presentation results of latent variables and observed variables in the three CMSAs.

The radar charts of the first- and second-order latent variables and observed variables illustrate the overall and detailed situations of all station areas. The results indicate that the performance graphs of Tianfu Square and Chunxi Road stations both exceed their importance graphs, suggesting that overall input should be maintained or reduced. Conversely, the performance graphs of 3rd Tianfu Street are significantly lower than the importance graphs, indicating a need for further optimization and improvement. In the first-order latent variables, field environment and urban aesthetics are identified as variables with high perceptual influence. Correspondingly, in the second-order latent variables, density, diversity, connection design, street characteristics, and order management are recognized as having significant perceptual effects. At Tianfu Square station, these variables fall into the "high importance–high performance" category, suggesting that maintaining the status quo is appropriate. At Chunxi Road station, the first-order variable of urban aesthetics falls into the "high importance–low performance" unmatched category. A detailed analysis reveals that while street characteristics at Chunxi Road are excellent, the evaluation of order management is poor and requires improvement. In the 3rd Tianfu Street station area, two important first-order variables also fall into the "high importance–low performance" mismatch category. Among the second-order variables, only order management performs well, while the others perform poorly and need significant improvement.

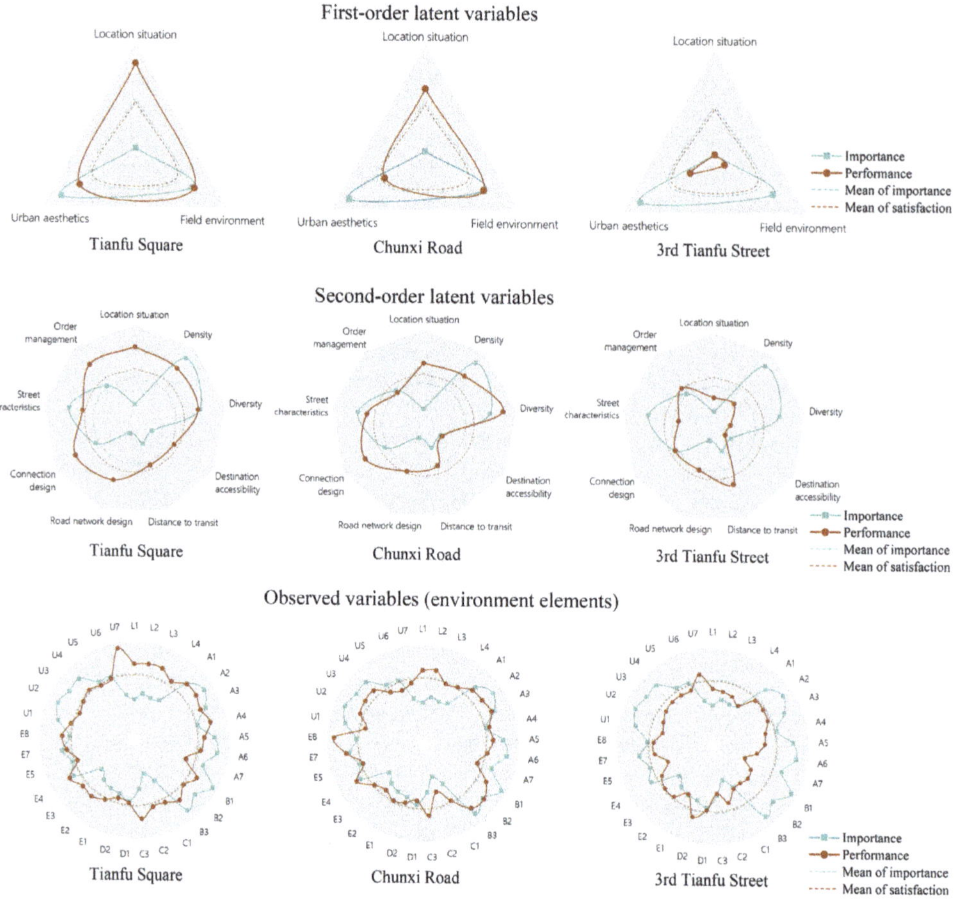

Figure 9. Comparison of the importance and performance of first-order, second-order, and observed variables.

The radar chart of observed variables can highlight specific environmental elements that need optimization. For example, at Chunxi Road station, the order management dimension includes U5 (street hygiene) and U6 (landscape maintenance), which are high-impact elements but exhibit low performance, categorizing them as "high importance–low performance" elements. There are more than two-thirds of such elements in the 3rd Tianfu Street station area. Improving these built elements will result in a noticeable enhancement in evaluations.

5. Discussion and Conclusions

5.1. Discussion

5.1.1. Field Environment

The field environment has the most significant direct effect on the satisfaction of the CMSA. Here, density, diversity, and connection design are the most important second-order latent variables. The density and diversity perception are related to the land use, while the connection design is the pedestrian planning and design with the metro entrance as the core. We discuss some of the key design points of land use and pedestrian design.

We listed a part of land use indicators for the three cases (Figure 10) and found that the apparent differences in density perception among the three stations are building height and open space. We infer that building height and open space can directly affect the user's perception of density. Another noticeable difference is the proportion of commerce. 3rd Tianfu Street needs a higher proportion of commercial function, which may reduce the diversity of users' activity choices. Based on the above analysis, we recommend the primary strategy to enhance land use satisfaction is to dispel the feeling of high-density development and increase the number of business-oriented functional types. High-density development does help improve land use efficiency and promote urban vitality; it is also considered a critical factor in the successful implementation of TOD, but a common problem central-type sites face is the poor sense of repression and congestion it brings [51]. One of the strategies to solve this problem is to appropriately add small-scale buildings in the station area so that the buildings have a difference of height and height. This strategy is conducive to creating a sparse and staggered urban spatial pattern in old cities and maximizing the agglomeration effect of land development [52]. Another suggestion is to increase the development area of the underground space and release the ground space, thus providing citizens with more open spaces like plazas and parks. Open space can also reduce the degree of interference between people and vehicles and enhance pedestrian safety and comfort.

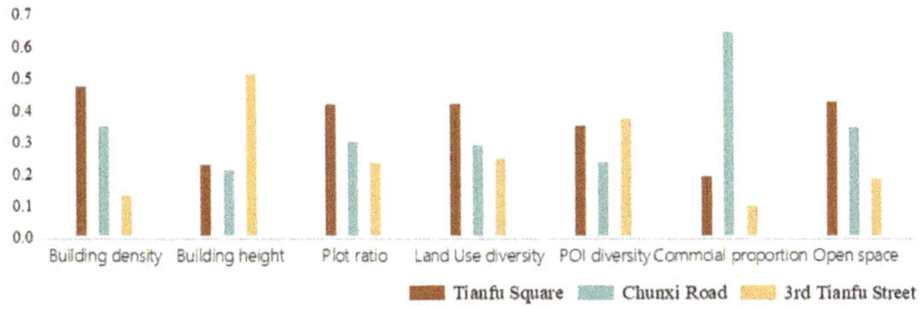

Figure 10. The indicators of land use of three catchment areas.

In terms of diversity, a study showed that neighborhoods with a relatively higher land-use mix have a significantly positive association with walking activity [53]. In the planning concept of the "15 min city", residents should be able to enjoy a higher quality of life within walking or cycling distance, where they can effectively fulfill six essential urban functions to sustain a decent urban life. These functions include living, working, commerce, healthcare, education, and entertainment [6]. These functions are integrated into TOD planning, expanding outward in concentric circles in decreasing order of economic value and development intensity, to enhance land-use diversity and the completeness of urban functions. Moreover, to maximize these benefits, diversity needs to be implemented at various scales within the city—not only at the urban level, across broader areas, but also at finer scales, such as at the building level.

A suggestion about pedestrian design is to strengthen the walking continuity and enrich the entrance's form. The planning and design of a station play a significant role in the construction of the station area walking system [54]. For example, the distance to the entrance of Tianfu Square station is more than 380 m, crossing several main roads to achieve passenger–vehicle separation and providing a comfortable underground walking space in extreme weather. Its entrance and exit of Tianfu Square take the form of a sunken square (Figure 11a) to organize passenger flow into the subway station and the commercial complex built on the station. Chunxi Road features a metro entrance that connects directly to the underground of the commercial center (Figure 11b). These designs not only facilitate the management of pedestrian traffic and enhance the integration of the station with the city but also help regulate the microclimate at the metro station entrance [55]. The form of the subway entrance is also an essential factor. Subway stations are usually built underground without building facades, and entrances are vital nodes for users to recognize and remember environmental information. Li's finding verifies that the traffic demand is the priority in the landscape design of the metro station entrance [56]. The entrances and exits of Chunxi Road adopt various forms of glass to give users a novel experience (Figure 11c). In conclusion, the planning of station entrances and exits should adopt a field-oriented approach, enhancing crowd diversion efficiency by integrating urban main roads and linking key areas and buildings. Additionally, the design of entrances and exits should be both memorable and aesthetically pleasing, while also reflecting the cultural characteristics of the city.

(**a**) (**b**) (**c**)

Figure 11. The station entrance design. (**a**) The open form entrance of Tianfu Square; (**b**) the entrance of Chunxi Road connects to the mall; (**c**) the artistic entrance of Chunxi Road.

5.1.2. Urban Aesthetics

Urban aesthetics had the strongest influence on satisfaction because of its direct and indirect effects. Here, street characteristics are the dominant second-order variable. Color and culture are two important elements that shape street characteristics. In the cases we studied, we found that 3rd Tianfu Street Station is located in the city's new district, which was completed in a short time, and the government uniformly planned the facade of the building. Therefore, it may lack characteristics compared to Tianfu Square, which has historical charm, and Chunxi Road, which has diversified colorful characteristics (Figure 12). The use of culture as a tool for urban regeneration is not recent; it has been used as a preferred concept to direct the revitalization of urban centers [57]. The color and form of architecture are the design methods that reflect a city's culture. It is of contemporary significance to retain regional characteristics, stimulate emotional resonance, and enhance the readability of the city image by portraying the urban ethnic image.

(**a**) (**b**) (**c**)

Figure 12. Street view of 3 catchment areas: (**a**)Tianfu Square; (**b**) Chunxi Road; (**c**) 3rd Tianfu Street.

Order management, as another second-order variable in urban aesthetics, is not more important than the average in our conclusion, but a variety of studies have shown that a clean, orderly, and safe environment is an essential condition of urban aesthetics and an important guarantee for citizens' quality of life [58]. This means that order management is also an indispensable part of urban aesthetics. Street hygiene and the pruning of plants are two significant factors. Keeping the streets clean requires adding garbage disposal facilities and regularly cleaning the ground. Vegetation is another fundamental element of urban design. Trees provide the shade necessary to walk and stay in the streets during the day's hottest hours. Also, vegetation has aesthetic qualities that are indispensable for the design of attractive and pleasing places [59]. The integration of the Smart City concept is also a key factor in improving management levels. For example, within the Smart City framework, factors such as inclusivity, resident participation, and real-time service delivery are promoted through various digital platforms [6].

5.1.3. Location Situation

The location situation does not have a direct impact on satisfaction, but it is not irrelevant. Instead, it indirectly influences satisfaction by affecting the field environment and urban aesthetics. The centers of new urban areas compete with the old urban centers, which tend to have better absolute geographical locations in the process of urban expansion [13]. It is worth noting that there is a negative correlation between location situation and urban aesthetics. For station areas with superior locations, which are more likely to represent the cultural and economic hubs of a city, people tend to have stricter evaluations of their urban aesthetics. From a psychological perspective, there is a tension between functionality and aesthetics. People typically expect high-density areas to not only fulfill basic functional needs but also provide a good visual and environmental experience.

Urban location is a multidimensional performance that is affected by natural and artificial urban areas [60]. Therefore, for stations located in less favorable areas, particularly in new districts, we recommend enhancing the relative centrality of the station area by increasing the number of subway lines at the central station of the new city and ensuring these lines traverse densely populated areas. These two strategies are also key components of our model findings. Studies indicated that increasing the number of subway lines at metro stations and enhancing transportation options to other central areas can improve urban compactness and promote more efficient living for residents [61,62]. There are many examples of urban planning in Japan that we can learn from, the construction goal of which is a "network compact city" [63].

It is worth noting that the relationship between the location situation and field environment is positively correlated, indicating that when the location situation is favorable, people tend to have a higher tolerance for the perceived aspects of the field environment. For instance, when a site is situated in a high-density development area, individuals are more accepting of crowded buildings and heavy pedestrian traffic. Conversely, the relationship between the location situation and urban aesthetics is negatively correlated. This can be understood as follows: These areas are typically located in the traditional center of the city or the old town, characterized by their mixed and chaotic streets. However, they often serve as urban public spaces characterized by gateway effects. People tend to apply more stringent evaluation criteria to these types of stations. Thus, evaluating the location situation provides insights into the design of the field environment and urban aesthetics. For stations with favorable location attributes, greater emphasis can be placed on the relationship between density, business type, and economic value, as people tend to be more accepting of these factors. Moreover, increased investment to enhance the beauty of the streets and elevate the overall image of the city is necessary.

5.2. Conclusions

This study explores the interaction between the built environment and public perception of CMSAs by combining SEM and IPA methods. The results revealed the effects of various variables (location, density, diversity, distance to transit, destination accessibility, road network, connection design, street characteristics, and order management) within the built environment (location, field environment, and urban aesthetics) on satisfaction. The main conclusions of the SEM model are as follows: First, perceptions of the field environment have a direct positive effect on satisfaction. Among these, density, diversity, and connection design are critical factors within the field environment dimension. Secondly, urban aesthetic perception directly affects satisfaction, with street characteristics being a critical factor. Thirdly, the location situation does not have a direct impact on satisfaction, but it indirectly influences satisfaction by positively affecting the field environment and negatively affecting urban aesthetics. Based on this, we used the value of the contribution of each element to perceived importance as the importance score, and the user satisfaction ratings of the elements as the performance score, to conduct an IPA analysis. We compared the importance and satisfaction alignment of various environmental elements across three sample CMSAs in Chengdu, and based on this, we proposed optimization suggestions.

This study aims to deeply understand the complex relationship between perceptions of the built environment and public satisfaction in the CMSAs. We provided an innovative approach to exploring environmental perception and assessing the effectiveness of the built environment. After the application of the proposed SEM-IPA model, the empirical results can provide valuable insights for urban designers and policymakers, as well as an empirical basis for the construction and renewal of central TOD in future cities.

This study has some limitations. Due to the limited sample size, sufficient data were not obtained for certain demographic characteristics, such as elderly individuals and children. Therefore, weight adjustments based on different sample proportions were not considered. In future research, we will aim to collect a broader range of samples using methods such as stratified sampling and targeted interviews with specific populations and further analyze whether significant differences in perceptions exist across various demographic groups.

Author Contributions: Conceptualization, J.M. and Z.S.; Investigation, Y.Z. and W.S.; Methodology, J.M. and P.L.; Software, J.M. and P.L.; Writing—Original Draft, J.M.; Writing—Review and Editing, Z.S., P.L. and Y.Z. All authors have read and agreed to the published version of the manuscript.

Funding: This research was funded by the National Natural Science Foundation of China [grant number 52378039 and 51978573].

Data Availability Statement: The data presented in this study are available on request from the corresponding author. The data are not publicly available due to privacy.

Acknowledgments: The authors wish to thank the people who filled out the survey in the study areas.

Conflicts of Interest: The authors declare no conflict of interest.

References

1. Choi, K.; Park, H.J.; Uribe, F.A. The impact of light rail transit station area development on residential property values in Calgary, Canada: Focus on land use diversity and activity opportunities. *Case Stud. Transp. Pol.* **2023**, *12*, 100924. [CrossRef]
2. Papa, E.; Bertolini, L. Accessibility and Transit-Oriented Development in European metropolitan areas. *J. Transp. Geogr.* **2015**, *47*, 70–83. [CrossRef]
3. Vale, D.S. Transit-oriented development, integration of land use and transport, and pedestrian accessibility: Combining node-place model with pedestrian shed ratio to evaluate and classify station areas in Lisbon. *J. Transp. Geogr.* **2015**, *45*, 70–80. [CrossRef]

4. Yang, L.; Chen, Y.; Xu, N.; Zhao, R.; Chau, K.W.; Hong, S. Place-varying impacts of urban rail transit on property prices in Shenzhen, China: Insights for value capture. *Sustain. Cities Soc.* **2020**, *58*, 102140. [CrossRef]
5. Mocak, P.; Matlovicova, K.; Matlovic, R.; Penzes, J.; Pachura, P.; Mishra, P.K.; Kostilnikova, K.; Demkova, M. 15-minute city concept as a sustainable urban development alternative: A brief outline of conceptual frameworks and slovak cities as a case. *Folia Geogr.-Slovak* **2022**, *64*, 69–89.
6. Moreno, C.; Allam, Z.; Chabaud, D.; Gall, C.; Pratlong, F. Introducing the "15-Minute City": Sustainability, Resilience and Place Identity in Future Post-Pandemic Cities. *Smart Cities* **2021**, *4*, 93–111. [CrossRef]
7. Mondschein, A.; Moga, S.T. New Directions in Cognitive-Environmental Research. *J. Am. Plan. Assoc.* **2018**, *84*, 263–275. [CrossRef]
8. Gao, Y.; Chen, Y.Y.; Mu, L.; Gong, S.Z.; Zhang, P.C.; Liu, Y. Measuring urban sentiments from social media data: A dual-polarity metric approach. *J. Geogr. Syst.* **2022**, *24*, 199–221. [CrossRef]
9. Olaru, D.; Curtis, C. Designing TOD precincts: Accessibility and travel patterns. *Eur. J. Transp. Infrastruct. Res.* **2015**, *15*, 6–26. [CrossRef]
10. Gao, W.; Sun, X.L.; Zhao, M.; Gao, Y.; Ding, H.R. Evaluate Human Perception of the Built Environment in the Metro Station Area. *Land* **2024**, *13*, 90. [CrossRef]
11. Li, C.N.; Lin, C.; Hsieh, T.K. TOD District Planning Based on Residents' Perspectives. *ISPRS Int. J. Geo-Inf.* **2016**, *5*, 52. [CrossRef]
12. Chen, T.; Chen, Y.; Zhou, Y.; Guo, J. Efficiency Assessment of Transit-Oriented Development Focusing on the 500-m Core Catchment of Metro Stations Based on the Concept of a Metro Microcenter in Beijing. *J. Transp. Eng. A-Syst.* **2023**, *149*, 04023117. [CrossRef]
13. Ma, J.X.; Shen, Z.W.; Xie, Y.; Liang, P.P.; Yu, B.J.; Chen, L. Node-place model extended by system support: Evaluation and classification of metro station areas in Tianfu new area of Chengdu. *Front. Environ. Sci.* **2022**, *10*, 990416. [CrossRef]
14. Wu, W.J.; Chen, W.Y.; Yun, Y.W.; Wang, F.L.; Gong, Z.Y. Urban greenness, mixed land-use, and life satisfaction: Evidence from residential locations and workplace settings in Beijing. *Landsc. Urban Plan.* **2022**, *224*, 104428. [CrossRef]
15. Lee, K.Y. Relationship between Physical Environment Satisfaction, Neighborhood Satisfaction, and Quality of Life in Gyeonggi, Korea. *Land* **2021**, *10*, 663. [CrossRef]
16. Bivina, G.R.; Gupta, A.; Parida, M. Walk Accessibility to Metro Stations: An analysis based on Meso- or Micro-scale Built Environment Factors. *Sustain. Cities Soc.* **2020**, *55*, 102047. [CrossRef]
17. Yang, L.; Yang, H.; Tang, J.; Lu, Y.; Liu, J. Assessing attribute performance and older adults' satisfaction with transit services: Implications for age-friendly planning. *Travel. Behav. Soc.* **2025**, *40*, 101026. [CrossRef]
18. Cervero, R.; Sarmiento, O.L.; Jacoby, E.; Gomez, L.F.; Neiman, A. Influences of Built Environments on Walking and Cycling: Lessons from Bogota. *Int. J. Sustain. Transp.* **2009**, *3*, 203–226. [CrossRef]
19. Banerjee, I. Production of Railwayscape in urban environment: Analysing railway heritage tourism potential in Siliguri City, India. *Reg. Sustain.* **2023**, *4*, 1–12. [CrossRef]
20. Bertolini, L. Nodes and places: Complexities of railway station redevelopment. *Eur. Plan. Stud.* **1996**, *4*, 331–345. [CrossRef]
21. Cao, Z.; Asakura, Y.; Tan, Z. Coordination between node, place, and ridership: Comparing three transit operators in Tokyo. *Transp. Res. Part D Transp. Environ.* **2020**, *87*, 102518. [CrossRef]
22. Deng, X.D.; Zhang, J.; Liao, S.Y.; Zhong, C.J.; Gao, F.; Teng, L. Interactive Impacts of Built Environment Factors on Metro Ridership Using GeoDetector: From the Perspective of TOD. *ISPRS Int. J. Geo-Inf.* **2022**, *11*, 623. [CrossRef]
23. Cao, J.; Ettema, D. Satisfaction with travel and residential self-selection: How do preferences moderate the impact of the Hiawatha Light Rail Transit line? *J. Transp. Land. Use* **2014**, *7*, 93–108. [CrossRef]
24. Huang, X.; Yuan, W.; White, M.; Langenheim, N. A Parametric Framework to Assess Generative Urban Design Proposals for Transit-Oriented Development. *Buildings* **2022**, *12*, 1971. [CrossRef]
25. Zhengyu, D.; Xuan, H.; Chuan, C. The Impacts of Urban Residential Location and Travel Behavior on Commuters' Quality of Life: A Case Study of Shanghai, China. *Mod. Urban Res.* **2023**, 88–93. [CrossRef]
26. Junyan, Y.; Yi, S. Research on spatial location mechanism and layout methods of urban district-wide center. *City Plan. Rev.* **2018**, *1*, 43–51.
27. Aghaabbasi, M.; Moeinaddini, M.; Shah, M.Z.; Asadi-Shekari, Z.; Kermani, M.A. Evaluating the capability of walkability audit tools for assessing sidewalks. *Sustain. Cities Soc.* **2018**, *37*, 475–484. [CrossRef]
28. Lynch, K. *Lynch Image of the City*; The MIT Press: Cambridge, MA, USA, 1960.
29. Cho, G.H.; Rodriguez, D. Location or design? Associations between neighbourhood location, built environment and walking. *Urban Stud.* **2015**, *52*, 1434–1453. [CrossRef]
30. Yin, C.H.; Liu, Y.F.; Wei, X.J.; Chen, W.Q. Road Centrality and Urban Landscape Patterns in Wuhan City, China. *J. Urban Plan. Dev.* **2018**, *144*, 05018009. [CrossRef]

31. Torquati, B.; Giacchè, G.; Tempesta, T. Landscapes and Services in Peri-Urban Areas and Choice of Housing Location: An Application of Discrete Choice Experiments. *Land* **2020**, *9*, 393. [CrossRef]
32. Yang, L.; Wang, X.; Sun, G.; Li, Y. Modeling the perception of walking environmental quality in a traffic-free tourist destination. *J. Travel. Tour. Mark.* **2020**, *37*, 608–623. [CrossRef]
33. Cetintahra, G.E.; Cubukcu, E. The influence of environmental aesthetics on economic value of housing: An empirical research on virtual environments. *J. Hous. Built Environ.* **2015**, *30*, 331–340. [CrossRef]
34. Fu, Y.; Zhang, X.L. Trajectory of urban sustainability concepts: A 35-year bibliometric analysis. *Cities* **2017**, *60*, 113–123. [CrossRef]
35. Julsrud, T.E.; Uteng, T.P. Technopolis, shared resources or controlled mobility? A net-based Delphi-study to explore visions of future urban daily mobility in Norway. *Eur. J. Futures Res.* **2015**, *3*, 1–13. [CrossRef]
36. Allen, J.; Eboli, L.; Forciniti, C.; Mazzulla, G.; de Dios Ortuzar, J. The role of critical incidents and involvement in transit satisfaction and loyalty. *Transp. Policy* **2019**, *75*, 57–69. [CrossRef]
37. Arbore, A.; Busacca, B. Rejuvenating importance-performance analysis. *J. Serv. Manag.* **2011**, *22*, 409–430. [CrossRef]
38. Rasovska, I.; Kubickova, M.; Ryglova, K. Importance-performance analysis approach to destination management. *Tour. Econ.* **2021**, *27*, 777–794. [CrossRef]
39. Ban, O.I.; Ban, A.I.; Tuşe, D.A. Importance–Performance Analysis by Fuzzy C-Means Algorithm. *Expert. Syst. Appl.* **2016**, *50*, 9–16. [CrossRef]
40. Martilla, J.A.; James, J.C. Importance-performance analysis. *J. Mark.* **1977**, *1*, 77–79. [CrossRef]
41. Kuby, M.; Barranda, A.; Upchurch, C. Factors influencing light-rail station boardings in the United States. *Transp. Res. Part A Policy Pract.* **2004**, *38*, 223–247. [CrossRef]
42. O'Sullivan, S.; Morrall, J. Walking Distances to and from Light-Rail Transit Stations. *Transp. Res. Rec. J. Transp. Res. Board.* **1996**, *1538*, 19–26. [CrossRef]
43. Field, A. *Discovering Statistics Using SPSS for Windows: Advanced Techniques for Beginners*; Springer: New York, NY, USA, 2000.
44. Hair, J.; Black, W.; Babin, B.; Anderson, R. *Multivariate Data Analysis*; Prentice Hall: Upper Saddle River, NJ, USA, 2009.
45. Wu, M. *Questionnaire Statistical Analysis: SPSS Operation and Application*; Chongqing University Press: Chongqing, China, 2010.
46. Hair, J.; Black, W.; Babin, B.; Anderson, R. *Multivariate Data Analysis*, 6th ed.; Prentice Hall: Upper Saddle River, NJ, USA, 2006.
47. Ong, A.K.S.; Prasetyo, Y.T.; Lagura, F.C.; Sigua, K.M.; Villas, J.A.; Nadlifatin, R.; Young, M.N.; Diaz, J.F.T. Determining tricycle service quality and satisfaction in the Philippine urban areas: A SERVQUAL approach. *Cities* **2023**, *137*, 104339. [CrossRef]
48. Steiger, J.H. Understanding the limitations of global fit assessment in structural equation modeling. *Personal. Individ. Differ.* **2007**, *42*, 893–898. [CrossRef]
49. Gefen, D.; Straub, D.W.; Boudreau, M. Structural Equation Modeling and Regression: Guidelines for Research Practice. *Commun. Assoc. Inf. Syst.* **2000**, *4*, 7. [CrossRef]
50. Hair, J.F. *Multivariate Data Analysis with Readings*; Macmillan, Collier Macmillan: New York, NY, USA, 1987.
51. Nematollahi, S.; Tiwari, R.; Hedgecock, D. Desirable Dense Neighbourhoods: An Environmental Psychological Approach for Understanding Community Resistance to Densification. *Urban Policy Res.* **2015**, *34*, 132–151. [CrossRef]
52. Dai, P.; Fu, H.; Yang, X.X.; Han, S.; Fu, G.N.; Wang, Y.J. Exploring the urban renewal strategy based on transit-oriented development concept--A case study of Japan and Hong Kong. *Front. Mater.* **2023**, *10*, 1098027. [CrossRef]
53. Sung, H.; Lee, S.; Jung, S. Identifying the relationship between the objectively measured built environment and walking activity in the high-density and transit-oriented city, Seoul, Korea. *Environ. Plan. B-Plan. Des.* **2014**, *41*, 637–660. [CrossRef]
54. Liang, X.; Lu, Z.R.; Ye, F.; Zhang, W.J. Investigation of design of independent metro station entrances in China. *Proc. Inst. Civ. Eng.-Munic. Eng.* **2023**, *176*, 10–31. [CrossRef]
55. Zhao, J.J.; Dong, J.K.; Zhang, X.H.; Na, Y.L.; Jiang, C.X.; He, F.; Cui, Q.J.; Liu, J. Field measurement of microclimate of sunken square and its effect on indoor environment of underground metro station in subtropical region. *Build. Environ.* **2023**, *228*, 109873. [CrossRef]
56. Li, Z.; Lin, X.S.; Han, X.; Lu, X.Y.; Zhao, H.Y. Landscape Efficiency Assessment of Urban Subway Station Entrance Based on Structural Equation Model: Case Study of Main Urban Area of Nanjing. *Buildings* **2022**, *12*, 294. [CrossRef]
57. Garza-Rodríguez, F. Identifying cultural clusters in Monterrey city center. A Case study. *Contexto-Rev. Fac. Arq.* **2019**, *13*, 41–56.
58. Mateo-Babiano, I. Pedestrian's needs matter: Examining Manila's walking environment. *Transp. Policy* **2016**, *45*, 107–115. [CrossRef]
59. Yang, L.; Yang, H.; Yu, B.; Lu, Y.; Cui, J.; Lin, D. Exploring non-linear and synergistic effects of green spaces on active travel using crowdsourced data and interpretable machine learning. *Travel. Behav. Soc.* **2024**, *34*, 100673. [CrossRef]
60. Portnov, B.A.; Schwartz, M. On the Importance of the 'Location Package' for Urban Growth. *Urban Stud.* **2009**, *46*, 1665–1679. [CrossRef]
61. Yang, L.; Lu, Y.; Cao, M.; Wang, R.; Chen, J. Assessing accessibility to peri-urban parks considering supply, demand, and traffic conditions. *Landsc. Urban Plan.* **2025**, *257*, 105313. [CrossRef]

62. Peng, P.P.; Liu, Z.; Guo, J.H.; Wang, C. Dynamic Metro Stations Importance Evaluation Based on Network Topology and Real-Time Passenger Flows. *Ksce J. Civ. Eng.* **2023**, *27*, 4459–4471. [CrossRef]
63. Liu, Q.; Qian, Z.; Lai, Y. Transit Oriented Shrinkage: Planning Methods of Location Optimization Plan in Japan from the Perspective of TOD. *Urban. Plan. Int.* **2023**, *38*, 112–122, 144.

land

MDPI

Article

The Role of Comprehensive Transportation in Shaping Spatial Expansion Patterns: A Case Study of the Yangtze River Middle Reaches Urban Agglomeration

Zaiyu Fan [1,2], Weiyang Luo [3], Chang Liu [4] and Mengyun Xie [5,*]

1 School of Civil Engineering, Architecture and Environment, Hubei University of Technology, Wuhan 430068, China; fzy@hbut.edu.cn
2 Key Laboratory of Intelligent Health Perception and Ecological Restoration of Rivers and Lakes, Ministry of Education, Hubei University of Technology, Wuhan 430068, China
3 CCCC Second Harbor Engineering Company Ltd., Wuhan 430040, China
4 School of Architecture & Urban Planning, Huazhong University of Science and Technology, Wuhan 430074, China
5 School of Urban Design, Wuhan University, Wuhan 430072, China
* Correspondence: 00033987@whu.edu.cn

Abstract: Regional comprehensive transportation infrastructures constitute the fundamental basis for the development of inland urban agglomerations. To elucidate the role of comprehensive transportation in shaping the spatial organization and expansion of urban agglomerations, this study takes the Yangtze River Middle Reaches Urban Agglomeration (YRMRUA) as a case example. It examines the spatial relationships between transportation network layout and spatial expansion patterns using fractal dimension based on traffic accessibility, traffic-weighted linear density, and Pearson correlation analysis. The key findings of this study are as follows: (1) The YRMRUA exhibits a partial fractal growth pattern influenced by transportation development, which indicates that the comprehensive transportation has a significant but limited impact on YRMRUA. (2) There is a moderate correlation between traffic-weighted linear density and spatial expansion intensity within YRMRUA. (3) Specific groups such as the Wuhan–Ezhou–Huanggang–Huangshi group, Changsha–Zhuzhou–Xiangtan group, and Nanchang–Yichun group have formed in areas where transportation development and spatial expansion are at the forefront. (4) Different modes of transportation, including waterway transportation, railway transportation, and road transportation, have varying effects on spatial expansion. The integration of these modes forms the fundamental framework of urban agglomerations.

Keywords: comprehensive transportation; cross-regional transportation infrastructure; fractal analysis; spatial expansion; Yangtze River Middle Reaches Urban Agglomeration

1. Introduction

The 14th Five-Year Plan of China has designated 19 urban agglomerations, some of which have released development planning that clarified the positioning of each urban agglomeration and outlined the development direction for them. The subject of urban agglomerations has been widely discussed, involving urban planning and policy [1,2], human geography [3], regional economics [4,5], transportation system, and other fields [6–8]. From the perspective of urban agglomeration formation, certain scholars argue that economic and industrial agglomeration serve as the foundational elements of urban agglomeration [9–11], and there are mutual feedback effects between them [12–14]. As for the further development and maturity of urban agglomeration, a growing body of scholarly opinion suggests that the evolution of it is influenced by a variety of factors, including geographical conditions, economic foundation, policy context, and more [15–17]. Furthermore, as an important support for regional urban areas, some significant advancements have been made in understanding the fundamental role of transportation in regional spatial development [18–20]. Researchers are concerned about the mode and degree of transportation construction for regional economic promotion and have conducted a thorough analysis [21–23]. This process has generated numerous insightful perspectives, including the enhancement of transportation infrastructure contributing to improved allocation efficiency of regional resources [24,25], and transportation networks fostering the sustainability of urban agglomeration spatial structures [26]. Nevertheless, transportation infrastructure may not always have a positive effect [27–29], especially in less developed areas, which leads to the imbalance in the development of urban areas [30–32].

At the regional macro level, the relationship between the spatial development of urban agglomerations and the layout of regional comprehensive transportation infrastructure necessitates further exploration [33]. First, the majority of studies endeavor to elucidate the mechanism in all factors for the macro-research object of urban agglomeration. This could lead to ambiguity in conclusions and is not conducive to construction and planning in the urbanization process. Second, studies that solely concentrate on the impact mechanism of regional traffic tend to overlook the exploration of spatial connections and hierarchies within the whole mega-region, while remaining confined to the internal spatial organization of built-up areas. Third, the relationship between individual cities and their internal transportation (such as the relationship between land use and transportation layout, and the impact of transportation infrastructure on urban economic development) has been extensively explored. However, due to the significant scale differences between urban agglomerations and individual cities, these conclusions cannot be directly applied to the spatial level of urban agglomerations. Therefore, the primary objective of this paper is to provide a confirmatory analysis of the link between comprehensive transportation and urban agglomerations from the local space perspective and propose corresponding explanatory mechanisms, complementing prior research on flowing space.

Yangtze River Middle Reaches Urban Agglomeration (YRMRUA), located in the core area of China's Yangtze River Economic Belt (Figure 1), is at the intersection of three horizontal and two vertical national traffic arteries. YRMRUA's significance is highlighted in the latest national comprehensive three-dimensional transportation network planning of China, with provincial capitals like Wuhan and Changsha recognized as crucial transportation hubs on a national scale [34,35].

The importance and research significance of YRMRUA are also reflected in the relevant planning. In 2010, China's Main Functional Zone Plan designated YRMRUA as one of the key development zones. Unlike the other 14 urban agglomerations in China's 14th Five-Year Plan [36], YRMRUA is one of the only five national-level urban agglomeration. In the development planning of YRMRUA itself, many development directions have been clarified. The planning requires YRMRUA to rely on key development axes such as Shanghai–Kunming, Beijing–Guangzhou, Beijing–Jiujiang, and others, to form a pattern of reasonable division of labor and coordinated development among large- and medium-sized cities and small towns along the line (Figure 2). These development axes are actually named after inter-regional transportation corridors. Therefore, this determines that, compared with other urban agglomeration, there is a close connection between the development of YRMRUA and comprehensive transportation.

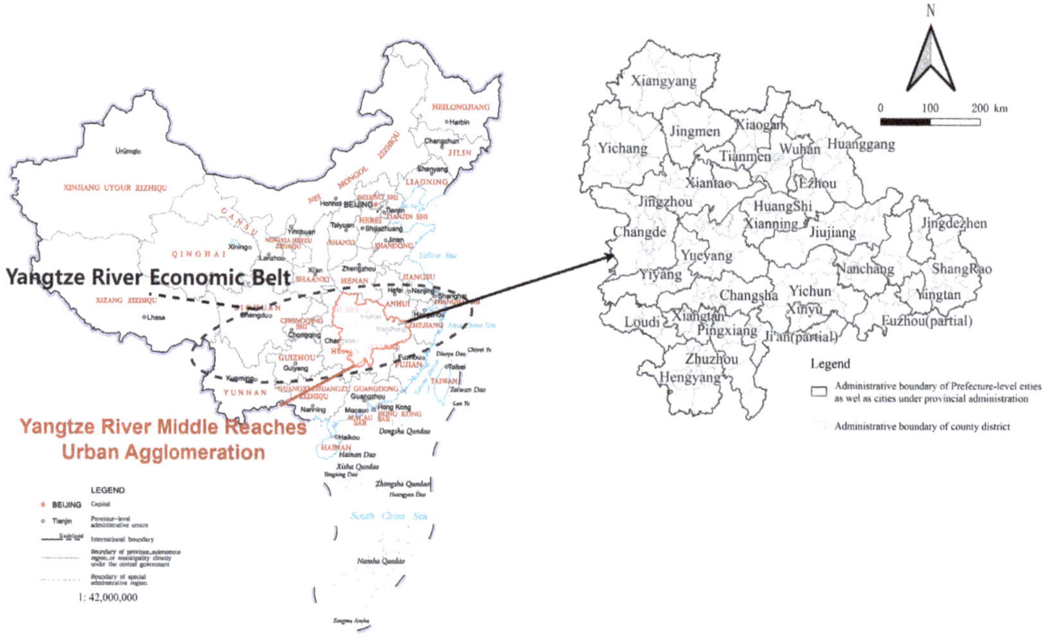

Figure 1. The research scope and location of YRMRUA.

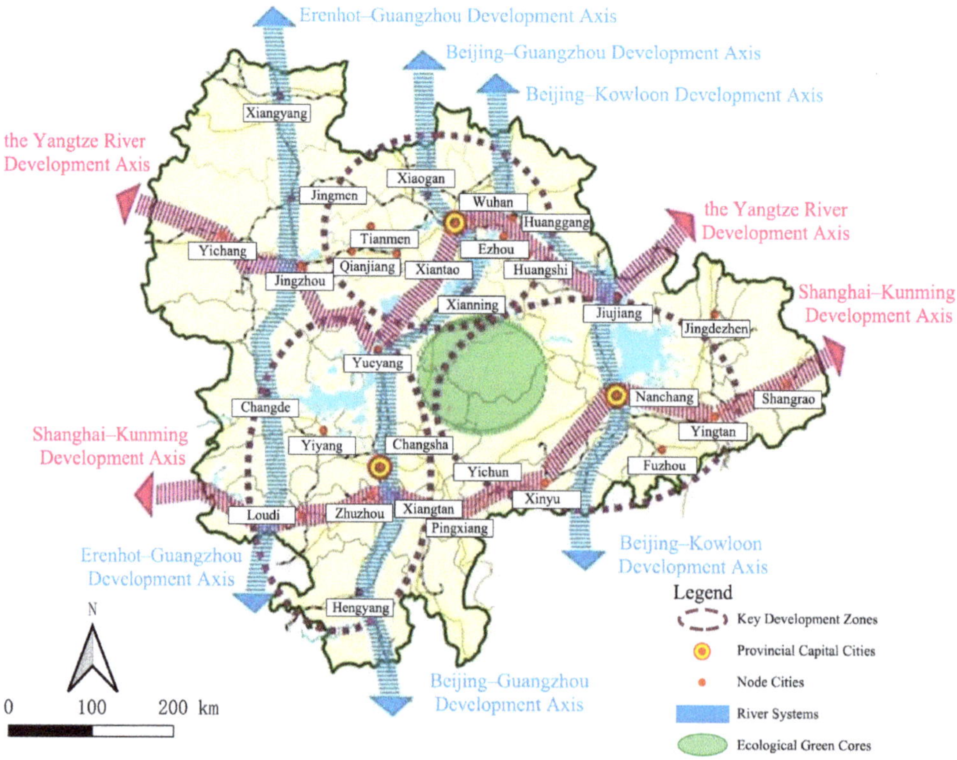

Figure 2. The strategic guidance of YRMRUA's development planning.

Consequently, this study utilizes YRMRUA as a case study to investigate the influence of transportation infrastructure development and its spatial expansion on the growth of urban agglomeration's built-up areas. This includes the extent and manner in which the former affects the latter. The research aims to provide insights for the development of growing urban agglomerations in Central and Western China, as well as for the construction of urban agglomerations in developing countries.

2. Materials and Methods

2.1. Research Framework

This study focuses on measuring the impact of comprehensive transportation on the spatial dynamic evolution of YRMRUA and explaining its mechanism. The research framework is illustrated in Figure 3. The research framework can be divided into three levels. The first level involves data processing and dataset creation, utilizing GEE (Google Earth Engine) and ArcGIS as tools. This process involves a two-step procedure of data pre-screening and confirmation [37]. After obtaining the aforementioned data, the required datasets for this study area are determined based on the defined research period. The second level involves data analysis. This study will sequentially employ methods such as fractal dimension based on traffic accessibility (FDTA), the traffic-weighted linear density (TLD), expansion intensity index (EII), and Pearson correlation coefficient (PCC). Specifically, we will employ the FDTA method to ascertain the fractal characteristics of spatial growth in the YRMRUA, that is, to clarify whether the built-up area's expansion of the YRMRUA exhibits a self-organizing pattern centered around comprehensive transportation infrastructure. Subsequently, we will comprehensively utilize TLD, EII, and PCC to obtain the results of correlational characteristics and group characteristics. These two types of results will, respectively, elucidate whether there is a correlation between TLD and EII within the YRMRUA at the grid cell level, as well as the clustering nature of their spatial distribution. Based on this, spatial groups of YRMRUA's spatial expansion and transportation resource clusters can be identified. The third level involves mechanism interpretation. Since the impact of comprehensive transportation on regional spatial expansion is a highly complex process, with its effects on regional scales demonstrably exceeding those on individual urban areas, it is difficult to determine the underlying mechanism solely through quantitative analysis. Therefore, this study endeavors to use graphical representation and illustrative examples in the Discussion section to supplement the explanation of the experimental results. This will contribute to a more in-depth understanding of the complex role that comprehensive transportation plays within the spatial context of urban agglomerations and help clarify the differences between the spatial-shaping effects of comprehensive transportation and those of intra-city transportation.

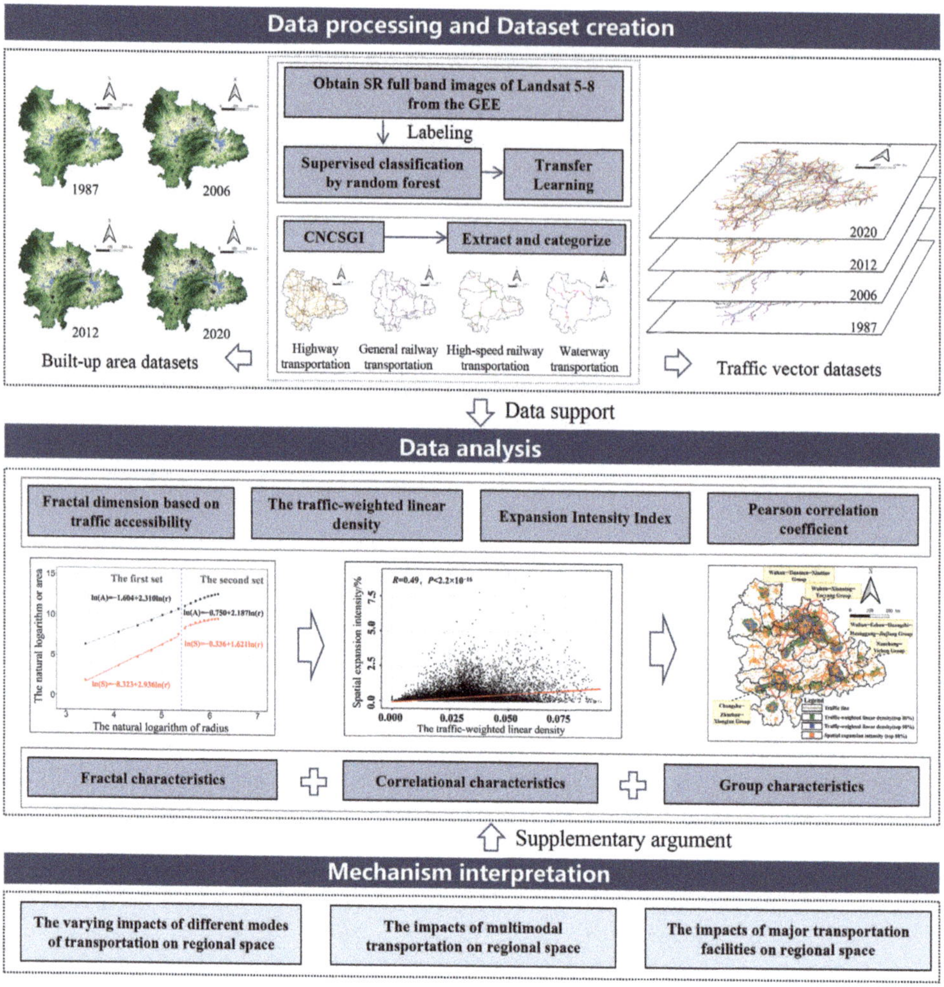

Figure 3. Research framework.

2.2. Data Sources

The data sources section of the paper includes data on urban built-up areas related to the spatial morphology of urban agglomerations, as well as data on major transportation infrastructure across the administrative area. The original data for built-up areas in this study come from LANDSAT 5 and 8 satellite imagery provided by the GEE (https://earthengine.google.org/ (accessed on 10th March 2024)) platform. The final data were obtained by processing with the random forest algorithm on GEE. These data will be subjected to multiple rounds of filtering and corrected using methods such as spatio-temporal trajectory maps [38–40]. The comprehensive transportation (highways, general railways, expressways, and waterways) network data used in this paper comes from China National Catalogue Service for Geographic Information (CNCSGI, https://www.webmap.cn/ (accessed on 10th March 2024)). In addition, regarding the scope of the study period, as 1987, 2006, 2012, and 2020 are crucial time nodes in the formation of the YRMRUA [37], this paper will conduct research in three segments, 1987–2006, 2006–2012, and 2012–2020, with the final year of each segment serving as a significant observation point.

2.3. Methods

2.3.1. Fractal Dimension Based on Traffic Accessibility

From the perspective of spatial fractal theory, the spatial growth of urban agglomerations has a certain tendency of self-organization. To validate whether certain regions of urban agglomerations are propelled by the development of comprehensive transportation infrastructure and network systems, traffic accessibility can be employed as a metric factor within this framework, thereby constructing an associated fractal dimension model [41]. This method helps to identify, at the macro level, whether the spatial expansion of the YRMRUA is unfolding along the layout of comprehensive transportation infrastructure. The construction process of the transportation network dimension constructed according to the fractal dimension is as follows (Equation (1)):

$$S^*(d_n) = kd_n^{D^*} \; d \in [M, N] \tag{1}$$

where S^* is the built-up area, k is the constant term, D^* is the corresponding agglomeration dimension of the built-up area, and $[M, N]$ is the scale-invariant of the solution. Take the derivative of this equation with d and we obtain the following equation (Equations (2)–(4)):

$$\frac{\partial S^*(d)}{\partial d} = D^* k d^{D^*-1} \; d \in [M, N] \tag{2}$$

$$A^*(d_n) = k_A d_n^{D^*_A} \tag{3}$$

$$\frac{\partial A^*(d)}{\partial d} = D^*_A k_A d^{D^*_A - 1} \; d \in [M, N] \tag{4}$$

where $A^*(d)$ is the service area of accessibility distance d; k_A is the constant term; D^*_A is the agglomeration dimension of the accessibility network, which is calculated in the scale invariant area; and the formula of the built-up areas' density of accessibility distance $\rho(d)$ is obtained by adjusting the above formula (Equation (5)).

$$\rho(d) = \frac{\partial S^*(d)}{\partial A^*(d)} = \frac{D^* k}{D^*_A k_A} d^{D^* - D^*_A} \; d \in [M, N] \tag{5}$$

The meanings of the variables in the above equation are consistent with those mentioned in the previous text. In this article, the starting point for calculating the fractal dimension is the tri-provincial border within the YRMRUA. The above model can be considered as an extended form of the Smeed model and judged according to the rules in the following table (Table 1).

Table 1. The judging standard of FDTA.

Judgment Standard	Status
$D^* < D_A$	The distribution of built-up areas along transportation networks decreases from the center to the periphery, exhibiting a centralized pattern.
$D^* = D_A$	There is no significant relationship between the density distribution of built-up areas and the distribution of transportation networks.
$D^* > D_A$	The distribution of built-up areas along transportation networks increases from the center to the periphery, exhibiting an outward diffusion pattern.

2.3.2. The Traffic-Weighted Linear Density

Linear density, along with point density and core density, is one of the measurement tools in ArcGIS. This method calculates based on linear vector elements within a unit area. The weighted density of a transportation network is defined by incorporating traffic-related variables based on linear density [42]. The method facilitates the assessment of the distributional characteristics of the comprehensive transportation development level in the YRMRUA and lays a solid foundation for subsequent correlation analysis and the identification of spatial groups. The definition of transportation network density in this empirical study is as follows (Equation (6)):

$$D = \frac{[(L_1 \times V_1) + (L_2 \times V_2)]}{A} \tag{6}$$

where L_1 and L_2 above are the length of the traffic network falling into the set-radius circle, V_1 and V_2 are the weight of calculating the density, respectively, A is the radius of the circle, and D represents the calculated value of TLD.

Based on the empirical research experience, different types and scales of transportation infrastructure generally have a direct impact on land use within a range of 3 to 30 km around them. Therefore, this study takes 30 km as the upper limit of the search radius and divides it into 10 groups, assigning values in a linear gradient from near to far. In addition, as for the weighted calculation, the weights of the railway, road, and water transportation modes are positioned as follows according to the mean value of the normalized passenger and freight turnover results (Table 2).

Table 2. Weight of linear density of different transportation modes.

Transportation Modes	High-Speed Railway	General Railway	Expressway	National Highway	Provincial Highway	Main Channel	Secondary Channel
weight	0.25	0.15	0.2	0.1	0.1	0.15	0.05

2.3.3. Expansion Intensity Index

To provide a quantitative description of the intensity of spatial expansion, the spatial EII can be utilized [43]. This method is capable of effectively abstracting the representation of numerous and tangible spatial expansion patches, with the formula presented as follows (Equation (7)):

$$I = \frac{S_t - S_0}{A_i t} \times 100\% \tag{7}$$

In the equation, I represents the intensity of growth and expansion of the built-up area within the specified study period, expressed as a percentage value. S_0 and S_t represent the areas of the built-up area at the beginning and end of the study period, respectively, with their difference indicating the change in the built-up areas during the study period. A_i denotes the spatial area of the study unit, and t represents the time interval of the study period.

2.3.4. Pearson Correlation Coefficient

The Pearson correlation coefficient (PCC) is widely used in the field of social sciences [44,45], which can intuitively show the distribution relationship between the two types of variables, and its formula is as follows (Equation (8)):

$$\rho_{xy} = \frac{Cov(X,Y)}{\sigma_x \sigma_y} = \frac{\sum_{i=1}^{n}(x_i - \overline{x})(y_i - \overline{y})}{\sum_{i=1}^{n}(x_i - \overline{x})^2 \sum_{i=1}^{n}(y_i - \overline{y})^2} \tag{8}$$

where ρ_{xy} is the calculated value of the correlation coefficient, $Cov(X,Y)$ is the covariance of variables X and Y, and σ_x and σ_y are the standard deviations of variables X and Y, respectively. In this study, the PCC is utilized to measure the relationship between the EII and the TLD.

3. Results

3.1. Spatial Fractal Characteristics Based on Traffic Network

In the calculation of FDTA, logarithmic operations are performed separately based on the service area formed by the accessibility radius and the built-up area within this service area to obtain the fitting results of $ln(A) - ln(r)$ and $ln(S) - ln(r)$. Concurrently, partitioning is conducted in combination with piecewise functions and R-squared R^2 to yield two distinct sets. The results obtained from the figure indicate that the built-up areas within YRMRUA exhibit a certain degree of association and growth characteristics with the transportation network. Furthermore, the construction of ln(A)−ln(r) shows a high goodness of fit with a value exceeding 0.9, indicating that FDTA is applicable to the YRMRUA (Figure 4).

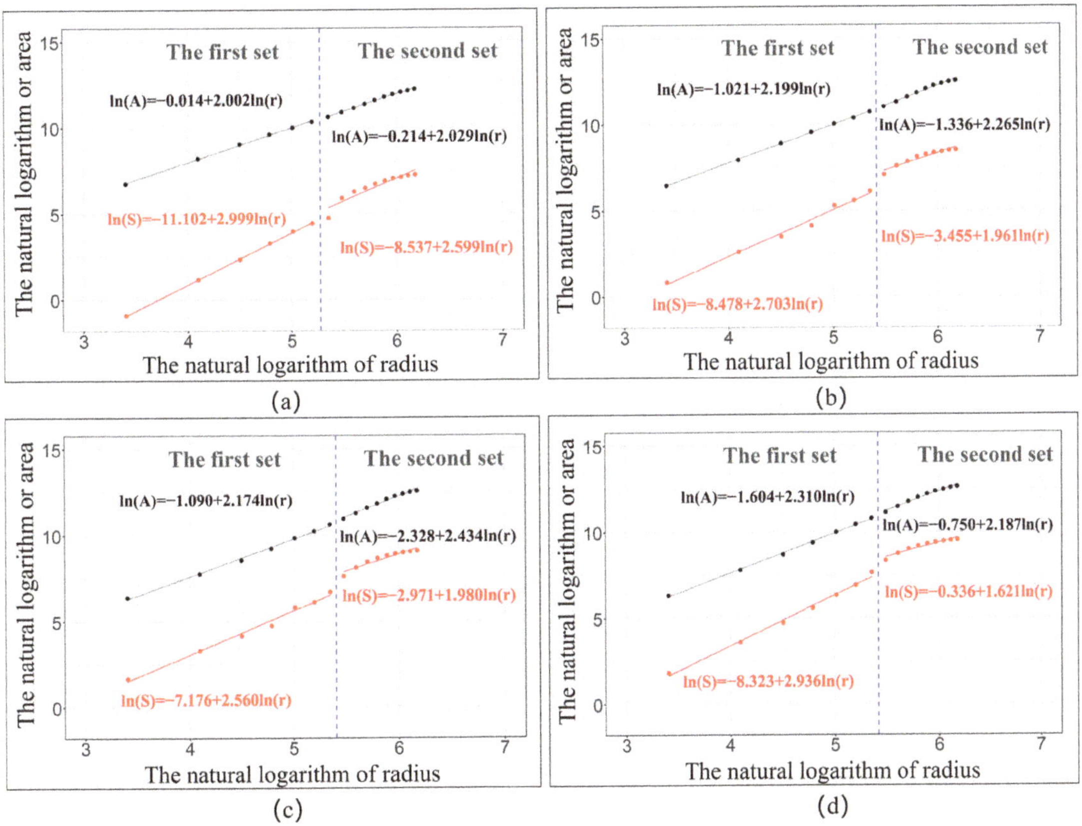

Figure 4. The calculated process of FDTA: (**a**) 1987; (**b**) 2006; (**c**) 2012; (**d**) 2020.

After calculating the fractal dimension based on traffic accessibility as shown in the table below (Table 3), the trend of the fractal dimension's variation becomes further clarified. The results reveal a general downward trend in both sets of dimensions, indicating a gradual shift in the overall morphological pattern of YRMRUA from spatial expansion to agglomeration. The fractal dimension in the first set remains higher than that in the second, with the former being greater than 2 and the latter less than 2, signifying a distinct boundary in growth patterns between the sets. This finding can be interpreted in the following ways: On the one hand, there is a close relationship between the expansion of built-up areas and the accessibility of comprehensive transportation. In particular, within a specific range starting from the provincial administrative boundary intersections, the fractal characteristics of urban built-up areas' expansion along the transportation network are particularly prominent. On the other hand, the role of comprehensive transportation in promoting the spatial growth of built-up areas is limited and also influenced by various factors such as topographical environment and economic radiation. The driving force of outward diffusion growth does not continue to increase with the extension of accessibility.

Table 3. The adjusted result of FDTA.

Sets	1987	2006	2012	2020
First set	2.99	2.97	2.35	2.54
Second set	2.56	1.73	1.63	1.48

3.2. *The Correlation Between TLD and EII*

The distributions of TLD at the ends of three stages, namely 2006, 2012, and 2020, as well as the correlational scatter plots of TLD and EII during the corresponding periods, are depicted below (Figure 5). The patches depicted in the figure represent grid cells measuring 30 m in size. In terms of the distribution of TLD, natural breakpoint classification was performed based on 2020 data and mapped onto the results of 2006 and 2012. The high-density areas gradually transformed from non-significant bands to a combined "band + area" pattern. In 2006, the TLD was concentrated along the "Wuhan-Ezhou-Huanggang-Huangshi-Jiujiang", "Changsha-Zhuzhou-Xiangtan", "Nanchang-Yichun", and "Nanchang-Fuzhou" urban zones. The TLD around Wuhan and its surrounding areas was significantly higher than in other regions. This trend was further reinforced by 2012 data, with the "Wuhan-Xianning-Yueyang", "Wuhan-Yichang", and the extension of the "Changsha-Zhuzhou-Xiangtan" axis towards "Yueyang-Changde-Yiyang-Loudi-Hengyang" also exhibiting prominent line densities (Figure 5a,b). The areal distribution of linear density became more pronounced in 2020, including urban clusters such as "Wuhan-Ezhou-Huanggang-Huangshi-Xianning" and "Changsha-Zhuzhou-Xiangtan-Pingxiang". Both relationships passed the significance test at the 0.01 level, with correlation coefficients of 0.47, 0.45, and 0.49 for the three time periods, respectively, indicating a moderate level of correlation. The correlation underwent a slight change, decreasing and then increasing by only 4%. Considering that the overall gridded units numbered over 35,000, this result aligns with statistical expectations and suggests, to some extent, a dependent relationship between TLD and EII (Figure 5d–f).

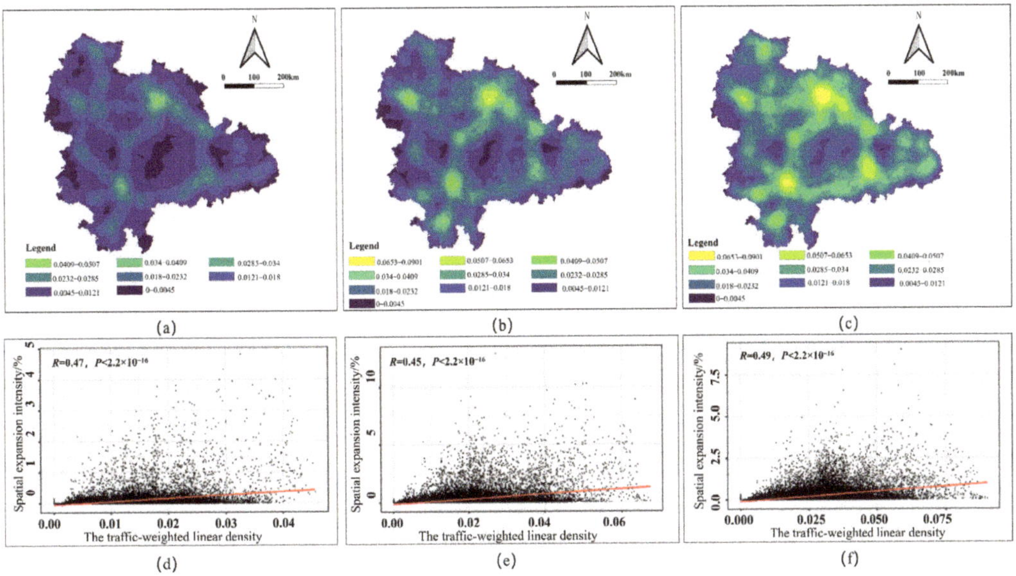

Figure 5. Distribution of TLD and its correlation with EII of built-up areas: (**a**) TLD in 2006; (**b**) TLD in 2012; (**c**) TLD in 2020; (**d**) correlation between TLD and EII, 1987–2006; (**e**) correlation between TLD and EII, 2006–2012; (**f**) correlation between TLD and EII, 2012–2020.

3.3. The Group Characteristics of Spatial Expansion Guided by Transportation

We selected grid cells with TLD and EII that ranked in the top 80% within their respective domains. Additionally, we segmented the grid cells with TLD in the top 90% to identify the source areas of outward linear density expansion (Figure 6a–c). During the study period, several group structures formed with the support of major transportation networks, such as the Wuhan–Ezhou–Huanggang–Huangshi group, the Changsha–Zhuzhou–Xiangtan group, and the Nanchang–Yichun group. Within the spatial system, key supporting elements included high-speed rail lines like Shanghai–Kunming and Beijing–Guangzhou, as well as major expressways such as Shanghai–Chongqing and Beijing–Hong Kong–Macau.

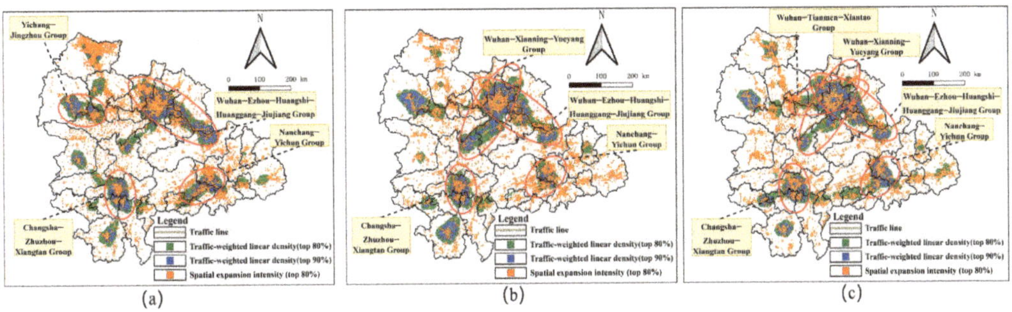

Figure 6. Cluster groups of TLD and EII: (**a**) group characteristics in 2006; (**b**) group characteristics in 2012; (**c**) group characteristics in 2020.

4. Discussions

4.1. Different Modes of Transportation Exhibit Diversified Spatial Supporting Effects

First and foremost, unlike waterways that merely serve as esthetic features within individual cities, water transportation functions as the fundamental driver of spatial formation in urban agglomerations. Despite the relatively weak overall spatial pattern observed in the YRMRUA region since 1987, the built-up areas of cities and their expansion ratios, while low, have consistently exhibited growth patterns aligned with water transportation routes. This is exemplified by the urban belts along the Yangtze River stretching from Yichang to Jiujiang, the Han River stretching from Xiangyang to Jingmen, and the Xiangjiang River stretching from Yueyang to Hengyang. These urban centers have maintained a profound relationship with water transportation throughout their development, from initial site selection to subsequent expansion (Figure 7). Under the guidance of past urban planning in China, it is noteworthy that the locations of municipal governments in cities are actually also the centers of built-up areas' expansion. Taking the main navigation channels of the Yangtze River as the central axis and expanding 30 km on either side to form an observation and buffer zone, we have identified the following: within the scope of the Yangtze River's Middle Reaches Urban Agglomeration, there are 15 prefecture-level urban growth centers along the river (accounting for 50% of the total), and 103 county-level urban growth centers (accounting for 48% of the total).

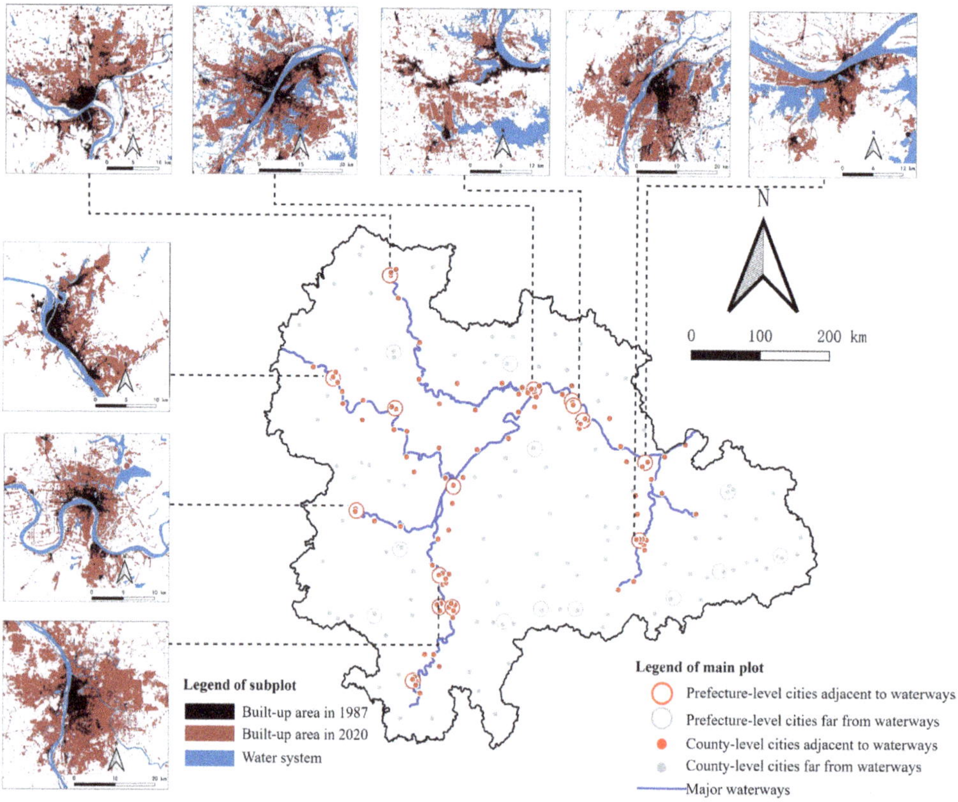

Figure 7. Urban growth centers along the Yangtze River waterways.

The horizontal extension of towns along riverbanks and the emergence of links between central city clusters, influenced by variations in urban planning, economic aggregate level and scale, have culminated in the formation of a composite belt network structure centered around waterways. From the spatial diagrams, we can gain a better understanding of these relationships (Figure 8). If we consider cities as point structures and transport routes as line structures, the "point—line" relationships in a region will differentiate into three types: ring-shaped form, belt-shaped form, and cone-shaped form. It is this diversity in types that constitutes the characteristics of YRMRUA spatial patterns.

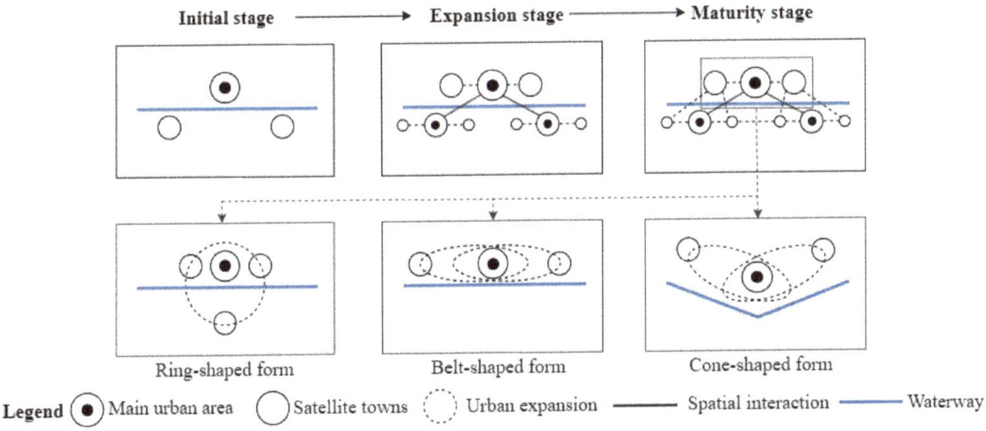

Figure 8. The correlation between the spatial system of urban agglomerations and waterways.

Secondly, rail transit, particularly in the form of high-speed and intercity railways, has emerged as a pivotal extension of spatial organization for urban agglomerations. Leveraging its advantages of high speed, punctuality, and superior service quality, rail transit has become a new cornerstone for the spatial organization of the YRMRUA within the transportation system. This is evident in cities aligned along the "Wuhan-Xianning-Changsha-Hengyang" corridor of the Beijing–Guangzhou Railway, where the integration of rail transit has facilitated the seamless stitching of spatial structure gaps and the enhancement of the overall spatial system. Since the inauguration of high-speed rail, spatial growth has demonstrated a notable "synchronous attribute", with the coefficient of variation for built-up area growth decreasing from 0.84 to 0.73, indicating a more consistent and coordinated development pattern. In alignment with the nationwide trend of constructing high-speed rail trunk lines, intercity high-speed rail projects have been incorporated into the key construction initiatives within the urban and land use planning of various regions in the YRMRUA. Cities along other intercity railway lines, such as "Wuhan-Xiaogan", "Wuhan-Huanggang", and "Nanchang-Jiujiang" have also exhibited increasing urbanization trends, with corresponding decreases in the coefficients of variation for built-up area growth. This indicates that the launch of high-speed rail within the YRMRUA has accelerated the integration process of cities along the railway lines. During this process, the economic spillover and feedback effects are transmitted through the HSR network, gradually guiding the spatial development of various cities within the YRMRUA.

Lastly, highway traffic serves as the essential supporting framework for spatial development, and together with the railway system, it has promoted the process of suburbanization in cities [46]. While rail transit has significantly reduced travel and logistics times, it alone cannot fully address the connectivity needs between central cities and peripheral counties. In this regard, roads offer greater flexibility, particularly in freight transportation, meeting specific time-sensitive delivery requirements and facilitating rapid point-to-point transportation to meet industrial customization needs. Previous studies have confirmed that the development of highway transportation exhibits adaptive characteristics to regional economic and social conditions [47,48]. Specifically, in urban studies, we typically examine the spatial structure formed by urban centers and transportation arteries, as well as how the construction of these transportation arteries shapes the unique characteristics of urban form. However, within the context of urban agglomerations, our focus shifts to the impact exerted by the extended areas along major highways on the entire regional system. Within YRMRUA, the spatial growth areas of the vast majority of cities have already been covered within a 3 km range along the highways (Figure 9). This indicates that the YRMRUA has developed a highly extensive expansion pattern along the highway network.

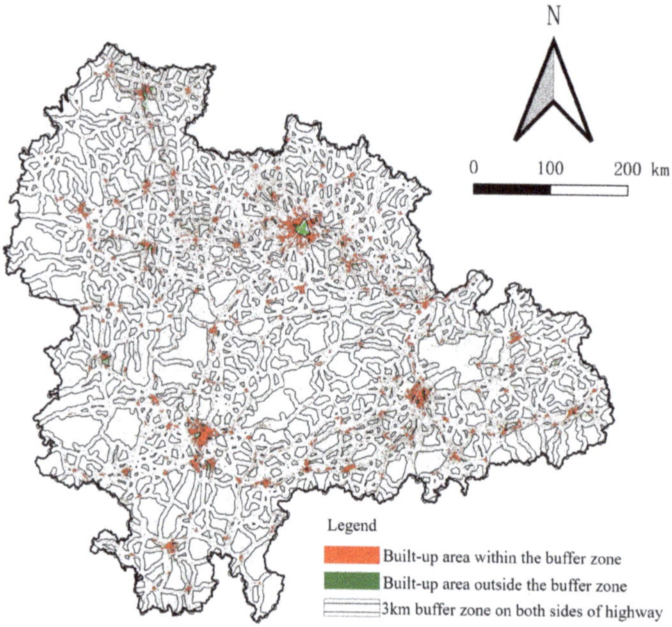

Figure 9. The urban built-up areas adjacent to highways within YRMRUA.

Furthermore, in regions where high-speed rail is not yet accessible, such as Jingmen and Yiyang, road transportation remains a highly dependent mode of transportation. The high flexibility and adaptability of road transportation make it an effective complement to other modes of transportation, playing a crucial supporting role in the regional transportation system. The formation of a three-tiered network combining "expressways, national highways, and provincial highways" has established a robust framework for city-to-city connectivity, with expressways serving as the backbone for passing through urban areas and gradually expanding to cover most counties and cities. Additionally, expressways function as spatial development boundaries and guides, with pre-delineated city-bypass routes often serving as urban development boundaries, defining multiple exits to direct and orient development, thus forging strong connections between various urban groups. The expansion pattern of cities in the YRMRUA, formed along road transportation routes, appears highly regular (Figure 10). For instance, the master planning of Wuhan and Changsha within the YRMRUA clearly delineated the location and alignment of ring expressways, which not only imposed effective constraints on urban sprawl but also strategically oriented the direction of spatial development. This finding also indicates that the spatial effects of highway transportation on the YRMRUA differ across the two levels of urban agglomerations and individual cities.

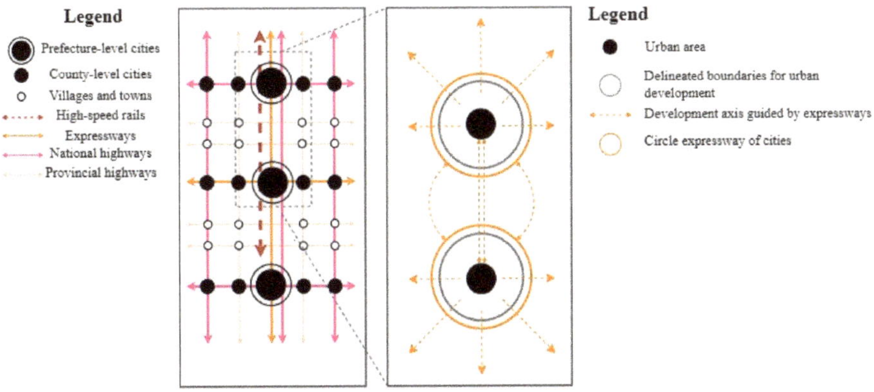

Figure 10. The supporting role of the highways system.

4.2. Multimodal Transportation Promotes Compact Outward Expansion of Region

In fact, there is a broad two-way interactive relationship between regional development and transportation infrastructure construction [49,50], and the former has also promoted the latter to become more mature. The mutual effects generated among regional transportation facilities can be understood as a form of transportation advantage, providing an explanatory framework for regional spatial structure [51]. Especially since the economic promotion effect of transportation often occurs in the early stages of transportation construction [52], it relies on the mode of comprehensive transportation to provide lasting impetus. The comprehensive transportation infrastructure and the spatial distribution of built-up areas align well with the development planning of the YRMRUA. Multimodal transportation supplements the traditional regional functions, solidifies the spatial configuration of urban agglomerations, and forms a spatial structure with regional characteristics (Figure 11).

The combination of roads, railways, waterways, and airports can form a variety of modes of multimodal transportation, which is also one of the characteristics of a modern transportation system. Including but not limited to the combined transportation of rail and water, and the integrated transportation of rail and water, air and water, etc. is essential [53,54]. Different multimodal transportation methods also have their own unique advantages and applicable scopes (Table 4).

Waterway combined transportation is the most common mode of multimodal transportation in YRMRUA and has profoundly shaped the regional space system (Figure 12a). In practical applications, this type of multimodal transportation is generally centered around the land use planning and specialized planning of the port. It guides the formation of a spatial system that relies on a banded and radial pattern along the waterway axis. The resulting reconstruction of the spatial system manifests in the improvement of spaces along both sides of the river and the emergence of distinct characteristics in the extended forms along the port access channels (Figure 12b). The figure reveals a complex "T"—shaped or dual—axis spatial expansion pattern formed by the interaction between towns and comprehensive transport systems. For instance, the Wuhan Yangluo Port has implemented a combined rail–water transportation system, utilizing the Yangtze River waterway, Jiangbei Railway, expressway, and the beltway to form a comprehensive "road-rail-water" logistics channel, resulting in a breakthrough efficiency of over 30% in its demonstration zone. Meanwhile, the Ezhou Sanjiang Port Area has also adopted a dedicated railway line and relies on the Huangxian Expressway to form a combined transportation mode. Similarly, the Huangshi New Port has integrated the Shannan Railway, Xingang Branch

Line, and Qijia Expressway to establish its own multimodal transportation system. In this process, certain areas demonstrate the comparative advantages of "water-rail" transportation. For example, in the "Nanchang-Jiujiang" corridor, goods from the Jiujiang Port along the Yangtze River are directly transported to the logistics park in the newly built area of Nanchang via railway, achieving point-to-point delivery compared to the shipping on Ganjiang River. Although tributary regions are constrained by waterway transportation turnover and remain on the fringes of the freight transportation system, they still retain distinct local characteristics. Specifically, the Zhuhui District in Hengyang has formed a combined transportation area centered around the Hongguang and Dongjiang logistics parks, connecting to the Beijing–Hong Kong line via a dedicated railway. This allows the construction materials, equipment manufacturing, and food processing industries along the line to participate in the regional division of labor and guides the urban area to expand towards the southeast and south.

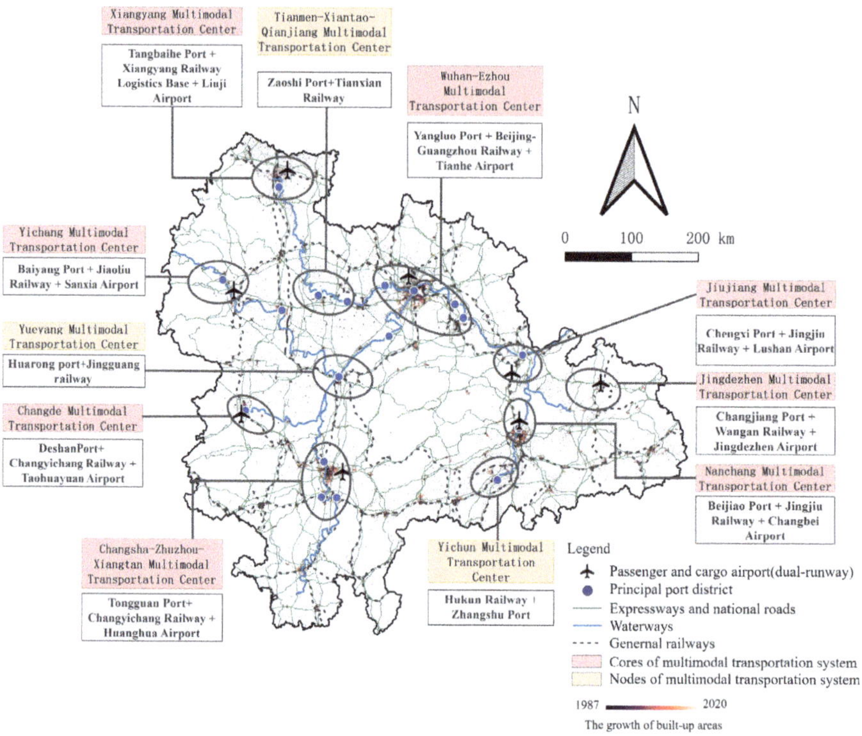

Figure 11. YRMRUA's multimodal transportation center distribution.

Table 4. Applicable scopes of different modes of multimodal transportation.

Form	Coverage	Speed	Freight Volume	Haul Distance	Costing
Road–rail multimodal transportation	●	○	●	●	○
Rail–water multimodal transportation	○	○	●	●	●
Air–road multimodal transportation	●	●	○	○	○
Air–rail multimodal transportation	●	○	●	●	○
Air–water multimodal transportation	○	○	○	●	●

Note: ● indicates that the multimodal transportation form performs well in the corresponding criterion. ○ indicates that the multimodal transportation form has limitations or disadvantages in the corresponding criterion.

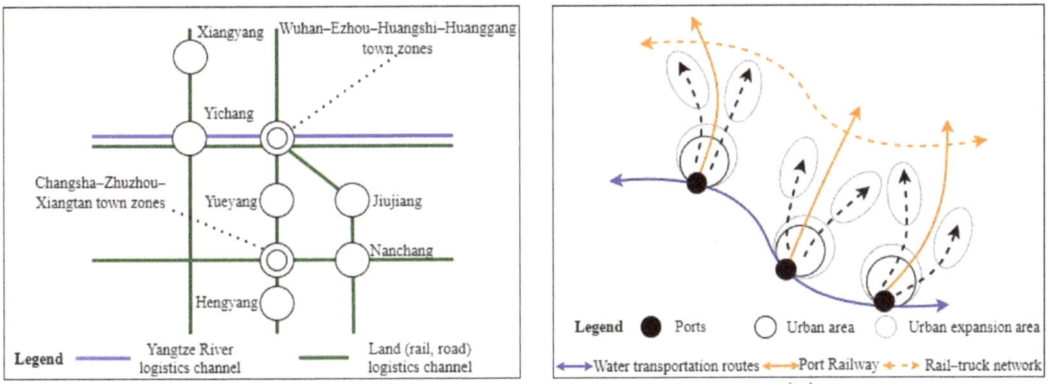

(a) (b)

Figure 12. Multimodal transportation promotes spatial reconstruction: (**a**) spatial structure of urban system; (**b**) spatial joint growth structure.

4.3. Major Transportation Facilities Guide the Inward Filling of Regional Space

Transportation hub stations have spearheaded the infill development of the built-up areas in YRMRUA. To better compare the differences in the driving effects of transportation hubs and networks, as well as their overall impact, we take the provincial capital cities of Wuhan, Changsha, and Nanchang as examples (Figure 13). The following diagrams illustrate the growth of built-up areas within the urban boundaries, along with rose diagrams, major transportation hubs, and associated networks. From the diagrams, it is evident that while the networks that play a guiding role at the macro level tend to express more restrictive effects on spatial growth at the micro level, the distribution of various hub stations shows a better correlation with the concentrated areas of spatial morphological growth.

From the perspective of transportation hubs, on one hand, the establishment of these hubs generates a strong aggregating force and integrates organically with the surrounding developed land, thereby enhancing vitality. Transportation infrastructure plays a pivotal role in economic development and cultural exchange [55]. On the other hand, the government often sets high-speed rail new cities, port new cities, and airport new cities as spatial development goals.

Newly built high-speed railway stations are often regarded as important means to guide new economic growth, promote industrial upgrading, and generate integration. In recent years, the impact of high-speed rail on economic development has been gradually expanding [56]. Therefore, in urban master planning, they are usually combined with business districts to form a "high-speed rail city" model for construction. In practice, this directly affects an area of over 1 square kilometer and guides the formation of spatial sequences within 3 km. The resulting effects include, on the one hand, the direct filling effect on idle land around the edge of the urban area, and on the other hand, due to the development of commercial and business functions, it further drives the value of surrounding land to increase and further attracts investment groups. Ultimately, a layered structure of "station core—surrounding commercial and trade service area—outer service area" is formed (Figure 14). Successful cases include the spatial combination of Wuhan High-speed Railway Station and its front business district, with a core construction area exceeding 1 million square meters. The large-scale spatial carrier serving as a landmark on the central axis in front of the station has been completed, and an industrial cluster focusing on "office buildings—hotel apartments" has gradually formed. In addition, after the completion of Changsha South Railway Station, a development axis for the high-speed rail new town running east to west was formed. To the east of the station is an ecological park, and 1 km across the Liuyang River is the Changsha International Convention and Exhibition Center and its conference hall. Two kilometers to the east is a series of commercial and business areas serving the exhibition, while within a range of 0.5 to 1 km to the southwest of the station core area, contiguous residential areas are used as a means of urban renewal. As for Nanchang West Railway Station, it adopts contiguous development as a means of spatial filling, forming a development axis symmetrical to the north–south station square. On both sides of the axis, hotel and shopping mall service clusters are formed in combination with residential areas, automobile passenger transportation, and parking stations, while to the south, contiguous residential development and reserved commercial axes are the main types.

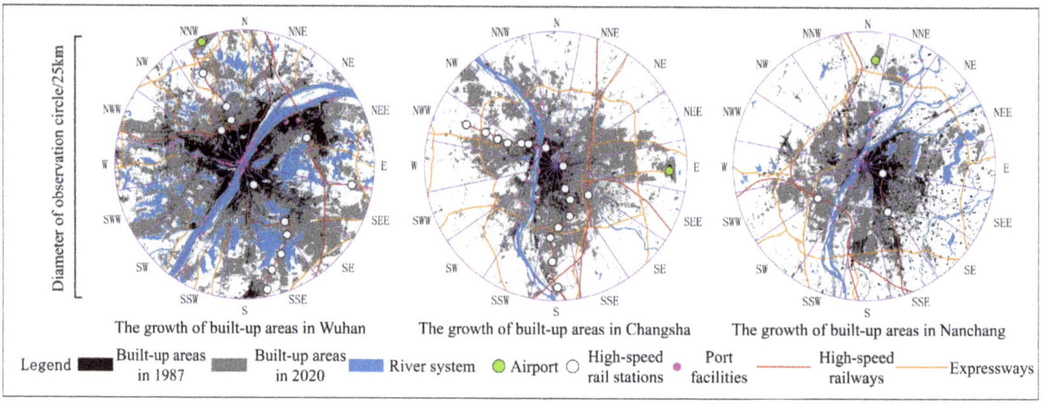

(a)

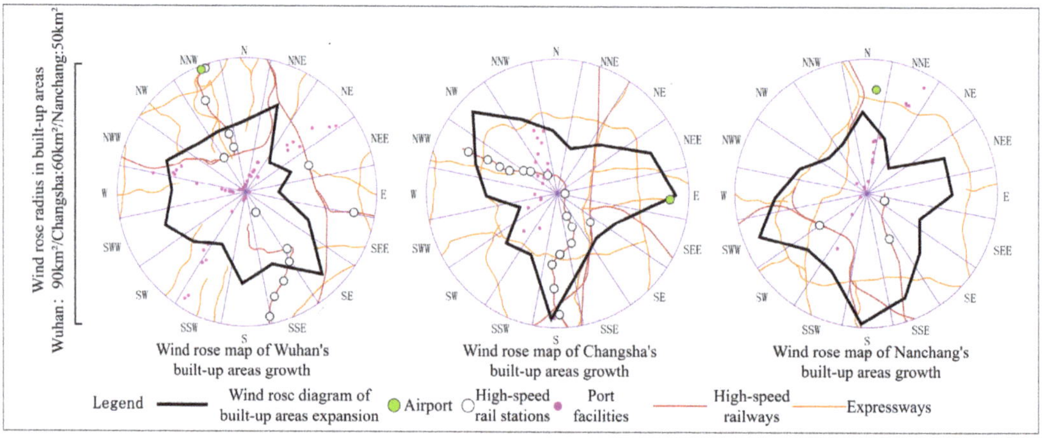

(b)

Figure 13. The correlation between transportation infrastructures and the distribution of built-up areas: (**a**) from the perspective of spatial form; (**b**) from the perspective of spatial distribution.

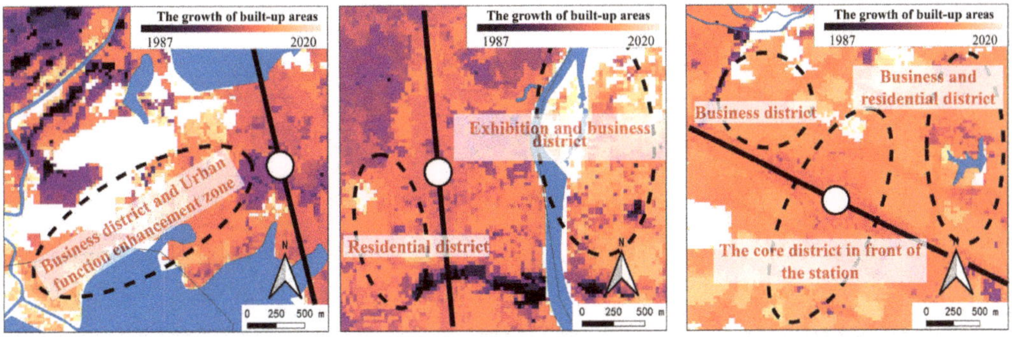

(a) (b) (c)

Figure 14. An illustration of the impact of high-speed rail stations on surrounding built-up areas: (**a**) Wuhan High-Speed Railway Station; (**b**) Changsha South High-Speed Railway Station; (**c**) Nanchang West High-Speed Railway Station.

In a broad sense, a port is a combination of a series of transportation facilities, including but not limited to waterway-level channels, berths, anchorages, as well as onshore facilities such as terminals, distribution yards, and production and storage facilities. Therefore, it is a complete and independent system in itself. The current development strategy of integration planning is to further integrate the dual system of "city-port", especially focusing on the construction of production space to improve bilateral efficiency and provide service support. In the integrated construction, it is usually guided to form outward concentric circles centered on the port area and facilitate the formation of multiple new urban sectors (Figure 15). Among urban agglomerations, the Wuhan Yangluo Port Area, Yueyang Chenglingji Port Area, and Jiujiang Chengxi Port Area stand out as exemplary cases, sharing a common trait of inward growth along the riverside industrial corridor. Specifically, the Yangluo Port Area fosters industrial linkages with the northwest and due north, extending and integrating a range of industrial development axes. Meanwhile, the eastern side's "tourism-residence-research-education" service cluster and the southern riverside industrial belt have also undergone renovation and enhancement. The Yueyang Chenglingji Port Area, built upon the foundation of a riverside industrial belt comprising river ports and the grain and oil industry, has subsequently expanded eastward to establish a "bioscience-new materials-building materials" industrial park and belt. As for the Jiujiang Chengxi Port Area, it has formed a horizontal westward expansion amidst the backdrop of port development. North of Jiurui Avenue lies the industrial development axis, centered around equipment manufacturing, new materials, and electrical equipment, giving rise to a segmented development approach.

In addition to the significant impact of the two types of facilities mentioned above on spatial expansion, cities in recent years have also been attempting development planning centered around airports [57]. The surrounding areas of airports possess both "functional restrictions" and "functional radiation" characteristics. As a result, they spatially form a layered and gradually increasing impact from the inside out, effectively filling the spaces in the near-port regions. More specifically, within a 5 km radius of the airport facilities, functions directly related to aviation, such as services, warehousing, and industry, will be arranged. Within a 20 km radius, there are height restrictions, reducing the intensity of development. Beyond this radius lies the economic radiation zone, which is typically developed in a combined form of "business—entertainment—residence" to guide and fill in the vacant areas of urbanized regions (Figure 16).

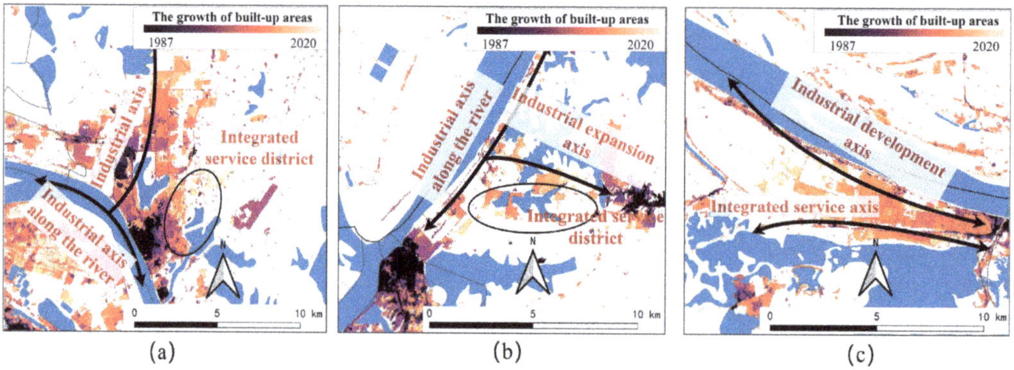

Figure 15. Illustration of impact of ports on surrounding built-up areas: (**a**) Wuhan Yangluo Port; (**b**) Yueyang Chenglingji Port; (**c**) Jiujiang Chengxi Port.

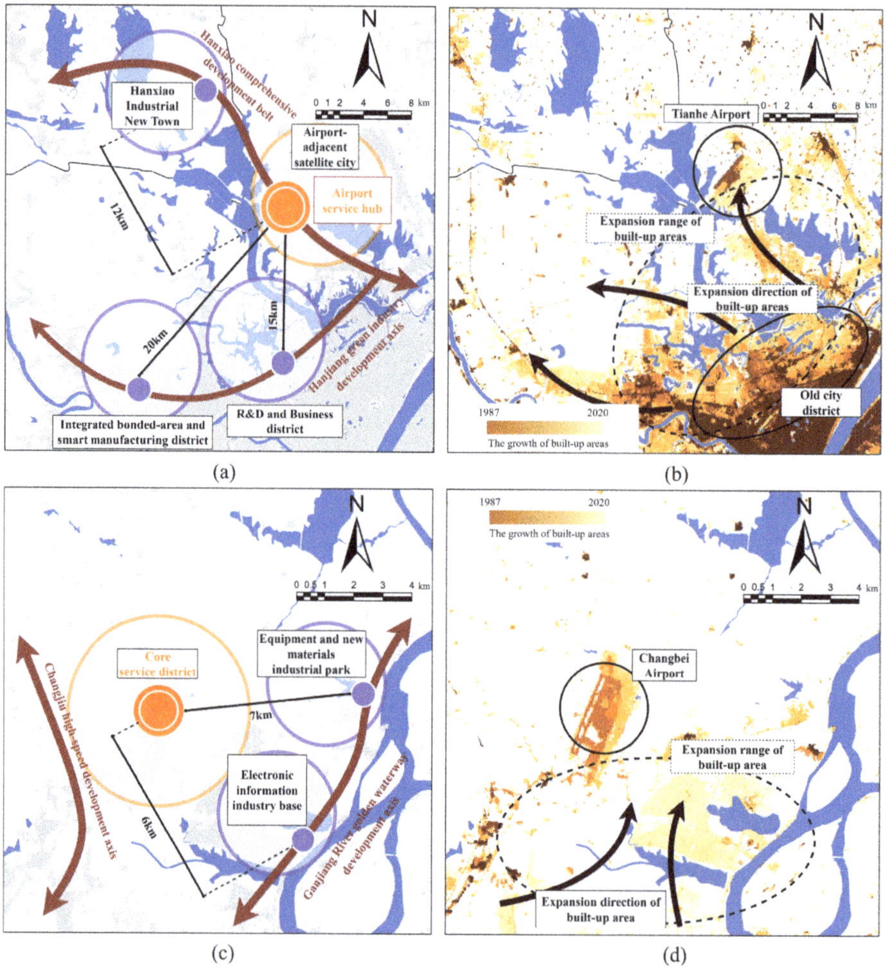

Figure 16. *Cont.*

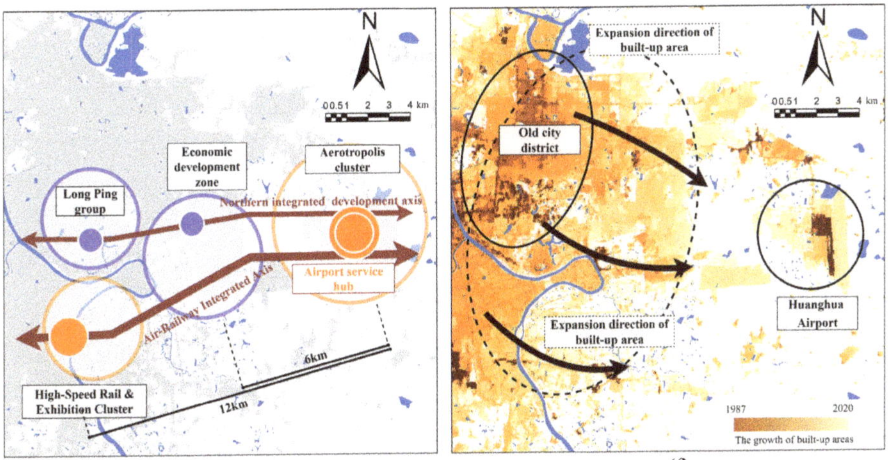

(e) (f)

Figure 16. An illustration of the impact of airports on surrounding built-up areas: (**a**) the structure of spatial expansion around Tianhe Airport; (**b**) the pattern of spatial expansion around Tianhe Airport; (**c**) the structure of spatial expansion around Changbei Airport; (**d**) the pattern of spatial expansion around Changbei Airport; (**e**) the structure of spatial expansion around Huanghua Airport; (**f**) the pattern of spatial expansion around Huanghua Airport.

5. Conclusions

Comprehensive transportation infrastructure can exert a characteristic shaping effect on the growth of built-up areas within urban agglomerations. For YRMRUA, regional transportation facilities not only serve as the fundamental basis for its spatial construction but also provide momentum for the formation of specific spatial expansion patterns at both the macro and micro levels of the urban agglomeration. This shaping effect on the agglomeration's space is distinct from the role of intra-city transportation in land use, and it is limited rather than determinative. It highlights the complexity of regional development dynamics and provides new insights into understanding spatial patterns at different scales between urban and urban agglomeration.

Through the use of GIS platforms and mathematical statistical analysis, this study observes YRMRUA in typical years and arrives at the following main conclusions:

(1) Constructing the FDTA model outward from the provincial administrative boundary intersections, the fractal dimension at each time node can be significantly divided into two sets. The first set consistently has a value greater than 2, while the second set evolves from being greater than 2 to less than 2. Overall, both sets show a decreasing trend in their dimensions. These results indicate that YRMRUA exhibits a certain fractal growth pattern under the influence of transportation development, with a fractal trend of becoming more cohesive along transportation routes. Nevertheless, the fractal trend of idealized urban built-up area growth exhibiting self-organization characteristics did not materialize.

(2) From the perspective of a 30 m grid cell, the pattern characteristics of YRMRUA's TLD gradually become clearer over time. High-density areas gradually transition from non-significant bands to a combined "band + area" form. The correlation coefficients between TLD and EII are 0.47, 0.45, and 0.49 at the end of the three time periods, respectively, indicating a moderate level of correlation and suggesting a certain spatial relationship between the two.

(3) By extracting areas with high TLD and EII, specific group structures are formed. These include the Wuhan–Ezhou–Huanggang–Huangshi group, the Changsha–Zhuzhou–Xiangtan group, and the Nanchang–Yichun group. Within these groups, regional transportation arteries such as the Shanghai–Kunming High-Speed Railway, Beijing–Guangzhou High-Speed Railway, and Beijing–Hong Kong–Macau Highway serve as critical supports.

(4) Different modes of transportation have varying effects on spatial expansion. Waterway transportation constitutes the basic form of space generation, railway transportation extends the space, and road transportation supports the space. The combined transportation of different modes enables the YRMRUA to form a more efficient and compact spatial structure.

The future research direction of this article will be to further investigate the interactive relationship between actual traffic volume and spatial expansion, as well as the relationship between factor movements and transportation facilities' layouts [58], in order to obtain the correlation characteristics from the dual perspectives of flowing space and local space. Additionally, different cities within the YRMRUA will be classified into various types and levels for in-depth discussion. Against the backdrop of big data development, we will propose comprehensive optimization strategies for the planning, construction, and regulation of urban and transportation systems for each type of city, aiming to enhance the universal applicability of our conclusions.

Author Contributions: Conceptualization, Z.F.; writing—original draft preparation, Z.F.; visualization, Z.F. and W.L.; investigation, Z.F. and C.L.; writing—review and editing, M.X.; funding acquisition, M.X. and Z.F. All authors have read and agreed to the published version of the manuscript.

Funding: This research was funded by the Post-Funded Projects of the National Social Science Fund (Grant No. 23FYSB047), the Doctoral Research Start-up Fund Project of Hubei University of Technology (Grant No. XJ2024003201), and the Fundamental Research Funds for the Central Universities (Grant No. 2021yjsCXCY044).

Data Availability Statement: The data presented in this study are available on request from the corresponding author; the data are not publicly available due to project confidentiality.

Conflicts of Interest: Author Weiyang Luo was employed by the CCCC Second Harbor Engineering Company Ltd. The remaining authors declare that the research was conducted in the absence of any commercial or financial relationships that could be construed as a potential conflict of interest.

Abbreviations

The following abbreviations are used in this manuscript:

EII	Expansion intensity index
FDTA	Fractal dimension based on traffic accessibility
PCC	Pearson correlation coefficient
TLD	Traffic-weighted linear density
YRMRUA	Yangtze River Middle Reaches Urban Agglomeration

References

1. Li, L.; Ma, S.; Zheng, Y.; Xiao, X. Integrated Regional Development: Comparison of Urban Agglomeration Policies in China. *Land Use Policy* **2022**, *114*, 105939. [CrossRef]
2. Lu, Y.; Xie, D.; Liu, H.; Zhong, Z. Impact of Urban Agglomeration Development Planning Policy on the Promotion of Efficiency of Urban Economic Development: Empirical Evidence from 283 Cities in China. *J. Urban Plan. Dev.* **2024**, *150*, 04024042. [CrossRef]
3. Shi, S.; Wong, S.K.; Zheng, C. Network Capital and Urban Development: An Inter-Urban Capital Flow Network Analysis. *Reg. Stud.* **2022**, *56*, 406–419. [CrossRef]

4. Maket, I.; Szakálné Kanó, I.; Vas, Z. Urban Agglomeration and Regional Economic Performance Connectedness: Thin Ice in Developing Regions. *Res. Glob.* **2024**, *8*, 100211. [CrossRef]
5. Chen, S.; Du, Y.; Liu, Y. Regional Integration and Urban Green and Low-Carbon Development: A Quasi-Natural Experiment Based on the Expansion of the Yangtze River Delta Urban Agglomeration. *Sustainability* **2025**, *17*, 3621. [CrossRef]
6. Zheng, W.; Du, N.; Wang, X. Understanding the City-Transport System of Urban Agglomeration through Improved Space Syntax Analysis. *Int. Reg. Sci. Rev.* **2022**, *45*, 161–187. [CrossRef]
7. Liu, X.; Huang, J.; Lai, J.; Zhang, J.; Zhao, P. Analysis of Urban Agglomeration Structure through Spatial Network Analysis and Mobile Phone Data. *Trans. GIS* **2021**, *25*, 1949–1969. [CrossRef]
8. Liu, Y.; Zhang, X.; Pan, X.; Ma, X.; Tang, M. The Spatial Integration and Coordinated Industrial Development of Urban Agglomerations in the Yangtze River Economic Belt, China. *Cities* **2020**, *104*, 102801. [CrossRef]
9. He, Y.; Zhou, G.; Tang, C.; Fan, S.; Guo, X. The Spatial Organization Pattern of Urban-Rural Integration in Urban Agglomerations in China: An Agglomeration-Diffusion Analysis of the Population and Firms. *Habitat Int.* **2019**, *87*, 54–65. [CrossRef]
10. Wu, F.; Fang, C.; Zhao, Y. PAF Model of Study on Urban Industrial Agglomeration Dynamic Mechanism and Patterns. *Geogr. Res.* **2011**, *30*, 71–82. [CrossRef]
11. Zhang, Y.; Wang, S.; Zhou, R. Spatial Sorting and Selection within Urban Agglomerations: A Tripartite Evolutionary Game Model Approach. *Humanit. Soc. Sci. Commun.* **2025**, *12*, 57. [CrossRef]
12. Thisse, J.-F. Human Capital and Agglomeration Economies in Urban Development. *Dev. Econ.* **2018**, *56*, 117–139. [CrossRef]
13. Pangarso, R.A.; Suharyadi, R.; Rijanta, R. The Spatial Patterns and Local Economic Determinant of Industrial Agglomeration in Semarang District, Indonesia. *Geoplanning J. Geomat. Plan.* **2020**, *6*, 99–112. [CrossRef]
14. Liu, Y.; Yang, M.; Cui, J. Urbanization, Economic Agglomeration and Economic Growth. *Heliyon* **2024**, *10*, e23772. [CrossRef]
15. Tan, R.; Zhou, K.; He, Q.; Xu, H. Analyzing the Effects of Spatial Interaction among City Clusters on Urban Growth—Case of Wuhan Urban Agglomeration. *Sustainability* **2016**, *8*, 759. [CrossRef]
16. Yeh, G.O.; Chen, Z. From Cities to Super Mega City Regions in China in a New Wave of Urbanisation and Economic Transition: Issues and Challenges. *Urban Stud.* **2020**, *57*. [CrossRef]
17. Cheng, M.; Li, L.; Zhou, Y. Exploring the Urban-Rural Development Differences and Influencing Factors in the Huang-Huai-Hai Plain of China. *J. Geogr. Sci.* **2020**, *30*, 1603–1616. [CrossRef]
18. Pokharel, R.; Bertolini, L.; Brmmelstroet, M.T.; Acharya, S.R. Spatio-Temporal Evolution of Cities and Regional Economic Development in Nepal: Does Transport Infrastructure Matter? *J. Transp. Geogr.* **2021**, *90*, 102904. [CrossRef]
19. Lu, Y.; Xu, X.; Zhu, G.; Peng, Y.; Li, Y.; Zhao, X. Land Space and High-Speed Transportation Coordinated Development Evaluation in the Beijing–Tianjin–Hebei Urban Agglomeration of China. *Land* **2024**, *13*, 1693. [CrossRef]
20. Bojan, V. Space, Agglomeration Economies and Urban Development. *Glas. Srp. Geogr. Društva* **2019**, *99*, 79–98. [CrossRef]
21. Zhou, H.; Gao, H. The Impact of Urban Morphology on Urban Transportation Mode: A Case Study of Tokyo. *Case Stud. Transp. Policy* **2018**, *8*, 197–205. [CrossRef]
22. Chen, Z.; Li, X. Economic Impact of Transportation Infrastructure Investment under the Belt and Road Initiative. *Asia Eur. J.* **2021**, *19*, 131–159. [CrossRef] [PubMed]
23. Ventura, P.; Zazzi, M.; Rossetti, S.; Carra, M. Urban Development and Airports in Northern and Central Italy: Main Trends and a Focus on Parma Giuseppe Verdi Case Study. *Transp. Res. Procedia* **2020**, *45*, 914–921. [CrossRef]
24. Cao, C.; Su, Y. Transportation Infrastructure and Regional Resource Allocation. *Cities* **2024**, *155*, 105433. [CrossRef]
25. Wang, L.; Wu, Z.; Ye, H.; Feng, W. Spatial Effect of Transportation Infrastructure on Regional Circular Economy: Evidence from Guangdong-Hong Kong-Macao Greater Bay Area. *Environ. Sci. Pollut. Res.* **2023**, *30*, 50620–50634. [CrossRef]
26. Sun, X.; Xiang, P.; Ngoduy, D.; Sui, H.; Zhang, B. Can Transportation Networks Contribute to the Sustainable Development of Urban Agglomeration Spatial Structures? *Sustain. Cities Soc.* **2024**, *117*, 105983. [CrossRef]
27. Rokicki, B.; Stpniak, M. Major Transport Infrastructure Investment and Regional Economic Development—An Accessibility-Based Approach. *J. Transp. Geogr.* **2018**, *72*, 36–49. [CrossRef]
28. Chan, H.-Y.; Xu, Y.; Wang, Z.; Chen, A. The Deeper and Wider Social Impacts of Transportation Infrastructure: From Travel Experience to Sense of Place and Academic Performance. *Transp. Policy* **2024**, *158*, 51–63. [CrossRef]
29. Wang, P.; Zhang, Q.; Cai, Y. Spatial Evolution of Water Transportation Industry Based on Multi-Source Data: Understanding the Structural Consolidation and Integration Demand in Coastal Cities. *Hum. Ecol. Risk Assess. Int. J.* **2023**, *29*, 817–835. [CrossRef]
30. Oswald Beiler, M.; Mohammed, M. Exploring Transportation Equity: Development and Application of a Transportation Justice Framework. *Transp. Res. Part Transp. Environ.* **2016**, *47*, 285–298. [CrossRef]
31. Wan, J.; Xie, Q.; Fan, X. The Impact of Transportation and Information Infrastructure on Urban Productivity: Evidence from 256 Cities in China. *Struct. Change Econ. Dyn.* **2024**, *68*, 384–392. [CrossRef]
32. Li, J.; Wang, P.; Ma, S. The Impact of Different Transportation Infrastructures on Urban Carbon Emissions: Evidence from China. *Energy* **2024**, *295*, 131041. [CrossRef]

33. Ren, B.; Tian, Y.; Xiao, X. Spatial Effects of Transportation Infrastructure on the Development of Urban Agglomeration Integration: Evidence from the Yangtze River Economic Belt. *J. Transp. Geogr.* **2022**, *104*, 103431. [CrossRef]
34. Zheng, Z.; Qingyun, H. Spatio-Temporal Evaluation of the Urban Agglomeration Expansion in the Middle Reaches of the Yangtze River and Its Impact on Ecological Lands. *Sci. Total Environ.* **2021**, *790*, 148150. [CrossRef]
35. Luo, Y.; Zhang, M.; Cao, Q.; Wang, L. An Evaluation of the WRF Physical Parameterizations for Extreme Rainfall Simulation in the Yangtze River Middle Reaches Urban Agglomeration. *Urban Clim.* **2024**, *58*, 102149. [CrossRef]
36. Zheng, H.; Hsiao, Y.L.; Ho, H.W.; Ji, C.; Zhang, S. How the Spatial Functional Division of Labor in Urban Clusters Affects Urban Population Size? Evidence from 19 Urban Clusters in China. *Humanit. Soc. Sci. Commun.* **2024**, *11*, 1459. [CrossRef]
37. Fan, Z.; Zhong, Z. Spatial Morphological Characteristics and Evolution of Policy-Oriented Urban Agglomerations—Take the Yangtze River Middle Reaches Urban Agglomeration as an Example. *Sustainability* **2023**, *15*, 13736. [CrossRef]
38. Li, X.; Gong, P.; Liang, L. A 30-Year (1984–2013) Record of Annual Urban Dynamics of Beijing City Derived from Landsat Data. *Remote Sens. Environ.* **2015**, *166*, 78–90. [CrossRef]
39. Zomlot, Z.; Verbeiren, B.; Huysmans, M.; Batelaan, O. Trajectory Analysis of Land Use and Land Cover Maps to Improve Spatial-Temporal Patterns, and Impact Assessment on Groundwater Recharge. *J. Hydrol.* **2017**, *554*, 558–569. [CrossRef]
40. Liu, D.; Chen, N. Satellite Monitoring of Urban Land Change in the Middle Yangtze River Basin Urban Agglomeration, China between 2000 and 2016. *Remote Sens.* **2017**, *9*, 1086. [CrossRef]
41. Zhang, C.; Gao, J.; LI, S.; Gao, S.; Zhao, J. Fractal Dimension Study of Urban Morphology Based on Network Accessibility (In Chinese). *Geogr. Res.* **2018**, *37*, 2528–2540.
42. Zhixin, F.; Yingbiao, C.; Qinglan, Q.; Shuaishuai, W. Relationship between the Structure of Urban Traffic Network and Urban Spatial Expansion: A Case Study of Dongguan City. *J. Geo-Inf. Sci.* **2014**, *16*, 79–86. (In Chinese)
43. Theres, L.; Radhakrishnan, S.; Ogwankwa, F.; Murali, G. Determining Region of Urban Expansion Based on Urban Growth Pattern and Intensity as a Driving Factor Using Regression Modelling Approach in Salem, India. *Earth Sci. Inform.* **2025**, *18*, 235. [CrossRef]
44. Mayoyo, N. Mental Health and Academic Engagement: The Influence of Cyber Dating Abuse Victimization on Undergraduate Students. *Open J. Soc. Sci.* **2025**, *13*, 16. [CrossRef]
45. Sang, H.; Zhang, D.; Zhang, C.; Xi, C.; Fang, K.; Shi, B.; Chang, L. Study on the Spatial Distribution Pattern of Correlation Between Surface Deformation and Reservoir Water Level in the Three Gorges Reservoir Area: Spatial Distribution Pattern of Correlation Study in the Three Gorges Reservoir Area: Sang et Al. *Landslides* **2025**, *22*, 77–93. [CrossRef]
46. Yudhistira, M.H.; Indriyani, W.; Pratama, A.P.; Sofiyandi, Y.; Kurniawan, Y.R. Transportation Network and changes in Urban Structure: Evidence from the Jakarta Metropolitan Area. *Res. Transp. Econ.* **2019**, *74*, 52–63. [CrossRef]
47. Hong, J.; Chu, Z.; Wang, Q. Transport Infrastructure and Regional Economic Growth: Evidence from China. *Transp. Infrastruct. Reg. Econ. Growth Evid. China* **2011**, *38*, 737–752. [CrossRef]
48. Wan, J.; Ma, C.; Jiang, T.; Phillips, A.; Wu, X.; Wang, Y.; Wang, Z.; Cao, Y. A Spatial Econometric Investigation into Road Traffic Accessibility and Economic Growth: Insights from the Chengdu-Chongqing Twin-City Economic Circle. *Palgrave Commun.* **2024**, *11*, 1–9. [CrossRef]
49. Magazzino, C.; Mele, M. On the Relationship between Transportation Infrastructure and Economic Development in China. *Res. Transp. Econ.* **2021**, *88*, 100947. [CrossRef]
50. Maparu, T.S.; Mazumder, T.N. Transport Infrastructure, Economic Development and Urbanization in India (1990–2011): Is There Any Causal Relationship? *Transp. Res. Part Policy Pract.* **2017**, *100*, 319–336. [CrossRef]
51. Wu, X.; Wu, P.; Guo, Z. The Correlations of Transportation with Regional Economy and Demographic Space. *J. Intell. Fuzzy Syst.* **2021**, *41*, 4813–4823. [CrossRef]
52. Komornicki, T.; Goliszek, S. New Transport Infrastructure and Regional Development of Central and Eastern Europe. *Sustainability* **2023**, *15*, 5263. [CrossRef]
53. Cherednichenko, K.; Ivannikova, V.; Sokolova, O.; Yanchuk, M. Model of Transport Safety Assessment in Multimodal Transportation Systems. *Transport* **2023**, *38*, 204–213. [CrossRef]
54. Liu, J.; Xu, H.; Chen, J. The Effects and Conflicts of Co-Opetition in a Rail-Water Multimodal Transport System. *Ann. Oper. Res.* **2023**, 1–46. [CrossRef]
55. Zhang, C.; Dong, X.; Tseng, C.-Y. From Ancient Silk Road to Contemporary Belt and Road: Different Cognition on Transportation Corridor Driven Regional Development. *China City Plan. Rev.* **2020**, *29*, 31–39. [CrossRef]
56. Shen, P.; Yin, P.; Niu, B. Assessing the Combined Effects of Transportation Infrastructure on Regional Tourism Development in China Using a Spatial Econometric Model (GWPR). *Land* **2023**, *12*, 216. [CrossRef]

57. Tang, K.; Wang, H.; Wang, N. The Relationship between the Airport Economy and Regional Development in China. *Emerg. Mark. Financ. Trade* **2022**, *58*, 812–822. [CrossRef]
58. Li, B.; Gao, S.; Liang, Y.; Kang, Y.; Prestby, T.; Gao, Y.; Xiao, R. Estimation of Regional Economic Development Indicator from Transportation Network Analytics. *Sci. Rep.* **2020**, *10*, 2647. [CrossRef]

land

MDPI

Article

Traffic–Tourism Spatial Interaction of Lai-Qu Expressway Based on the Traffic Flow Data

Yujian Gao [1,2], Long Bai [3,*], Shengqiang Liu [1,2], Hongjuan Zheng [3], Jie Liu [1,2], Jinxiang Cheng [1,2], Haiyuan Yao [1,2] and Qing Zhao [3,*]

1 Transport Planning and Research Institute, Ministry of Transport, Beijing 100028, China
2 Laboratory of Transport Pollution Control and Monitoring Technology, Beijing 100028, China
3 School of Geographic Sciences and Environment, Shijiazhuang University, Shijiazhuang 050035, China
* Correspondence: 2309017@sjzc.edu.cn (L.B.); 1310023@sjzc.edu.cn (Q.Z.)

Abstract: In the Taihang Mountain Tourism Development Plan (2020–2035), the Taihang Mountain Expressway is included in the construction of the National Tourism Scenic Road around Taihang Mountain to promote the integrated development of regional transportation and tourism. The Lai-Qu Expressway is part of the Baoding section of the Taihang Mountain Expressway. Based on the data of traffic flow on the Lai-Qu Expressway, data of regional tourism resources, and data of regional economic and social development, this paper studies the interaction between the traffic and tourism space of the Lai-Qu Expressway by using spatial interaction, geographically weighted regression (GWR), and other geospatial analysis theories and methods. The results show that the traffic flow of the Baishishan Tollgate is directly correlated with the passenger flow of the Baishishan scenic spot. The spatial pattern of two tourism resource cluster centers and one sub-center, and one residential cluster center and one sub-center is expected to be formed along the Lai-Qu Expressway. The newly built traffic routes extend the influence of the traffic space and overlaps with the regional tourism space, not only providing new opportunities and possibilities for the development of regional tourism, but also promoting the change in the regional tourism spatial pattern and the cluster form of tourism resources. The research on the interaction between the traffic–tourism space in this paper can help to enrich the theoretical connotation of the research on the integration of transport and tourism, and can also be used to evaluate the tourism impact of newly built transport routes and serve the regional tourism development.

Keywords: integration of transportation and tourism; traffic flow; spatial interaction; geographically weighted regression; Lai-Qu Expressway

1. Introduction

Driven by the integration of transportation and tourism modes, high-speed tourism transportation is transitioning from a single transportation function to offering a rich travel experience and enhanced tourism services, so as to meet diverse and personalized travel needs [1]. With the improvement in the diversified and comprehensive transportation system, road transportation plays an increasingly obvious driving role in the development of the tourism industry [2,3], which is mainly manifested in location optimization, spatial reorganization, resource integration, and other aspects [4]. For example, Park et al. [5] found that road traffic infrastructure facilitates travel to tourist destinations, thereby having a positive impact on community living standards and tourism development. Nunkoo [6] argued that the development of road transportation creates employment opportunities,

https://doi.org/10.3390/land14061197

thus generating active support for the community and the regional tourism industry. Kanwal et al. [7] proposed that road traffic construction would have dual impacts on the development of the tourism industry and the changing concepts of local residents. In addition, some studies have shown that road traffic often plays a framework and supporting role in the formation of the regional tourism spatial pattern. For example, Novianti et al. [8] studied the planning and development of tourism destinations in the Ocean Road region of Australia and the dispersion and mobility of tourists. It is found that tourism transportation is the strategic basis for a region tourism destination and will continue to influence the development direction and spatial composition of tourism destinations. Best [9] took the No. 11 Highway in Canada as an example with which to study the role of the tourist highway in shaping regional lifestyles and developing natural tourism resources. It is found that road traffic has a profound impact on the business development mode of tourist destinations and the composition of the tourist communities. Ishikura et al. [10] studied the impact of urban high-speed traffic on the tourism economy and transportation network by taking Ring Road in Tokyo as an example and found that high-speed traffic has a significant impact on reducing transportation costs, enhancing tourism agglomeration, and avoiding monopolistic competition. It can be seen that most studies on the integration of transport and tourism are about the framework and conceptual spatial impact, and it is difficult to effectively correlate traffic data indicators with the regional tourism spatial impact and tourism industry development, and the impact evaluation of newly built tourism roads is still insufficient.

The "Taihang Mountain Tourism Development Plan (2020–2035)" integrates the Taihang Mountain Expressway into the construction category of the National Tourism Scenic Byway around Taihang Mountain, and suggests that the Taihang Mountain Expressway should be taken as an important basis for building a tourism scenic road with an environment-friendly feature from an ecological perspective, a rich cultural landscape, and the orderly separation of passengers and goods. The Lai-Qu Expressway is a part of the Baoding section of the Taihang Mountain Expressway, which was completed and opened to traffic at the end of 2018. It is a representative example of transportation infrastructure construction promoting the development of the tourism industry. Its completion and opening have changed the original pattern of traffic flow in the region. In the process of statistics and collation of traffic flow data of the Lai-Qu Expressway, it has been found that the traffic flow changes with the seasonal changes in tourism, the traffic flow at the exit of the same tollgate is greater than the traffic flow at the entrance, and the traffic flow gap between tollgates is large.

This study aims to explore the interaction between the influence space of expressways and the tourism space. To achieve this research goal, the newly built Lai-Qu Expressway is taken as the research area, the expressway traffic data and regional tourism data are used as the research data, and social exchange theory and geographical enhanced regression are used as the main theory and method. The application of the geographical spatial analysis method will help to enrich the ideas, methods, and theoretical connotation of the research on the integration of transportation and tourism. The results will serve the regional tourism development of the East Taihang Mountain and the construction of the regional tourism transportation system with a fast–slow combination, and can also be used to evaluate the tourism impact of newly built transport routes.

2. Theoretical Basis and Research Methods

2.1. Theoretical Basis

Tourism spatial interaction (TSI) and social exchange theory (SET) serve as the theoretical underpinnings of the research presented in this paper.

The spatial expression of tourism spatial interaction on tourism behavior encompasses the spatial attributes of tourists' behaviors, the spatial configurations of tourism resources, and other dimensions [11]. It represents the concept of elucidating the relationship between man and land in the region and the correlation, interaction, and dynamic response among each subspace within the framework of tourism geography [12]. In the gravity model, travel time is often used to replace travel distance, and space extension is expressed by space–time compression, agglomeration effect, and regional cooperation, so as to study the interaction between spaces [13]. In this theory, the newly built tourism highways break the original traffic framework structure of the region, improve the accessibility of the whole region [14], reduce the accessibility differences of the region, and are more likely to guide tourists to the tourist resources with high attraction and high agglomeration in the reachable region [15], and then promote the agglomeration development of tourism resources and the linkage of scenic spot marketing [16]. Then, the spatial pattern of tourism is restructured with the extension of traffic-influenced space.

From the perspective of social interaction, social exchange theory establishes the relationship between people such as interest exchange and demand acquisition, and emphasizes the social relations formed by individuals and groups based on resource exchange [17]. It has been widely used in the analysis and research of community support, population mobility, resource interaction, and individual–group social relations in regional tourism development [18]. For instance, the establishment of the China–Pakistan Economic and Cultural Corridor (CPEC) is underpinned by transportation infrastructure, which facilitates the integration of critical elements such as population dynamics, resource allocation, environmental sustainability, and cultural exchange [17]. Social exchange theory can provide interaction possibility, travel motivation, and benefit analysis for regional tourism development on the needs of different groups. And it can also include infrastructure factors [7], social and economic factors [19], environmental protection factors [20], and other multiple influencing factors into regional tourism analysis and research.

Based on spatial interaction theory and social exchange theory, this paper regards tourism resource points as an important attraction of traffic flow, uses traffic volume to reflect the intensity of spatial interaction, and constructs the transportation–tourism spatial interaction relationship of "tourism resource points attract traffic flow and traffic flow promotes tourism spatial agglomeration". In view of the dual significance of regional tourism to the development of tourism resources and community development, it is necessary that we regard residential area as an important factor affecting regional traffic flow, and link transportation–tourism–community all together when analyzing the impact of road traffic on regional tourism development. Finally, the multi-type space composed of traffic routes, tourism resources, and residential areas in the study area is analyzed by superposition and interaction.

2.2. Research Methods

Geographically weighted regression (GWR) is a local linear spatial regression model that utilizes a correction algorithm to assess the variation in influencing parameters in response to the spatial heterogeneity of factors [21]. It has obvious advantages in the study of spatial parameter difference and spatial differentiation of influencing factors and the analysis of difference and imbalance of influencing factors in different geographical Spaces. Local spatial reinforcement regression based on global regression has been widely used in regional spatial analysis [22], and regional traffic volume assessment. The geographical weighted regression model is planned to be used to analyze the spatial impact of traffic flow on tourism resources and residential areas in the study area in this paper. The model is constructed as follows: $y_i = \beta_0(u_i, v_i) + \sum_k \beta_k(u_i, v_i)x_{k,i} + \varepsilon_i$, where y_i is the dependent variable of the i element unit, (u_i, v_i) is the geographical coordinates of the i element unit, $\beta_k(u_i, v_i)$ is the value of the continuous function $\beta_k(u, v)$ on the i element unit (that is, the coefficient of the independent variable x_k on the i element unit), $x_{k,i}$ is the independent variable, and ε_i is the random error. Finally, the spatial expression and influence characteristics analysis of the non-linear and heterogeneous impacts of regional tourism development of the high-speed transportation obtained by geographically weighted regression are analyzed to construct the self-organizing space of tourism destinations, in the continuous plane [15], and show the influence process of newly built expressway on regional tourism traffic flow from disorder to new dynamic order.

In addition, the Tyson polygon rule, a geographical method that can localize the discrete sampling points, is used to determine the tourism resource points and residential areas of the actual influence of each tollgate along the Lai-Qu expressway before regression calculation [23]; the correlation analysis, a method that can be used to identify, analyze, and quantify the correlation between elements, is used to determine regression variables before regression calculation [7]; and the fuzzy density calculation method, which can be used for spatial geographic information extraction and clustering calculation, is used to process the regression results after regression calculation [24].

3. Case Area and Data

The Lai-Qu Expressway (S31), a provincial-level expressway, originates from Yansuya Village in Laiyuan County, Baoding City, and extends southward through Laiyuan County and Tang County to Quyang County. It connects to the Rongwu Expressway in the north and the Beijing–Kunming Expressway and Quyang–Huanghuagang Expressway in the south. The expressway has a total length of 72.8 km and a design speed of 100 km/h. There are five tollgates along it, and its traffic impact covers more than 20 towns. The completion and opening of the expressway have altered the traffic location of the surrounding areas, resulting in complex changes in regional traffic flow [7]. At the same time, many tourism entities in the region that have no hierarchical relationship have been effectively organized and coordinated to form a self-organizing space of tourism destinations that are not restricted by administrative regions [15]. By improving the regional tourism traffic location and optimizing the regional tourism road traffic, it has an impact on the surrounding areas; that is, there is a spatial interaction between the Lai-Qu Expressway and the tourism resource points and residential areas along the route.

The real traffic data along the Lai-Qu Expressway from May 2019 to December 2020 are obtained through the actual daily statistical data report form of the five tollgates, including the export traffic flow, the entrance traffic flow, the vehicle category, and the tolls. The traffic flow data are counted by day. A total of 948,573 entrance traffic flows and 1,434,876 export traffic flows are investigated and counted, and the annual total traffic toll is about 47.31 million yuan (Table 1). The traffic flow data of the section is obtained through the local database of the Transportation Management Center, and a total of 175,213 vehicle flows of the section were counted. Comprehensive statistical data are obtained through the website of the Transportation Department of Hebei Province as a supplement to other necessary data (http://jtt.hebei.gov.cn/jtyst/zwgk/jcxxgk/jttj/zhtjfx/, accessed on 7 March 2021).

Through database screening and a field investigation, the data of tourism resources in this study area are obtained, including the names of completed scenic spots, the available development tourism resources, the longitude and latitude, and the counties to which they belong. According to the national standard [15], the classification, single investigation, and evaluation of tourism resources (1–5) are carried out, and 109 tourism resources points that may be affected by the Lai-Qu Expressway are initially identified and investigated

(Figure 1a). Through the database screening and field investigation, the detailed data of residential areas along the route are obtained, including the name of settlements, latitude and longitude, and administrative districts. Among them, the township point coordinates the location of the government, and the village-level settlements coordinate the village-level administrative boundary polygon geometric center. According to the population, economy, and other factors of the settlements, 17 township-level settlements and 237 villages that may be affected by the Lai-Qu Expressway are initially identified and graded (Figure 1a).

Table 1. Traffic flow data of Lai-Qu Expressway in 2019.

Name of Tollgate	Total Annual Traffic Flow	Annual Entrance Traffic Flow	Annual Export Traffic Flow	Total Annual Tolls	Location of Section Point	Section Traffic Flow
Baishishan	306,714	116,556	190,158	6,549,596.79	Yansuya interchange	/
Chuanli	171,197	70,434	100,763	2,458,500.18	Bao-Fu interchange	85,062
Gubeiyue	88,184	28,845	59,339	2,340,639.21	Dongwang interchange	90,151
Lingshannan	280,674	109,371	171,303	2,823,291.08		
Quyangbei	1,536,680	623,367	913,313	33,133,209.25		

The Yansuya interchange monitoring site was completed in September 2020, so there is no statistical data in 2019.

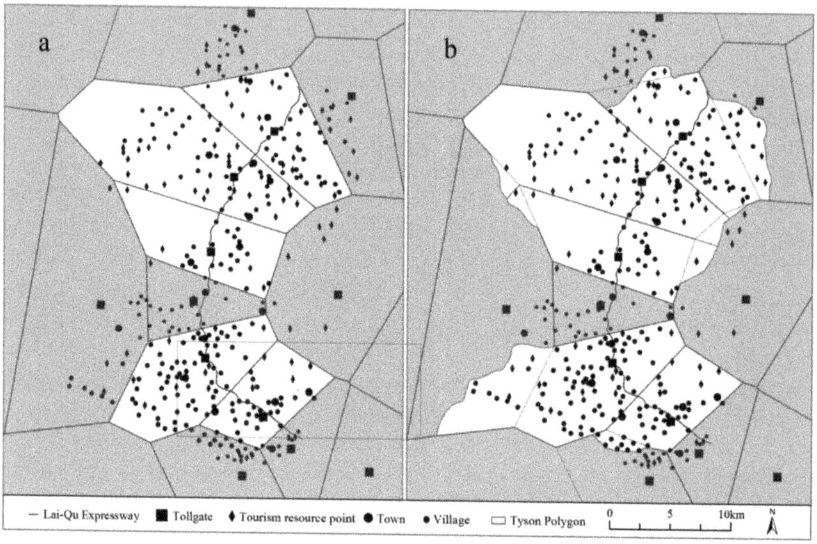

Figure 1. The influence areas before adjustment (**a**) and after adjustment (**b**) of each tollgate along Lai-Qu Expressway based on Tyson polygon law.

The data information of other expressways around the Lai-Qu Expressway, including expressway line data and tollgate data, are continuously collected, and all the collected tollgate data (including tollgates along the Lai-Qu Expressway) are divided into polygon regions by Arcgis Pro according to the Tyson polygon rule to initially identify the influence area of each tollgate along the Lai-Qu Expressway (Figure 1a). According to the topography, traffic flow direction, residents' travel habits, administrative district management, and other factors, combined with the results of the field research, the Tyson polygon boundary is adjusted to obtain the actual influence area of each tollgate of the Lai-Qu Expressway (Figure 1b). According to the statistics, 81 tourism resource points, 13 towns-level residential areas, and 181 villages are affected by the Lai-Qu Expressway (Table 2). The statistical data of 13 towns are obtained by querying the Statistical Yearbook, including the area of jurisdiction, the number of village committees, the household registration population, the number of enterprises, and the number of comprehensive stores (Table 2), which are used to reflect the basic situation of regional economic, social, and tourism development, and to conduct the following spatial impact analysis of the Lai-Qu Expressway.

Table 2. Overview of areas along Lai-Qu Expressway in 2019.

Name of Tollgate	Number of Tourist Resource Points	Number of Villages	Number of Township-Level Settlements	Overview of Township-Level Settlements					
				The Serial Number	Area	Number of Village Committees	The Registered Population	Enterprise Number	Number of General Stores
Baishishan	29	29	2	1	5930	25	33,720	6	69
				2	6802	25	33,215	4	32
Chuanli	29	31	4	3	12,120	35	70,123	50	107
				4	5039	14	9889	0	0
				5	5400	17	40,404	0	22
				6	9200	14	12,230	5	22
Gubeiyue	6	18	2	7	9700	23	23,327	45	27
				8	11,800	22	17,204	0	20
Lingshannan	12	69	2	9	10,300	8	4328	0	9
				10	23,200	21	14,502	24	4
Quyangbei	5	34	3	11	15,700	20	19,546	26	50
				12	10,100	14	8961	3	22
				13	5723	14	26,720	356	29

In order to study the impact of the traffic flow of the Lai-Qu Expressway on regional tourism development, the annual traffic flow of the Baishishan Tollgate and the tourist flow of the Baishishan Scenic Spot in 2019 were statistically sorted out (Table 3). As a scenic spot with the highest level (5A) and popularity in the case area, its main tourist source market is the Beijing–Tianjin–Hebei region, followed by Inner Mongolia, Shanxi, and other regions. The total number tourists of the Baishishan Scenic Spot in 2019 was 1.5026 million person-times, accounting for nearly three-quarters of the total tourists in the area affected by the Baishishan Tollgate (http://www.laiyuan.gov.cn/ldjh/1246.jhtml, accessed on 17 November 2021).

Table 3. Statistics of traffic flow at Baishishan Tollgate of Lai-Qu Expressway and visitors of Baishishan Scenic Area in 2019.

	January	February	March	April	May	June	July	August	September	October	November	December	Total
Baishishan Tollgate	2.67	2.08	2.77	3.08	3.68	2.78	2.91	2.58	2.00	2.26	2.04	1.81	30.67
Baishishan Scenic Area	7.08	6.43	9.67	16.42	22.52	14.04	15.72	14.98	12.74	16.25	8.03	6.38	150.26

The completion and opening of the Lai-Qu Expressway effectively communicates the tourist flow of Shijiazhuang, Xingtai, Handan, and other cities in the central and southern part of Hebei Province, which can better reflect the interaction between traffic flow and tourism flow, and traffic space and tourism space. That is, the completion and opening of the Lai-Qu Expressway has changed the traffic location of the case area and the Baishishan Scenic Spot, and provided more diversified and direct route choices for tourists from the source to the destination. Therefore, this paper takes the tourism flow of the Baishishan Scenic Spot as the representative of regional tourism flow research, and studies the correlation between the transportation and tourism industry development based on the traffic flow data of the Lai-Qu Expressway and the tourism flow data of the Baishishan Scenic Spot.

4. Results

4.1. The Correlation Between Lai-Qu Expressway Traffic Flow and Tourist Flow of the Scenic Spots

A correlation analysis (Table 4) and a curve comparison analysis (Figure 2) are conducted between the tourism flow of the Baishishan Scenic Spot and the traffic flow of the Baishishan Tollgate on the Lai-Qu Expressway. Simultaneously, the traffic flow data of the Baofu Interchange (noting that there is no traffic flow data available for 2019 at the Yansuya Interchange), which is in close proximity to the Baishishan Tollgate, are incorporated into the comparative study. The findings are as follows: (1) The tourism flow of the Baishishan Scenic Spot is correlated with the traffic flow of the Baishishan Tollgate of the Lai-Qu Expressway and the correlation is significant, but the correlation with the traffic flow of the interchange section is not obvious, whereas the traffic flow of the interchange section is correlated with the upward traffic flow. (2) Although there is no obvious parallel structure between the tourism flow of the Baishishan Scenic Spot and the traffic flow curve of the Baishishan Tollgate, there is still a strong correlation. The peak–valley value of tourism flow and traffic flow almost appear at the same time, especially in May, when the annual maximum tourism flow statistics of the Baishishan Scenic Spot (28,000 tourists on May 2) and the annual maximum traffic flow statistics of the Baishishan Tollgate (2009 vehicles on May 2) appear. (3) The upward traffic flow is directly related to the amount of interchange section traffic flow, but the curve synchronization between the interchange section traffic flow and tourism flow is not obvious. Although the peak–valley values of the two are also obvious, there is a certain lag or buffer in the interchange section data, which may be related to the traffic flow from Beijing, Tianjin, and other directions (the Yansuya interchange data may be better verified), the non-one-day travel traffic flow, and the local residents' travel and regional development.

Table 4. Correlation between traffic flow data of Lai-Qu Expressway and visitors of Baishishan Scenic Spot in 2019.

	a	*b*	*c*	*d*
a	/	/	/	/
b	0.737/0.006	/	/	/
c	0.312/0.323	0.416/0.179	/	/
d	−0.030/0.092	0.105/0.746	0.860/0.000	/

a: traffic flow of Baishishan Tollgate. b: tourism flow of Baishishan Scenic Spot. c: traffic flow of Bao-Fu interchange section. d: uplink traffic flow of Bao-Fu interchange section.

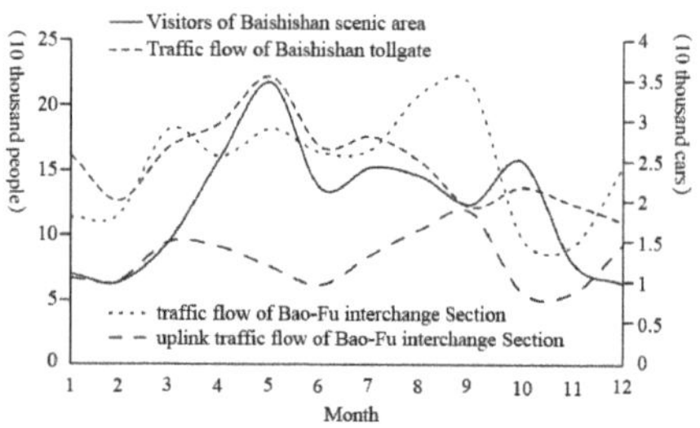

Figure 2. Comparison of traffic flow data of Lai-Qu Expressway and visitors of Baishishan Scenic Spot in 2019.

4.2. The Correlation Between Toll Charges and Regional Development Index

To verify the correlation between traffic flow and regional development, the traffic flow data of the Lai-Qu Expressway were subdivided, and statistical analyses were conducted according to the categories of entrance, exit, cars/buses, and vans to assess the correlation between the regional economy, social development, the regional impacts of the Lai-Qu Expressway, and its development significance. SPSS (27.0) software was employed to analyze the correlation between the aforementioned statistical data and regional development data in the affected area (Table 5). It was found that the import traffic flow (0.769) and the export total traffic flow (0.753) has a strong correlation with the number of village committees in the region, the export car/bus (0.897) had the strongest correlation with the regional household registration population, the charge data (0.715/0.871/0.687) all had the strongest correlation with the number of village committees, but the export van charges had a weak correlation and poor significance compared with the other two indicators. The number of village committees (0.871) has the strongest correlation with the export car/bus charge data, and the household registration population (0.897) has the strongest correlation with the export car/bus data. The administrative area and the number of enterprises are negatively correlated with the traffic flow, and the correlation is weak. The number of general stores (0.788) has the strongest correlation with the export car/bus data.

Table 5. Correlation between the traffic flow of Lai-Qu Expressway and statistical data of regional economic and social development.

	Entrance Traffic Flow	Export Traffic Flow	Export Passenger Cars	Export Vans	Total Export Charges	Export Passenger Car Charges	Export Van Charges
Number of village committees	0.769/0.129	0.753/0.141	0.795/0.108	0.724/0.166	0.715/0.175	0.871/0.055	0.687/0.200
The registered population	0.728/0.163	0.725/0.165	0.897/0.039	0.656/0.229	0.641/0.243	0.841/0.074	0.609/0.276
Area	−0.296/0.696	−0.305/0.618	−0.356/0.557	−0.282/0.646	−0.249/0.687	−0.310/0.612	−0.238/0.700
Number of enterprises	−0.141/0.696	−0.238/0.700	0.056/0.929	−0.326/0.593	−0.333/0.585	−0.090/0.886	−0.361/0.551
Number of general stores	0.563/0.323	0.557/0.329	0.788/0.113	0.472/0.422	0.466/0.429	0.720/0.170	0.428/0.472

In summary, the number of settlements, the regional population, and the level of the agriculture and economy are directly related to the traffic flow of the Lai-Qu Expressway, in addition to other indicators that failed to be reflected. In other words, settlements with a high population density, high per capita income, and high level of agricultural development along the Lai-Qu Expressway are more likely to build up a close relationship between transportation, tourism, and community, and are more prominent in areas with a high concentration of settlements.

4.3. *Characteristics of Tourism Spatial Impact of Lai-Qu Expressway*

The GWR model is employed to investigate the spatial relationship between the traffic flow at each tollgate and the tourism resource points, as well as residential areas along the Lai-Qu Expressway. Based on the analysis results presented in Section 4.1, and drawing upon the screening methodologies for the explanatory variables of the GWR model utilized in previous studies, the grade, quantity, and distance between tourism resource points and the nearest toll station are initially selected as the explanatory variables of the GWR model pertaining to tourism resource points. Similarly, in accordance with the analysis results in Section 4.2, and referring to the screening approaches for the explanatory variables of the GWR model adopted in other research, the number of residential areas, the distance to the nearest tollgate, the regional area, and the regional population are initially designated as explanatory variables of the GWR model associated with residential points. SPSS software was utilized to perform standardized processing and collinearity tests on the aforementioned explanatory variables. Subsequently, the explanatory variable "regional area", which had a variance inflation factor (VIF) exceeding 10, was eliminated. Finally, three explanatory variables related to tourism resource points (grade, quantity, and distance to the nearest tollgate) and three explanatory variables related to residential areas (quantity, distance to the nearest tollgate, and household population) were determined, respectively.

The GWR calculation is carried out for tourism resource points and residential areas, respectively, and the regression formula of tourism resource points is constructed as $y_i = \beta_0(i) + \beta_1(i)RG(i) + \beta_2(i)RN(i) + \beta_3(i)RD(i) + +\varepsilon(i)$, where y_i is the traffic flow of each tollgate, $RG(i)$ is the grade of tourist resource points, $RN(i)$ is the number of tourist resource points, and $RD(i)$ is the distance between tourism resource points and the nearest expressway tollgate. With the help of the Arcgis GWR analysis tool for data calculation, the AICc value is 343.94, the Neighbors value is 10.55 km, the R^2 value is 0.9206, and the adjusted R^2 is 0.9046. The regression formula of the residential area is constructed as $y_{ii} = \beta_0(ii) + \beta_1(ii)PN(ii) + \beta_2(ii)PD(ii) + \beta_3(ii)PP(ii) + \varepsilon(ii)$, where y_{ii} is the traffic flow of each tollgate, $PN(ii)$ is the number of residential areas, $PD(ii)$ is the distance between residential area and the nearest expressway tollgate, and $PP(ii)$ is the registered population of the residential area. With the help of the Arcgis GWR analysis tool for data calculation, the AICc value is 209.65, the Neighbors value is 14.19 km, the R^2 value is 0.9083, and the adjusted R^2 is 0.8889. The rule is that the value of R^2 tends to be 1 and, the better the fitting degree is, the better the regression results of tourism resource points are compared to those of residential areas; that is, the regression results of tourism resource points can better reflect the actual situation of the spatial impact of the Lai-Qu Expressway.

4.3.1. Characteristics of Spatial Influence of Traffic Flow on Tourism Resource Points

As illustrated in Figure 3, the regression results of the spatial relationship between the traffic flow of the Lai-Qu Expressway and the tourism resource points show that the above attributes of the tourism resource points (β_1–β_3) have an impact on the traffic flow from big to small in order of the tourism resource point grade (β_1), the distance to the nearest tollgate station (β_2), and the number of tourism resource points (β_3). The significance of its impact is in order of the tourism resource point grade (β_1), the number of tourism resource points (β_3), and the distance to the nearest tollgate (β_2).

Specifically, the Baishishan Tollgate and Chuanli Tollgate are associated with more tourism resource points (including the highest level of the Baishishan Scenic Spot in the region). Under the influence of the Baishishan Tollgate and Chuanli Tollgate, several tourism resource points in the northern section of the Lai-Qu expressway show a close relationship with the traffic flow, and the interaction between the two tollgates and tourism resource points is obvious. The Gubeiyue Tollgate has the fewest tourism resource points, and the correlation between tourism resource points and tollgate traffic flow is poor. There are few tourism resource points associated with the Lingshannan Tollgate and Quyangbei Tollgate, but the interaction characteristics between the distance to the nearest tollgate and the traffic flow is obvious, and the influence is significant. In terms of the spatial characteristics, the two tollgates in the northern section of the Lai-Qu Expressway provide opportunities and possibilities for the formation of high-level and high-density tourism resource cluster centers. The spatial interaction between the Lai-Qu Expressway and tourism resource points is obvious, and the impact is significant. The Gubeiyue Tollgate in the middle section is associated with fewer tourism resource points, and its spatial interaction with tourist resource points is insufficient. The two tollgates in the southern section have obvious interaction characteristics with the associated tourist resource points of distance attenuation.

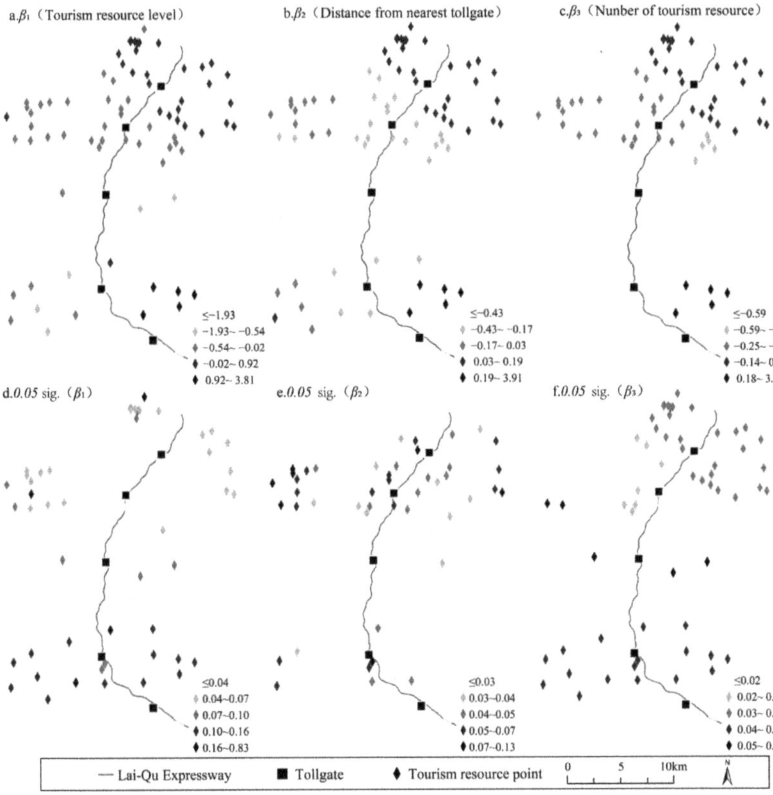

Figure 3. Spatial characteristics of the impacts of tourism resource point level, number, and distance to the nearest tollgate on the traffic flow of Lai-Qu Expressway.

4.3.2. Characteristics of Spatial Influence of Traffic Flow on Residential Area

As shown in Figure 4, the regression results of the spatial relationship between the traffic flow of the Lai-Qu Expressway and the residential areas show that the above attributes of the residential areas ($\beta_4 - \beta_6$) have an impact on the traffic flow from big to small in order of the number of residential areas (β_5), the household population (β_6), and the distance to the nearest tollgate (β_4). The significance of its impact is in order of the distance to the nearest tollgate station (β_4), the household population (β_6), and the number of residential points (β_5).

Specifically, the Baishishan Tollgate and Chuanli Tollgate are associated with a large number of scattered residential areas, and the interaction between each property of the settlements and the Lai-Qu Expressway is positive. The Gubeiyue Tollgate has the fewest residential areas, and the correlation between the residential areas and the traffic flow is not strong. The Lingshannan Tollgate and Quyangbei Tollgate are associated with a large number of dense settlements, but the interaction between them and the associated residential areas is different. The interaction between the Lingshannan Tollgate and the associated residential areas is poor and only one attribute (β_4) is significant, while the interaction between the Quyangbei Tollgate and the associated residential areas is very obvious and significant. In terms of the spatial characteristics, the interaction between the two tollgates in the northern section and the associated residential areas is mostly positive, but not significant; the distance decay characteristics between the Gubeiyue Tollgate and Lingshannan Tollgate and the associated residential areas are very obvious; the agglomeration characteristics of the associated residential areas in the southern section of the Lai-Qu Expressway are obvious, and the influence of the number of residential areas and the household population on the traffic flow of the Lai-Qu Expressway is more significant.

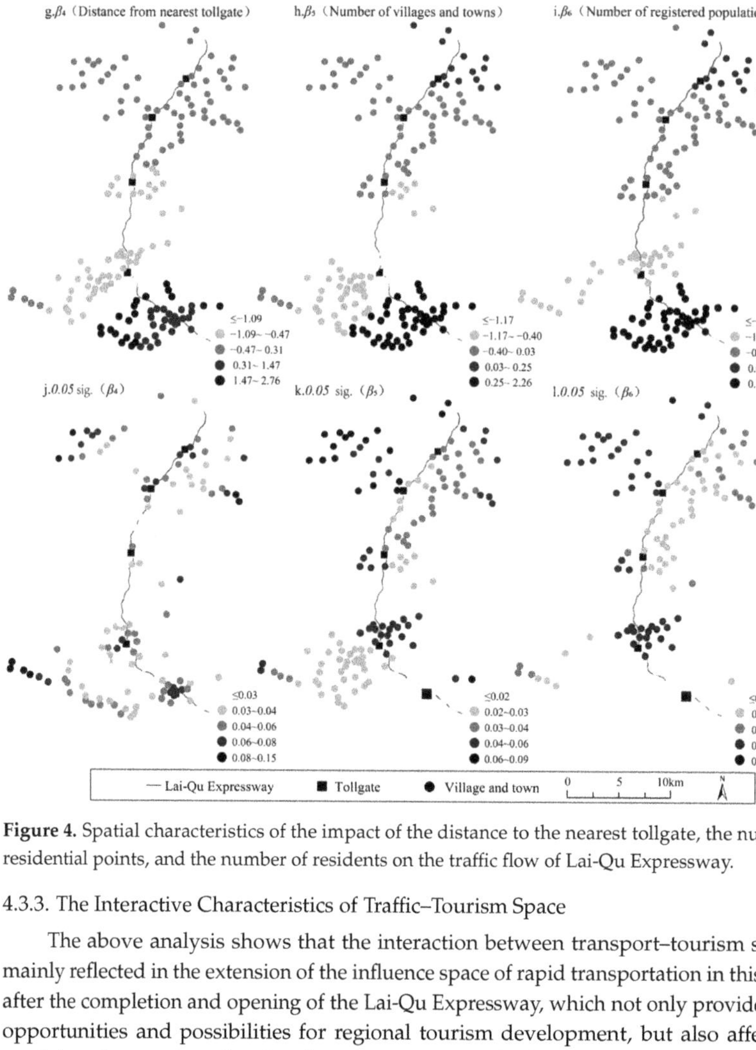

Figure 4. Spatial characteristics of the impact of the distance to the nearest tollgate, the number of residential points, and the number of residents on the traffic flow of Lai-Qu Expressway.

4.3.3. The Interactive Characteristics of Traffic–Tourism Space

The above analysis shows that the interaction between transport–tourism space is mainly reflected in the extension of the influence space of rapid transportation in this region after the completion and opening of the Lai-Qu Expressway, which not only provides more opportunities and possibilities for regional tourism development, but also affects the spatial agglomeration of regional tourism resources. In the regression results of $\beta_1 - \beta_6$, the grade of the tourist resource points (β_1) and number of residential areas (β_5) have important effects on the traffic flow of each tollgate along the Lai-Qu Expressway. The regression results are rasterized, and the raster output image size is set to 1.70; then, the fuzzy density calculation method is used to calculate and analyze the regression results of $\beta_1 - \beta_6$, respectively, and the spatial interaction map of transportation–tourism and transportation–residential areas are obtained (Figure 5); and the distance decay curves of the influence spatial range of each tollgates are also obtained (Figure 6).

It can be seen from the figure that the Lai-Qu Expressway provides opportunities and possibilities for the formation of two tourism resource clustering centers in the northern section and one tourism resource clustering sub-center in the southern section (Figure 5a). It also provides opportunities and possibilities for the formation of one residential area clustering sub-center in the northern section and one in the southern section (Figure 5b). In terms of regional development, the transportation–tourism interaction is obvious in the northern section, and the completion and opening of the Lai-Qu Expressway provides the possibility for the development of regional tourism and the construction of regional tourism destinations. The transportation–residential interaction is obvious in the southern section, and the completion and opening of the Lai-Qu Expressway improves the traffic location of regional economic and social development along the expressway. The whole expressway effectively links the tourist resource points in the northern section and the residential areas in the southern section, creating conditions for close tourism travel, and optimizing the route selection for linking with the tourism source market in central and southern Hebei Province. In terms of spatial impact characteristics, the spatial impact of each tollgate attenuates differently with distance (Figure 6). The northern section of the Lai-Qu Expressway mainly slows down its distance attenuation effect due to the existence of a large number of tourism resource points, and the influence range of the Lai-Qu Expressway is extended due to the traffic–tourism interaction (Figure 6a). The southern section of the Lai-Qu Expressway mainly slows down its distance attenuation effect due to the agglomeration of a large number of residential areas, and the interaction between traffic and residential areas extends the influence range of the Lai-Qu Expressway (Figure 6b).

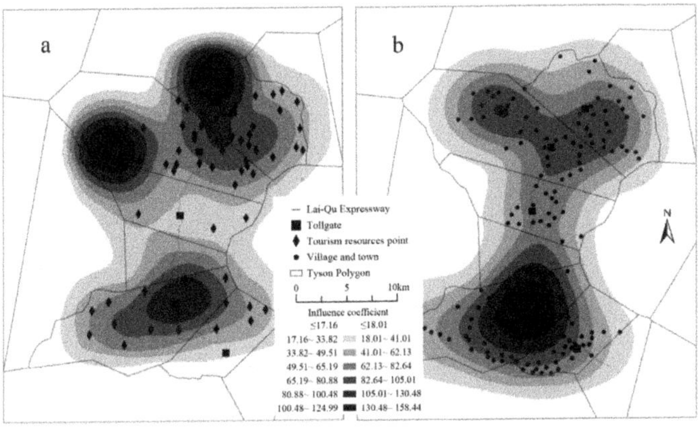

Figure 5. The spatial interaction map of traffic–tourism (**a**) and traffic–residential points (**b**) on Lai-Qu Expressway.

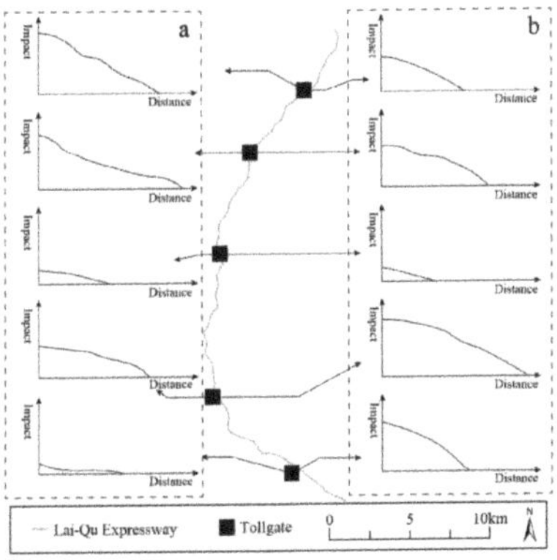

Figure 6. The distance attenuation curve of each tollgate along Lai-Qu Expressway under the interaction of traffic–tourism (**a**) and traffic–residential points (**b**).

5. Conclusions and Discussion

The findings of the study are as follows:

(1) The influence space of high-speed transportation extends along with the newly built expressway, and realizes the spatial interaction between transportation and tourism in the process of overlapping of the transportation-affected space and regional tourism space, which provides new opportunities and possibilities for regional tourism, and will also promote the change in the regional tourism spatial pattern and agglomeration form of tourism resources.

(2) The peak–valley value data of the Baishishan Tollgate traffic flow and Baishishan Scenic Spot tourist flow almost simultaneously show that they are directly correlated. The GWR results of tourism resource points show that the grade of tourism resource points (β_1) has the greatest impact and the strongest significance on the traffic flow of the Lai-Qu Expressway. In the interaction between the traffic and tourism space, it is shown that the northern section of the Lai-Qu Expressway is expected to form a high-grade and high-density tourism resource agglomeration area.

(3) The traffic flow of the Lai-Qu Expressway has the strongest correlation with the number of residential areas (β_5). The results of the GWR of residential areas show that the number of residential areas (β_5) has the greatest impact on the traffic flow at each tollgate of the Lai-Qu Expressway, but the most significant explanatory variable is the distance between settlements and the nearest tollgate (β_4); that is, the impact of residential areas on the traffic flow of the Lai-Qu Expressway is obviously attenuating with distance, which is most typical in the Lingshannan Tollgate.

(4) The interaction map of the transportation–tourism space of the Lai-Qu Expressway is calculated by the fuzzy density calculation method and the GWR results shows that it is expected to form a spatial pattern of two tourism resource cluster centers and one sub-center, and one residential cluster center and one sub-center under the influence of the Lai-Qu Expressway. The distance attenuation curve shows that the tollgates in the northern section of the Lai-Qu Expressway are associated with a large number of tourism resource points, and the transportation–tourism spatial interaction extends the spatial impact of the Lai-Qu Expressway. A large number of residential areas associated with the southern section of the Lai-Qu Expressway slows down the distance attenuation effect, and the interaction between traffic and residential areas extends the spatial influence of the Lai-Qu Expressway.

The regional tourism spatial impact of the Lai-Qu Expressway starts from the optimization of the traffic location of tourism resource points, and provides opportunities and possibilities for the development of regional tourism along the route. Based on the integration of transportation and tourism, we hold the following beliefs:

(1) With regard to the research on the spatial interaction of transportation and tourism at a small scale and a small sample size, this paper adopts the form of eigenvalue sampling similar to that in statistics, that is, to compare and analyze the traffic flow of the Baishishan Scenic Spot and the traffic flow data of the Baishishan Tollgate, to obtain the interaction between regional transportation and tourism development. The evaluation method of the tourism efficiency and high-speed traffic coordination degree provided by Guo et al. [25] can also better reflect the regional transportation–tourism interaction and spatial interaction, but this method is not adopted in this paper. The reason is that this method is more suitable for a large-scale and multi-sample analysis, and needs the support of multiple regional development statistics. And the analytical method for the spatial interaction between transportation and tourism under small-scale spatial and limited sample conditions, as proposed in this study, demonstrates a greater suitability for county-

level spatial planning and tourism policy formulation. In other words, this approach internalizes externalities such as transportation networks, regional policies, and ecological development into regional development "responses" [26]. The findings offer practical cases for enhancing transportation networks and optimizing regional spatial development.

(2) With regard to the positioning of residential areas in the study of the transportation–tourism spatial interaction and the development of regional tourism destinations, many achievements of the social exchange theory generally believe that communities (residential areas) play an important role in regional tourism development [19,20]. Considering the regional conflict of functional–structural space units and the dynamic space–time evolution of the "production–ecological" space in territorial spatial planning, the boundary between scenic spots and communities in areas where tourism resources are concentrated becomes more unclear. Therefore, this paper also tends to regard residential areas as the elements to be considered in regional tourism development under the guidance of social exchange theory. Residential areas can serve regional tourism development from the aspects of characteristic dwellings, labor force, infrastructure construction, and so on. Simultaneously, the regional development role of the Lai-Qu Expressway is "optional" rather than "alternative". Under the framework of the social exchange theory, this perspective integrates external factors (including transportation network integrity and policy coordination between tourism and ecological development) into a comprehensive regional development framework. This integration enables a holistic analysis of the expressway's regional impacts, making the methodology better suited to small-scale spatial contexts for transportation planning and evaluation, as well as tourism strategy development.

(3) With regard to the construction of tourism resource agglomeration areas and regional tourism destinations, with the opening up of newly built highways and the continuous effect on regional tourism development, the status quo of the "isolated island" development and integrated development of low-concentration industries will be broken, and replaced by the organic series of scattered tourism resources, the contiguous development of tourism resources along highways, and the development of cross-administrative region boundaries. For example, in the research results of this paper, the northern section of the Lai-Qu Expressway may form two tourism resource agglomeration centers. Gao et al. [24] mentioned in their research on tourism agglomeration areas in Hebei Province that the formation of agglomeration areas needs the support of traffic conditions and takes high-level scenic spots as the core attraction of agglomeration areas. As time goes on, the influence of the Lai-Qu Expressway on the development of the regional tourism industry and the spatial pattern of tourism will be more obvious.

Author Contributions: Conceptualization, L.B. and J.L.; methodology, L.B., H.Z. and Q.Z.; software, Y.G.; validation, S.L. and Q.Z.; formal analysis, H.Y.; investigation, J.L.; resources, S.L., J.C. and Q.Z.; data curation, Y.G., L.B. and J.C.; writing—original draft, L.B.; writing—review and editing, Y.G. and Q.Z.; supervision, H.Z. and H.Y.; project administration, L.B.; funding acquisition, Y.G. All authors have read and agreed to the published version of the manuscript.

Funding: This research was funded by the Key Science and Technology Projects in the Transportation Industry (grant number: 2020-MS4-113); and the National Natural Science Foundation of China (grant number: U23A2007).

Data Availability Statement: The original contributions presented in the study are included in the article, further inquiries can be directed to the corresponding authors.

Conflicts of Interest: The authors declare no conflicts of interest.

References

1. Van, N.; Shimizu, T. The effect of transportation on tourism promotion: Literature review on application of the Computable General Equilibrium (CGE) Model. *Transp. Res. Procedia* **2017**, *25*, 3096–3115.
2. Palhares, L. The role of transport in tourism development: Nodal functions and management practices. *Int. J. Tour. Res.* **2003**, *5*, 403–407. [CrossRef]
3. Huang, T.; Xi, J.; Ge, Q. Spatial differentiation and integration optimization of an urban agglomeration tourism system under the influence of high-speed railway network evolution. *Appl. Spat. Anal. Policy* **2019**, *12*, 349–376. [CrossRef]
4. Fisher, M.; Wood, A.; Roh, H.; Kim, C.K. The geographic spread and preferences of tourists revealed by user-generated information on Jeju Island, South Korea. *Land* **2019**, *8*, 73. [CrossRef]
5. Park, B.; Nunkoo, R.; Yoon, Y.S. Rural residents' attitudes to tourism and the moderating effects of social capital. *Tour. Geogr.* **2015**, *17*, 112–133. [CrossRef]
6. Nunkoo, R.; Ramkissoon, H. Developing a community support model for tourism. *Ann. Tour. Res.* **2011**, *38*, 964–988. [CrossRef]
7. Kanwal, S.; Rasheed, I.; Pitafi, H.; Pitafi, A.; Ren, M. Road and transport infrastructure development and community support for tourism: The role of perceived benefits, and community satisfaction. *Tour. Manag.* **2020**, *77*, 104014. [CrossRef]
8. Novianti, S.; Fauzi, C.; Suhartanto, D. Spatial analysis of tourist dispersal and mobility for tourism destination planning and development: A case study of great ocean road region, Australia. *IOP Conf. Ser. Mater. Sci. Eng.* **2020**, *830*, 032081. [CrossRef]
9. Best, K. *Highways and Lifeways: Highway 11 and the Shaping of the Ways of Life and Senses of Place of Nature-Based Tourism Operators in South-Central Almaguin, Ontario, Canada*; University of Northern British Columbia: Prince George, BC, Canada, 2016.
10. Ishikura, T.; Yoshikawa, H.; Yokoyama, F. Spatial economic impacts of ring road highway development in Greater Tokyo Area. In Proceedings of the 22nd Annual Conference on Global Economic Analysis, Warsaw, Poland, 19–20 June 2019; pp. 1–16.
11. Lee, J. Conflict mapping toward ecotourism facility foundation using spatial Q methodology. *Tour. Manag.* **2019**, *72*, 69–77. [CrossRef]
12. Lin, R.; Chen, K. Design and Research of Traffic-Tourism Integration Service System Based on Scenario Theory. In *International Conference on Green Intelligent Transportation System and Safety*; Springer Nature: Singapore, 2022; pp. 539–548.
13. Feng, B. Coupling and coordinated development of traffic accessibility and regional tourism economy. *Res. Transp. Bus. Manag.* **2023**, *49*, 101010. [CrossRef]
14. Ravazzoli, E.; Streifeneder, T.; Cavallaro, F. The effects of the planned high-speed rail system on travel times and spatial development in the European Alps. *Mt. Res. Dev.* **2017**, *37*, 131–140. [CrossRef]
15. Bai, L.; Lu, Z.; Gao, Y.; Gao, W. Study on Spatial Pattern of Regional Tourism under Influence of High-Speed Tourism Corridor: Case Study of Taihang Mountain Expressway. *J. Highw. Transp. Res. Dev.* **2021**, *15*, 102–110. [CrossRef]
16. Masson, S.; Petiot, R. Can the high speed rail reinforce tourism attractiveness? The case of the high speed rail between Perpignan (France) and Barcelona (Spain). *Technovation* **2009**, *29*, 611–617. [CrossRef]
17. Ali, L.; Mi, J.; Shah, M.; Khan, A.; Imran, M. Transport culture akin to the China-Pakistan economic corridor. *Hum. Syst. Manag.* **2017**, *36*, 381–396. [CrossRef]
18. Sun, F.; Jia, Y. Influence factors of residents' political trust from perception of tourism community residents: Based on explanation of social exchange theory. *J. Beijing Inst. Technol. (Soc. Sci. Ed.)* **2020**, *22*, 90–99. (In Chinese)
19. Gursoy, D.; Chi, G.; Dyer, P. Locals' attitudes toward mass and alternative tourism: The case of Sunshine Coast, Australia. *J. Travel Res.* **2010**, *49*, 381–394. [CrossRef]
20. Kanwal, S.; Pitafi, H.; Rasheed, I.; Pitafi, A.; Iqbal, J. Assessment of residents' perceptions and support toward development projects: A study of the China-Pakistan Economic Corridor. *Soc. Sci. J.* **2022**, *59*, 102–118. [CrossRef]
21. Brunsdon, C.; Fotheringham, S.; Charlton, M. Geographically weighted regression. *J. R. Stat. Soc.* **1998**, *47*, 431–443. [CrossRef]
22. Yu, H.; Wang, Q.; Zhang, B.; Liu, J. Driving mechanism and the spatial differentiation of coupling coordinated development of tourism supply and demand in China. *Sci. Geogr. Sin.* **2020**, *40*, 1889–1898. (In Chinese)
23. Napitupulu, D.; Rahmaitria, F.; Rosita, R. The Effect of tourism accessibility perception towards tourists visiting intention to Toba lake in samosir district. *J. Indones. Tour. Hosp. Recreat.* **2021**, *4*, 39–52. [CrossRef]
24. Gao, W.; Zhang, Q.; Lu, Z.; Wu, D.; Du, X. Modelling and application of fuzzy adaptive minimum spanning tree in tourism agglomeration area division. *Knowl.-Based Syst.* **2018**, *143*, 317–326. [CrossRef]

25. Guo, X.; Mu, X.; Ding, Z.; Ming, Q. The coordination pattern of tourism efficiency and high-speed transportation: A case study of 41 cities in the Yangtze River Delta. *Geogr. Res.* **2021**, *40*, 1042–1063. (In Chinese)
26. Du, X.; Li, Z.; Bai, L.; Pan, L. Navigating sustainable development: Empirical analysis of corporate social responsibility, eco-friendly production, and stakeholder green commitment through the lens of stakeholder theory. *Sustain. Dev.* **2024**, *32*, 4250–4260. [CrossRef]

land

MDPI

Article

Exploring the Relationship Between the Built Environment and Bike-Sharing Usage as a Feeder Mode Across Different Metro Station Types in Shenzhen

Yiting Li [1], Jingwei Li [1,*], Ziyue Yu [1], Siying Li [1] and Aoyong Li [2]

[1] School of Architecture and Design, Beijing Jiaotong University, Beijing 100044, China; 21322056@bjtu.edu.cn (Y.L.); 21311023@bjtu.edu.cn (Z.Y.); 21311008@bjtu.edu.cn (S.L.)
[2] School of Transportation Science and Engineering, Beihang University, Beijing 100191, China; liaoyong@buaa.edu.cn
* Correspondence: jwli3@bjtu.edu.cn

Abstract: Bike-sharing has been widely recognized for addressing the "last-mile" problem and improving commuting efficiency. While prior studies emphasize how the built environment shapes feeder trips, the effects of station types and spatial heterogeneity on bike-sharing and metro integration remain insufficiently explored. Taking the urban core area of Shenzhen as a case study, this paper examines how the built environment influences such integration during morning peak hours and how these impacts differ across station types. First, we proposed a "3Cs" (convenience, comfort, and caution) framework to capture key built environment factors. Metro stations were classified into commercial, residential, and office types via K-means clustering. Subsequently, the ordinary least squares (OLS) regression model and the multiscale geographically weighted regression (MGWR) model were employed to identify significant factors and explore the spatial heterogeneity of these effects. Results reveal that factors influencing bike-sharing–metro integration vary by station type. While land-use mix and enclosure affect bike-sharing usage across all stations, employment and intersection density are only significant for commercial stations. Furthermore, these influences exhibit spatial heterogeneity. For instance, at office-oriented stations, population shows both positive and negative effects across areas, while residential density has a generally negative impact. These findings enhance our understanding of how the built environment shapes bike-sharing–metro integration patterns and support more targeted planning interventions.

Keywords: bike-sharing; metro station types; feeder mode; built environment; spatial heterogeneity

1. Introduction

With the acceleration of China's urbanization, issues such as traffic congestion and high energy consumption have become increasingly severe. As an effective way to ease environmental pressure and improve commuting efficiency, shared bikes help meet residents' mobility needs and support multimodal transfers [1]. Their speed advantage over walking makes them a better solution to the first-and-last-mile problem, thereby enhancing transit accessibility and equity [2]. As one of China's first low-carbon pilot cities and a model for public transportation development, Shenzhen has witnessed shared bicycles emerge as a vital mode of transportation for commuters. According to data from the Shenzhen Municipal Transportation Bureau, approximately 52.3% of daily bike-sharing trips serve as

https://doi.org/10.3390/land14061291

a feeder mode to public transit, with usage primarily concentrated during weekday peak hours [3].

Given the context above, increasing attention has been given to feeder trips between bike-sharing and public transportation in recent years [4–9]. Among the factors shaping people's travel behavior, the built environment has been identified as a key determinant influencing residents' choice of feeder modes [7,8]. In this context, a growing number of studies have explored the relationship between the built environment and the integration of bike-sharing with metro systems, confirming that factors such as population density, location conditions, and land use around metro stations significantly influence the choice of bike-sharing as a feeder mode [4,6,7,10]. These studies have often measured built environment characteristics within the service areas of metro stations, analyzing them as a unified whole. However, as the direct environment in which cycling behavior occurs, the impact of both macro and micro road features on bike-sharing usage is significant and should not be overlooked [11,12], and effective urban design along cycling routes plays a key role in enhancing the comfort and convenience of cycling [13]. While many studies have considered road-related built environment variables, such as road density and intersection density [6,14], few have focused on the impact of road characteristics (i.e., sky openness, traffic safety infrastructure, greenery) surrounding metro stations on commuters' feeder behaviors.

Furthermore, given that different types of metro stations, such as commercial-oriented, residential-oriented, and employment-oriented stations, exhibit variations in location, interchange level, passenger volume, and travel patterns [15,16], the surrounding built environment, such as land use, road pattern, and streetscape, also differs across station types [17,18]. Existing studies have proved that both station characteristics and its surrounding environment are associated with commuters' feeder choice [5,7,17]. Therefore, categorizing metro stations within the study area is essential, as it allows for a more comprehensive consideration of how station-specific characteristics mediate the effects of the built environment on feeder behavior. However, there is limited research on how built environment effects vary across different stations, which limits the ability to provide targeted urban design and transport planning recommendations that address the specific needs of different station contexts.

To fill these gaps, we focus on the urban core area of Shenzhen, China, which is the city's concentration area for metro stations and bike-sharing usage. This study examines how the surrounding road environment influences bike-sharing and metro integration during morning peak hours across different types of metro stations by identifying key influencing factors and their spatial heterogeneity. The main research questions are as follows: (1) How do the built environment and feeder characteristics around different metro stations vary? (2) What built environment factors around various types of stations have a significant impact on the integration of bike-sharing and metro usage? (3) What is the spatial heterogeneity of these influential features? To solve these problems, we first develop a "3Cs" framework (convenience, comfort, and caution) to comprehensively assess the impact of the road built environment around metro stations on the integration of bike-sharing and metro. Second, we use K-means clustering to categorize metro stations based on their surrounding transportation characteristics, population density, and land use. Third, the OLS regression model is applied to identify significant factors influencing bike-sharing around different station types. Finally, we employ a multiscale geographically weighted regression (MGWR) model to further examine the spatial heterogeneity of these factors.

The remainder of this paper is organized as follows: Section 2 presents a literature review on bike-sharing–metro integrated use and the measurement of the built environment; Section 3 describes the study area, data collection, and major methods; Section 4 introduces the results of the metro station classification, OLS regression model, and the MGWR model; Section 5 presents the discussion of the results. Finally, Section 6 concludes the paper with the main findings, policy recommendations, and limitations.

2. Literature Review

2.1. Built Environment Determinants of Bike-Sharing–Metro Integrated Use

2.1.1. Built Environment Correlates of Cycling Behavior

The integration of bike-sharing and metro systems is influenced not only by urban infrastructure and economic policies but also closely associated with the built environment characteristics surrounding metro stations [19,20]. In terms of built environment research, Cervero et al. initially explored the impact of the built environment on travel behavior through the "3Ds" framework, including density, diversity, and design, in 1997 [21]. Building upon this foundation, Cervero et al. later expanded the framework to the "5Ds" by incorporating distance to transit and destination accessibility [22]. By 2010, Ewing and Cervero further developed the widely cited "6Ds" framework, which added the category of demand management and has since become one of the most commonly used theories in built environment studies [23,24].

In terms of convenience-related factors, Ni and Chen confirmed that residential and employment density can promote bike-sharing usage [25]. However, some studies have also found that these factors may inhibit bike-sharing in certain contexts [6,26]. Apart from specific land use types, the overall degree of land-use mix can also encourage residents' cycling behavior [26,27]. Population size has also been shown to significantly affect cycling, but there is no consensus on the specific direction of this effect [4,8]. Furthermore, road-related variables such as road and intersection density can also influence cycling behavior, as supported by earlier studies [6,14].

Although the aforementioned studies have provided a relatively comprehensive consideration of the impact of the built environment, they remain focused on objective factors, often overlooking cyclists' perceptual experiences on the cycling route, including comfort and safety [8,12,28]. In recent years, an increasing number of scholars have incorporated perceived built environment factors into their research. Specifically, a study in India found that the comfort of the in-car environment and journey safety can affect commuters' choices between using cars or public transportation [29,30]. Meanwhile, a more aesthetically pleasing and safer street environment can stimulate people's desire to walk [31]. Furthermore, existing studies have demonstrated that perceived factors such as cycling safety [8,10] and street design (e.g., greenery, enclosure, and shade) [12,32] also significantly shape residents' cycling behavior, indicating that the impact of psychological perception factors on people's travel behavior cannot be ignored. For example, Bai et al. discovered that greater enclosure could promote cycling behavior [33], since well-enclosed spaces often suggest a sense of safety and vitality [32], and thus encourage more physical activity [34]. While Song et al. suggested that enclosure could adversely affect the cycling experience [12]. Moreover, sky openness and greenery also have psychological value [33], by improving the mood of residents [35,36]. Previous studies have shown a positive correlation between sky openness, street greenness, and cycling [12,33,37]. In addition, with the emergence of big data methods such as street view imagery and its integration with machine learning (ML), the correlation between these perceived built environment factors and cycling can be further confirmed. For example, Ito and Biljecki combined street view imagery and computer vision (CV) techniques to incorporate micro-level factors into an assessment system for bikeability [38]. Bai et al. used street view imagery and ML methods to confirm that streetscapes promote the use of bike-sharing as a feeder mode among students [37]. Although the existing literature has considered the influence of objective and subjective built environment on cycling behaviors, few studies have integrated them to provide a comprehensive assessment of their impact on feeder trips.

As for built environment data extraction, when analyzing the built environment around metro stations, previous studies tend to establish a buffer centered on the metro station entrance. This buffer is then analyzed as a whole to examine how built environment factors such as population density, road density, and the land-use mix affect feeder trips [5–7]. The research on how both the subjective and objective built environments of roads affect cycling efficiency and integrated travel remains limited [11,12].

2.1.2. Metro Station Types and Travel Behavior

Existing studies often classify metro stations based on their inherent characteristics, such as location [17], passenger volume [16,39], and interchange level [16,17]. These characteristics have been shown to be closely associated with the travel behavior of residents in the surrounding areas [17,40,41]. Due to variations in location and urban functions, different types of stations are often associated with significant differences in the surrounding built environment, such as land use, facility density, and street network structure [15,17,18,42]. Previous studies have shown a high degree of consistency between residents' feeder behavior and the built environment factors around metro stations [15,17]. In particular, according to typical riding distance for shared bike users [43], for bike-sharing and metro integration, the area within 800–2000 m of a metro station is typically regarded as the major attractive space that generates a large number of potential demands [9,10,44], and studies have found that land use, population density, road network characteristics, and perceived built environment factors within this area are significantly associated with bike-sharing usage. For example, Ma and Dill found that within a half-mile radius (approximately 804.67 m) around metro stations, the number of retail stores and road density can promote bike-sharing usage; Sun et al. found that within a 2000 m radius around metro stations, road density, POI characteristics, and land-use mix are closely associated with bike-sharing–metro feeder use [7]; Guo and He studied the 800 m buffer area, finding a strong correlation between a perceived built environment factor (traffic safety) and bike-sharing volume around the station [8]. However, these studies analyze all metro stations in the study area as a homogeneous whole, without categorizing metro stations by functional type. Therefore, this may overlook both the influence of the metro station itself on bike-sharing–metro integration usage and the differences in the surrounding built environment across different types of metro stations, making it difficult to implement more targeted policies around each station type.

To better explore the impact of the built environment around different types of stations on the use of bike-sharing–metro integrated usage, it is essential to first classify the metro stations in the study area. Clustering techniques such as K-means, expectation maximization, and hierarchical clustering are commonly employed to categorize metro stations within a region [17]. For instance, Wu et al. identified five TOD types in Wuhan based on the Place–Node model and K-means clustering [18]. Yang et al. categorized Beijing metro stations into seven categories using the K-means method by taking the station passenger flow characteristics and land use characteristics as the clustering criteria [45]. Xia and Gai, based on the distribution of population distribution, development intensity, public transportation connection, road network length distribution, station location attributes, and station passenger flow data of six dimensions of clustering factors, applied the K-means method to categorize Shenyang metro stations into five categories: residential, commercial and business, comprehensive development, industrial, and transportation hub [46].

2.2. Evaluation for Impact of Built Environment

When exploring the impact of built environment elements on bike-sharing usage, it is essential to first identify the significant factors that affect the dependent variable. Specifically, Wang et al. and Guo et al. used a regression model to test the relationship between the built environment and the cycling volume around metro stations [44,47]. However, using only the traditional regression model would ignore the spatial heterogeneity of the effects and could not further explore the relationship between variables that would be affected by different geographic locations [45].

Therefore, studies have started to introduce geographically weighted regression (GWR). For example, Zheng et al., when studying the impact of the built environment on ride-hailing in Chengdu, in addition to applying the OLS regression model to select the variables, also established a GWR model to investigate the impact of variables in different geographical locations [48]. Li et al. similarly combined the OLS and GWR models to explore the effects of factors such as the proportion of transportation facility coverage, population, and land use on the use of shared bicycles in Shanghai and its spatial heterogeneity [49]. However, GWR uses the same bandwidth for each variable [50], which may reduce the robustness of the model [45], while multiscale geographically weighted regression (MGWR) uses variable bandwidths and different geographic weighting matrices for each geographic object, which avoids the degradation of the performance of the dependent variable and its correlations due to overfitting and underfitting of the variables, as compared to the GWR method using fixed bandwidths and differentiated weighting matrices [51].

3. Methods

3.1. Study Area

The urban core area of Shenzhen was selected as the study area (Figure 1). Shenzhen is a mega-city in southern China, located in the Pearl River Delta near Hong Kong. Its urban core area comprises the three central districts of Luohu, Nanshan, and Futian, as well as Xin'an and Xixiang Streets of Bao'an District, Minzhi and Longhua Streets of Longhua District, and Bantian, Buji, Jihua, and Nanwan Streets of Longgang District. These areas cover four key functional centers: Qianhai, Nanshan, Futian, and Luohu, which are densely populated and have a high demand for bike-sharing usage.

As a transit-oriented city, Shenzhen possesses a highly modernized rail network, with over 40% of daily trips made via the metro system [7]. It is also one of the largest cities that have experienced an increase in DBS usage [10]. According to the Shenzhen Bike-Sharing Development Assessment Report, as of April 2019, the city had two major bike-sharing operators ("ofo" and "Mobike") managing a fleet of 480,000 bikes, serving 26.47 million registered users with an average of 849,000 daily rides. Of these rides, 54.43% were for commuting, and 45% took place during morning and evening peak hours, underscoring the role of dockless bike-sharing (DBS) in complementing public transit for first- and last-mile connectivity. This integration is further highlighted by the fact that 98.6% of shared bicycles are concentrated in metro-dense areas. The two main operators charge between 0.6 yuan (approximately $0.08) and 2.5 yuan (approximately $0.35) per hour.

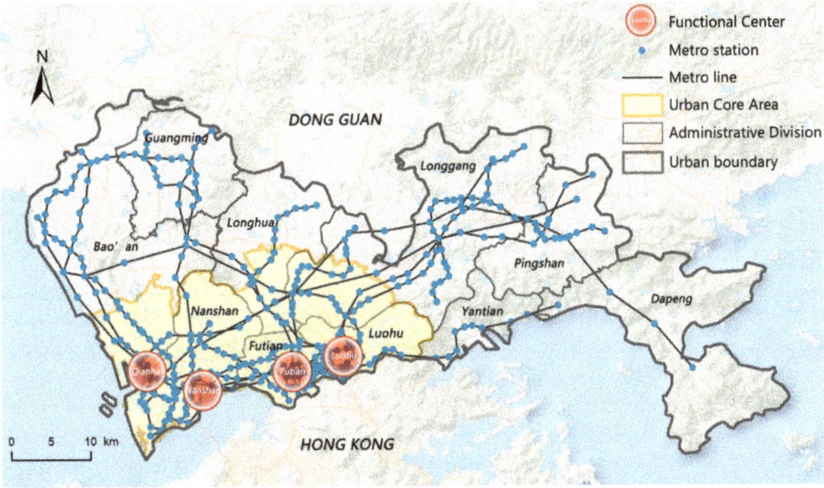

Figure 1. Study area.

3.2. Data

This study includes two main datasets: the bike-sharing datasets and the built environment datasets.

The bike-sharing records data were obtained from the Shenzhen Government open data platform, which included vehicle IDs, start and end locations (X and Y coordinates), and start and end times (accurate to the second). The data were collected during the morning peak hour (07:00–09:00 AM) over three weekdays from 24 August to 26 August 2021, which were considered representative of typical weekday bike-sharing usage patterns [8]. During this period, the average temperature ranged between 29 °C and 30 °C under cloudy conditions, which were suitable for cycling. A total of 1,186,362 records of bike-sharing trips were selected, offering a comprehensive view of cycling activity in the study area.

As for the source of built environment data, population data were obtained from Worldpop (https://hub.worldpop.org/), which provides a 100 m precision raster dataset of the population. POIs data, including residential communities, restaurants, companies, and shopping and entertainment facilities, were sourced from GaodeMap (https://lbs.amap.com/) using an API interface; metro line and station data were obtained from BaiduMap via an API interface (https://lbsyun.baidu.com/). Road information that covers all levels of roads in the study area was collected from OpenStreetMap (https://www.openstreetmap.org/), which is a reliable open data source for road-related elements [8,37]. Finally, we sourced the street-view image (SVI) data from BaiduMap with Python3.9 to measure the micro-level built environment elements of the cycling route.

3.3. Variables

3.3.1. Measure of the Dependent Variable: Bike-Sharing–Metro Integrated Use

The dependent variable was the volume of bike-sharing usage as a feeder mode on the road segment over three days, which was measured in the following three steps:

Step1: Identifying integrated usage

In this phase, a 100 m buffer was established for each metro station entry, since parking or picking up vehicles within 100 m of the station entry is considered effective feeder use [9,44]. GIS was then applied to crop out the bike-sharing data with the start and end points within the 100 m buffer, and duplicated data were deleted. Additionally, empirical studies suggest that shared bikes around metro stations are usually used for short-distance feeder trips, with travel distances usually ranging from 200 to 3000 m [52,53]. Since the original bike-sharing data did not provide direct information on travel distance, usage time was used as a proxy. Based on the average cycling speed in Shenzhen [3], trips with durations between 96 and 1440 s were selected.

Step2: Cleaning road data

According to previous travel behavior studies, the 85th percentile value of the cumulative distribution of transfer distance is often used to define the search radius. Guo and He demonstrated that the distribution of 85% cumulative transfer distances for access and egress usage of shared bikes in Shenzhen refers to 1960 m and 2040 m, respectively [9]. Therefore, in this study, an average value of 2000 m was chosen as the radius to measure the built environment elements of roads around the metro station. Based on this buffer, bike-accessible roads were then extracted and segmented at intersections, and road centerlines were derived, resulting in a total of 4568 road segments for subsequent analysis.

Step3: Calculating bike-sharing volume

Subsequently, the selected data were then imported into GIS based on the start and end points. Using the shortest path-matching tool in network analysis, the OD trajectories of bike-sharing users were reconstructed based on the actual road network. The frequency of shared bicycle passages on each road was then summarized, representing the bike-sharing flow associated with feeder trips.

3.3.2. Measure of the Independent Variable: The Built Environment

This study employs built environment factors as independent variables, which are conceptualized based on the "3Cs" framework. To comprehensively identify the built environment features across the road, a 50 m linear buffer was created for each road segment, which basically includes the roadway and its surrounding buildings, open spaces, traffic facilities, and green areas that may affect cycling behavior [54]. For population density, road density, and intersection density, we created 500 m buffers around the road centerline because we needed to consider the surrounding contexts [38]. Before the analysis, the built environment data were coordinate transformed to ensure that all the data were under the same coordinate system, before being clipped based on the mentioned 50 m or 500 m buffer.

We used the summarized built environment and travel behavior studies as a basis to construct the "3Cs" framework, which comprehensively combines both objective and subjective built environment factors [8,55,56] with commuters' demand [57]. Under this framework, the built environment variables in this study were categorized into convenience, comfort, and caution, containing 13 variables in total (Table 1).

(1) Convenience

The category of convenience includes variables related to urban density, POIs, and location, which reflect the commuting efficiency and practicality of using shared bikes as a feeder mode.

Population density serves as a direct indicator of the potential demand for feeder services [1]; and the density of employment, residential, catering, shopping, and entertainment facilities, whose spatial distributions were quantified by POI data, is expected to enhance the cycling convenience and integrated use [9]. The formula is as follows:

$$\rho = \frac{N_i}{A} \tag{1}$$

$$\rho_i = \frac{N_i}{L} \tag{2}$$

where ρ is the density of population; N_i is the number of people or the POI in the buffer; A is the area of the corresponding buffer; and L is the length of the road segment.

Table 1. The "3Cs" built environment framework.

Category	Variable	Definition
Convenience	Population density	Population density within a 500 m circular buffer around the road centroid (persons/km^2)
	Residential density	Number of residential facilities within a 50 m linear buffer per kilometer of road length (facilities/km)
	Employment density	Number of employment facilities within a 50 m linear buffer per kilometer of road length (facilities/km)
	Catering density	Number of catering facilities within a 50 m linear buffer per kilometer of road length (facilities/km)
	Shopping and entertainment density	Number of shopping and entertainment facilities within a 50 m linear buffer per kilometer of road length (facilities/km)
	Land-use mix	POI diversity within a 50 m linear buffer
	MHD	Road network centrality and accessibility calculated by the sDNA model
	Road density	Total road length within a 500 m circular buffer around the road centroid divided by the buffer area (km/km^2)
Comfort	Sky openness	% of pixels classified as the sky
	Greenery	% of pixels classified as trees
	Enclosure	% of pixels classified as buildings, columns, walls, and trees
Caution	Intersection density	Number of intersections within a 500 m circular buffer around the road centroid (intersections/km^2)
	Traffic safety infrastructure density	% of pixels classified as pillars, guardrails, and street lights

Land-use mix has been widely used to measure the functional complexity in built environment studies and is expected to promote the integrated usage of bike-sharing [58,59]. In this study, 4 primary types and 31 subtypes of POI data were used to measure this variable. Specifically, the number and type of POI in each buffer were calculated using GIS. Subsequently, the proportion of each type in each buffer was calculated to arrive at a measure of the land-use mix, also known as the Shannon–Wiener diversity index (SHDI) [33], the formula is as follows:

$$H = -\sum\nolimits_{m=1}^{n} (P_m \cdot \ln P_m) \quad (3)$$

where H is the land-use mix degree; n is the number of all POI types contained in the buffer; and P_m is the proportion of type m in the whole buffer.

Road density can reflect the optionality for cycling path [60], the formula is as follows:

$$d_{road} = \frac{L}{A} \quad (4)$$

where d_{road} is road density; L is the total length of roads in the buffer; and A is the buffer area.

MHD is computed based on the sDNA model, reflecting the centrality and accessibility of each road segment [54].

(2) Comfort

Comfort refers to cyclists' subjective perception of the built environment, specifically including sky openness, greenery, and enclosure.

Sky openness represents the portion of the sky visible from the ground [33]; low openness can lead to a sense of oppression. Greenery enhances the attractiveness of the landscape [28] and provides shade for the bike lane. Enclosure describes the sense of confinement created by vertical elements like buildings, walls, and trees that define the street edges [12,61]. In line with Jia et al., we calculated the proportion of pixels classified as buildings, columns, walls, and trees to quantify this variable [62].

To effectively measure the comfort-related variables discussed above, we extracted street-view sampling points at 100 m intervals along the road segments and retrieved street view images (SVI) from BaiduMap for each point in four directions: 0°, 90°, 180°, and 270° [37]. The size of each SVI was 512 × 512 pixels, and a total of 47,684 valid SVIs were selected. To extract these elements and calculate the ratios of the pixels classified as them over the total number of pixels in the image [38], we applied the widely used Pyramid Scene Parsing Network (PSPNet) [12,37,63]. Using the model pre-trained on the ADE20K dataset, we quantified 150 objective elements in each street segment, ranging from natural objects such as the sky, grass, and trees, to public infrastructure elements like buildings, walls, and columns [12]. Elements unrelated to the variables were excluded, and a total of 11 elements were ultimately selected. Each variable was calculated as the sum of pixel proportions for all relevant visual elements according to its definition, and the pixel proportion was averaged across all sampling points within each street segment to obtain representative values. The formula for each point is as follows:

$$\mathrm{Pr} = \frac{\sum_{i=1}^{4} \mathrm{Pixel}_{\mathrm{element}}}{\sum_{i=1}^{4} \mathrm{Pixel}_{\mathrm{total}}} \tag{5}$$

where Pr is the pixel ratio of each element, such as trees, grass, columns, and sky; $\mathrm{Pixel}_{\mathrm{element}}$ is the pixels of the element in one SVI; and $\mathrm{Pixel}_{\mathrm{total}}$ is the total pixels in this SVI.

(3) Caution

Caution refers to the objective facility configurations and road characteristics, including two variables: traffic safety infrastructure density and intersection density. Traffic infrastructure facilities such as pillars, guardrails, and street lights can regulate traffic order and ensure cycling safety, which were calculated using the same method as comfort-related variables in this study. However, too many intersections can increase the risk of conflict between cyclists and other traffic flows [9], and the calculation formula for its density is as follows:

$$\mathrm{d}_{\mathrm{Intersection}} = \frac{\mathrm{N}_{\mathrm{Intersection}}}{\mathrm{A}} \tag{6}$$

where $\mathrm{d}_{\mathrm{Intersection}}$ is the intersection density; $\mathrm{N}_{\mathrm{Intersection}}$ is the number of intersections inside the buffer; and A is the buffer area.

3.4. Models

The study first categorized all metro stations in the urban core area of Shenzhen using K-means clustering into commercial-oriented, residential-oriented, and office-oriented stations. Next, we examined the built environment and identified significant factors influencing bike-sharing usage by constructing an OLS regression model. Finally, an MGWR model was employed to further examine the spatial heterogeneity of these factors across station types.

3.4.1. K-Means Clustering Algorithm

The K-means clustering method is widely adopted for categorizing data points into discrete groups according to their similarity [64]. The process of metro station classification in this study consists of four main steps: (1) selecting and calculating the clustering variables; (2) standardizing the variables; (3) performing factor analysis; and (4) applying the K-means algorithm.

First, we classified the clustering variables into three categories: transportation characteristics, population density, and land use [17,42]. Transportation characteristics include four factors: interchange level, road length, station location, and bus stop density, while land use includes POI-based factors representing residential, commercial, and office functions. In line with Xia and Gai, we used the number of transfer routes at each station to represent the interchange level and the distance from the city center to characterize station location [46], while other variables were calculated based on an 800 m buffer around the station.

After computing the variables, the data were standardized using the Z-score method in SPSS Statistics 26 software. Factor analysis was then conducted, extracting three common factors from the original eight variables to serve as input for the K-means clustering algorithm. Based on practical considerations, the clustering result with K = 3 was ultimately selected as the classification of metro stations. Finally, as stated by Abdullahi et al. [64], the K-means clustering algorithm was applied.

3.4.2. OLS Regression Model

To identify the built environment factors that significantly influence bike-sharing–metro integrated usage around different types of metro stations, an ordinary least squares (OLS) regression model was employed. This model is widely used in built environment and travel behavior studies to explore the relationship between dependent and independent variables [7,49]. Before conducting the regression analysis, multicollinearity among the built environment variables was assessed using the variance inflation factor (VIF) to ensure the reliability of the model and avoid the influence of collinearity [7].

Subsequently, the OLS model was employed to analyze the impact of 13 built environment factors on bike-sharing and metro integrated use around three types of metro stations. The formula is as follows:

$$Y_i = \beta_0 + \sum_i \beta_i x_i + \varepsilon_i \tag{7}$$

where Y_i represents the dependent variable; x_i is the independent variable; β_0 is a constant; and β_i is the regression coefficient; and ε_i is the random error.

3.4.3. MGWR Model

The MGWR model is widely adopted when capturing spatial non-stationarity, as it accounts for the spatial heterogeneity associated with sample locations. Compared to the geographically weighted regression (GWR) model, MGWR enables the relationship between independent and dependent variables to change at different spatial scales [65,66], making it more suitable for revealing the spatial distribution of impact factors [67]. In this study, we applied MGWR2.2 software to perform the model, investigating the effects of the "3Cs" built environment factors on bike-sharing and metro trips in different geographic spaces. The specific formula is as follows:

$$y_i = \sum_{j=1}^{k} \beta_{bwj}(u_i, v_i)x_{ij} + \varepsilon_i \tag{8}$$

where in sample road segment i, y_i represents the response variable. k is the number of road segments; x_{ij} is the predictor variable j; (u_i, v_i) is centroid coordinates of street segment i; β_{bwj} is the bandwidth used in (u_i, v_i); and ε_i is the random error.

4. Results

4.1. Statistics and Spatial Characteristics of Built Environment Variables

4.1.1. Metro Station Clustering Results

Based on the K-means results, metro stations in the study area can be categorized into 120 commercial-oriented stations, 28 residential-oriented stations, and 32 office-oriented stations (Figure 2), with distinct characteristics identified for each station type as follows (Table 2).

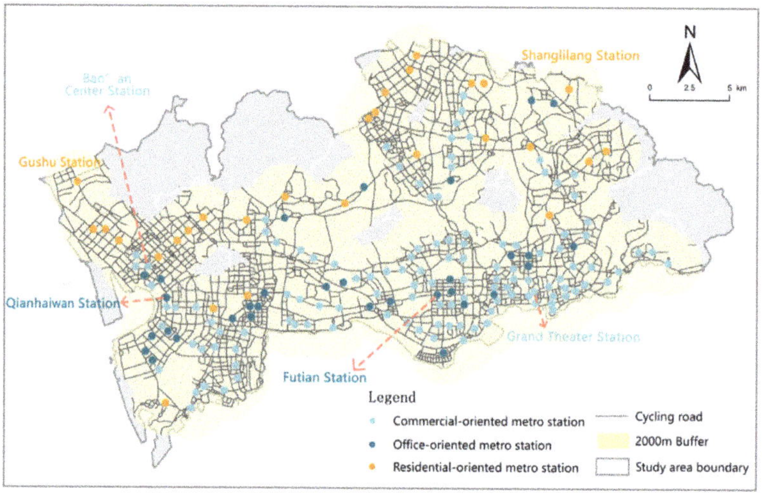

Figure 2. Spatial distribution of different station types.

Table 2. Z-scores for each cluster.

Clustering Factor	Cluster 1 (Commercial-Oriented Station)	Cluster 2 (Residential-Oriented Station)	Cluster 3 (Office-Oriented Station)
Interchange level	0.03	−0.43	0.28
Road length	0.01	−0.55	0.44
Location	0.25	−0.85	−0.19
Bus stop density	−0.28	1.72	−0.44
Population density	−0.031	1.67	−0.29
Employment density	−0.49	0.39	1.50
Shopping density	0.39	−0.16	−1.33
Residential density	0.43	−0.47	−1.20

Commercial-oriented stations are the most common type in the study area, mainly located in the southern part, including key interchange hubs such as Grand Theater Station and Bao'an Center Station, with high surrounding road density and strong transport connectivity. Surrounding these stations, POI categories such as commercial facilities (z-score: 0.40) and residential facilities (z-score: 0.42), which include both commercial and residential properties, exhibit comparatively higher z-score values.

Residential-oriented stations (e.g., Shanglilang Station and Gushu Station) are primarily located beyond the outer ring of the study area, characterized by fewer interchange stations, lower road density, and higher concentration of bus stops. The residential population (z-score: 1.67) is focused in these areas, with access flows during the morning peak. POI categories such as residential facilities (z-score: −0.46) and employment facilities (z-score: 0.39) exhibit comparatively higher values.

Office-oriented stations are mainly located around functional centers, close to commercial-oriented stations. These stations are characterized by higher interchange stations (e.g., Qianhaiwan Station and Futian Station), and dense road networks. Employment facilities (z-score: 1.50) are prominent in these regions, with large egress flows during the morning peak.

4.1.2. Spatial Characteristics of Bike-Sharing Volume and Built Environment Factors

Figure 3 presents the spatial distribution of bike-sharing–metro integrated usage across different station types. Spatially, high bike-sharing volume during the morning peak is concentrated in the Futian, Luohu, Nanshan, and Qianhai functional centers, primarily around commercial- and office-oriented stations. Furthermore, there are differences in bike-sharing usage around different types of metro stations. Specifically, during the study period, the average bike-sharing volume per road segment around office-oriented stations was 227 trips, followed closely by commercial-oriented stations with 221 trips, while residential-oriented stations had only 157 trips.

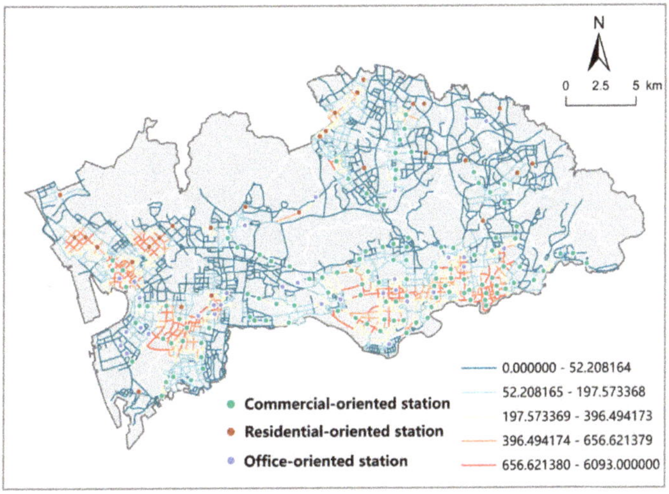

Figure 3. Bike-sharing volume around different types of metro stations.

Figures 4–6 show the spatial distribution of built environment variables under convenience, comfort, and caution categories (only some variables are shown; others are referred to in the Supplementary Materials). From the perspective of spatial distribution, most built environment variables exhibit significant spatial clustering.

Specifically, high population density and land-use mix are mainly distributed around the functional centers and the northeast of the study area, which closely resembles the spatial distribution of bike-sharing volume (Figure 4). Regarding comfort-related variables, while the spatial distribution of enclosure is relatively dispersed, areas with high sky openness are mainly located at the outer edges of the study area, where bike-sharing volume is low (Figure 5). For caution-related variables, high values of intersection density are similarly clustered around the functional centers, whereas the spatial distribution of traffic safety infrastructures does not show significant clustering (Figure 6).

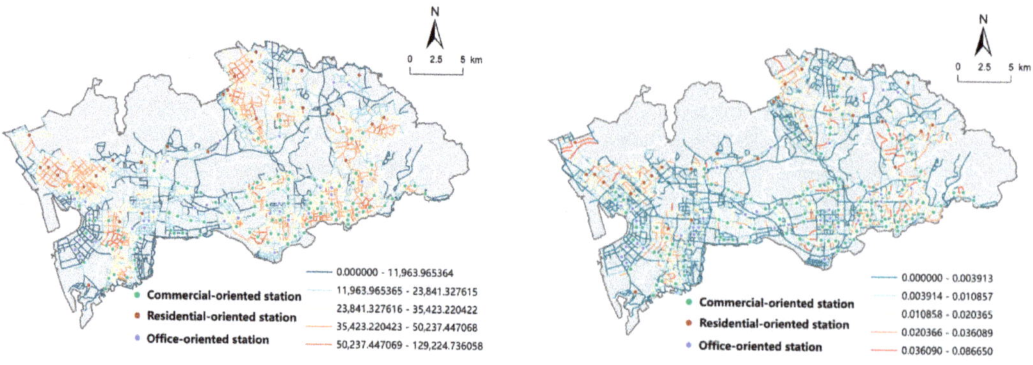

(**a**) (**b**)

Figure 4. Spatial distribution of convenience variables: (**a**) population density; (**b**) land-use mix.

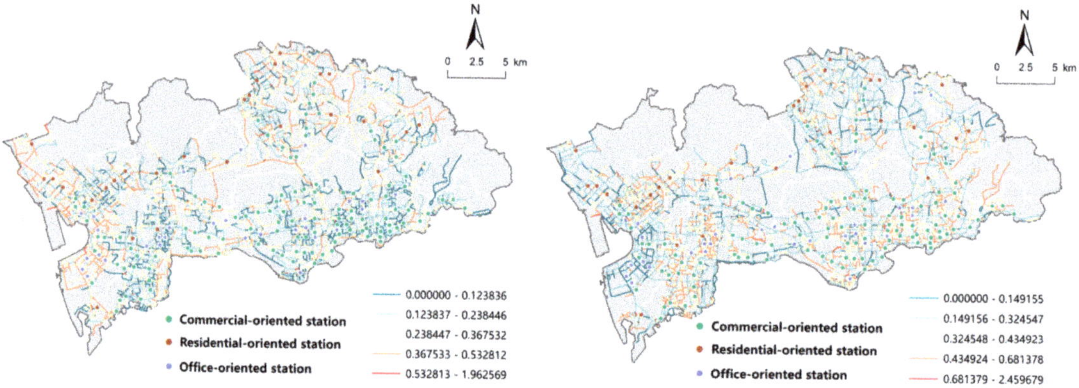

(**a**) (**b**)

Figure 5. Spatial distribution of comfort-related variables: (**a**) sky openness; (**b**) enclosure.

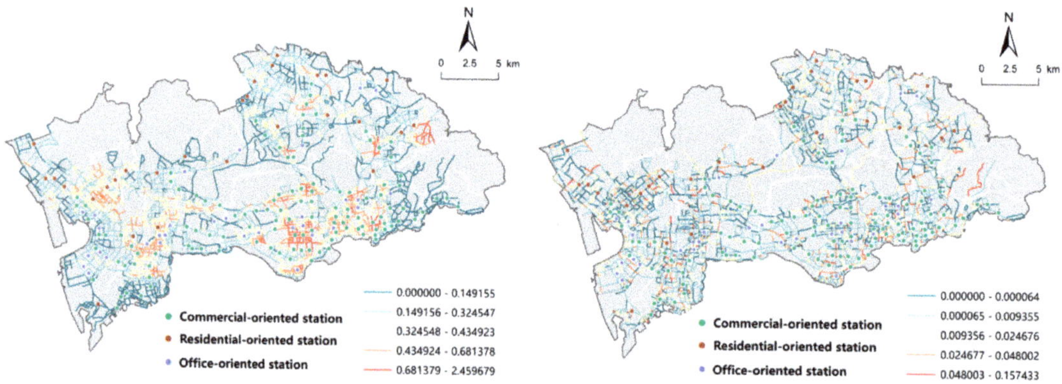

(**a**) (**b**)

Figure 6. Spatial distribution of caution-related variables: (**a**) intersection density; (**b**) traffic safety infrastructure density.

4.2. Results of the OLS Model

The OLS regression model was constructed to examine the effect of the built environment variables on bike-sharing–metro integrated usage. Before employing the model, the variable inflation factors (VIF) were used to test multicollinearity among the independent variables. Generally, a VIF value of 10 or higher indicates the presence of severe multicollinearity among the variables. In this study, all VIF values were below 4, indicating no significant multicollinearity.

Table 3 displays the results of the OLS regression model. Results show that population density, land-use mix, MHD, and road density are significant in all types of stations, indicating that the convenience variables have a strong attraction for bike-sharing feeder use, while enclosure has a significant negative impact on all stations. Meanwhile, sky openness, greenery, and traffic safety infrastructure density are not significantly related to any station type. In addition, for commercial- and office-oriented stations, most convenience-related variables have a significant impact on bike-sharing usage around the station, with eight and seven significant variables, respectively. Most POI-related variables (residential, employment, and shopping and entertainment density) exhibit negative effects on bike-sharing usage, which deviates from our previous expectations. The number of convenience-related significant variables for residential-oriented stations is notably lower, with only four showing a significant effect.

Table 3. Results of the OLS regression model.

	Commercial-Oriented Station		**Residential-Oriented Station**		**Office-Oriented Station**	
	Coef.	*p*	**Coef.**	*p*	**Coef.**	*p*
Population density	0.156 ***	0.000	0.104 ***	0.000	0.216 ***	0.000
Residential density	−0.052 ***	0.010	−0.041	0.097	−0.063 **	0.012
Employment density	0.047 **	0.014	−0.033	0.177	−0.014	0.573
Catering density	−0.065 ***	0.003	−0.003	0.905	−0.088 ***	0.001
Shopping and entertainment density	−0.046 **	0.022	−0.016	0.565	−0.071 ***	0.004
Land-use mix	0.081 ***	0.004	0.088 **	0.011	0.125 ***	0.000
MHD	0.151 ***	0.000	0.103 ***	0.000	0.165 ***	0.000
Road density	0.133 ***	0.000	0.226 ***	0.000	0.145 ***	0.000
Sky openness	−0.027	0.155	0.013	0.573	−0.006	0.804
Greenery	−0.024	0.173	−0.039	0.081	−0.042	0.056
Enclosure	−0.036 **	0.050	−0.069 ***	0.004	−0.061 ***	0.008
Intersection density	0.084 ***	0.000	−0.012	0.597	0.045	0.063
Traffic safety infrastructure density	0.012	0.477	0.017	0.402	0.01	0.618

Note: *** and ** represent the significance levels of 1% and 5%, respectively.

4.3. Results of the MGWR Model

Based on the OLS results above, we then introduced the significant variables for each station type into the MGWR model along with the bike-sharing volume to further explore the spatial heterogeneity of the effects. Before introducing the model, a spatial autocorrelation analysis was conducted on the built environment factors. The results showed that all 13 independent variables had Moran's I values greater than 0 and *p*-values less than 0.001, indicating a strong positive spatial correlation that meets the requirement for MGWR model construction.

The MGWR models of commercial-oriented stations, residential-oriented stations, and office-oriented stations are presented in Figures 7–9, respectively. The standard deviation of regression coefficients reflects spatial heterogeneity, with larger values indicating stronger heterogeneity. Overall, the MGWR results reveal spatial heterogeneity in the effects of built environment variables around metro stations, with varying impacts across different areas and station types. As for convenience-related factors, population and road density show the highest spatial heterogeneity, while land-use mix and MHD exhibit minimal variation. Enclosure shows low heterogeneity across station types and intersection density significantly affects only commercial stations, with clear spatial variation. The results for each type of station are as follows:

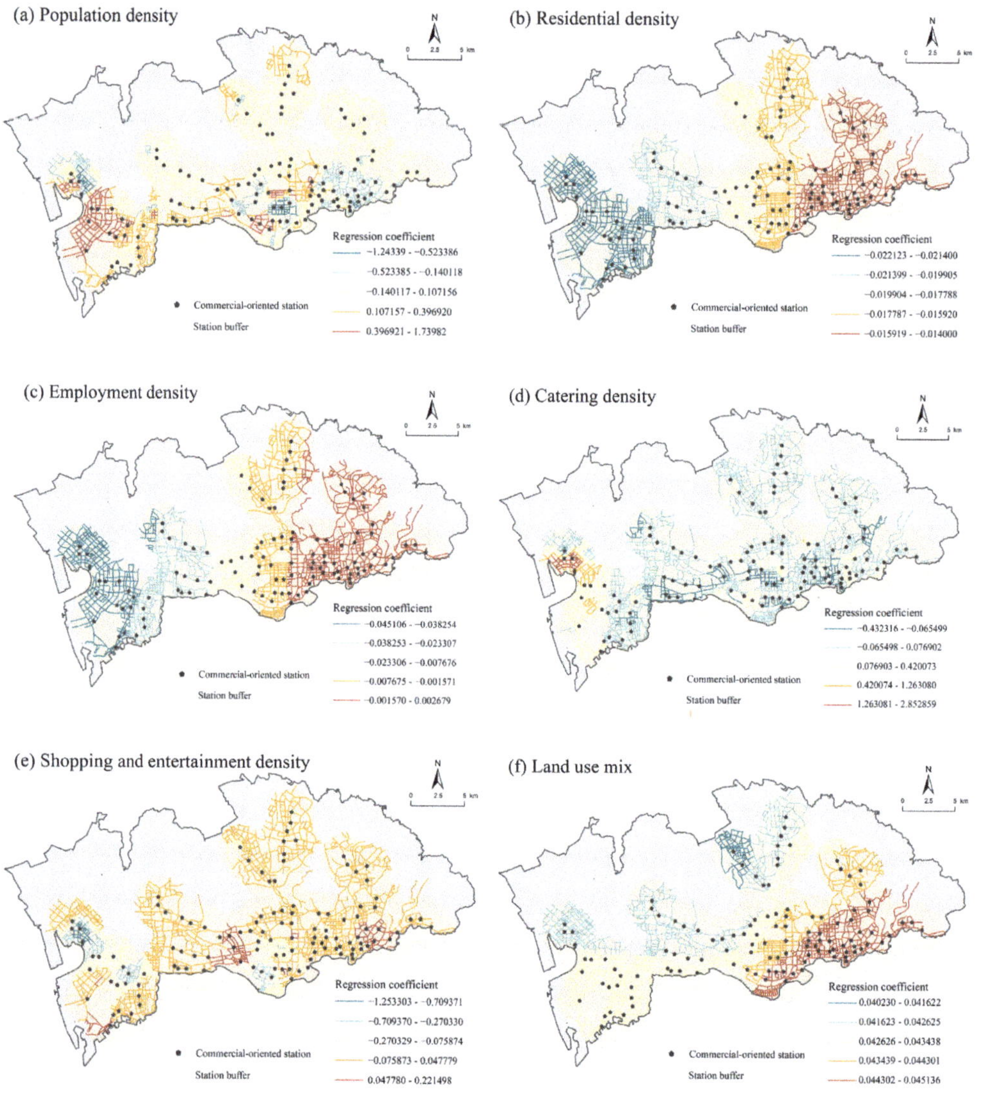

Figure 7. *Cont.*

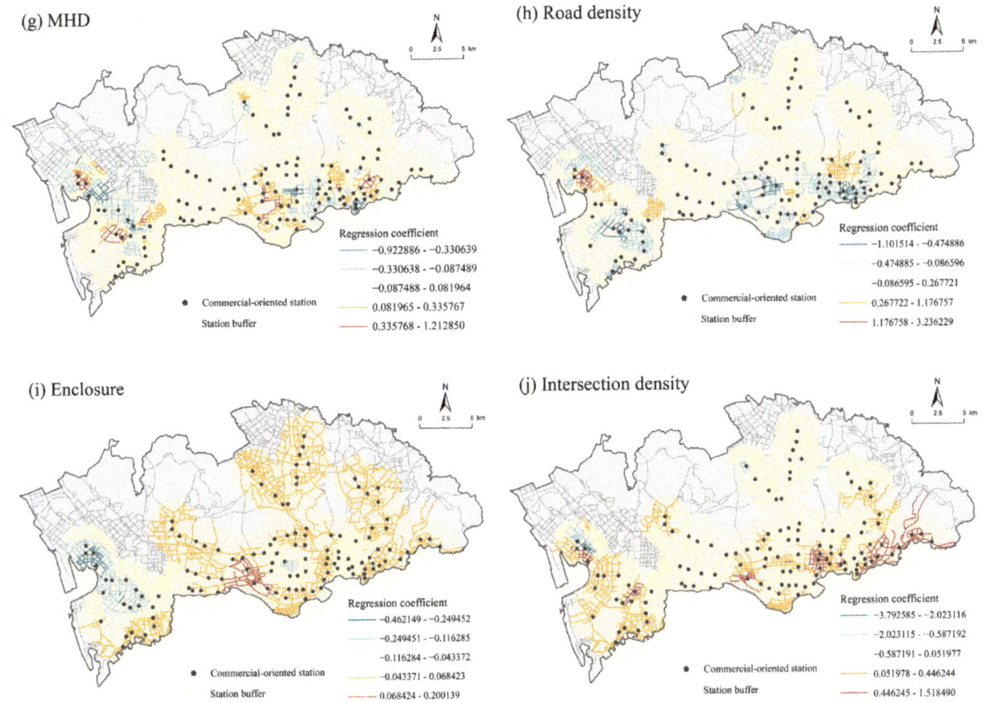

Figure 7. MGWR regression coefficient distribution for commercial-oriented stations.

Figure 7 shows the spatial distribution of regression coefficients for commercial-oriented stations. The effect of intersection density is the most significant and bidirectional, with coefficients ranging from −3.793 to 1.519, concentrated around Qianhai Center in the west and Luohu and Futian centers in the southeast. The effects of population density, catering density, and road density also show notable spatial heterogeneity.

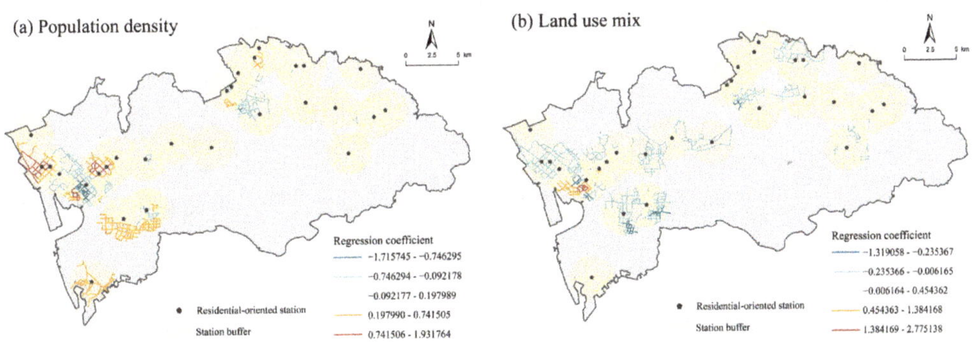

Figure 8. *Cont.*

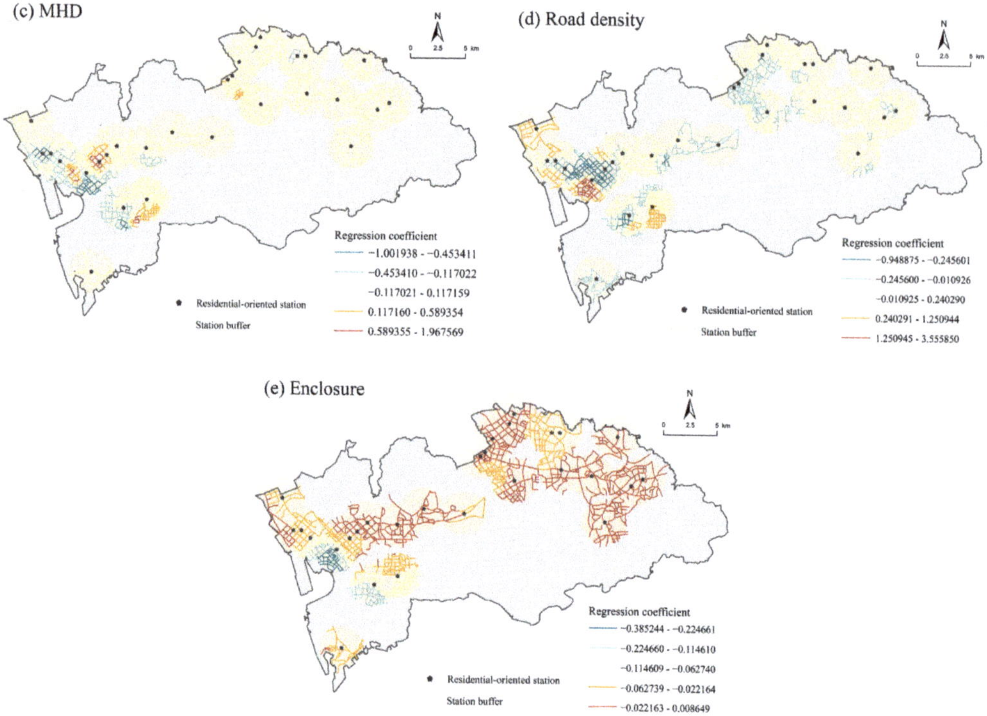

Figure 8. MGWR regression coefficient distribution for residential-oriented stations.

Figure 8 shows the spatial distribution of regression coefficients for residential-oriented stations. Road density has the most significant spatial heterogeneity, with coefficients ranging from −0.949 to 3.556, and extreme values clustered around Qianhai Center in the western part of the study area. Population density follows with coefficients between −1.716 and 1.932, showing a spatial distribution similar to road density.

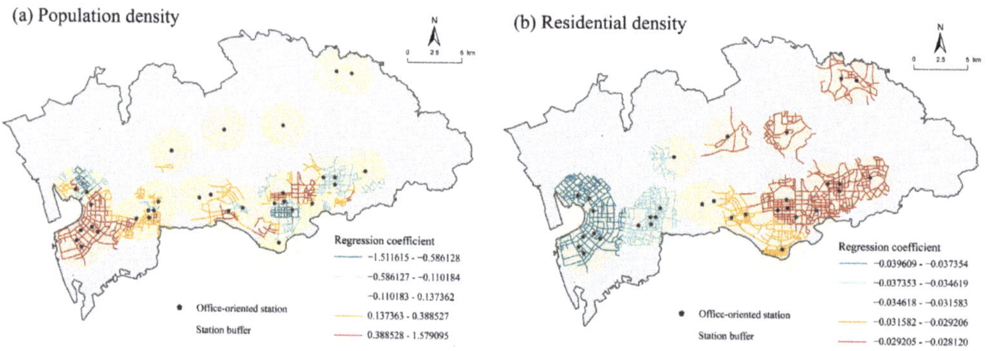

Figure 9. *Cont.*

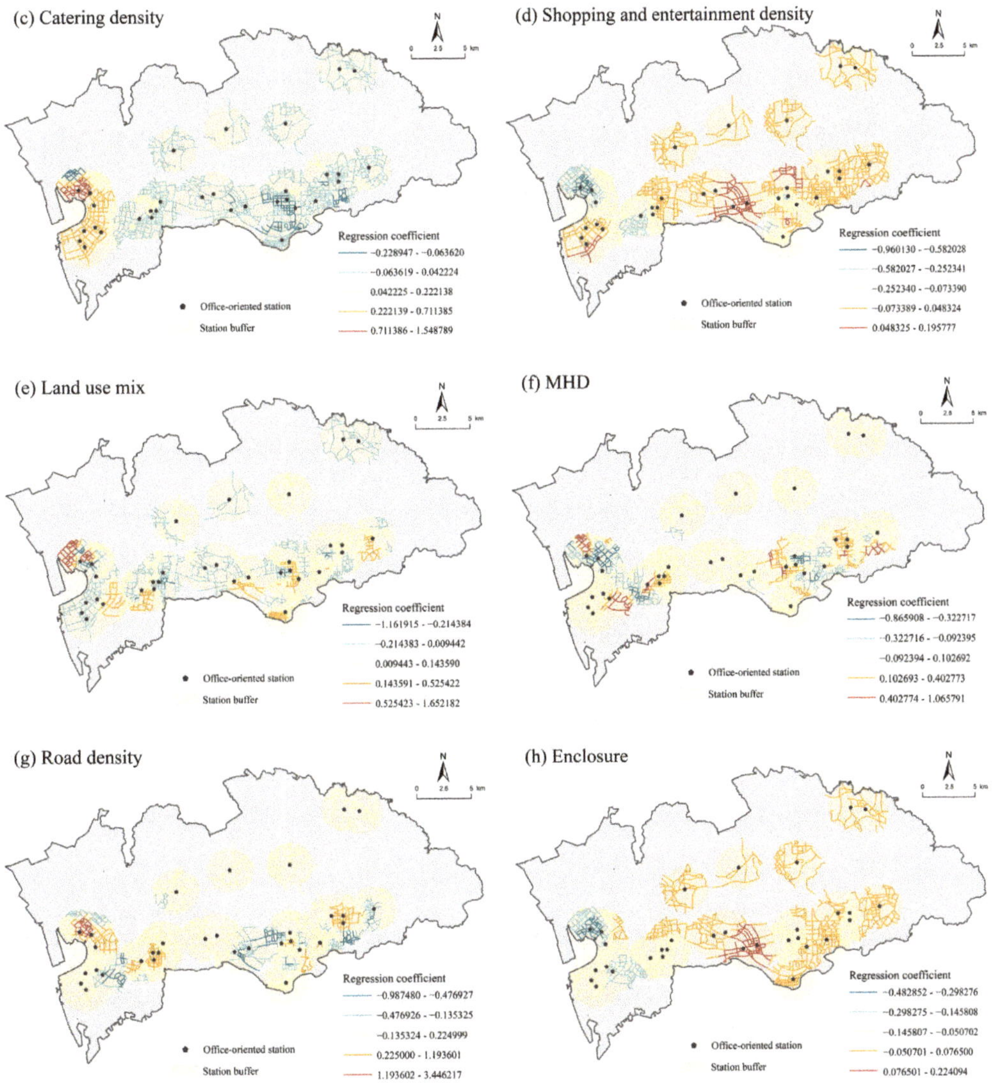

Figure 9. MGWR regression coefficient distribution for office-oriented stations.

Figure 9 shows the spatial distribution of regression coefficients for office-oriented metro stations. Among all factors, road density exhibits the most significant spatial heterogeneity, with regression coefficients ranging from −0.987 to 3.446. The low and high values are mainly concentrated around functional centers. In contrast, residential density and enclosure show the weakest spatial heterogeneity, with coefficient ranges of −0.040 to −0.028 and −0.483 to 0.224, respectively. The spatial distribution of these coefficients demonstrates a "low-in-the-west, high-in-the-east" pattern.

5. Discussion

5.1. Impact of the Built Environment on the Integrated Use

As shown in Figure 10, among three categories of built environment factors, convenience-related variables have a greater impact on the integrated use of bike-sharing and metro, which is consistent with the earlier findings [9,68]. In high-population areas, strong demand for metro services increases the use of bike-sharing for "last-mile" travel. Contrary to previous expectations [7,60], most facility indicators, including catering, shopping and entertainment, and residential density have a negative impact on bike-sharing usage. On one hand, a high density of facilities may exacerbate road congestion during the morning peak hours, thereby reducing residents' willingness to cycle. Another possible explanation is that, in this study, POI data were used to capture the distribution and attributes of facilities around metro stations. However, these data have limitations, as they do not capture the scale, quality, or operational status of facilities, causing low-attractiveness or inactive sites to be equally counted, which may misrepresent the true appeal of the surrounding environment. Finally, better spatial proximity and a well-connected street network enhance connectivity, providing cyclists with more route options.

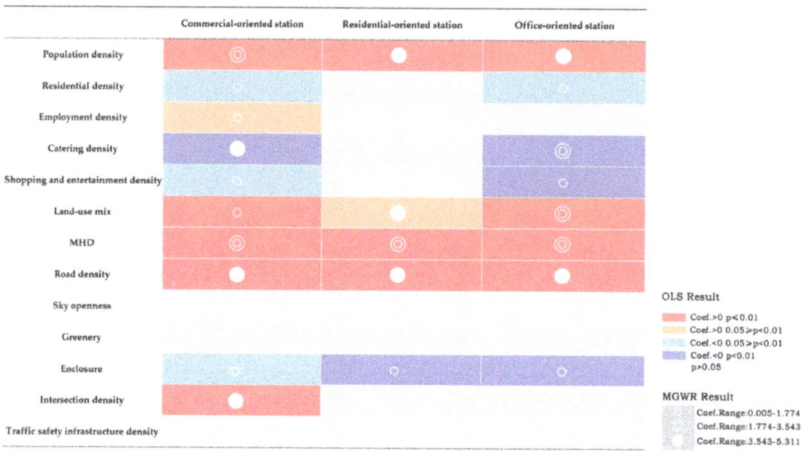

Figure 10. OLS and MGWR model results.

Conversely, comfort and caution-related factors have a minimal impact on bike-sharing use for feeder trips. Enclosure exerts a significant negative impact on shared bike usage across all station types. While a certain level of enclosure may enhance the sense of safety and vitality [32], excessive enclosure during peak hours tends to increase the sense of crowding, thereby reducing the willingness to cycle. In contrast to some prior studies, most comfort-related factors, including sky openness and greenery, have no significant effect on the dependent variable [12,37]. Although the openness and greenery of streets can enhance the attractiveness of the streetscape and improve cyclists' comfort and sense of well-being [33], during weekday morning peak hours, cycling is primarily for commuting purposes [69], with a greater emphasis on efficiency. Additionally, the bike-sharing data were collected during August, with an average temperature of approximately 29.5 °C. In such hot weather, the demand for open sky exposure among cyclists tends to be suppressed. Our study also revealed differences in the impact of built environment variables across metro station types. As supported by earlier studies [9,26], excessive residential, shopping and entertainment, and catering facilities reduce bike use around office-oriented and commercial-oriented stations during the morning peak, while having no significant impact on bike-sharing usage around residential-oriented stations. This may be attributed to differences in user travel behavior: commuters near office and commercial stations typically prioritize efficiency and time savings [17], as bike-sharing is often used for the "last mile" of their commute. Moreover, high residential density may intensify the pressure on cycling roads or cause a shortage of shared bikes, leading more people to choose walking or bus connections instead. This finding suggests that while mixed land uses along cycling routes can support access and vitality, excessive facility density may reduce travel efficiency. Finally, intersection density significantly enhances bike-sharing–metro integrated usage only around commercial-oriented stations. A possible explanation is that commercial areas generate diverse travel demands (e.g., commuting, early shopping, and deliveries), where a higher number of intersections provides more route options and better network connectivity.

5.2. Spatial Heterogeneity of the Impact of Built Environment on the Integrated Use

Figure 10 demonstrates the spatial heterogeneity of the impact of built environment variables on different types of stations. For commercial-oriented stations, intersection density exhibits the strongest spatial heterogeneity in its effect on the dependent variable. Although an earlier study demonstrated that intersections can attract more bike-sharing users [56], the MGWR result revealed spatial non-stationarity. Specifically, intersection density shows a strong positive effect around the four functional centers in the southern region, while in the northern areas, the promoting effect is relatively weak and may even become suppressive on some roads. The reason for this is that for high commercial density areas, intersections can improve roadway access and commuting efficiency; however, in lower traffic areas, excessive intersections can increase the length of unnecessary waiting time to cross the street.

For residential-oriented stations, road density exhibits the most noticeable spatial variation among all variables, with both significant positive and negative impacts occurring in areas with high road density (around the Qianhai functional center). These stations are located near densely populated communities with numerous schools, hospitals, and service facilities, resulting in strong demand for cycling-friendly environments. In areas with sparse road networks, higher road density can enhance connectivity and accessibility, while in traffic-congested zones, it may also lead to more intersections, resulting in longer waiting times and greater safety risks for cyclists [10]. Similarly, population density, land-use mix, and MHD show spatial non-stationarity, with the strongest positive and negative effects also clustered around the Qianhai functional center. This may be because high facility density draws large populations, resulting in congestion that simultaneously increases cycling demand and impedes its efficiency. Compared to other station types, the effect of enclosure on residential-oriented stations shows less spatial variation. For most residential stations, enclosure has a suppressive effect on bike-sharing usage, possibly indicating that commuters in residential areas place a higher value on cycling comfort and that highly enclosed streets may intensify the feeling of congestion, thus reducing their willingness to cycle.

For office-oriented stations, the results for population density, road density, and POI-related variables are similar to those for commercial-oriented stations. However, it is interesting to note that, the effect of land-use mix on feeder trips varies spatially among office-oriented stations, which is inconsistent with previous studies [26,70]. Specifically, this factor shows a significant positive effect around the functional centers of Futian, Luohu, Nanshan, and Qianhai, but exhibits a negative impact in the northern areas with relatively low bike-sharing volume. For core regions, a high functional mix allows residents' living, working, and service needs to be met, reducing commuting distances and attracting more commuters to ride to their workplaces. In contrast, in non-core areas, a high functional mix may not generate strong biking demand due to insufficient infrastructure, such as limited bike-sharing stations and poor cycling roads [49], preventing POIs from converting into cycling needs.

6. Conclusions

This study explored how the built environment affects bike-sharing–metro integration during morning peak hours and the spatial heterogeneity of these influences, by employing the OLS regression model and the MGWR model. Particular emphasis has been given to exploring this relationship with a focus on various metro station types that are clustered based on transportation characteristics, population density, and land use. The key findings of the study are summarized as follows: (1) Bike-sharing as a metro feeder mode shows significant spatial agglomeration in Shenzhen's urban core area. There are notable differences in bike-sharing volume across different station types. Specifically, during the morning peak period, bike-sharing volume is significantly higher around office-oriented and commercial-oriented stations compared to residential-oriented stations. (2) The spatial distribution of built environment variables is also closely related to metro station types and shows a high degree of correlation with the spatial pattern of bike-sharing usage. Convenience-related variables, such as population density, POI-related factors, and land-use mix, are concentrated around commercial- and office-oriented stations, where bike volume is high. Comfort-related variables are mainly centered around residential stations, while caution-related variables are more dispersed. (3) The built environment around different types of metro stations has different impacts on cycling behavior. Population density, road density, MHD, and functional mixing show significant positive effects for all types of stations, while enclosure generally inhibits feeder behavior. Unlike expectations, POI-related variables do not promote feeder rides around all types of stations and most comfort-related factors have a weak effect on morning peak bike-sharing feeder use. (4) The influence of the built environment on bike-sharing usage exhibits significant spatial heterogeneity. Specifically, population density and road density exert strong spatial non-stationarity across all types of metro stations, showing both positive and negative effects in different areas, while enclosure consistently has a negative effect in most regions. The areas with significant impacts are generally concentrated around the four functional centers mentioned above.

This study also presents categorized planning recommendations for various types of metro stations, emphasizing differentiated strategies for central areas and low bike-sharing volume areas. For commercial-oriented stations, enhancing the functional mix while ensuring appropriate facility density is essential. Commuter-focused amenities, such as convenience stores and restaurants, can be added without overcrowding the street. Additionally, intersection and road management in core zones is crucial for maintaining traffic order, along with ensuring an adequate supply of shared bicycles. For residential-oriented stations, planning efforts can focus on improving road density and connectivity. Furthermore, attention should also be paid to the psychological comfort of cyclists in the areas surrounding residential neighborhoods by reducing the sense of enclosure created by walls, buildings, and fences and enhancing micro-level street design to support a more comfortable cycling environment. Finally, as for office-oriented stations, it is recommended to improve traffic organization in areas with dense road networks, plan surrounding dining, shopping, and entertainment facilities more rationally, and enhance road connectivity in northern areas to improve the efficiency of last-mile cycling.

It is important to acknowledge several limitations of this study. First, certain variables within the "3Cs" framework may have been overlooked. Future research is encouraged to incorporate factors such as competitiveness and commuters' socioeconomic characteristics. Furthermore, greater emphasis should be placed on psychological indicators, including cyclists' perceived comfort, security, and aesthetic appreciation of the cycling environment. Second, this study used POI data to analyze the distribution of catering, shopping and entertainment, employment and residential facilities as well as land-use mix around metro stations. However, POI data are static and lack information on temporal changes, facility size, or quality, providing only a partial view of the built environment's influence. Future research could incorporate more detailed and dynamic datasets, including land use, qualitative factors, and environmental audits, to better capture the characteristics of the facilities. Finally, in terms of methodology, this study primarily analyzes the linear relationships between the independent and dependent variables, as well as the spatial heterogeneity of these effects. The exploration of potential nonlinear relationships remains an area for further investigation in future research.

Supplementary Materials: The following supporting information can be downloaded at: https://www.mdpi.com/article/10.3390/land14061291/s1, Figure S1: Spatial characteristics of built environment factors; Table S1: MGWR model fitting goodness; Table S2: MGWR results for commercial-oriented station; Table S3: MGWR results for residential-oriented station; Table S4: MGWR results for office-oriented station.

Author Contributions: Conceptualization, Y.L. and J.L.; methodology, Y.L., J.L., Z.Y., S.L. and A.L.; software, Y.L. and Z.Y.; validation, J.L.; formal analysis, Y.L., Z.Y. and S.L.; data curation, Z.Y.; writing—original draft preparation, Y.L., Z.Y., S.L. and J.L.; writing—review and editing, J.L., A.L. and Y.L.; visualization, Y.L., Z.Y. and S.L.; supervision, J.L.; project administration, J.L.; funding acquisition, J.L. and A.L. All authors have read and agreed to the published version of the manuscript.

Funding: This research was jointly supported by the Fundamental Research Funds for the Central Universities (No. 2024XKRCW007) and the Young Scientists Fund of the National Natural Science Foundation of China (No. 52202389).

Data Availability Statement: Data are available from the corresponding authors upon request.

Conflicts of Interest: The authors declare no conflicts of interest.

References

1. Guo, Y.; Yang, L.; Chen, Y. Bike Share Usage and the Built Environment: A Review. *Front. Public Health* **2022**, *10*, 848169. [CrossRef] [PubMed]
2. Zuo, T.; Wei, H.; Chen, N.; Zhang, C. First-and-last mile solution via bicycling to improving transit accessibility and advancing transportation equity. *Cities* **2020**, *99*, 102614. [CrossRef]
3. Shenzhen Municipal Transportation Bureau. *Shenzhen Bicycle Traffic Development Plan (2021–2035)*; Shenzhen Municipal Transportation Bureau: Shenzhen, China, 2022.
4. Cheng, L.; Jin, T.; Wang, K.; Lee, Y.; Witlox, F. Promoting the integrated use of bikeshare and metro: A focus on the nonlinearity of built environment effects. *Multimodal Transp.* **2022**, *1*, 100004. [CrossRef]
5. Zhan, Z.; Guo, Y.; Noland, R.B.; He, S.Y.; Wang, Y. Analysis of links between dockless bikeshare and metro trips in Beijing. *Transp. Res. Part A Policy Pract.* **2023**, *175*, 103784. [CrossRef]
6. Cheng, L.; Huang, J.; Jin, T.; Chen, W.; Li, A.; Witlox, F. Comparison of station-based and free-floating bikeshare systems as feeder modes to the metro. *J. Transp. Geogr.* **2023**, *107*, 103545. [CrossRef]
7. Sun, Y.; Wang, Y.; Wu, H. How does the urban built environment affect dockless bikesharing-metro integration cycling? Analysis from a nonlinear comprehensive perspective. *J. Clean. Prod.* **2024**, *449*, 141770. [CrossRef]
8. Guo, Y.; He, S.Y. The role of objective and perceived built environments in affecting dockless bike-sharing as a feeder mode choice of metro commuting. *Transp. Res. Part A Policy Pract.* **2021**, *149*, 377–396. [CrossRef]
9. Guo, Y.; He, S.Y. Built environment effects on the integration of dockless bike-sharing and the metro. *Transp. Res. Part D Transp. Environ.* **2020**, *83*, 102335. [CrossRef]
10. Guo, Y.; He, S.Y. Perceived built environment and dockless bikeshare as a feeder mode of metro. *Transp. Res. Part D Transp. Environ.* **2021**, *92*, 102693. [CrossRef]
11. Yin, Z.; Guo, Y.; Zhou, M.; Wang, Y.; Tang, F. Integration between Dockless Bike-Sharing and Buses: The Effect of Urban Road Network Characteristics. *Land* **2024**, *13*, 1209. [CrossRef]
12. Song, Q.; Huang, Y.; Li, W.; Chen, F.; Qiu, W. Unraveling the effects of micro-level street environment on dockless bikeshare in Ithaca. *Transp. Res. Part D Transp. Environ.* **2024**, *132*, 104256. [CrossRef]
13. Caigang, Z.; Shaoying, L.; Zhangzhi, T.; Feng, G.; Zhifeng, W. Nonlinear and threshold effects of traffic condition and built environment on dockless bike sharing at street level. *J. Transp. Geogr.* **2022**, *102*, 103375. [CrossRef]

14. Ethier, B.G.; Wilson, J.S.; Camhi, S.M.; Shi, L.; Troped, P.J. An analysis of built environment characteristics in daily activity spaces and associations with bike share use. *Travel Behav. Soc.* **2024**, *37*, 100850. [CrossRef]
15. Dong, S.; Wang, Y.; Dou, M.; Wang, C.; Gong, J. Uncovering the similarity and heterogeneity of metro stations: From passenger mobility, land use, and streetscapes semantics. *Appl. Geogr.* **2025**, *174*, 103450. [CrossRef]
16. Chen, L.; Chen, Y.; Wang, Y.; Li, Y. Research on the Classification of Rail Transit Stations and Passenger Flow Patterns—A Case from Xi'an, China. *Buildings* **2024**, *14*, 1068. [CrossRef]
17. Shi, Y.; Zeng, L. How do built environment characteristics influence metro-bus transfer patterns across metro station types in Shanghai? *J. Transp. Geogr.* **2025**, *123*, 104137. [CrossRef]
18. Wu, T.; Li, M.; Gao, L.; Zhou, Y. Integrating spatial vitality and node-place model to evaluate and classify metro station areas in Wuhan. *Front. Archit. Res.* **2024**, *13*, 1363–1376. [CrossRef]
19. Guan, H.; JI, X.; LI, W.; Chen, F.; Deng, R. Influence of Built Environment on Integrated Use of Bike Sharing and Metro. *J. Transp. Syst. Eng. Inf. Technol.* **2024**, *24*, 200–211. [CrossRef]
20. Zhao, P.; Li, S. Bicycle-metro integration in a growing city: The determinants of cycling as a transfer mode in metro station areas in Beijing. *Transp. Res. Part A Policy Pract.* **2017**, *99*, 46–60. [CrossRef]
21. Cervero, R.; Kockelman, K. Travel demand and the 3Ds: Density, diversity, and design. *Transp. Res. Part D Transp. Environ.* **1997**, *2*, 199–219. [CrossRef]
22. Cervero, R.; Sarmiento, O.L.; Jacoby, E.; Gomez, L.F.; Neiman, A. Influences of Built Environments on Walking and Cycling: Lessons from Bogotá. *Int. J. Sustain. Transp.* **2009**, *3*, 203–226. [CrossRef]
23. Munshi, T. Built environment and mode choice relationship for commute travel in the city of Rajkot, India. *Transp. Res. Part D Transp. Environ.* **2016**, *44*, 239–253. [CrossRef]
24. Reid, E.; Robert, C. Travel and the Built Environment: A Meta-Analysis. *J. Am. Plan. Assoc.* **2010**, *76*, 265–294. [CrossRef]
25. Ni, Y.; Chen, J. Exploring the Effects of the Built Environment on Two Transfer Modes for Metros: Dockless Bike Sharing and Taxis. *Sustainability* **2020**, *12*, 2034. [CrossRef]
26. Bi, H.; Li, A.; Hua, M.; Zhu, H.; Ye, Z. Examining the varying influences of built environment on bike-sharing commuting: Empirical evidence from Shanghai. *Transp. Policy* **2022**, *129*, 51–65. [CrossRef]
27. Zhang, Y.; Thomas, T.; Brussel, M.; van Maarseveen, M. Exploring the impact of built environment factors on the use of public bikes at bike stations: Case study in Zhongshan, China. *J. Transp. Geogr.* **2017**, *58*, 59–70. [CrossRef]
28. Jeon, J.; Woo, A. The effects of built environments on bicycle accidents around bike-sharing program stations using street view images and deep learning techniques: The moderating role of streetscape features. *J. Transp. Geogr.* **2024**, *121*, 103992. [CrossRef]
29. Sarkar, P.P.; Mallikarjuna, C. Effect of perception and attitudinal variables on mode choice behavior: A case study of Indian city, Agartala. *Travel Behav. Soc.* **2018**, *12*, 108–114. [CrossRef]
30. He, S.Y.; Thøgersen, J. The impact of attitudes and perceptions on travel mode choice and car ownership in a Chinese megacity: The case of Guangzhou. *Res. Transp. Econ.* **2017**, *62*, 57–67. [CrossRef]
31. Chan, E.T.H.; Schwanen, T.; Banister, D. The role of perceived environment, neighbourhood characteristics, and attitudes in walking behaviour: Evidence from a rapidly developing city in China. *Transportation* **2019**, *48*, 431–454. [CrossRef]
32. Wang, R.; Lu, Y.; Zhang, J.; Liu, P.; Yao, Y.; Liu, Y. The relationship between visual enclosure for neighbourhood street walkability and elders' mental health in China: Using street view images. *J. Transp. Health* **2019**, *13*, 90–102. [CrossRef]
33. Bai, Y.; Bai, Y.; Wang, R.; Yang, T.; Song, X.; Bai, B. Exploring Associations between the Built Environment and Cycling Behaviour around Urban Greenways from a Human-Scale Perspective. *Land* **2023**, *12*, 619. [CrossRef]
34. Dai, L.; Zheng, C.; Dong, Z.; Yao, Y.; Wang, R.; Zhang, X.; Ren, S.; Zhang, J.; Song, X.; Guan, Q. Analyzing the correlation between visual space and residents' psychology in Wuhan, China using street-view images and deep-learning technique. *City Environ. Interact.* **2021**, *11*, 100069. [CrossRef]
35. Dong, R.; Zhang, Y.; Zhao, J. How Green Are the Streets Within the Sixth Ring Road of Beijing? An Analysis Based on Tencent Street View Pictures and the Green View Index. *Int. J. Environ. Res. Public Health* **2018**, *15*, 1367. [CrossRef]
36. Lambert, G.W.; Reid, C.; Kaye, D.M.; Jennings, G.L.; Esler, M.D. Effect of sunlight and season on serotonin turnover in the brain. *Lancet* **2002**, *360*, 1840–1842. [CrossRef] [PubMed]
37. Bai, Y.; Cao, M.; Wang, R.; Liu, Y.; Wang, S. How street greenery facilitates active travel for university students. *J. Transp. Health* **2022**, *26*, 101393. [CrossRef]
38. Ito, K.; Biljecki, F. Assessing bikeability with street view imagery and computer vision. *Transp. Res. Part C Emerg. Technol.* **2021**, *132*, 103371. [CrossRef]
39. Feng, H.; Chen, Y.; Wu, J.; Zhao, Z.; Wang, Y.; Wang, Z. Urban Rail Transit Station Type Identification Based on "Passenger Flow—Land Use—Job-Housing". *Sustainability* **2023**, *15*, 15103. [CrossRef]
40. Wang, Y.; Li, J.; Yang, X.; Guo, Y.; Ren, J.; Zhan, Z. Examining the impact of station location on dockless bikesharing-metro integration: Evidence from Beijing. *Travel Behav. Soc.* **2024**, *37*, 100835. [CrossRef]

41. Li, X.; Yan, Q.; Ma, Y.; Luo, C. Spatially Varying Impacts of Built Environment on Transfer Ridership of Metro and Bus Systems. *Sustainability* **2023**, *15*, 7891. [CrossRef]
42. Wang, H.; Ma, H. Classification Method of Urban Rail Transit Stations Based on POI. *Traffic Transp.* **2020**, *36*, 33–37. [CrossRef]
43. Pan, H.; Shen, Q.; Xue, S. Intermodal Transfer Between Bicycles and Rail Transit in Shanghai, China. *Transp. Res. Rec. J. Transp. Res. Board* **2010**, *2144*, 181–188. [CrossRef]
44. Guo, Y.; Yang, L.; Lu, Y.; Zhao, R. Dockless bike-sharing as a feeder mode of metro commute? The role of the feeder-related built environment: Analytical framework and empirical evidence. *Sustain. Cities Soc.* **2021**, *65*, 102594. [CrossRef]
45. Yang, S.; Peng, C.; Hu, S.; Zhang, P. Geospatial modelling of housing rents from TOD using MGWR and implications on integrated transportation-land use planning. *Appl. Geogr.* **2024**, *170*, 103356. [CrossRef]
46. Xia, X.; Gai, J. Classification of urban rail transit stations and points and analysis of passenger flow characteristics based on K-Means clustering algorithm. *Mod. Urban Transit.* **2021**, *4*, 112–118. [CrossRef]
47. Wang, R.; Lu, Y.; Wu, X.; Liu, Y.; Yao, Y. Relationship between eye-level greenness and cycling frequency around metro stations in Shenzhen, China: A big data approach. *Sustain. Cities Soc.* **2020**, *59*, 102201. [CrossRef]
48. Zheng, Z.; Zhang, J.; Zhang, L.; Li, M.; Rong, P.; Qin, Y. Understanding the impact of the built environment on ride-hailing from a spatio-temporal perspective: A fine-scale empirical study from China. *Cities* **2022**, *126*, 103706. [CrossRef]
49. Li, A.; Zhao, P.; Huang, Y.; Gao, K.; Axhausen, K.W. An empirical analysis of dockless bike-sharing utilization and its explanatory factors: Case study from Shanghai, China. *J. Transp. Geogr.* **2020**, *88*, 102828. [CrossRef]
50. Brunsdon, C.; Fotheringham, A.S.; Charlton, M.E. Geographically Weighted Regression: A Method for Exploring Spatial Nonstationarity. *Geogr. Anal.* **1996**, *28*, 281–298. [CrossRef]
51. Fatima, M.; Ünsal, Ö. Geospatial analysis of diarrhoea determinants among children under five in Pakistan using Multiscale Geographically Weighted Regression (MGWR). *Appl. Geogr.* **2024**, *172*, 103434. [CrossRef]
52. Ma, X.; Ji, Y.; Yuan, Y.; Van Oort, N.; Jin, Y.; Hoogendoorn, S. A comparison in travel patterns and determinants of user demand between docked and dockless bike-sharing systems using multi-sourced data. *Transp. Res. Part A Policy Pract.* **2020**, *139*, 148–173. [CrossRef]
53. Liang, M.; Wang, G. A Study of Parking Space of Shared Bikes Near Rail Transit Stations, Chengdu. *Planners* **2019**, *35*, 54–61. [CrossRef]
54. Xiang, Z.; Li, Q.; Hong, L.; Sheng, J.; Ban, P. Effects of Built Environment on the Spatio-Temporal Trajectories of Shared Bicycles: A Case Study of Shenzhen. *Trop. Geogr.* **2024**, *44*, 236–247. [CrossRef]
55. Shen, H.; Weng, J.; Lin, P. Exploring the nuanced correlation between built environment and the integrated travel of dockless bike-sharing and metro at origin-route-destination level. *Sustain. Cities Soc.* **2025**, *119*, 106090. [CrossRef]
56. Zhang, Y.; Chen, X.-J.; Gao, S.; Gong, Y.; Liu, Y. Integrating smart card records and dockless bike-sharing data to understand the effect of the built environment on cycling as a feeder mode for metro trips. *J. Transp. Geogr.* **2024**, *121*, 103995. [CrossRef]
57. Mitra, R.; Schofield, J. Biking the First Mile: Exploring a Cyclist Typology and Potential for Cycling to Transit Stations by Suburban Commuters. *Transp. Res. Rec. J. Transp. Res. Board* **2019**, *2673*, 951–962. [CrossRef]
58. Mateo-Babiano, I.; Bean, R.; Corcoran, J.; Pojani, D. How does our natural and built environment affect the use of bicycle sharing? *Transp. Res. Part A Policy Pract.* **2016**, *94*, 295–307. [CrossRef]
59. Jiao, J.; Rollo, J.; Fu, B.; Liu, C. Exploring Effective Built Environment Factors for Evaluating Pedestrian Volume in High-Density Areas: A New Finding for the Central Business District in Melbourne, Australia. *Land* **2021**, *10*, 655. [CrossRef]
60. Chen, E.; Ye, Z. Identifying the nonlinear relationship between free-floating bike sharing usage and built environment. *J. Clean. Prod.* **2021**, *280*, 124281. [CrossRef]
61. Ewing, R.; Handy, S. Measuring the Unmeasurable: Urban Design Qualities Related to Walkability. *J. Urban Des.* **2009**, *14*, 65–84. [CrossRef]
62. Jia, J.; Wang, Z.; Li, L. Research on the Correlation between Road Visual Features and Cycling Traffic Based on Multi-source Big Data and Deep Learning. *Chin. Landsc. Archit.* **2023**, *39*, 84–90. [CrossRef]
63. Helbich, M.; Yao, Y.; Liu, Y.; Zhang, J.; Liu, P.; Wang, R. Using deep learning to examine street view green and blue spaces and their associations with geriatric depression in Beijing, China. *Environ. Int.* **2019**, *126*, 107–117. [CrossRef] [PubMed]
64. Abdullahi, M.R.; Lu, Q.-C.; Hussain, A.; Tripura, S.; Xu, P.-C.; Wang, S. Location optimization of EV charging stations: A custom K-means cluster algorithm approach. *Energy Rep.* **2024**, *12*, 5367–5382. [CrossRef]
65. Zhao, J.; Guo, F.; Zhang, H.; Dong, J. Mechanisms of non-stationary influence of urban form on the diurnal thermal environment based on machine learning and MGWR analysis. *Sustain. Cities Soc.* **2024**, *101*, 105194. [CrossRef]
66. Liu, H.; Huang, B.; Gao, S.; Wang, J.; Yang, C.; Li, R. Impacts of the evolving urban development on intra-urban surface thermal environment: Evidence from 323 Chinese cities. *Sci. Total Environ.* **2021**, *771*, 144810. [CrossRef]
67. Shabrina, Z.; Buyuklieva, B.; Ng, M.K.M. Short-Term Rental Platform in the Urban Tourism Context: A Geographically Weighted Regression (GWR) and a Multiscale GWR (MGWR) Approaches. *Geogr. Anal.* **2020**, *53*, 686–707. [CrossRef]

68. Lin, J.-J.; Zhao, P.; Takada, K.; Li, S.; Yai, T.; Chen, C.-H. Built environment and public bike usage for metro access: A comparison of neighborhoods in Beijing, Taipei, and Tokyo. *Transp. Res. Part D Transp. Environ.* **2018**, *63*, 209–221. [CrossRef]
69. Liu, Y.; Zhang, Y.; Jin, S.T.; Liu, Y. Spatial pattern of leisure activities among residents in Beijing, China: Exploring the impacts of urban environment. *Sustain. Cities Soc.* **2020**, *52*, 101806. [CrossRef]
70. Shen, Y.; Zhang, X.; Zhao, J. Understanding the usage of dockless bike sharing in Singapore. *Int. J. Sustain. Transp.* **2018**, *12*, 686–700. [CrossRef]

land

MDPI

Article

Urban Renewal Strategy Guided by Rail Transit Development Based on the "Node–Place–Revenue" Model: Case Study of Shenyang Metro Line 1

Xu Lu *, Mengqin Zhu, Zeting Li, Qingyu Li and Shan Huang

School of Architecture and Urban Planning, Shenyang Jianzhu University, Shenyang 110168, China; huangshan@stu.sjzu.edu.cn (S.H.)

* Correspondence: luxupku@163.com

Abstract: Under the backdrop of urban renewal, harmonizing transit-oriented development (TOD) with urban renewal to maximize rail value has emerged as a critical focus in contemporary planning. Based on this, this paper proposes the node–place–revenue (NPR) model, which constructs evaluation indexes from the three dimensions of the node, place, and revenue. It determines the weights of each index by using expert scoring and the Analytic Hierarchy Process (AHP). Taking Shenyang Metro Line 1 as an example, the study first used the model to measure the node value, place value, and revenue value of each sample TOD station area. Secondly, K-means clustering analysis was used to form a spatial classification of five station areas. Finally, this paper proposes one differentiated urban renewal strategy for each type of station area. It is found that (1) the NPR model classifies stations into five categories: stress and high revenue, balanced, unbalanced node, unbalanced place, and dependence and low revenue and (2) the differentiated urban renewal strategies for each type of station area can be explored in terms of precise decongestion, node upgrading, function expansion, endogenous optimization, and infill quality improvement. This paper examines the economic driving effect of Shenyang Metro Line 1 stations on the renewal of the surrounding areas from the perspective of the economic balance of payments, providing a new reference for Shenyang-rail-transit-guided urban renewal work.

Keywords: rail transit; TOD; urban renewal; node–place model; Shenyang

1. Introduction

China's rapid urbanization has resulted in tight urban land resources and an urgent need for urban renewal efforts to further achieve high-quality urban development. In 2021, the "Implementation of Urban Renewal Initiatives" was included in China's Government Work Report and the 14th Five-Year Plan for the first time, marking a pivotal shift from an incremental to a stock-based urban development era. Urban renewal has thus become a central challenge in China's urban transformation agenda [1].

In the context of urban renewal, leveraging rail transit construction to achieve efficient land use and optimize urban spatial structure has emerged as a key strategy for significant cities [2]. Currently, China is experiencing a period of rapid development in rail transit. By the end of 2024, 58 cities in mainland China will have opened urban rail transit lines, totaling 361 rail transit lines. Capitalizing on this momentum is crucial for guiding transit-oriented development (TOD), which can accelerate urban renewal and promote compact, sustainable urban growth [3].

In recent years, some cities in China have successfully integrated rail transit with the development of surrounding land, achieving notable results [4]. Since 2010, cities such as Shenzhen, Shanghai, Guangzhou, Beijing, Chengdu, Hangzhou, Nanjing, Xiamen, and others have issued relevant implementing regulations, management measures, opinions, rules, and other restrictions. In 2015, the Chinese Ministry of Housing and Urban–Rural Development launched the "Guidelines for Planning and Design of Areas along the Urban Rail Transit", focusing on the core of railway stations to build an intensive, efficient, humanized urban environment and activity space [5]. In 2021, the Urban Planning Society of China issued the group standard "Guidelines for Spatial Planning and Design of Facilities Around Urban Rail Transit Stations", which puts forward corresponding optimized design guidance and control requirements for the transport connection and auxiliary facilities, distribution, and public space in the areas surrounding rail transit stations. Transit-oriented urban renewal has thus become a common development strategy for an increasing number of cities [6].

Although China's design guidelines and standards for rail transit station areas are relatively complete at this stage, and the relevant integration practices have also achieved specific results, there is still a significant gap between China's rail-transit-guided urban regeneration and that of advanced international cities. On the one hand, Tokyo and other international cases have demonstrated that successful renewal necessitates the establishment of a land value capture mechanism to internalize the external benefits generated by rail transit, thereby providing the impetus for renewal [7]. In China, the practice of value capture for urban rail transit development is far from mature. Unlike the U.S. urban rail transit system, which receives significant federal government subsidies, the construction and operation of urban rail transit in China typically do not receive any financial support from central and provincial governments [8]. Therefore, Chinese municipal governments have strong economic incentives to adopt viable and effective urban rail transit financing models. On the other hand, Singapore's experience shows that integrated station development requires a collaborative model of risk-sharing and benefit-sharing among developers, operators, and owners. China's integrated station development currently lacks institutional innovation in balancing the interests of multiple parties, and these practices mostly rely on the government to take the lead, failing to fully activate the synergistic effect of market capital and community participation [9].

Existing studies have shown that urban renewal guided by rail transit is completed under the background of multiple interests; rail transit has a positive premium effect on the surrounding areas of a station, stimulating urban rail construction and operation enterprises and real estate developers to enter the field of development of rail transit station areas, deriving from the "rail + property", "rail + community", and other modes [10]. At the same time, the government can utilize metro property joint development, land asset pricing financing, and other premium recovery policies to policies capture the land and real-estate value-added gains brought about by rail transit, serving as a financial source to support both rail transit construction and operation as well as site area development [11]. This also requires that the process of rail-transit-guided land development or renewal can be achieved to alleviate traffic congestion and improve land-use efficiency while also achieving capital gains with an additional premium. Additionally, it needs to consider the synergistic effect of multiple subjects.

Based on this, this paper tries to consider the transportation conditions, land use conditions, and external economic effects of the station area. The "node–place–revenue"(NPR) model is proposed, and the evaluation indicators are constructed from the three dimensions of the "node–place–revenue" model, and the weights of the indicators are determined by expert scoring and Analytic Hierarchy Process (AHP). Then, we choose Shenyang

Metro Line 1 as the research object through the quantitative assessment of the node value, place value, and income value of the station area, using the K-means clustering method to summarize the station type, and then put forward both the urban renewal strategy under different station types as well as the exploration of the synergistic governance mechanism of multiple subjects in the process.

The rest of the paper is organized as follows. Section 2 reviews the previous literature and describes the research methodology. Section 3 analyzes the results of the station clustering model. Section 4 discusses the model station clustering results and proposes a differentiated regeneration strategy. Section 5 summarizes the main findings and recommendations for future research.

2. Materials and Methods

2.1. Previous Literature

TOD was proposed in 1993, aiming to create pedestrian-friendly neighborhoods around public transportation stations, reduce dependence on private cars, and promote sustainable development through high-density, mixed-use development patterns [12]. Since then, domestic and international scholars and academics have proposed implementation guidelines such as 3D principles (density, diversity, design) [13] and 5D principles (increased distance to transit, destination accessibility) [14] and have gradually derived a multidimensional covering transportation, economic, and environmental evaluation system [15]. Early studies focused on physical space design, and with the continued deepening of research and integration of multiple disciplines, scholars began to pay attention to the importance of the coordinated relationship between land use and transportation [16,17]. Lowry established the Land-Use and Transportation Model (LULM) earlier, which considered urban transportation and land use as a system. Handy [18] studied the integrated development planning of both urban land use and transportation systems from the perspective of urban smart growth; Lawrence et al. [19] studied the relationship between urban land use and transportation systems from the perspective of improving the quality of life; Davis et al. [20] addressed the sustainability of both of these from the viewpoint of technology, finance, land use, and transportation system integration.

With urban renewal gradually becoming the direction and goal of most urban development planning, TOD research has begun to expand to the trend of "Transit-Oriented-Renewal" [21], and some studies have explored the interactive relationship between rail transit and urban renewal from different angles, enriching the theoretical connotation of TOD-led urban renewal. Shin [22] analyzed the methods and characteristics of urban regeneration to explore urban planning methods applicable to the development of station impact zones in small- and medium-sized local cities, which are often excluded from domestic urban regeneration efforts. Dai et al. [23] explained the concept and principles of TOD, exploring the logic behind TOD and urban renewal. Sho et al. [24] focused on urban regeneration in Taipei City, Taiwan. They found that promoting connectivity between modes of public transportation can be effective, especially in residential land use areas, as it can improve policies based on the urban regeneration TOD model. The 15 min city concept, as mentioned in the studies by Mocák et al. [25] and Moreno et al. [26], emphasizes proximity, diversity, density, and ubiquity, which overlap with the TOD approach, and provides an analytical framework for studying the integration of transportation and urban renewal.

The basis for researching transit-oriented urban renewal is to evaluate each TOD accurately. Bertolini (1998) [27] developed a node–place model based on station redevelopment, which includes two key conclusions (Figure 1a). The first is to take the matching degree of node value and place value as an indicator to evaluate the effectiveness of spatial development in the station area. The second is that the model can be based on the degree of coordination between nodes and places and can classify the subway station area into five states: stress, unbalanced place, unbalanced node, balanced, and dependence. The "node–place" model is a standard model for evaluating subway station areas. It provides planning guidance based on the coordinated development statuses of station areas. To the author's study, considering the influence of the feedback value on station area revenue, we add a revenue dimension to the original "node–place" model. Scholars have improved and extended various NP models for different application scenarios. Lyu et al. [28] refer to the concept of TOD and use "Transit" and "Development" to represent node and place indicators and add "Oriented" indicators to describe the interaction between transportation and urban development. Vale et al. [29] examine the urban conditions of pedestrian accessibility in station areas and introduce the "Design" dimension to enhance the understanding of the integration of transportation and land use. Cao et al. [30] included the "Ridership" indicator to analyze the interaction and coordination between the passenger flow of rail transit and the degree of node–place integration. Rodriguez et al. [31] extended the dimensions of "integration" and "value" to form a multi-axis model that planners and policymakers can use to understand the development of public transportation stations.

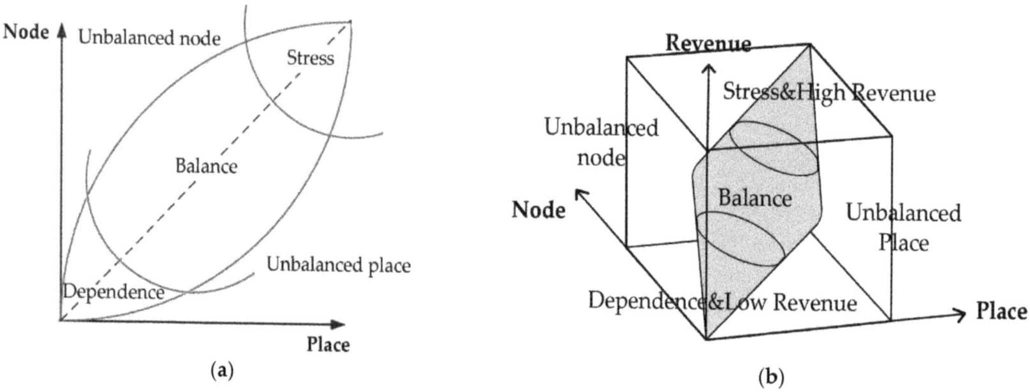

Figure 1. Comparison diagram of node–place (NP) model and node–place–revenue (NPR) model. (**a**) The NP model. (**b**) The NPR model.

The importance of transit premium feedback has received much attention in existing studies of transit-led urban regeneration [32–36]. Turbay et al. [32] have shown that the Bus Rapid Transit (BRT) corridor in Curitiba, Brazil has benefited the high-income class and high-quality real estate along the corridor. Mulley et al. [33] and Cervero et al. [34] point out that an urban transit system can increase land and property values and create economic benefits. The value-added portion can fund existing and new urban transit systems. Yang et al. [35] further pointed out that urban rail transit has a positive impact on surrounding house prices; this positive impact diminishes rapidly as the distance from urban rail transit stations increases; interchange stations have a more significant positive impact on surrounding house prices than ordinary stations; urban rail transit accessibility has a more substantial effect on house prices in the suburbs. Mittal [36] and others summarize the land value capture implementation process and present current debates and controversies, providing an initial decision-making "toolkit" for cities seeking innovative financing solutions for infrastructure improvements.

Therefore, based on the "node–place" model, this paper considers the spatial and temporal economic effects of rail transit on the development of surrounding land, adds the element of revenue feedback to the 3rd axis, and constructs the "node–place–revenue" classification method to provide a quantitative analysis tool for transit-guided urban renewal. The K-means clustering method is used to summarize the development advantages and shortcomings of different types of site areas, identify the key directions of urban renewal, and formulate differentiated renewal strategies and multiple synergistic guidance. Sites can be classified into five theoretical situations based on the differences in 3-dimensional indicators: (1) stress and high revenue—node, place, and revenue value are all strong; (2) unbalanced place—place value is much higher than node value; (3) unbalanced node—node value is much higher than place value; (4) balanced—located in the middle of the diagonal area, nodes, place, and revenue value promote each other and develop in a balanced way; and (5) dependence and low revenue—node, place, and revenue value are all relatively absent (Figure 1b).

2.2. Research Method

This study was divided into four stages: sample data collection, indicator selection and processing, analysis of assessment results, and guidance for updating differences (Figure 2). In the first stage, we selected the station spatial research samples and calibrated the scope. We then systematically collected multi-source data related to the samples and integrated the multi-source data to construct the fundamental database.

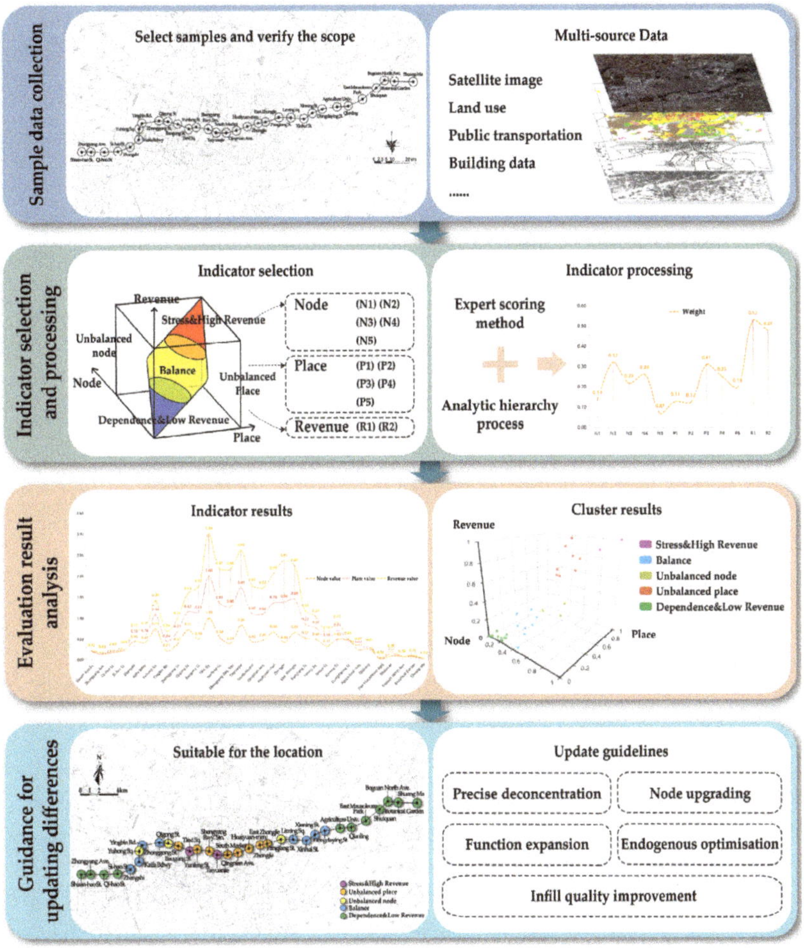

Figure 2. Flowchart of this research.

In the second stage, we referred to relevant studies, constructed evaluation indicators from the three dimensions of "node–place–revenue", determined the weights of each indicator using expert scoring and AHP, assigned values to the fundamental evaluation unit indicators, and chose the positive and negative attributes of the indicators. The positive and negative characteristics of the indicators were also determined.

In the third stage, we quantitatively evaluated each TOD site's node value, place value, and revenue value based on the "node–place–revenue" model. Based on this, we used the K-means clustering algorithm to classify the station area types and finally formed five types of station areas with significant differences.

In the fourth stage, based on the research results of the first three stages, differentiated urban renewal strategies were proposed for the value characteristics of different types of station space. The research results can provide a scientific basis for decision-making and planning guidelines for rail-transit-oriented urban renewal in Shenyang.

2.2.1. Research Object and Scope

As the first city in Northeast China to open a subway, Shenyang's rail transit development can cover multiple levels of impactful renewal of the station area after the completion of rail transit construction, renewal of the station periphery triggered by the intersection of old and new lines, and urban renewal initiated by the new construction of rail transit. It is helpful to analyze the prevailing problems comprehensively; therefore, this paper chooses to take Shenyang Metro Line 1 as the research object (Figure 3).

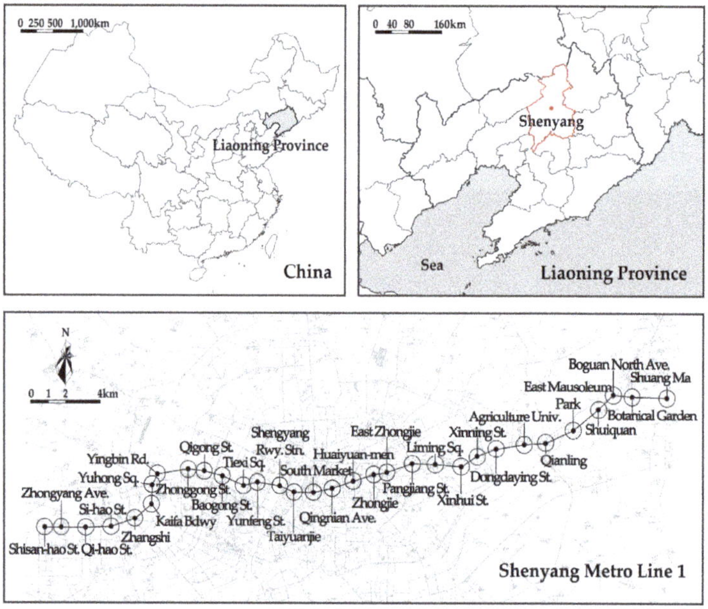

Figure 3. Location and research area of Shenyang Metro Line 1.

The first phase of Shenyang Metro Line 1 was put into operation on 27 September 2010, marking both the first subway line to be built and operated in Shenyang City as well as the first subway line to be put into operation in Northeast China. Line 1 Phase I project, starting from the west at Shisan-hao Street Station and ending at the east at Liming Square Station, passes through Yuhong District, Tiexi District, Heping District, Shenhe District, Dadong District, connecting Shenyang Economic Development Zone, Tiexi Industrial Zone and Shenyang Railway Station, Taiyuan Street Shopping District, Middle Street Square City, and other old city passenger flow distribution areas, facing the situation of micro-renewal within the site area after the completion of construction.

Shenyang Metro Line 1 Phase II East Extension Project, from Line 1 Liming Square Station in the west to Shuangma Station in the east, passes through Shenhe District, Hunnan District, and other new urban areas. It faces the urban renewal triggered by the new railway construction and the renewal situation around the site initiated by the intersection of the latest and original lines (Table A1).

Overall, the diversity of TOD station types in Shenyang Metro Line 1 provides us with a rich and diverse dataset [37]. This diversity offers ample opportunities to study TOD-led urban renewal strategies, which help analyze prevailing problems and develop comprehensive targeted renewal strategies.

In studies on the built environment around rail transit stations, the Pedestrian Catchment Area (PCA) is commonly used to represent the service coverage of rail transit stations. Most existing studies employ circular buffer zones to represent the PCA of the stations [38]. Considering the concept of the 15 min city, according to the relevant guidelines [39], this paper selects an 800 m buffer as the PCA of metro stations.

2.2.2. Data Sources

The applied multi-source big data mainly includes railway station data, railway line data, bus station data, bus line data, point of interest (POI) data around the station, building data, land data, property data, and population data. The above multi-source data primarily come from various data sources, including electronic map operators and public data (Table 1). A spatial database was built on the geographic information system (GIS) platform for integration and measurement. This approach could comprehensively analyze the urban development factors related to TOD and effectively combine qualitative and quantitative data, thus supporting our research findings [40].

Table 1. Details of multi-source data usage.

Name	Field	Source
Rail Transit Station Data	Station Name, Latitude, and Longitude	Open street map [41]
Rail Transit Line Data	Line Name, Latitude, and Longitude	Amap [42]
Bus Station Data	Station Name, Latitude, and Longitude	Amap
Bus Line Data	Line Name, Latitude, and Longitude	Amap
POIs Around Stations	Name, Category, Latitude, and Longitude	Open street map
Building Data	Height, Type, Area, Location	Open street map
Land Use Data	Type, Area	Open street map
Real Estate Data	Name, Price	Lianjia [43]
Population Data	Population, Location	Swguancha [44]

2.2.3. Indicator Selection and Processing

The indicator system of the "node–place–revenue" model is based on the various indicators of the classical "node–place" model, as described in the literature [45–48]. We have introduced the benefits of land and property premiums to construct an indicator system based on three dimensions with 12 indicators (Table 2). The selection of indicators is directly or indirectly related to the urban renewal of subway stations, and the indicator system of the multidimensional "node–place–revenue" model has been constructed to measure the renewal potential of the station area under different circumstances.

1. Nodal value reflects the transportation service function and value presented by metro stations as urban transportation nodes. The number of station entrances and exits

(N1) (Reusser et al. [45]), the number of directions served by the station (N2) (Reusser et al. [45]), and the number of bus stops in the station domain (N3) (Reusser et al. [45], Kim et al. [46]), ease of bus transfer (N4) (Bertolini et al. [47]), and accessibility to the city center (N5) (Kim et al. [46], Bertolini et al. [47]) are five indicators.

2. Place value reflects whether the built environment of the station area where the metro station is located is conducive to urban renewal. Building development intensity (P1) (Kim et al. [46]), building density (P2) (Lu et al. [48]), a functional mix of land use (P3) (Reusser et al. [45], Lu et al. [48]), number of place facilities (P4) (Lu et al. [48]), and population density (P5) (Reusser et al. [45]) are five indicators.
3. Revenue value reflects the ability of the land and property value added generated in the process of urban renewal to feed back the investment in rail transit construction in the form of rent or sales profit in the future. Two indicators, land premium benefit (I1) (Salat et al. [7]) and property premium benefit (I2) (Salat et al. [7]), have been selected.

Table 2. Evaluation index system and weights of the "node–place–revenue" model.

Target Layer	Indicator Layer	Indicator Description	Weight	Attribute
Node Value	Number of Station Entrances/Exits (N1)	The number of entrances and exits at the station.	0.14	Forward direction
	Number of Service Directions (N2)	Terminal stations have 1 direction; additional transfer lines increase number of directions by 2.	0.32	Forward direction
	Number of Bus Stops in the Area (N3)	The count of bus stops within the station's vicinity.	0.22	Forward direction
	Bus Transfer Convenience (N4)	The number of bus transfer routes within 200 m of the station.	0.26	Forward direction
	Accessibility to City Center (N5)	Distance from each station to the city center (Zhongjie).	0.07	Positive direction
Place Value	Building Development Intensity (P1)	The ratio of total building floor area to the total land area.	0.13	Forward direction
	Building Density (P2)	The ratio of building footprint area to the total land area.	0.12	Forward direction
	Land Use Function Mix (P3)	Calculated using the entropy method for POI land use mix. ($H = \sum_{i=1}^{n} p_i \ln p_i$)	0.31	Forward direction
	Number of Facilities (P4)	The count of POIs within the station area.	0.25	Forward direction
	Population Density (P5)	The ratio of the population in the station area to the total land area.	0.19	Forward direction
Revenue Value	Land Premium Benefit (R1)	Calculated based on empirical values from urban rail transit premium effect assessments.	0.53	Forward direction
	Property Premium Benefit (R2)	The ratio of average housing prices within 500 m to those within 500–800 m of the station.	0.47	Forward direction

In this paper, combined with the development stage of Shenyang urban rail transit and station space characteristics, firstly, the evaluation index system is constructed through hierarchical analysis, and the expert scoring method is used to compare the relative importance of the relationship between the two and to determine the weight of each index [49]. Secondly, the evaluation results of each indicator are weighted, with negative indicators receiving a positive weighting, allowing them to be integrated with the positive indicators to reflect the overall evaluation results. Finally, to eliminate the influence of different scales in each index data, this paper, based on SPSS 24, uses normalization to process the data of each index so that all the data are re-adjusted to a range between 0 and 1.

In terms of weight selection for node value, Li et al.'s [50] weights are the same as ours. Both set the station service capacity and bus transfer convenience as higher weights and set the city center accessibility as lower weights. In the selection of weights for place value, the weights of Li et al. [51] are the same as ours. Both set the land-use-function mixing degree as a higher weight and the building development intensity as a lower weight. In the weight selection of revenue value, the weights are distributed relatively evenly to avoid subjective bias.

2.2.4. Site Clustering Method

Existing studies have shown that the station evaluation method based on the extended "node–place" model can effectively identify the characteristics of TOD development and provide a scientific basis for urban planning. Guo et al. [52] classified the state of urban development into nine modes by constructing a three-dimensional evaluation system of nodes, places, and green spaces. The experimental results in Harbin showed that this method can provide specific guidance for urban planning and sustainable development policies. Similarly, Yang et al. [53] also extended the "node–place" model to classify Ningbo metro stations into four categories based on their suitability for TOD improvement. The practice in Ningbo demonstrated that the method is effective in identifying station locations suitable for TOD improvement.

Based on this, to further explore the synergistic effect between node, place, and revenue dimensions and to evaluate the stations in more detail, this paper adopts the K-means cluster analysis method to validate the results and form the empirical findings of station spatial typing. This method automatically divides the data into K clusters with high cohesion by calculating the Euclidean distance between samples [54], which is suitable for identifying potential divergence laws in station spatial features. Specifically, we first normalize the results of node, place, and revenue data to eliminate the influence of magnitude; secondly, we determine the optimal number of K clusters through the elbow rule to ensure the maximization of interclass differences with the minimization of intraclass differences; finally, we combine the profile coefficients to validate the reasonableness of the clustering results. This method not only reveals the spatial differentiation characteristics of TOD but also identifies the dominant dimensions of each type of station area through the eigenvalues of the clusters, providing empirical evidence for the subsequent development of differentiated renewal strategies.

3. Results

3.1. Indicator Results

Based on the above index system and calculation method, the weights and attributes of each basic index were determined and obtained. Then, the weighted calculation yielded the node, place, and revenue values for 32 metro TOD station areas along Shenyang Metro Line 1 (Table A2).

3.1.1. Node Value

The node value reflects the comprehensive transportation advantages of the TOD station area, and the higher its value is, the stronger the centrality of the station area is. From the perspective of the spatial differentiation of the node value of Shenyang Metro Line 1 (Figure 4), the node value of station areas did not show a single decrease from the center to the periphery but instead formed a hierarchical differentiation of high, medium, and low values within the center and peri-urban areas, respectively. High-quality station areas are primarily concentrated in the city's central area, such as Tiexi Square Station and Taiyuanjie Station, which have well-developed transportation and significant location advantages. Medium-value station areas are primarily located in central and suburban areas, such as Huaiyuan-men, East Zhongjie, and Pangjiang Streets, which are essential subway intersections and have obvious transportation advantages. Additionally, Zhangshi and Xinhui Streets are considered low-value station areas. Railways and other low-value station areas are typically mixed-use areas, combining residential, industrial, and commercial uses with some transportation value. Low-value station areas are primarily located in suburban areas, such as around the Boguan North Avenue and Botanical Garden stations, where transportation support has not yet been fully established and ecological or cultural tourism functions predominate.

Figure 4. Node value spatial differentiation map.

3.1.2. Place Value

The place value reflects the station area's development intensity and functional diversity. From the perspective of the spatial differentiation of the place value of Shenyang Metro Line 1 (Figure 5), the place value of the station area decreased from the city center to the surrounding area, showing the apparent spatial differentiation feature of "high center and low periphery". High-value station areas are primarily located in the city center, mostly in mature business districts or residential areas such as Zhongjie, Taiyuanjie, and Yunfeng Street. These areas are saturated with land development, and urban functions are well supported, forming self-contained urban vitality centers. Medium-value station areas are primarily located in the city center and suburban areas, with mixed-function areas such as South Market, Qingnian Avenue, and other medium-high-value station areas, which are highly mixed with functions and have relatively perfect support facilities, and medium-low-value station areas such as Yuhong Square and Qigong Street, which are relatively low in function mixing but have formed basic support facilities. Although the mix of functions is relatively low, an essential support system has been formed. The low-value station areas are mainly located in the outlying suburbs such as around East Mausoleum Park Station and Boguan North Avenue Station. These areas are restricted by the ecological protection policy or the stage of urban development and have a relatively single type of function with low development intensity and an insufficient coverage of public service facilities, which shows the typical characteristics of urban peripheral areas.

Figure 5. Place value spatial differentiation map.

3.1.3. Revenue Value

The revenue value measures the economic and social benefits generated by urban renewal and its ability to repay the investment in railway construction from the perspective of spatial differentiation in the revenue value of Shenyang Metro Line 1 (Figure 6). High-value station areas are primarily located in the city center, primarily in the development of mature commercial areas, such as Tiexi Square, Taiyuanjie, and Zhongjie, which feature a mature transportation network and commercial agglomeration effect. This leads to the urban renewal process producing land premium benefits and property premium benefits that are higher. Medium-value station areas are mainly distributed in the city center and suburban areas, mostly in mixed residential and industrial areas or new expansion areas, such as Baogong Street, Qingnian Avenue, and other medium-to-high-value station areas, as well as Yuhong Square and Liming Square and other medium-to-low-value station areas, where the income value matches the development intensity. Low-value station areas are concentrated in the remote suburbs of the city, primarily in ecological zones or undeveloped areas, such as around Shuiquan Station and Shuang Ma Station, where land and property premium benefits are low. Still, they have huge development potential due to the large amount of inefficient land.

3.2. Cluster Results

To further identify the type of station space, clustering was utilized using the K-means clustering algorithm. The elbow and contour maps verified the plausibility of contour clustering (Figure A1). The clustering results showed that 34% of the station spaces were categorized into Cluster 5 (11), followed by Cluster 2 and Cluster 4 (8), Cluster 3 (3), and Cluster 1 (2) (Table 3). These five categories of stations were characterized as follows: Category 1 had high node, place, and revenue values; Category 2 stations had low node values but high place and revenue values; Category 3 had high node values and medium place and revenue values; Category 4 had medium node, place, and revenue values; Category 5 had low node, place, and revenue values. According to the positions of the categories in the model, combined with the clustering of the sample node, the place and revenue values are summarized as follows: stress and high revenue, unbalanced place, unbalanced node, balanced, and dependence and low-revenue station space, respectively (Figure 7). The categories' spatial differentiation in Shenyang City is shown in Figure 8.

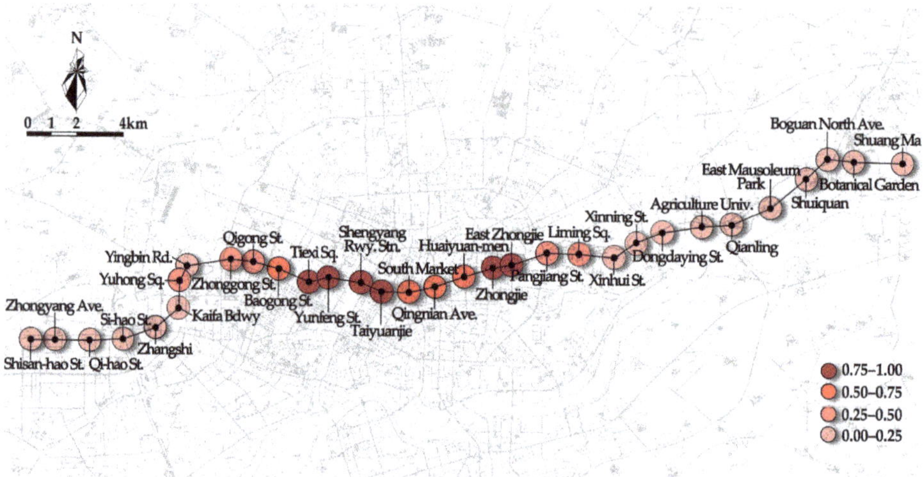

Figure 6. Revenue value spatial differentiation map.

Table 3. The centroid features and descriptive statistics table of NPR model clustering results.

Cluster	Centroid						Cluster Distribution	
	Node		Place		Revenue		Number of Cases (Units)	Proportion (%)
	Average Value	Standard Deviation	Average Value	Standard Deviation	Average Value	Standard Deviation		
Cluster 1	0.91	0.13	0.95	0.08	0.95	0.08	2.00	0.06
Cluster 2	0.50	0.11	0.76	0.13	0.76	0.13	8.00	0.25
Cluster 3	0.66	0.12	0.37	0.06	0.37	0.06	3.00	0.09
Cluster 4	0.39	0.09	0.19	0.09	0.19	0.09	8.00	0.25
Cluster 5	0.14	0.09	0.03	0.02	0.03	0.02	11.00	0.34

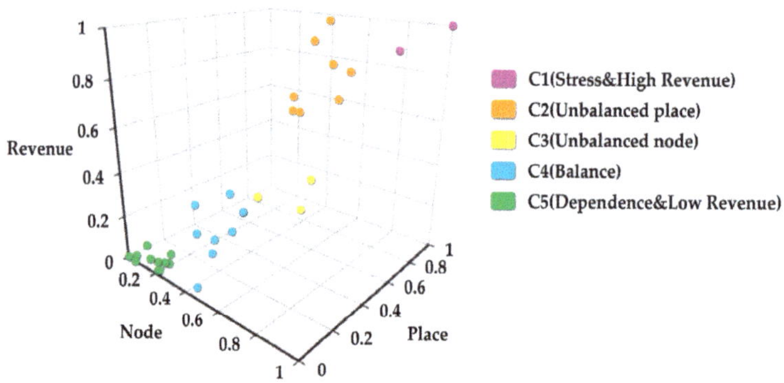

Figure 7. NPR model clustering results.

3.2.1. Stress and High Revenue

Category 1 is characterized by high pressure and revenue and is defined by a state of pressure in terms of node and place value and simultaneously exhibits a high revenue feedback value. It is typically situated in the city's core business district, offering convenient transportation, a functional composition, and high revenue-generating value. However, due to space development, it tends to be saturated, and its further renewal and development may be limited. For example, Taiyuanjie Station is the traditional commercial core area of

Shenyang, containing both Line 1 and Line 4 of the interchange station, but also connects the Shenyang station rail passenger transport and has a high node value, and the surrounding collection of the central commercial body, medical and cultural facilities, and place function is highly complex. The Tiexi Square station also belongs to this category as it is the core business district and transportation hub, and development has been saturated.

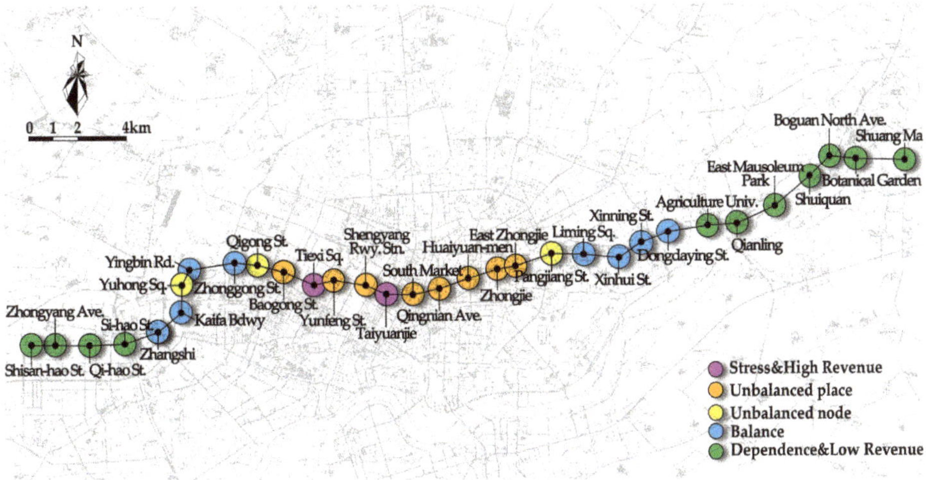

Figure 8. Spatial distribution map of station domain classification.

3.2.2. Unbalanced Place

Category 2 is summarized as being of the uneven place type, where the place function significantly exceeds the node function, and the development intensity is high, but the transportation support is inadequate. Members of this category are primarily located in urban central business districts (CBDs), core business districts, and mature residential areas. This area type features mixed land use, high development intensity, dense population, rich commercial resources, and high property value. However, there are problems, such as a larger neighborhood scale, lower traffic accessibility, and poor road network penetration. Its transportation system struggles to accommodate the large number of people generated by the high-density development of the area, which is prone to traffic congestion. At the same time, the land price of this type of station area is relatively high, with strong market development momentum. For example, Zhongjie Station is a traditional commercial pedestrian street with high development intensity; however, the connection between the station and the pedestrian street is not optimal, and the utilization rate of the underground passage is low.

3.2.3. Unbalanced Node

Category 3 is summarized as being of an unbalanced node type; the node value of this type of site is significantly higher than the place value, and the transportation location advantage is outstanding; however, development is lagging. This site area is primarily located in residential clusters near the city center, and it is served by two rail transit lines and numerous conventional bus lines, offering obvious transportation advantages. However, the level and scale of commercial and public service facilities available at these sites are relatively low, and the range of services offered is limited. There is some revenue feeder value, making it a hotspot for investment by developers and members of the public. For example, Pangjiang Street Station has obvious transportation advantages as a subway interchange station and is near the East First Ring Expressway. Still, excessive transportation functions detract from the area's urban functions, and the site's low development intensity and poor mixed land use create an imbalance in the value of the sites.

3.2.4. Balance

Category 4 is characterized as a balanced type, where the balance between node value, place value, and revenue is relatively high. Usually located in the city's suburbs, these indicators are relatively balanced, with a certain foundation of transportation facilities, functional layout, and revenue-generating value; however, they still have the potential for further improvement. For example, the Liming Square Station is mainly for residential functions, and its node value and place value are more balanced; as the connecting station of Line 1 Phase I and Phase II, it has a certain scale of developable land or land to be renewed, and it is a potential zone for upgrading the urban functions and redeveloping the land at this stage.

3.2.5. Dependence and Low Revenue

We summarize Category 5 as dependent and low-revenue, a generally low composite indicator of locations. The areas in this category are typically situated on urban fringes or in undeveloped areas. The current situation is characterized by predominantly non-urban built-up areas or ecological protection zones, with a lack of road and public transportation facilities, as well as commercial and public services, resulting in a weak potential for market renewal. For example, Shuangma Station, as the terminus of Metro Line 1 East Extension, has a low significance in the rail transportation network. Regarding land use, the surrounding area consists of almost undeveloped land that has not yet been planned or constructed and lacks the foundation of transportation and service facilities.

4. Discussion

4.1. The Spatial Differentiation Characteristics of Multidimensional Value in Station Domain

The study found that the nodes, places, and revenue value of the station areas of Shenyang Metro Line 1 exhibit spatial differentiation characteristics characterized by a "high center and low periphery" pattern. The phenomenon of unbalanced development in the station area is prominent, and the differences between regions are significant. The issue of unbalanced growth in the railway station area has been widely discussed in numerous studies both domestically and abroad. Yang et al. [55], Zhang et al. [56], and Zhang et al. [57] also showed, in their studies, that there is a certain degree of imbalance in station development among regions. In the study by Zhu et al. [58], it was noted that transportation–land mismatch, edge lag, and the influence of value dimensions can lead to significant imbalance. In the studies by Guo et al. [52] and Yang et al. [59], it was noted that station areas with lower comprehensive development indexes are primarily located in the city's suburbs. In comparison, those with higher indexes are typically located in the center of the town, where socio-economic conditions are generally better.

4.2. Characteristics and Update Potential Evaluation of Station Domain Clustering

In this paper, the station areas of Shenyang Metro Line 1 are categorized into five types using the clustering method: pressure and high revenue, unbalanced place, unbalanced node, balanced, and dependence and low revenue. By analyzing the characteristics of the five types of subway station areas and examining their renewal potential, it has been found that areas with high pressure and revenue are usually located in the city's core business districts, and further development is limited, resulting in low renewal potential. The unbalanced areas are generally located in the CBD, core business district, and mature residential areas, with strong renewal potential. The unbalanced node is typically situated in residential clusters near the central business district (CBD) and has strong renewal potential. Balance nodes are usually located in the city's suburbs and have some potential for regeneration. Dependence and low-income areas are generally located on the urban fringe or in undeveloped areas and have weak regeneration potential. Dou et al. [60], Nemroudi et al. [61], and Dai et al. [62] also showed that based on the composite scores of the NP model, the geographic distribution of clustering results roughly decreases from the city center to the suburbs.

4.3. Differentiated Update Strategy for Site Domain Types

Based on the characteristics and renewal needs of different types of station spaces, this paper proposes five differentiated urban renewal strategies [63] and further discusses the corresponding policy tools, synergistic benefit mechanisms, and dynamic effectiveness indicators to ensure the effective implementation of the strategy and the evaluation of their effectiveness (Table 4).

Table 4. Differentiated governance strategies for different types of site areas.

Station Type	Stress and High Revenue	Unbalanced Place	Unbalanced Node	Balance	Dependence and Low Revenue
Key Characteristics	Urban core with high pedestrian and vehicular traffic, peak land value	Urban centers with strong place functions but weak node functions	City center fringe, with strong transportation but weak service functions	Suburban areas with a balanced work and residence profile but with demand for upgrading	Urban fringe, lagging in development, low financial returns but high social demand
Renewal Strategy	Precise deconcentration	Node upgrades	Function expansion	Endogenous optimization	Infill quality improvement
Policy Tools	Congestion charging, plot ratio transfer, air rights sale	Microcirculation public transportation subsidy	Functional replacement tax deduction, temporary land use permit	Incremental tax financing and flexible control of mixed land use	Land allocation + franchising, tax exemption period + small and micro loans
Stakeholder Collaboration Mechanism	Developer: Underground Link Construction (Hong Kong) Rail Company: Integrated Interchange Hub (Shanghai)	Neighborhood Council: Slow Path Optimization (Shanghai) Bus Company: Microcirculation Routes (Singapore)	Artists: Public Space Revitalization (Tokyo) Community: Shared Facility Operation (Beijing)	Merchants Union: Street Renewal (Barcelona) Transportation Bureau: Bus Optimization (Chengdu)	Government: subsidized schools/hospitals (Guangzhou) Social organizations: tourism training (Kyoto)
Performance Metrics	Short-term: 20% reduction in congestion Long-term: commercial rents balanced with commuter efficiency	Short-term: 30% increase in feeder satisfaction Long-term: ≥40% share of non-commuter commercial revenue	Short-term: 50% increase in non-commuter traffic Long-term: destination station conversion rate achieved	Short-term: job-to-life ratio of 1.2 to 1.0 Long-term: 30% increase in business diversity index	Short-term: 200 new jobs Long-term: ≥30% of earnings from specialty industries

Firstly, for stress and high-revenue station spaces, in order to avoid over-development, it is appropriate to adopt the strategy of "fine deconcentration". The capacity of transportation facilities should be able to meet the demand of regional peak traffic distribution, avoid large-scale construction behavior, and pay more attention to the enhancement of the degree of functional mixing and refinement of urban design. Specific policy tools can be adopted, such as congestion charging, the transfer of plot ratio, the sale of air rights, etc. For the synergistic mechanism, one can refer to Hong Kong's model of underground corridor construction, or Shanghai's case of an integrated interchange hub built by the government in cooperation with a railroad company [34]. The long-term goal should be able to achieve a balance between commercial rents and commuting efficiency.

Second, for unbalanced place station spaces, it is appropriate to adopt a "node upgrade" renewal strategy. Priority should be given to upgrading the value of the nodes to match the value of the place to clarify the regional traffic demand and traffic supply level, assess the necessity and feasibility of connecting commercial and other facilities in the station space, attract the market to make up for the gap in transportation facilities in a targeted manner, and improve the level of integration and connectivity of transportation connections. Policy tools such as micro-circulation bus subsidies can be used; benefit synergies can learn from Shanghai's slow path optimization practice or Singapore's micro-circulation line operation model. The long-term goal should be to make the share of non-commuter commercial revenue ≥40%.

Thirdly, for unbalanced node station spaces, it is appropriate to adopt the "function expansion" renewal strategy. Scientific planning is needed to guide the functional differentiation and complementary development of the station area, expand diversified functional nodes, attract new business development, and strengthen the construction of commercial facilities and public facilities in the station area through public investment to stimulate the market demand and realize the synergistic and high-quality development of the station area. Policy tools, such as tax credits for functional replacement and temporary land use permits, etc., can be used. For the synergy of interests, one can refer to Tokyo's public space revitalization or Beijing's shared-facilities operation case [55]. The long-term goal should be to achieve the conversion rate of "destination stations".

Fourth, for balanced station spaces, it is appropriate to adopt an "endogenous optimization" renewal strategy. One must study the station type and explore the development potential of the station area. This type of station area typically still retains a particular scale of urban renewal land, offering a relatively considerable investment return and high market development potential. In the future, it is advisable to adopt appropriate incentives in conjunction with market development to guide market forces toward simultaneously enhancing the node value and place value. Policy tools include tax incremental financing and the flexible control of mixed-use land among others. The synergy of interests could be based on Barcelona's merchant alliance street renewal or Chengdu's transit optimization practice. The long-term goal should be to increase the business diversity index by 30%.

Fifth, for dependence and low-revenue station spaces, it is appropriate to adopt the "infill and improve quality" type of renewal strategy. For this type of station area, it is relevant to limit its disorderly development shortly, retain the ecological base, and stimulate the vitality of the station area through financial, planning, and other multi-faceted policies in the long term; analyze the mismatch between the advantages of the regional development and the functional positioning of the area on the regional scale; and promote the development of such an area in an orderly and concentrated manner. Policy tools can include land allocation and franchising, as well as tax holidays and small- and medium-sized loans. The synergy of interests can be exemplified by the government-subsidized school and hospital model in Guangzhou or the tourism training program for social organizations in Kyoto. The long-term goal is to achieve a 30% share of revenue from specialty industries.

4.4. Limitations

However, this paper still has some limitations. Firstly, the "node–place–revenue" model proposed in this paper primarily aims to elucidate the guiding effect of rail transit on the revitalization of the surrounding areas from the perspective of economic balance of payments. The paper has selected Shenyang Metro Line 1 as the research object, which limits the generalizability of the paper's conclusions due to the geographic environment, financial structure, and policy context of a single city. Therefore, in the future, the sample coverage will be expanded, and a comparative multi-case survey will be conducted to refine the common patterns and adaptive adjustment strategies across regions.

Second, in the current study, the indicators of the "node–place–revenue" model were static data, which represented the current stage. It is necessary to incorporate dynamic variables that reflect the city's stage of development, policy cycle, and market environment into future research. In particular, the measurement of land premium benefits and property premium benefits may vary depending on the market environment, and an in-depth discussion of land premium and property premium benefits will be added in the future.

Finally, the clustering analysis results are limited by the available data size and the limitations of algorithm selection. Future research can explore the introduction of machine learning algorithms, combined with richer, multi-source, heterogeneous data, to improve classification accuracy and the model's explanatory power and to further investigate the intrinsic correlations of complex systems.

5. Conclusions

This study conducted systematic research on rail-TOD-guided urban renewal, and the main conclusions were as follows:

1. A three-dimensional evaluation system of TOD-guided urban renewal was constructed at the theoretical level. By interpreting the mechanism of rail-TOD-guided urban renewal, it was transformed into three dimensions and 12 indicators, among which the node dimension reflected the transportation service function and value of the subway station as an urban transportation node, the place dimension reflected whether the built environment of the station area where the subway station was located was conducive to urban renewal, and the revenue dimension demonstrated the economic and social benefits generated by urban renewal that could feed back into the investment in rail transit construction.
2. We developed a quantitative analysis framework for urban renewal guided by rail TOD at the methodological level. Based on multi-source data, the node value, place value, and revenue value of 32 Shenyang Metro Line 1 stations were quantified. The K-means clustering method was applied to divide the stations into five categories: pressure and high revenue, unbalanced place, unbalanced node, balanced, and dependence and low revenue. The different types of stations exhibited significant spatial differentiation characteristics, providing a scientific basis for the development of differentiated renewal strategies.
3. At the practical level, differentiated regeneration strategies were proposed. Based on the differences in the characteristics of five types of stations—stress and high revenue, balanced, unbalanced node, unbalanced place, and dependence and low revenue—it was proposed that targeted regeneration strategies can be explored from the perspectives of precise dissolution, node upgrading, function expansion, endogenous optimization, and infill and quality improvement. We also incorporated the exploration of policy instruments, synergistic mechanisms, and dynamic effectiveness indicators to enhance the relevance of the proposed strategies.

This article can serve as a practical reference for urban renewal in areas with sizable urban rail transit stations. However, this study still had some limitations, including the selection of cases limited to a single type of city, a lack of dynamic adaptability in the construction of the indicator system, and limitations in the application of data and algorithms for clustering methods. Future research can further explore the applicability of the NPR model in other cities. Secondly, it can construct a dynamic indicator system that allows the evaluation model to be adjusted according to the stage of urban development. Ultimately, it can incorporate machine learning methods to enhance classification accuracy.

Author Contributions: Conceptualization, X.L. and M.Z.; methodology, X.L. and M.Z.; software, X.L. and M.Z.; validation, X.L. and M.Z.; formal analysis, X.L., M.Z., Z.L. and Q.L.; investigation, M.Z., Z.L., Q.L. and S.H.; resources, X.L. and M.Z.; data curation, X.L., M.Z. and Z.L.; writing—original draft preparation, X.L., M.Z. and Z.L.; writing—review and editing, X.L., M.Z., Z.L., Q.L. and S.H.; visualization, X.L. and M.Z.; supervision, X.L.; funding acquisition, X.L. All authors have read and agreed to the published version of the manuscript.

Funding: This research was funded by the National Natural Science Foundation of China (grant number 51778375).

Data Availability Statement: The original contributions presented in this study are included in the article. Further inquiries can be directed to the corresponding author.

Acknowledgments: We would like to thank the journal experts who edited this paper. We also appreciate the constructive suggestions and comments on the manuscript from the reviewers and editors.

Conflicts of Interest: The authors declare no conflicts of interest.

Abbreviations

The following abbreviations have been used in this manuscript:

TOD	Transit-oriented development
NP	Node–Place
NPR	Node–Place–Revenue
LULM	Land-Use and Transportation Model
AHP	Analytic Hierarchy Process
POI	Point of interest
GIS	Geographic information system

Appendix A

Appendix A.1

The following is the basic information of the stations on Shenyang Metro Line 1.

Table A1. Basic information of stations on Shenyang Metro Line 1.

Number	Station Name	Administrative District	Platform Type	Transfer Route
1	Shisan-hao St.	Tiexi District	Underground Side	-
2	Zhongyang Ave.			
3	Qi-hao St.			
4	Si-hao St.			
5	Zhangshi	Yu Hong District	Underground Island	
6	Kaifa Bdwy			
7	Yuhong Sq.			
8	Yingbin Rd.			
9	Zhonggong St.	Tiexi District		
10	Qigong St.			
11	Baogong St.			
12	Tiexi Sq.			Shenyang Metro Line 9
13	Yunfeng St.			-
14	Shengyang Rwy. Stn.	Heping District		Shenyang Metro Line 5 (Planned)
15	Taiyuanjie			Shenyang Metro Line 4
16	South Market			-
17	Qingnian Ave.	Shenhe District		Shenyang Metro Line 2
18	Huaiyuan-men			-
19	Zhongjie			Shenyang Metro Line 6 (Planned)

Table A1. *Cont.*

Number	Station Name	Administrative District	Platform Type	Transfer Route
20	East Zhongjie	Dadong District	Underground Island	-
21	Pangjiang St.			Shenyang Metro Line 10
22	Liming Sq.		Underground Side	
23	Xinhui St.	Shenhe District		
24	Xinning St.			
25	Dongdaying St.			
26	Agriculture Univ.			
27	Qianling		Eastern Extension (Not Yet Opened)	-
28	East Mausoleum Park			
29	Shuiquan	Hunnan District		
30	Boguan North Ave.			
31	Botanical Garden			
32	Shuang Ma			

Appendix A.2

This study employed the elbow rule to determine the optimal number of clusters (K), ensuring maximum inter-class differences and minimum intra-class differences. It also verified, intra-classes, the rationality of the clustering results by combining the contour coefficients. In the elbow diagram (Figure A1a), the SSE decreases significantly during the stages of k = 1 to k = 5. After k = 5, the decrease in SSE sharply decreases. Therefore, k = 5 is the "elbow" position, representing the optimal balance between data interpretability and model simplicity. The contour coefficient method (Figure A1b) further verifies this conclusion. When k = 5, the contour coefficient is higher, reflecting the compactness within clusters and clear separation between clusters, and the clustering results are reasonable.

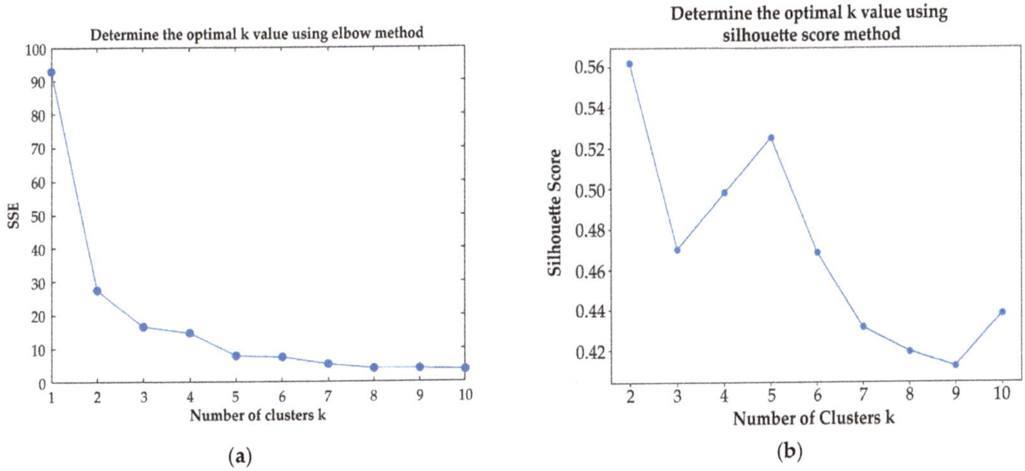

Figure A1. Comparison chart of elbow method and silhouette score method for determining the optimal K value: (**a**) elbow method; (**b**) silhouette score method.

Appendix B

The following is the normalized sample station data. To eliminate the influence of different dimensions of each indicator data, this article uses SPSS 24 to normalize the indicator results, so that all data are within the range of 0–1.

Table A2. Normalized sample station data.

Station Name	N1	N2	N3	N4	N5	P1	P2	P3	P4	P5	R1	R2
Shisan-hao St.	0.50	0.00	0.12	0.31	0.00	0.03	0.21	0.70	0.01	0.37	0.00	0.50
Zhongyang Ave.	0.50	0.33	0.21	0.04	0.04	0.13	0.44	0.89	0.03	0.37	0.09	0.41
Qi-hao St.	0.50	0.33	0.33	0.04	0.11	0.01	0.13	0.87	0.03	0.49	0.08	0.45
Si-hao St.	0.50	0.33	0.17	0.23	0.17	0.12	0.30	0.87	0.04	0.60	0.05	0.46
Zhangshi	0.50	0.33	0.67	0.27	0.24	0.51	1.00	1.00	0.12	0.49	0.16	0.54
Kaifa Bdwy	0.26	0.33	0.67	0.23	0.30	0.39	0.67	1.00	0.16	0.54	0.22	0.50
Yuhong Sq.	0.26	0.33	0.79	0.77	0.37	0.45	0.62	0.91	0.34	0.51	0.33	0.38
Yingbin Rd.	0.26	0.33	0.37	0.04	0.41	0.35	0.74	0.89	0.23	0.31	0.16	0.18
Zhonggong St.	0.26	0.33	0.46	0.27	0.49	0.38	0.56	0.93	0.31	0.43	0.23	0.33
Qigong St.	0.26	0.33	0.96	0.58	0.54	0.57	0.56	0.91	0.43	0.40	0.46	0.23
Baogong St.	0.26	0.33	0.83	0.23	0.59	0.65	0.80	0.87	0.62	0.40	0.52	0.27
Tiexi Sq.	1.00	1.00	0.92	1.00	0.65	0.82	0.79	0.87	1.00	0.40	0.74	0.15
Yunfeng St.	0.00	0.33	1.00	0.19	0.69	0.84	0.73	0.96	0.81	0.71	0.70	0.35
Shengyang Rwy. Stn.	1.00	0.33	0.71	0.00	0.77	0.87	0.73	0.93	0.89	0.49	0.77	0.13
Taiyuanjie	0.26	1.00	1.00	0.62	0.81	1.00	0.76	1.00	0.89	0.60	1.00	0.13
South Market	0.00	0.33	0.92	0.00	0.86	0.81	0.84	0.96	0.67	0.66	0.56	0.21
Qingnian Ave.	0.50	1.00	0.67	0.12	0.91	0.66	0.71	0.91	0.63	0.46	0.76	0.17
Huaiyuan-men	0.26	0.33	0.92	0.42	0.97	0.74	0.85	0.89	0.70	0.34	0.75	0.07
Zhongjie	0.76	0.33	0.75	0.00	1.00	0.71	0.84	0.87	0.98	0.54	0.71	0.08
East Zhongjie	0.50	0.33	0.96	0.35	0.99	0.73	0.82	0.89	0.80	0.77	0.70	0.44
Pangjiang St.	0.26	1.00	0.83	0.12	0.92	0.49	0.68	0.91	0.33	0.40	0.41	0.30
Liming Sq.	0.50	0.33	0.71	0.23	0.86	0.43	0.66	0.87	0.27	0.37	0.25	0.35
Xinhui St.	0.26	0.33	0.50	0.08	0.79	0.42	0.84	1.00	0.16	0.54	0.34	0.61
Xinning St.	0.50	0.33	0.54	0.38	0.66	0.42	0.66	1.00	0.21	0.49	0.39	0.45
Dongdaying St.	0.26	0.33	0.25	0.58	0.66	0.23	0.60	0.96	0.02	0.29	0.12	0.22
Agriculture Univ.	0.26	0.33	0.08	0.27	0.60	0.18	0.49	1.00	0.07	0.29	0.12	0.16
Qianling	0.26	0.33	0.12	0.27	0.55	0.22	0.37	1.00	0.04	0.23	0.10	0.10
East Mausoleum Park	0.26	0.33	0.17	0.19	0.47	0.00	0.01	0.57	0.01	0.00	0.01	0.00
Shuiquan	0.50	0.33	0.04	0.00	0.36	0.10	0.18	0.43	0.06	0.66	0.07	1.00
Boguan North Ave.	0.50	0.33	0.00	0.00	0.28	0.12	0.17	0.39	0.00	0.43	0.07	0.53
Botanical Garden	0.26	0.33	0.04	0.00	0.21	0.19	0.28	0.46	0.01	1.00	0.03	0.94
Shuang Ma	0.26	0.00	0.00	0.00	0.11	0.00	0.00	0.00	0.00	0.00	0.00	0.79

References

1. Cao, K.; Deng, Y.; Song, C. Exploring the drivers of urban renewal through comparative modeling of multiple types in Shenzhen. *Cities* **2023**, *137*, 104294. [CrossRef]
2. Suzuki, H.; Cervero, R.; Iuchi, K. *Transforming Cities with Transit: Transit and Land-Use Integration for Sustainable Urban Development*; World Bank Publications: Washington, DC, USA, 2013.
3. Xu, W.; Guthrie, A.; Fan, Y.; Li, Y. Transit-oriented development in China: Literature review and evaluation of TOD potential across 50 Chinese cities. *J. Transp. Land Use* **2017**, *10*, 743–762. [CrossRef]
4. Wang, L.; Xia, H. A Comprehensive Review of the Development Characteristics and Future Trends of TOD in Chinese Urban Rail Transit. *Urban Rail Transit* **2024**, *10*, 335–349. [CrossRef]
5. Chinese Ministry of Housing and Urban-Rural Development. *Guidelines for Planning and Design of Areas Along the Urban Rail Transit*; Chinese Ministry of Housing and Urban-Rural Development: Beijing, China, 2015.
6. Hua, X.Y.; Cheng, L.; Bai, H. Thinking on application of transit oriented development in Chinese cities. *Transp. Stand.* **2009**, *200*, 80–84.
7. Salat, S.; Ollivier, G. *Transforming the Urban Space Through Transit-Oriented Development: The 3V Approach*; World Bank: Washington, DC, USA, 2017.
8. Sharma, R.; Newman, P. Can land value capture make PPP's competitive in fares? A Mumbai case study. *Transp. Policy* **2018**, *64*, 123–131. [CrossRef]
9. Li, X.; Yuan, J.; Liu, X.; Ke, Y.; Jia, S. Identifying critical influencing factors of the value creation of urban rail transit PPP projects in China. *Buildings* **2022**, *12*, 1080. [CrossRef]
10. Douglas, G.C. Rail transit identification and neighbourhood identity. Exploring the potential for 'community-supportive transit'. *J. Urban Des.* **2010**, *15*, 175–193. [CrossRef]
11. Ibeas, Á.; Cordera, R.; dell'Olio, L.; Coppola, P.; Dominguez, A. Modelling transport and real-estate values interactions in urban systems. *J. Transp. Geogr.* **2012**, *24*, 370–382. [CrossRef]

12. Calthorpe, P. *The Next American Metropolis: Ecology, Community, and the American Dream*; Princeton Architectural Press: Princeton, NJ, USA, 1993; ISBN 1-878271-68-7.
13. Cervero, R.; Kockelman, K. Travel demand and the 3Ds: Density, diversity, and design. *Transp. Res. Part D* **1997**, *2*, 199–219. [CrossRef]
14. Ewing, R.; Cervero, R. Travel and the Built Environment. *J. Am. Plan. Assoc.* **2010**, *76*, 265–294. [CrossRef]
15. Jamme, H.-T.; Rodriguez, J.; Bahl, D.; Banerjee, T. A twenty-five-year biography of the TOD concept: From design to policy, planning, and implementation. *J. Plan. Educ. Res.* **2019**, *39*, 409–428. [CrossRef]
16. Aston, L.; Currie, G.; Pavkova, K. Does transit mode influence the transit-orientation of urban development?—An empirical study. *Transp. Geogr.* **2016**, *55*, 83–91. [CrossRef]
17. Stover, V.G. *Transportation and Land Development—Transportation Research Board Special Report*; National Academy of Science: Washington, DC, USA, 1987.
18. Handy, S. Smart growth and the transportation-land use connection: What does the research tell us? *Int. Reg. Sci. Rev.* **2025**, *28*, 146–167. [CrossRef]
19. Frank, L.D. Land Use and Transportation Interaction: Implications on Public Health and Quality of Life. *Plan. Educ. Res.* **2000**, *20*, 6–22. [CrossRef]
20. Dittmar, H. A broader context for transportation planning: Not just an end in itself. *J. Am. Plan. Assoc.* **1995**, *61*, 7–13. [CrossRef]
21. Olawepo, R.A. Perspectives on urban renewal and transportation development in Lagos: Implications for urban development in Nigeria. *Afr. Res. Rev.* **2010**, *4*, 273–287. [CrossRef]
22. Shin, Y. Urban Regeneration Strategies of local Metropolitan Cities' Railway Stations and the Surroundings in Japan. *J. Asia* **2011**, *14*, 79–99.
23. Dai, P.; Fu, H.; Yang, X.; Han, S.; Fu, G.; Wang, Y. Exploring the urban renewal strategy based on transit-oriented development concept—A case study of Japan and Hong Kong. *Front. Mater.* **2023**, *10*, 1098027. [CrossRef]
24. Sho, K.; Zhu, J. Are Urban Renewal Projects in Taipei City Transit Oriented? *Jpn. Archit. Rev.* **2025**, *8*, e70016. [CrossRef]
25. Mocák, P.; Matlovicova, K.; Matlovic, R.; Pénzes, J.; Pachura, P.; Mishra, P.K.; Kostilnikova, K.; Demková, M. 15-minute City Concept as a Sustainable Urban Development Alternative: A Brief Outline of Conceptual Frameworks and Slovak Cities as a Case. *Folia Geogr.* **2022**, *64*, 69–89.
26. Moreno, C.; Allam, Z.; Chabaud, D.; Gall, C.; Pratlong, F. Introducing the "15-Minute City": Sustainability, resilience and place identity in future post-pandemic cities. *Smart Cities* **2021**, *4*, 93–111. [CrossRef]
27. Bertolini, L.; Spit, T. *Cities on Rails: The Redevelopment of Railway Stations and Their Surroundings*; Routledge: London, UK, 1998; ISBN 1135811253.
28. Lyu, G.; Bertolini, L.; Pfeffer, K. Developing a TOD typology for Beijing metro station areas. *Transp. Geogr.* **2016**, *55*, 40–50. [CrossRef]
29. Vale, D.S.; Viana, C.M.; Pereira, M. The extended node-place model at the local scale: Evaluating the integration of land use and transport for Lisbon's subway network. *Transp. Geogr.* **2018**, *69*, 282–293. [CrossRef]
30. Cao, Z.; Asakura, Y.; Tan, Z. Coordination between node, place, and ridership: Comparing three transit operators in Tokyo. *Transp. Res. Part D* **2020**, *87*, 102518. [CrossRef]
31. Rodriguez, D.A.; Kang, C.D. A typology of the built environment around rail stops in the global transit-oriented city of Seoul, Korea. *Cities* **2020**, *100*, 102663. [CrossRef]
32. Turbay, A.L.; Pereira, R.H.; Firmino, R. The equity implications of TOD in Curitiba. *Case Stud. Transp. Policy* **2024**, *16*, 101211. [CrossRef]
33. Mulley, C.; Tsai, C.H.P. When and how much does new transport infrastructure add to property values? Evidence from the bus rapid transit system in Sydney, Australia. *Transp. Policy* **2016**, *51*, 15–23. [CrossRef]
34. Cervero, R.; Murakami, J. Rail and property development in Hong Kong: Experiences and extensions. *Urban Stud.* **2009**, *46*, 2019–2043. [CrossRef]
35. Yang, L.; Chen, Y.; Xu, N.; Zhao, R.; Chau, K.W.; Hong, S. Place-varying impacts of urban rail transit on property prices in Shenzhen, China: Insights for value capture. *Sustain. Cities Soc.* **2020**, *58*, 102140. [CrossRef]
36. Mittal, J.; Berson, S. *Sustainable Urbanism in Developing Countries*, 3rd ed.; CRC Press: Boca Raton, FL, USA, 2022; pp. 143–165.
37. Yi, Z.; Zhang, Z.; Wang, H. Measuring metro station area's TODness: An exploratory study of Shenyang based on multi-source urban data. *Trans. GIS* **2024**, *28*, 623–653. [CrossRef]
38. Wu, W.; Divigalpitiya, P. Assessment of Accessibility and Activity Intensity to Identify Future Development Priority TODs in Hefei City. *Land* **2022**, *11*, 1590. [CrossRef]
39. Zhao, J.; Deng, W.; Song, Y.; Zhu, Y. What influences Metro station ridership in China? Insights from Nanjing. *Cities* **2013**, *35*, 114–124. [CrossRef]
40. Chen, H.; Zhao, K.; Zhang, Z.; Zhang, H.; Lu, L. Methods for the Performance Evaluation and Design Optimization of Metro Transit-Oriented Development Sites Based on Urban Big Data. *Land* **2024**, *13*, 1233. [CrossRef]

41. OpenStreetMap. Available online: https://www.openstreetmap.org (accessed on 4 May 2025).
42. Amap. Available online: https://www.amap.com (accessed on 4 May 2025).
43. Lianjia. Available online: https://bj.lianjia.com (accessed on 4 May 2025).
44. Swguancha. Available online: https://www.swguancha.com (accessed on 4 May 2025).
45. Reusser, D.E.; Loukopoulos, P.; Stauffacher, M.; Scholz, R.W. Classifying railway stations for sustainable transitions–balancing node and place functions. *Transp. Geogr.* **2008**, *16*, 191–202. [CrossRef]
46. Kim, J.S.; Shin, N. Planning for Railway Station Network Sustainability Based on Node–Place Analysis of Local Stations. *Sustainability* **2021**, *13*, 4778. [CrossRef]
47. Bertolini, L.; Le Clercq, F.; Kapoen, L. Sustainable accessibility: A conceptual framework to integrate transport and land use plan-making. Two test-applications in the Netherlands and a reflection on the way forward. *Transp. Policy* **2005**, *12*, 207–220. [CrossRef]
48. Lu, X.; Wang, J.; Zhang, Z.; Hu, X.; Guo, H. Evaluation and classification of Beijing rail microcenter subway stations based on extended node-place model. *Sci. Technol. Rev.* **2023**, *41*, 41–51.
49. Saardchom, N. The validation of analytic hierarchy process (AHP) scoring model. *Int. J. Liab. Sci. Enq.* **2012**, *5*, 163–179. [CrossRef]
50. Li, J.; Shen, Z.; Ma, J.; Hao, Y.; Zeng, R.; Lin, Z. Evaluation of Sustainable Development Potential of Railway Station Areas in Guangzhou Metropolitan Area: Based on the Extended "Node–Place–Support" Model. *Land* **2025**, *14*, 369. [CrossRef]
51. Teng, L.; Zhong, C.; Cai, D. Study on the Spatial Type Differentiation of Guangzhou Metro TOD Zones: Based on" Node-Place-Linkage" Coupling Model. *Econ. Geogr.* **2023**, *42*, 103–111.
52. Guo, R.; Song, X.; Li, P.; Wu, G.; Guo, Z. Large-Scale and Refined Green Space Identification-Based Sustainable Urban Renewal Mode Assessment. *Math. Probl. Eng.* **2020**, *2020*, 2043019. [CrossRef]
53. Yang, L.; Song, X. TOD typology based on urban renewal: A classification of metro stations for Ningbo city. *Urban Rail Transit* **2021**, *7*, 240–255. [CrossRef] [PubMed]
54. Arthur, D.; Vassilvitskii, S. *K-Means++: The Advantages of Careful Seeding*; Stanford University: Stanford, CA, USA, 2006.
55. Yang, W.; Yan, W.; Chen, L.; Li, H. From city center to suburbs: Developing a timeline-based TOD assessment model to explore the dynamic changes in station areas of Tokyo metropolitan area. *Environ. Plan. B Urban Anal. City Sci.* **2024**, *52*, 214–230. [CrossRef]
56. Zhang, S.; Huang, Z.; Zhao, K. Research on the Coordinated Development of "Node-Place" in Intercity Railway Station Areas: A Case Study of the Guangdong–Hong Kong–Macao Greater Bay Area, China. *ISPRS Int. J. Geo-Inf.* **2025**, *14*, 121.
57. Zhang, Y.; Qin, M. Applying "Node-Place-Value (NPV)" Model on the Renewal of Suburban Areas Along Rail Transit Lines. In Proceedings of the 2024 2nd International Conference on Big Data and Privacy Computing (BDPC), Macau, China, 10–12 January 2024; pp. 148–153.
58. Zhu, K.; Chen, W.; Zhang, Y. Development Coordination of Chinese Megacities Using the Node–Place–Value Model: A Case Study of Changsha. *Urban Sci.* **2025**, *9*, 121. [CrossRef]
59. Yang, Y.; Zhong, C.; Gao, Q.L. An extended node-place model for comparative studies of transit-oriented development. *Transp. Res. Part D Transp. Environ.* **2022**, *113*, 103514. [CrossRef]
60. Dou, M.; Wang, Y.; Dong, S. Integrating network centrality and node-place model to evaluate and classify station areas in Shanghai. *ISPRS Int. J. Geo-Inf.* **2021**, *10*, 414. [CrossRef]
61. Nemroudi, R.; Ortuño, A.; Flor, M.; Guirao, B. Application of the Node–Place Model in Algiers (Algeria). *Sustainability* **2024**, *16*, 6428. [CrossRef]
62. Dai, P.; Bi, L.; Cen, S.; Fu, G.; Fang, H.; Wang, Y. Research on the Coupling Relationship Between Park and Metro Station Space in Qingdao. *Sustainability* **2025**, *17*, 3029. [CrossRef]
63. Reggiani, M. Urban Regeneration Strategies and Place Development in Contemporary Tokyo: The Case of Shibuya Station Area. *J. Place Manag. Dev.* **2022**, *15*, 40–54. [CrossRef]

land

MDPI

Article

Street View-Enabled Explainable Machine Learning for Spatial Optimization of Non-Motorized Transportation-Oriented Urban Design

Yichen Ruan [1], Xiaoyi Zhang [1,*], Shaohua Wang [2,3], Xiuxiu Chen [1,4] and Qiuxiao Chen [1]

1 School of Spatial Planning and Design, Hangzhou City University, Hangzhou 310015, China; ruanyc@hzcu.edu.cn (Y.R.); cxxtribal@zju.edu.cn (X.C.); chenqx@hzcu.edu.cn (Q.C.)
2 State Key Laboratory of Remote Sensing and Digital Earth, Aerospace Information Research Institute, Chinese Academy of Sciences, Beijing 100101, China; wangshaohua@aircas.ac.cn
3 University of Chinese Academy of Sciences, Beijing 101408, China
4 Department of Regional and Urban Planning, College of Civil Engineering and Architecture, Zhejiang University, Hangzhou 310058, China
* Correspondence: zhangxiaoyi@hzcu.edu.cn

Abstract: To advance evidence-based urban design prioritizing non-motorized mobility, this study proposes a street view-enabled explainable machine learning framework that systematically links built environment semantics to non-motorized transportation vitality optimization. By integrating Baidu Street View images with deep learning-based object detection (Faster R-CNN), we quantify fine-grained human-powered and mechanically assisted mobility vitality. These features are fused with multi-source geospatial data encompassing 23 built environment variables into an interpretable machine learning pipeline using SHAP-optimized random forest models. The key findings reveal distinct nonlinear response patterns between HP and MA modes to built environment factors; for instance, a notable promotion in mechanically assisted NMT vitality is observed as enterprise density increases beyond 0.2 facilities per ha. Emergent synergistic and threshold effects are evident from variable interactions requiring multidimensional planning consideration, as demonstrated in phenomena such as the peaking of human-powered NMT vitality occurring at public facility densities of 0.2–0.8 facilities per ha, enterprise densities of 0.6–1 facilities per ha, and spatial heterogeneity patterns identified through Bivariate Local Moran's I clustering. This research contributes an innovative technical framework combining street view image recognition with explainable AI, while practically informing urban planning through evidence-based mobility zone classification and targeted strategy formulation, enabling more precise optimization of pedestrian-/cyclist-oriented urban spaces.

Keywords: street view imagery (SVI); object detection; explainable machine learning (XAI); Shapley additive explanation (SHAP); non-motorized transportation (NMT)

1. Introduction

Many cities around the globe are increasingly recognizing the critical importance of non-motorized transportation (NMT) in promoting sustainable urban living. The vitality of NMT systems, encompassing *human-powered modes* such as walking and *mechanically assisted modes* like cycling and e-bikes, directly impacts public health, carbon neutrality, and social equity [1,2]. As a key determinant of travel behavior, the built environment (BE) profoundly shapes NMT adoption by mediating accessibility, safety, and comfort [3,4]. However, understanding how multidimensional BE features synergistically drive NMT

 https://doi.org/10.3390/land14071347

vitality remains challenging, requiring granular analysis of both physical infrastructure and behavioral dynamics [5,6].

The associations between the built environment and NMT travel behavior have been confirmed by numerous studies [3–8]. A general conclusion is that a compact area with mixed land use, a high population density, and high accessibility tends to enlarge the probability of NMT travel and reduce car travel [9–11]. Yet conceptual fragmentation persists; some studies treat NMT as a single mode within broader modal choice frameworks [12,13], while others isolate specific subtypes like pedestrian activity [14,15], cycling [16,17], or emerging e-bike usage [18,19]. However, analyses differentiating human-powered (HP) and mechanically assisted (MA) NMT subtypes remain scarce and constrained by methodological reliance on subjective survey data [20,21].

Recent advances in street view imagery (SVI) analytics now enable objective, large-scale classification of NMT activities through computer vision [22], offering unprecedented spatial resolution to decode BE-NMT interactions. Computer vision algorithms, from early feature extraction to deep learning-based object detection such as Faster R-CNN enable the automated identification of NMT agents like pedestrians and cyclists with high precision [23], as well as micro-scale urban elements, including vegetation [24], sidewalks and buildings [25], etc. Unlike satellite imagery or field surveys, SVI captures the ground-level perspectives essential for evaluating pedestrian-scale environments [26], and its temporal coverage supports longitudinal BE change tracking [27]. This SVI-centric approach addresses the resolution limitations of survey-based studies and satellite-derived land use data [28,29], providing a new quantitative means for analyzing interactions between BE factors and NMT choice.

While SVI technologies provide unprecedented observational granularity, conventional approaches to modeling built environment (BE) impacts—including multinomial logit models [11,12], spatial regressions [30], and factor analysis [31]—have established foundational relationships; their reliance on linear in-parameter assumptions oversimplifies the complex, interdependent nature of urban systems [32,33]. Emerging evidence demonstrates that nonlinear frameworks better capture threshold effects in BE–mobility interactions [34] and subtype-specific response patterns [35] critical for NMT optimization [36]. For instance, machine learning models have outperformed linear methods in explaining active travel variance and electric bike usage dynamics [37], revealing BE factor interactions that traditional approaches mask [38]. Recent studies have used SVI data to explore these relationships, leveraging the high-quality and widespread use of SVI to identify nonlinear correlations between built environment features and travel choice [39]. However, there is insufficient exploration of how nonlinear BE interactions differentially shape subtype-specific mobility patterns. This oversight impedes the design of tailored infrastructure interventions, as human-powered and mechanically assisted NMT modes exhibit divergent dependencies on gradients, connectivity, and safety features [33].

This paper advances a novel integration of SVI analytics and explainable machine learning to decode the nonlinear influence of BE factors on NMT dynamics in Hangzhou's Binjiang District. Utilizing Faster R-CNN [24,25] for the automated detection of NMT agents and BE features from Baidu SVI, we quantify 23 geospatial variables influencing NMT vitality. By leveraging SHAP (SHapley Additive exPlanations) values [40], this study reveals not only the relative importance of each factor but also interaction effects between urban form variables. Spatial heterogeneity in these relationships is further mapped via Bivariate Local Moran's I [41], which has important implications for the construction of NMT-friendly city planning. By bridging SVI-driven NMT observation with interpretable nonlinear modeling, our framework unfolds a deeper understanding of the diverse urban factors that drive NMT use and shows why planners must consider multiple environmental aspects for sustainable cities.

In the following section, the paper presents the research area and the data collection process used in the study. Section 3 discusses the distribution characteristics of NMT objects and BE factors, presents the model-fitting results, and conducts an analysis of the findings. Section 4 summarizes the spatial heterogeneity of different NMT types and the potential BE factors behind them. The final section discusses the novelty of the research outcomes and strategies for applying them in practical applications, combining the research findings with real-world practices.

2. Data and Methods

2.1. Study Area

The study area, Binjiang District in Hangzhou (Figure 1), serves as a representative urban case for examining the nonlinear relationship between BE with NMT vitality. As a central urban district established in 1996, Binjiang later integrated with the Hangzhou National High-Tech Industrial Development Zone in 2002. With the support of this policy, it has undergone rapid urbanization and infrastructure development. By 2023, it had a population density of 7341 persons/km^2 and covered an administrative area of 72.2 km^2 [42].

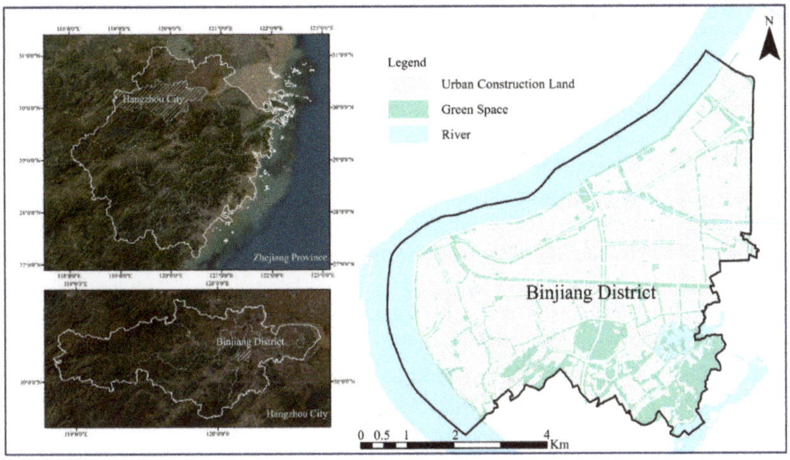

Figure 1. Study area: Binjiang District in the city of Hangzhou.

The district is renowned for its comprehensive non-motorized transport network, including a decade-old public bicycle system [43] and an extensive riverside greenway system that integrates recreational spaces, residential areas, and commercial hubs [44]. This infrastructure provides a unique opportunity to analyze how urban design elements influence walking, cycling, and e-bike behavior across space and time.

2.2. Research Framework

A quantitative analysis framework (Figure 2) is established to assess the nonlinear impact and synergistic effects of the built environment on NMT vitality, grounded in the concept of environmental amenities. As defined in the framework, amenities encompass desirable features of built environments that enhance comfort, convenience, or enjoyment through four key dimensions: esthetic comfort (visual appeal), functional comfort (utility optimization), safety comfort (risk mitigation), and social comfort (interpersonal interaction space). These environmental stimuli interact with human responses—including perceived comfort levels and behavioral intentions—to jointly determine NMT vitality.

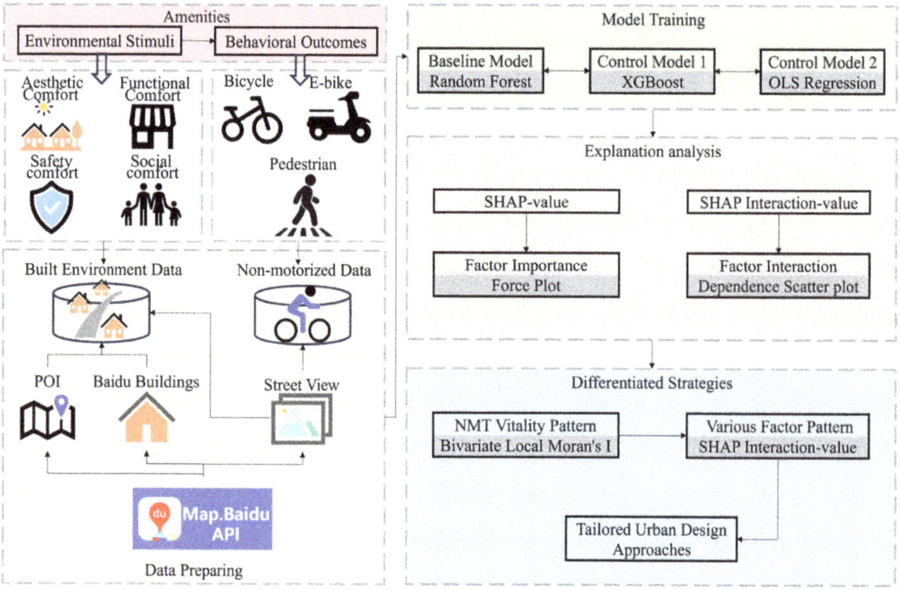

Figure 2. Research framework and three interconnected modules.

The framework integrates three interconnected modules, an environmental stimuli module, a model training module, and an explanation analysis module. The environmental stimuli module uses multi-source geospatial data, such as building footprints and Points of Interest (POIs) from the Baidu Maps open platform (https://lbsyun.baidu.com/ accessed on 18 April 2023) and street view imagery, and extracts factors of the built environment in four comfort dimensions as well as NMT vitality. The model training module employs a hybrid validation strategy, including a random forest model [45] as a baseline model, and two control models using XGBoost and OLS regression. And the explanation analysis module leverages SHAP values and integration terms to hierarchically rank factor importance, identify nonlinear relationships, and visualize spatial heterogeneity through Bivariate Local Moran's I and dependence scatter plots.

By coupling built environment factor prioritization with interaction effect detection, this framework enables targeted optimization strategies. The findings inform differentiated design interventions and tailored urban design guidelines for enhancing NMT vitality through evidence-based spatial reconfiguration.

2.3. Object Detection for Non-Motorized Transportation via Street View Images

The street view images for this study were collected through Baidu Street View API in 2022, and the capture time was from June 2019 to August 2021. The street view images were systematically collected across Binjiang District, covering all administrative subdistricts through a geographically stratified sampling framework. The sampling framework is composed of the street network of the study area obtained by OpenStreetMap (OSM), and the interval between sampling points is 50 m. Through multi-task object detection, we quantified distributions of eight mobility-related agents as proxy indicators of non-motorized transport vitality (Figure 3), including pedestrians, bicycles, e-bikes, traffic lights, fire hydrants, parking signs, parking meters, and animals like cats and dogs. We implemented a modified Faster R-CNN framework with a ResNet-50 backbone pre-trained on the COCO 2017 dataset. The model achieved an 89.2% mean average precision across eight classes, with pedestrian detection showing an exceptional precision of 93.7%. Notably, animal detection served as a social comfort indicator of the built environment. Previous studies have shown that this model also performs well for built environments in Chinese streetscapes such as Guangzhou and Wuhan [25,35], demonstrating satisfactory generalization.

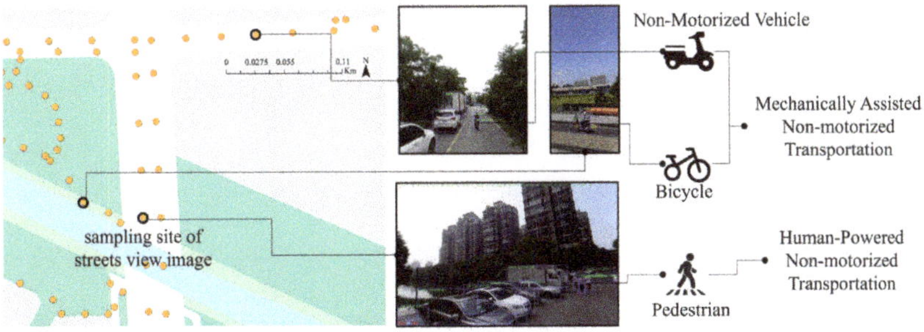

Figure 3. Object detection from Street View Images.

For each SVI coordinate point, we constructed 400 m Euclidean buffers to represent neighborhood-scale mobility catchments. The buffer corresponds to roughly a 5 min walking distance (assuming an average walking speed of about 1.3 m/s) and is a commonly used neighborhood-scale metric in urban planning and transit research [30]. This buffer both encompasses the typical catchment area for residents' daily transfers and leisure trips and remains computationally efficient, effectively characterizing the spatial reach of NMT. Two NMT vitality metrics were derived through kernel density estimation. Human-powered NMT vitality is calculated as the mean kernel density intensity of pedestrian detections, weighted by occlusion-corrected visibility scores. Mechanically assisted NMT vitality is quantified as the sum of bicycle and e-bike kernel density intensities.

2.4. *Quantifying the Nonlinear Influences of the Built Environment*

2.4.1. Built Environment Factors

Urban NMT vitality emphasizes residents' perceptions and experiences of urban space when walking or cycling. The choice to travel by foot or bicycle depends not only on destination accessibility but also on factors such as environmental comfort, pedestrian environment quality, visual esthetics, opportunities for social interaction, and perceived safety [11]. Amenity theory highlights the role of environmental factors—livability, convenience, safety, and esthetics—in an area's attractiveness. This theory offers a framework to analyze spatial variations in urban NMT vitality and its driving factors. Recent studies apply amenity theory to NMT vitality: using streetscape data to assess the esthetic comfort of urban spaces and demonstrating its positive effect on walking activity [4]; examining how the density of different facility types (commercial, transportation, and public services) promotes walking/cycling [20]; and showing that safety infrastructure (crosswalks, traffic signals, and lighting) increases pedestrian vitality [46].

At present, scholars typically divide urban amenities into four dimensions: esthetic comfort, functional comfort, safety comfort, and social comfort. The esthetic comfort dimension measures the visual and architectural quality of urban landscapes. Previous research confirms that visual esthetics significantly affect pedestrian/cyclist choices [47]. Based on existing studies, we selected proximity to parks and plazas (PKS), the building height-to-width ratio (BHR), and building density (BD) [47,48]. BHR and BD data derive from Baidu Buildings, a vector spatial dataset via Baidu Maps API, which provides footprints, heights, floor counts, and functional data for our study area. BHR is defined as the building height divided by the street frontage width per analysis unit (400 m Euclidean buffers); BD is the ratio of total building footprint area to analysis unit area.

Functional comfort evaluates the accessibility and convenience of facilities for NMT [5]. We selected key pedestrian destinations informed by prior research [49]: parking meters (PMs), food and beverage services (FBs), public facilities (PFs), enterprises (ENTs), shopping venues (Ss), transportation services (TSs), financial and insurance services (FISs), commercial residences (CHs), daily life services (DLSs), sports and recreation facilities (SRs), and accommodation services (ASs).

Safety comfort assesses safety infrastructure for pedestrians and cyclists. Safety factors strongly influence individuals' willingness to engage in NMT [15]. Evidence confirms that facilities such as traffic lights (TL), parking signs (PGSs), fire hydrants (FHs), and medical services (MSs) improve actual and perceived safety, encouraging non-motorized activity [50].

The social comfort dimension encompasses factors related to social interaction opportunities and population vitality, grounded in social capital theory and human-centered urban design principles [51]. The presence of pets (cats and dogs, CDs), resident population (R), employed population (EMP), science and educational and cultural service facilities (SESs), and government and social organizations (GOSGs) have all been widely recognized as important indicators affecting social interaction, community cohesion, and walking vitality [5]. All built environment factors with the associated data sources can be seen in Table 1.

Table 1. Built-environment factors and data sources associated with the four comfort states.

Category	Factors	Abbreviation	Data Source
Esthetic comfort	Park and Square	PKS	POI
	Building Height-to-Width Ratio	BHR	Baidu buildings
	Building Density	BD	Baidu buildings
Functional comfort	Parking Meter	PM	Street view
	Food and Beverages	FB	POI
	Public Facility	PF	POI
	Enterprises	ENTs	POI
	Shopping	S	POI
	Transportation Service	TS	POI
	Finance and Insurance Service	FIS	POI
	Commercial House	CH	POI
	Daily Life Service	DLS	POI
	Sports and Recreation	SR	POI
	Accommodation Service	AS	POI
Safety comfort	Traffic Light	TL	Street view
	Fire Hydrant	FH	Street view
	Parking Sign	PGS	Street view
	Medical Service	MS	POI
Social comfort	Cats and Dogs	CDs	Street view
	Residents	Rs	Mobile data
	Employees	EMPs	Mobile data
	Science/Culture and Education Service	SES	POI
	Governmental Organization and Social Group	GOSG	POI

2.4.2. Regression Model Construction

To comprehensively assess model performance and validate the robustness of the nonlinear relationships, we implemented a hybrid analytical framework. Random forest (RF) served as the core interpretable machine learning model, XGBoost worked as a high-performance gradient-boosting counterpart, and ordinary least squares (OLS) worked as a linear baseline.

Random forest employs dual stochastic mechanisms to enhance generalizability. First, each decision tree is trained on a bootstrapped subset of the data (sampled with replacement), allowing ensemble aggregation to mitigate overfitting. Second, at every node split, a random subset of features is evaluated to decorrelate trees and improve collective predictive stability.

As a state-of-the-art gradient boosting implementation, XGBoost optimizes differentiable loss functions through additive tree construction. Its sequential error-correction mechanism contrasts with RF's parallelized bagging approach, providing insights into nonlinear pattern capture efficiency. OLS with standardized coefficients assumes fixed marginal effects between BE factors and NMT vitality. This traditional parametric method highlights the limitations of linear additivity in capturing threshold behaviors and variable interactions.

2.4.3. Nonlinear Influences Interpretation

To decode the nonlinear mechanisms through which built environment (BE) factors shape non-motorized transportation vitality, we employed SHAP [52], a game theory-based interpretability framework uniquely suited to analyzing complex spatial interactions in urban systems. Unlike global feature importance metrics, SHAP quantifies localized contributions of individual BE variables to specific NMT vitality predictions. This granular perspective aligns with our focus on spatially heterogeneous BE-NMT dynamics in high-density urban cores.

Using the SHAP v0.43.0 Python package, we applied TreeExplainer, optimized for ensemble tree models like random forests, to calculate two critical metrics. The Shapley value (Equation (1)) quantifies the marginal impact of each BE variable on localized NMT vitality predictions relative to the dataset mean. Shapley interaction values (Equation (2)) identify synergistic or antagonistic effects between BE variable pairs.

$$\phi_k(f,x) = \sum_{S\in\tau} \frac{1}{K!}[f_x\left(P_k^S \bigcup k\right) - f_x(P_k^S)] \tag{1}$$

where $\phi_k(f,x)$ is the Shapley value, τ is the set of possible variable orderings, k is the number of variables, P_k^S is the variables that come before the variable k in the ordering S, x is the value of explanatory variables, and f means a single prediction.

$$\phi_{i,j}(f,x) = \sum_{T\in\xi\backslash\{i,j\}} \frac{|T|!(K-|T|-2)!}{2(k-1)} \nabla_{i,j}(f,x,T) \tag{2}$$

where $\phi_{i,j}(f,x)$ is the Shapley interaction value that reflects the interaction effect between variable i and variable j on a single prediction, K is the number of variables, ξ is the set of input variables, T is the possible variable coalitions, and x is the value of input variable for the prediction.

Synergism emerges when a cooperation or interaction causes an increase in the whole so that the whole is greater than the sum of parts [53]. For our NMT context, positive interactions $\phi_{i,j} > 0$ indicate BE factor synergies amplifying subtype vitality, while negative values $\phi_{i,j} < 0$ suggest conflicting design priorities.

This approach advances traditional linear paradigms by addressing three core challenges in BE-NMT modeling, the spatial heterogeneity of nonlinear effects, context-dependent variable interactions, and subtype-specific infrastructure optimization thresholds.

3. Results

3.1. NMT Target Objects and BE Factors in the Study Area

A total of 31,302 street view images were collected across Binjiang District via Baidu Map API, achieving a spatial density of approximately 433 images per square kilometer. These images comprehensively cover both arterial roads and secondary street networks. Using Faster R-CNN-based object detection, we identified eight critical street-level elements (Table 2), including pedestrians, bicycles, and e-bikes (response variables, *Y*), as well as infrastructure features such as traffic lights and fire hydrants (explanatory variables, *X*).

Table 2. Detected elements from street view imagery and descriptive statistics.

Elements	Mean	Range	Median	Standard Deviation	Variable Type	Statistic Method
Pedestrian	1243.87	[0, 5000]	994.13	979.24	*Y*	kernel density
Bicycle	443.78	[0, 3828]	274.82	547.80	*Y*	kernel density
E-Bikes	218.21	[0, 868]	218.21	166.24	*Y*	kernel density
Traffic Light	0.21	[0, 8]	0.00	0.66	*X*	count
Fire Hydrant	0.02	[0, 3]	0.00	0.14	*X*	count
Parking Sign	0.02	[0, 2]	0.00	0.13	*X*	count
Parking Meter	0.00	[0, 1]	0.00	0.03	*X*	count
Cats and Dogs	0.00	[0, 1]	0.00	0.04	*X*	count

NMT vitality was measured through kernel density estimation within 400 m buffers. Pedestrian vitality has an average number of 1243.87 (SD = 979.24), exhibiting a polycentric clustering pattern. Pedestrians typically travel at speeds ranging from 3 to 4 km per hour, offering a high degree of freedom and spontaneity that allows for thorough observation of the streetscape during transit. Bicycle vitality has an average number of 443.78 (SD = 547.80), concentrated along greenways and metro station corridors. E-bike vitality has an average number of 218.21 (SD = 166.24), primarily aggregated near arterial road connectors. Bicycles and e-bikes, generally travel at speeds between 15 and 20 km per hour. These travel modes are more constrained by non-motorized lanes and topographical conditions, leading to less spontaneity and more superficial streetscape observation. Additionally, they require specific parking spaces.

The other five built environment elements are all in a low mean count, while having different distributions. The average number of traffic lights within the 400 m buffer is 0.21 ± 0.66, a quite large SD compared to the mean. It suggests limited signalization in secondary streets, potentially prioritizing vehicle flow over NMT safety in peripheral areas. Proximity to hydrants may signal well-maintained public spaces conducive to walking, though sparse distribution (mean = 0.02 ± 0.14) limits broad applicability. Clustered parking signage (mean = 0.02 ± 0.13) often coincides with commercial hubs, where vehicular encroachment on sidewalks may deter walking comfort. A near absence of parking meters reflects Binjiang's reliance on off-street parking facilities, suggesting a segregated rather than shared street design. Regarding the situation for animals, however, even a minimal presence (max = 1) could signal micro-scale social comfort zones attractive to pedestrians.

Fire hydrants (FHs) and cats and dogs (CDs) exhibit zero-inflated distributions, yet they each represent critical dimensions of safety and social comfort. Even a single fire hydrant in a neighborhood can signal effective curb-side management and emergency preparedness, factors closely linked to pedestrian confidence and route choice in prior research. Likewise, the occasional presence of pets on the street enhances the "eyes on the street" effect, fostering social ambiance and encouraging walking, so we have retained these variables.

A comprehensive analysis of built environment (BE) factors was conducted using 70,661 Points of Interest (POIs), supplemented by two variables derived from Baidu building data and two mobile data-driven metrics. Pearson correlation coefficients were calculated to assess interdependencies among all 23 BE variables, with results ranging from 0 to 0.7 (Figure 4). This moderate correlation spectrum suggests generally independent explanatory power across most variables, aligning with the study's goal of minimizing multicollinearity in predictive modeling. However, the coefficient for residents and employees is 0.99, indicating a clear collinearity between them. Including both variables in the model may lead to endogeneity issues due to explanatory redundancy. So, we combined the resident population and the employed population into a new variable called population by calculating its average. The retained correlation range ensures stable regression coefficient estimation while preserving granular BE dimensionality.

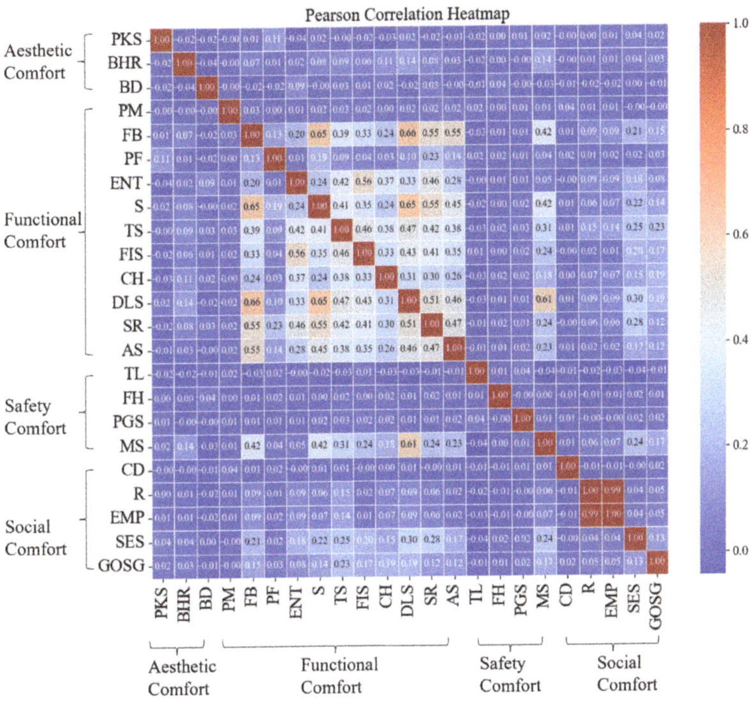

Figure 4. Pearson correlation coefficients of factors.

3.2. Regression Model Optimization and Comparative Analysis

The comparative analysis of regression models systematically evaluated the predictive performance of nonlinear machine learning algorithms against traditional linear regression for NMT vitality prediction. Hyperparameter optimization through grid search identified optimal configurations: random forest (RF) achieved a maximum accuracy with max_depth = None, min_samples_leaf = 1, min_samples_split = 2, and n_estimators = 200, while XGBoost demonstrated superior calibration under constraints of gamma = 0, learning_rate = 0.2, max_depth = 5, min_child_weight = 3, and n_estimators = 300. Both models were rigorously validated against ordinary least squares (OLS) regression using root mean squared error (RMSE) for neighborhood-scale error quantification and adjusted R^2 for variance explanation relative to model complexity. The results in Table 3 reveal consistent nonlinear advantages: XGBoost achieves a 75.16% lower test RMSE than OLS, while RF reduces it by 79.15%, on average. Adjusted R^2 improvements were equally striking, rising from 0.438 (OLS) to 0.960 (XGB) and 0.970 (RF) in the test sets, confirming strong nonlinear relationships between built environment variables and NMT vitality.

Table 3. The performance of the XGBoost, RF, and OLS models.

Model		Random Forest		XGBoost		OLS	
		HP	**MA**	**HP**	**MA**	**HP**	**MA**
RMSE	Training Set	55	42	125	86	617	522
	Test Set	148	102	171	126	643	546
Adjusted R^2	Training Set	0.996	0.995	0.978	0.978	0.459	0.216
	Test Set	0.970	0.965	0.960	0.957	0.438	0.200

Notably, human-powered mobility vitality (HPMV) exhibited superior modeling performance compared to mechanically assisted mobility vitality (MAMV), with pedestrians showing higher adjusted R^2 scores than bicycles/e-bikes across all models. This divergence stems from fundamental differences in mobility patterns; pedestrian activity demonstrates strong spatial autocorrelation due to fixed infrastructure and micro-scale environmental factors, which align with both the linear and tree-based models' capacity to capture localized interactions. In contrast, MAMV's heterogeneity—spanning bicycle lane preferences, e-bike speed dynamics, and route choice hierarchies—introduces spatial non-stationarity that challenges conventional modeling assumptions. The persistent performance gap underscores the importance of differentiated analytical frameworks, particularly the need for interpretability-driven methods to decipher complex nonlinear interactions while maintaining actionable linkages to urban design interventions.

3.3. *Influence of BE Factors on NMT Vitality*

To address the black box nature of machine learning models while preserving their capacity to capture complex built environment dynamics, we calculated mean SHAP values to reveal significant disparities in the relative importance of built environment factors between HP and MA NMT modes (Figure 5). The larger the value, the stronger the global influence of the feature on the model's prediction.

Nine of the top ten built environment factors overlap between the HP and MA models—enterprises, transportation services, population density, governmental organizations and social groups, parks and squares, commercial establishments, food and beverage outlets, public facilities, and finance and insurance services—yet their rank order differs markedly. Daily life services, for instance, rank 4th in the HP model but fall to 12th in the MA predictions, while medical services climb from 11th in HP to 7th in MA. These rank shifts reflect distinct spatial dependencies: pedestrian trips hinge on micro-scale accessibility to daily needs—shorter distances minimize spatial friction and directly boost walking frequency—whereas cyclists and e-bike users tend toward longer, utility-driven journeys. For them, access to medical facilities often depends on dedicated infrastructure that enhances multimodal connectivity. In other words, HP networks prioritize dense, local amenity clusters, while MA networks emphasize key infrastructure placements to serve longer-distance travel demands.

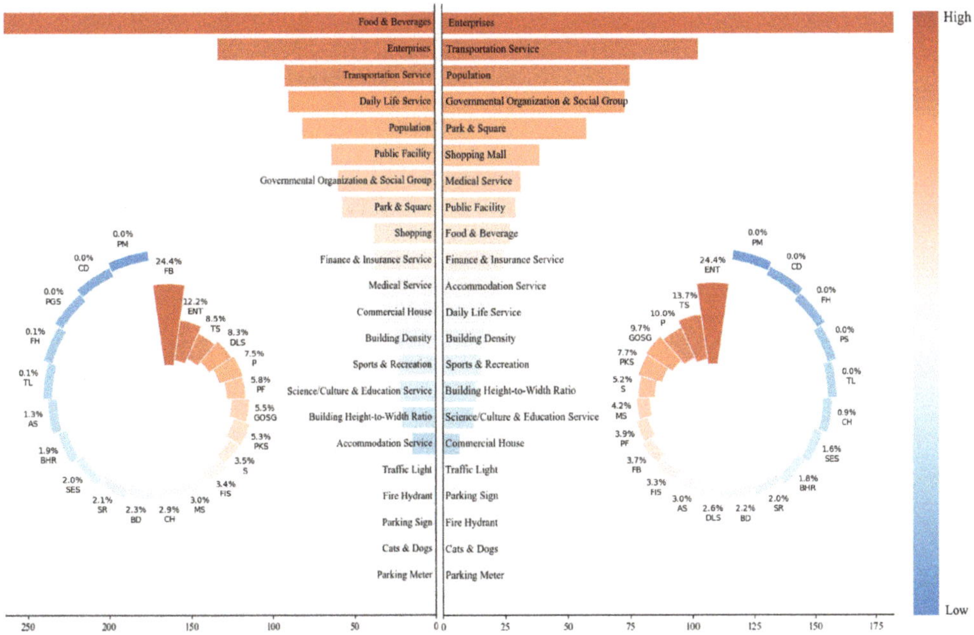

Figure 5. SHAP values of built environment factors for HP and MA scenarios.

On the other hand, factors like public facilities, parks and squares, and shopping amenities show a strong influence on HP but are less significant for MA, revealing critical insights into mobility mode-specific environmental dependencies. High concentrations of public facilities (e.g., libraries or community centers) within 400 m walking catchments reduce spatial friction for pedestrians, creating gravity-driven trip generation patterns. In contrast, MA users prioritize route efficiency over proximity, often bypassing localized amenities to optimize travel time. Parks and squares enhance pedestrian activity through sensory stimulation and social facilitation [54], which are less relevant to MA riders focused on utility-oriented navigation. The spatial configuration of commercial areas further amplifies this dichotomy—compact retail clusters with wide sidewalks boost walking trips but show negligible impact on bicycle route choices.

Overall, the SHAP value shows the nuanced needs of different non-motorized transportation modes and the importance of tailoring mechanism analyses especially nonlinear effects and the interaction effects of factors, as well as tailoring urban design and policy interventions to support the varied requirements of urban NMT vitality.

3.4. Interaction of BE Factors for NMT Vitality

Beyond evaluating each variable's overall importance, we leverage Shapley interaction values to uncover deeper insights into their nonlinear effects. Figures 6 and 7 show the six influential factors driving human-powered and mechanically assisted NMT vitality and their SHAP dependency graphs with the most relevant factors. Each subplot corresponds to a specific BE factor; the Y-axis remains the total SHAP value for the specific BE factors, representing its total contribution to the NMT vitality model, including the main effect and the interaction effect. Color represents the value of an interactive factor, and color mapping helps to intuitively analyze the synergistic effect between features. The trend line fits the relationship between the SHAP value and the BE factors through a Generalized Additive Model (GAM; a GAM fits each predictor with its own smooth function allowing us to capture and visualize nonlinear effects), revealing the average marginal contribution of this feature to the model's output under different values. It removes noise and nonlinear fluctuations, showing the overall trend between BE factors and SHAP values.

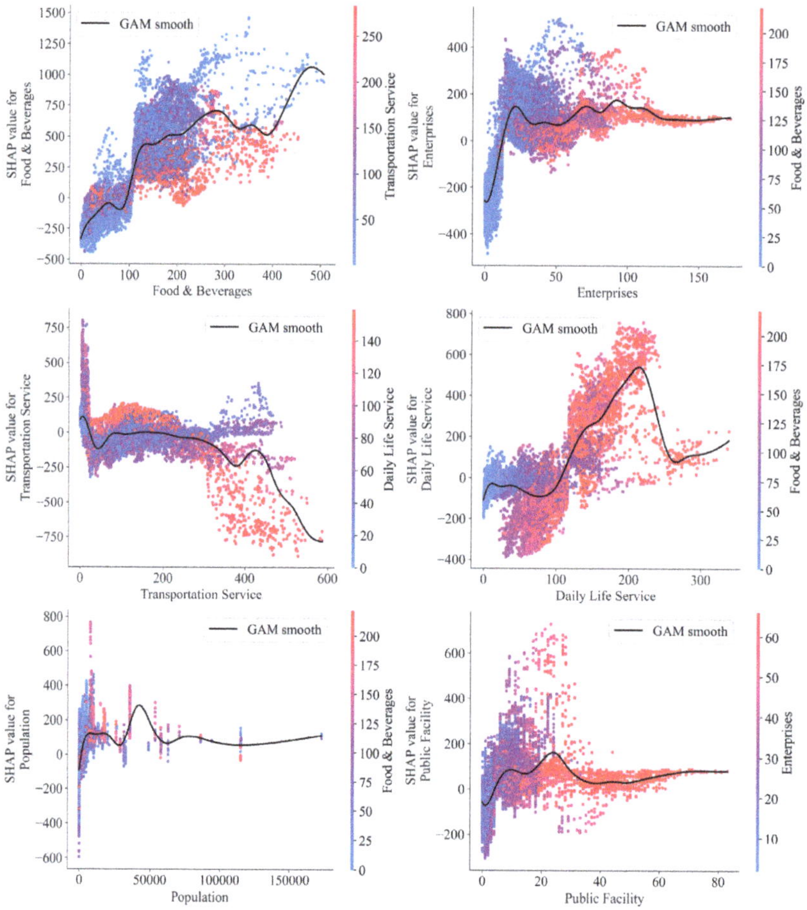

Figure 6. Interaction importance of top six built environment factors on HP.

The SHAP values for food and beverages demonstrate a nonlinear impact on human-powered vitality; when FB density exceeds a threshold of 100, SHAP values become positive. This indicates that dense concentrations of food and beverage services enhance pedestrian vitality[1]. The most relevant factor for food and beverages is transport services. Transportation services have some antagonistic effects on the food and beverage factor. When the food and beverage factor is at a low level—that is, when the SHAP value is negative—the SHAP value gradually increases as the transport service density increases. When the food and beverage factor is at a high level, the SHAP value gradually decreases with the increase in transportation service density.

Enterprise density has a similar threshold effect. When the density exceeds 20, equal to 0.5 enterprises/ha, the SHAP value becomes positive and enterprise density begins to have a positive impact on HP NMT vitality. When the density is in the range of 50 to 125—that is, 1 to 2.5 enterprises/ha—the promotion effect reaches its peak. Beyond this peak range, the promotion effect declines and tends to stabilize. An excessive density of enterprises indicates a higher occupation of space in the area, which can exert a repulsive force on other urban functions. The highest interaction factor for enterprises is food and beverages. There is also a certain antagonistic effect between the two factors. When the enterprise density is the same, the larger the food and beverage store density is, the smaller the SHAP value is. When the SHAP value of enterprise density reaches a stable level, the effect between food and beverages and enterprises disappears.

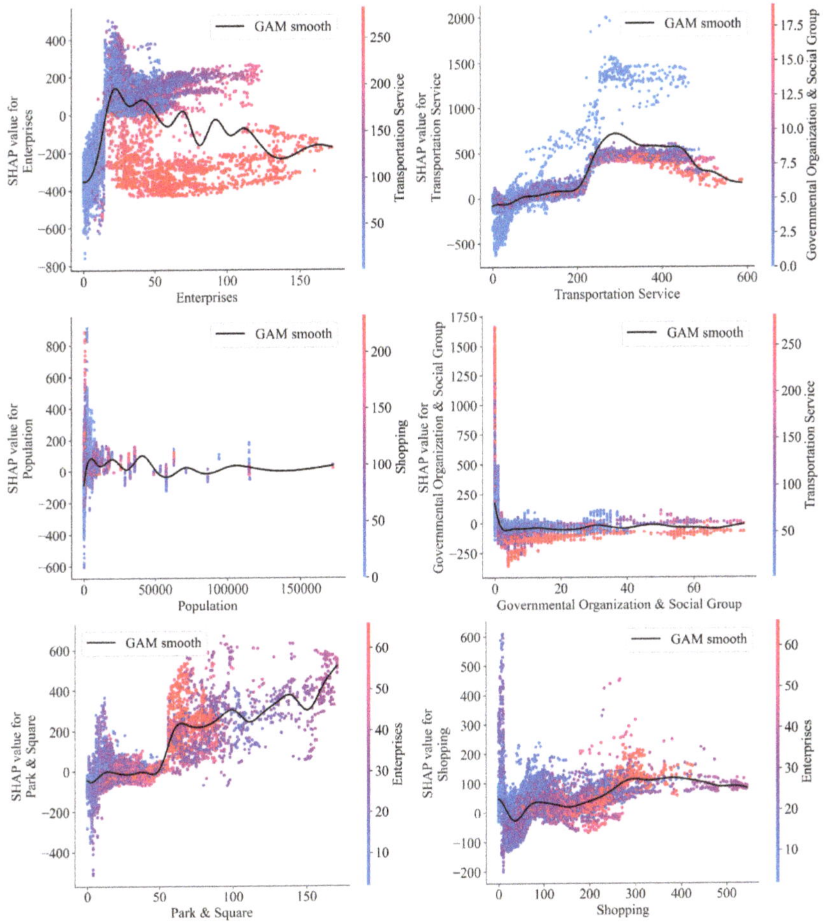

Figure 7. Interaction importance of top six built environment factors on MA.

The transportation service factor exhibits a non-monotonic effect on NMT vitality. Its SHAP value is highest at very low TS levels, then steadily declines as TS density increases, reaching zero around 50 facilities (≈1/ha) and turning negative past 300 facilities (≈6/ha). This pattern makes intuitive sense: when transport options are scarce, walking predominates and contributes positively to NMT vitality; at moderate TS levels, walking serves efficiently as a transfer mode; but when TS is abundant, alternative modes dilute its positive impact on NMT. Furthermore, TS shows its strongest synergy with daily life services. This interaction peaks at medium TS densities (50–250 facilities, or 1–5/ha), where each additional DLS facility significantly boosts the SHAP value of TS. Beyond 250 TS facilities (>5/ha); however, further increases in DLS coincide with a reduction in TS's SHAP contribution.

The SHAP distribution for daily life services is similar to the food and beverage factor, with a threshold of 120, or 2.4 stores/ha. After the threshold is exceeded, the SHAP value increases with the increase in the service level of daily life. At the same time, it has the most significant synergies with the factor of food and beverages. Especially when the level of daily life service exceeds the threshold, there is an obvious enhancement between the two factors. The SHAP value curve of population reaches its maximum promotive effect at approximately 10, equal to 2000 people/ha, after which it tends to stabilize. The optimal interactive factor for population is the food and beverage factor, further indicating that community activities within living scenes are more conducive to promoting HP. The SHAP distribution for public facilities is similar to the enterprise factor, with a threshold of 10 to 40, or 0.2 to 0.8 facilities/ha. Exceeding the range, the SHAP value begins to diminish and stabilize. The enterprises factor shows an additive effect. When the level of public facilities is the same, the increase in enterprise factors is accompanied by the increase in SHAP value.

In the context of MA (Figure 7), the factor of enterprises emerges as the most impactful among the top six factors, exhibiting a nonlinear relationship in its SHAP value analysis. A significant threshold is observed as enterprise density exceeds 15, equaling 0.3 enterprises/ha. When this value is exceeded, the synergistic effect of the transportation service factor is obvious. When the enterprise density is the same, the lower the level of transportation services, the larger the SHAP value is and the value is positive; while when the level of traffic facilities is higher, the SHAP value becomes smaller and negative. For Transportation Service, an initial surge in SHAP values corresponds to an increase in facility count, indicating a potent positive influence on MA transportation. This effect plateaus upon reaching 8 facilities/ha, suggesting a saturation point. Transportation Service shows two similar trend lines with comparable shapes and thresholds, indicating a stable operational mechanism, potentially with spatial heterogeneity contributing to divergent impacts. The strongest interaction factor for transportation service is governmental organization and social group, with the synergistic effect becoming apparent below 0.1 facilities/ha.

Population density affects MA NMT vitality in a nonlinear way: it bolsters MA usage at low to moderate densities but plateaus—and even declines—when areas become overly crowded, suggesting that extreme density offers diminishing returns for cycling and e-biking. The strongest synergy for population is with shopping density, highlighting how access to retail amenities amplifies functional comfort and MA uptake. Governmental organizations and social groups also drive MA vitality at lower presence levels, likely because institutional hubs concentrate trip demand. Moreover, under a fixed level of government infrastructure, increasing transportation service slightly reduces population's SHAP contribution—implying that when transit options are plentiful, travelers may shift toward private cars, buses/metro, or walking rather than MA modes. The SHAP values for parks and squares unveil a distinct pattern where beyond 60, equal to 1.2 facilities/ha, there is a positive impact on MA. The most important interaction factor of parks and squares is enterprises, indicating that internal interactions between functional factors can also generate significant synergistic effects and promote MA. A high concentration of Shopping tends to positively affect MA transportation, yet beyond 300, equal to 6 shops/ha, the promotional effect plateaus. There is also a suppressive effect observed in local samples between 200 and 300, that is 4 to 6 shops/ha, suggesting a possible oversaturation effect.

Overall, the SHAP values for both HP and MA highlight the nonlinear actions and interactions between factors affecting non-motorized transportation. Most factors are dynamically reversed before and after a certain threshold. For example, when the density exceeds 0.3~0.5 enterprises/ha, the promotion of HP vitality turns to inhibition; When the density of transportation services reaches 8 facilities/ha, the marginal benefit of MA is saturated, and when the density exceeds 300 facilities/ha, the marginal benefit of MA is negative. There are trade-offs and complementary effects among built environment factors. Food and beverages and daily life services exert synergistic gains on each other in HP NMT vitality, but a high enterprise density will inhibit the contribution of food and beverage services. The positive effect of transportation services on MA increases first and then decreases with the increase in the number of facilities, while its promotion of HP competes with the population density. The intensity of the interaction effect is closely related to the combination of factors and the range of densities. For example, the synergistic effect of government organizations and social groups on MA is more significant at low transportation service levels, while parks and squares are synergistic with enterprise density in areas of medium enterprise concentration. Another characteristic is the space crowding effect at high densities. For example, a high density of enterprises inhibits HP vitality, and a high density of transportation services weakens MA attraction. However, the threshold points and influence paths of different functional areas are different.

4. Discussion

4.1. Bivariate Local Moran's I and Strategies

By examining the spatial heterogeneity of key factors, we can formulate more targeted urban planning strategies. We employed a multi-stage analytical framework combining spatial statistics and machine learning interpretability to guide context-sensitive urban design. First, using Bivariate Local Moran's I (Figure 8), we classified spatial heterogeneity in NMT vitality into four distinct zones. H-L clusters means human-powered mobility hotspots with below-average mechanically assisted mobility, L-H clusters represent MA-dominant zones with suppressed HP vitality, H-H clusters have dual-mode vitality hotspots, and L-L clusters have dual-mode underperformance areas.

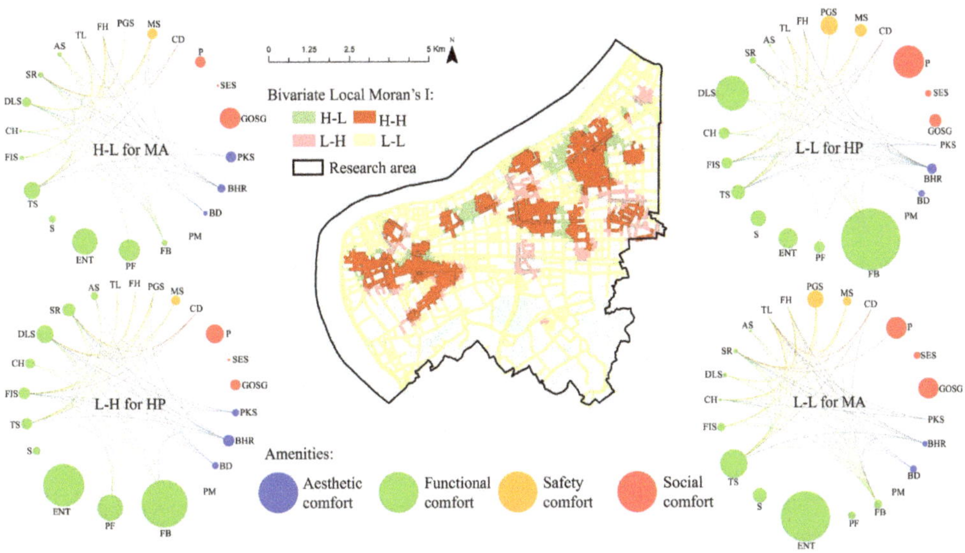

Figure 8. Factor effects and interactions in three types of low-vitality clusters.

Subsequently, we conducted weighted factor analysis within each zone, leveraging SHAP values and interaction metrics to identify context-specific drivers. For visualization, Figure 8 integrates spatial clustering results with SHAP-based feature importance. The diameter of the circles corresponds to the average absolute SHAP magnitude, and the line connections denote interaction effects with the most relevant factors. This integrated approach enables evidence-based, spatially differentiated planning that resolves the paradoxical coexistence of HP and MA deficiencies through tailored interventions. By coupling geostatistical diagnostics with explainable AI techniques, the framework advances precision urbanism in NMT system optimization.

4.1.1. H-L Clusters

H-L clusters are primarily found near the H-H cluster. These zones maintain significant HP vitality, though MA vitality has notably declined. Spatially, potential areas for reinforcement are some H-L clusters along the river. These regions boast well-developed greenways and are well-equipped with infrastructure supporting non-motorized transportation. Therefore, we conduct an interpretable analysis of MA for this segment of the sample to facilitate the formulation of targeted planning strategies.

The most significant factors influencing this trend are enterprises, public facilities, and governmental organizations and social groups, which suggests that MA predominantly involves purpose-driven, short- to medium-distance trips. Transportation services also play a crucial role, with transit exchanges being one of the primary purposes of NMT. This suggests that functionality is paramount in MA non-motorized transportation. In urban planning, emphasis should be placed on achieving a balance between work and residential areas, promoting the development of parcels under a transit-oriented development (TOD) strategy, and enhancing the accessibility and mobility of the NMT system.

The strongest interaction effect occurs between cats and dogs and shopping, indicating that pet-accompanied shopping expeditions constitute a stable NMT micro-mobility pattern. The significance of these factors is currently limited by the urban pet ownership ratio. While its current expression is constrained by urban pet ownership rates, this pattern exhibits high growth potential as pet-friendly infrastructure expands. The factor of enterprises and the factor of governmental organizations and social groups exert significant effects but they lack strong interactions; one possible reason for this might be that these trips are more unidirectional and work-related, being characterized by point-to-point transit. The most frequent strong interactions are between social comfort and functional comfort, followed by interactions between esthetic comfort, safety comfort, and functional comfort. Therefore, emphasizing mixed-use land strategies that blend urban functions can enhance the vitality of NMT activities.

4.1.2. L-H Clusters

L-H clusters are similarly distributed around the H-H cluster, but unlike H-L clusters, they are located in the southern parts of the study area. These regions lack the well-developed NMT infrastructure such as the riverside and offer opportunities for simultaneous strengthening of planning and spatial design.

The most influential BE factor is food and beverages. Unlike MA, HP includes more lifestyle-oriented or non-purposeful dwelling with higher randomness and urban service interaction, making it more susceptible to the influence of the density of food and beverages. Enhancing this type of NMT movement could emphasize commercial route design. Enterprises remain important for balancing work and residence. Following that, public facilities shift from 6 in HP importance ranking to 3 in L-H importance ranking, indicating that walkable public services are a critical urban function. Population and daily life services are also significant, underscoring the promotional effect of residential functions on HP. This suggests that community-building strategies focused on walkability, such as the 15 min living circle and neighborhood units, are viable strategies for enhancing HP non-motorized transportation.

The strongest interactions are between shopping and transportation services, reinforcing that TOD combining transit and shopping remains an effective strategy for developing NMT levels. Enterprises have significant importance but lack crucial interactions, similar to H-L clusters for MA. Also, there are strong interactions between social comfort and functional comfort factors, such as cats and dogs and medical services, building height-to-width ratio, and food and beverages. As anticipated, the higher time cost of walking necessitates comprehensive benefits including safety, esthetics, and functional comfort.

4.1.3. L-L Clusters

L-L clusters are prevalent in most of the study areas, especially in the southern region, despite it having more green space. For HP, significant factors include food and beverages, daily life services, and population, highlighting the critical role of lifestyle services in stimulating pedestrian activity in low-vitality areas. Therefore, urban revitalization efforts in these zones should prioritize the planning of commercial spaces tailored to retail activities to enhance walkability. The most crucial correlations within this cluster arise between public facilities and transportation services, underscoring the importance of transit-oriented public services as a foundation for HP. Consequently, improving accessibility to these facilities should be a focal point.

For MA, significant factors include enterprises, followed by transportation services and population. Similarly to the H-L clusters, the focus of urban renewal should be on achieving a work–residence balance and implementing TOD strategies to enhance MA. The most important interactions for MA involve commercial houses with transportation services, and with cats and dogs. A possible explanation is that commercial housing serves as a significant source of NMT travel, thus making transit-related and pet-accompanied dwelling impact NMT vitality. Therefore, enhancing NMT infrastructure remains essential.

4.2. Nonlinear Relationships Between BE Factors and NMT Vitality

The results of our study demonstrate that the built environment significantly influences the non-motorized traffic patterns of both human-powered and mechanically assisted transportation, and the effects exhibit differentiated and complex local characteristics.

4.2.1. Threshold Effects

The nonlinear impact of BE factors exhibits a turning point where their local effects switch from negative to positive or vice versa. Changes in the rate of increase or decrease in local effects before and after the threshold are often observed, aligning with studies on the nonlinear impact of the built environment on urban vitality [55]. For example, a notable promotion of MA vitality is observed as enterprise density increases beyond 0.2 enterprises/ha.

In our study, we identify two kinds of thresholds: Lower limits, below which a factor has little effect. For example, HP's NMT vitality only rises once food and beverage density exceeds two stores/ha. Upper limits, beyond which extra increases no longer help. For example, MA's NMT vitality plateaus once transport services exceed eight facilities/ha. Recognizing these bounds helps planners target the detailed rules where each factor most effectively boosts non-motorized transport.

4.2.2. Synergistic Effects

Nonlinear effects exist when a built environment factor can be amplified or diminished by changes in another factor. In particular, when factors from different dimensions meet under specific conditions, they produce significant synergistic effects, supporting the amenities theory [56–58]. For instance, for HP's NMT vitality, a synergistic effect is observed when the density of public facilities is within the range of 0.2 to 0.8 facilities/ha, combined with enterprises at a moderate density of around 0.6 to 1 enterprise/ha. This confirmed that considering only one dimension is often insufficient when implementing NMT facilities or road planning to enhance vitality. Instead, policy-makers should take into account the circumstances of other built environment factors to achieve synergy and avoid negative interactions.

4.2.3. Spatial Clustering Effects

H-L clusters reveal asymmetric suppression effects, where dominant HP infrastructure (e.g., pedestrian zones or bike lanes) inadvertently restricts MA integration, creating trade-offs between accessibility and sustainability. Conversely, L-H clusters demonstrate over-reliance inertia, where car-centric design displaces walkability, reflecting nonlinear thresholds where MA saturation erodes opportunities for HP reclamation. H-H clusters highlight synergistic scaling, where the optimized coexistence of HP and MA amplifies network efficiency through multiplicative gains. L-L clusters, despite their balanced infrastructure potential, suffer from latent interaction deficits—neither mode thrives due to fragmented land use policies or insufficient demand triggers.

These dynamics underscore that HP and MA are nonlinearly interdependent, not merely competing forces. For example, in H-L clusters, HP vitality plateaus once MA infrastructure exceeds a critical threshold, triggering congestion that degrades pedestrian appeal. Similarly, L-H clusters exhibit tipping points where MA density surpasses a threshold, nonlinearly suppressing pedestrian activity via safety declines or route fragmentation. H-H clusters, by contrast, achieve super-linear returns when HP and MA networks intersect strategically. This nonlinear framework challenges linear planning models, emphasizing that mobility outcomes depend on threshold-crossing, context-dependent interactions, and phase shifts between modes rather than isolated investments.

5. Conclusions

This study demonstrates that built environment factors and non-motorized transportation vitality interact in highly nonlinear ways, with clear threshold and synergistic effects, and distinct spatial clusters that demand tailored interventions. Leveraging street view images and geospatial big data, twenty-three BE factors were systematically extracted, and human-powered NMT was distinguished from mechanically assisted NMT. Through comparative modeling—random forest, XGBoost, and OLS regressions—the analysis revealed that nonlinear models outperformed linear counterparts, underscoring the non-additive nature of relationships between the built environment and NMT. Specifically, threshold effects emerged. For instance, a notable promotion in MA NMT vitality is observed as enterprise density increases beyond 0.2 enterprises/ha, while HP NMT requires more than two stores/ha of food and beverages to boost its vitality. Synergistic effects also dominated, such as the density of public facilities that has a range from 0.2 to 0.8 facilities/ha, when combined with enterprises, which has a range from 0.6 to 1 enterprise/ha, can largely boost the vitality of HP NMT.

These findings challenge linear planning paradigms, advocating instead for adaptive, context-sensitive interventions that account for nonlinear thresholds and cross-mode synergies. For instance, areas identified as H-L clusters, with high human-powered mobility but suppressed mechanically assisted mobility, require recalibration of land use configurations to balance pedestrian dominance with strategic MA integration, such as shared-use pathways alongside transit hubs. Conversely, in L-H clusters, MA-dominant but HP-suppressed zones, demand de-prioritizing car-centric infrastructure while incentivizing compact mixed-use developments to reignite walkability. L-L clusters, marked by dual-mode underperformance, necessitate holistic interventions that simultaneously address fragmented zoning, inadequate transit frequency, and underutilized green spaces. By coupling bivariate Local Moran's I analysis with SHAP interaction mapping, this study identifies actionable leverage points. Crucially, the spatial heterogeneity of SHAP interactions underscores the need for tailored, cluster-specific strategies rather than uniform policies.

In our study, the street-view images we obtained were captured between June 2019 and August 2021, all during daytime hours. As these images are static and time-specific snapshots, the numbers of pedestrians, cyclists, and e-bikers may fluctuate depending on the capture date, season, weekdays or weekend, and specific time of day (e.g., morning peak, midday, evening peak), and may deviate from the actual NMT activity level. In addition, changes in camera angle, lighting conditions, and occlusions between images may affect the accuracy of the target detection algorithm. These limitations in time and image quality may affect our estimation of NMT vitality. Future work could incorporate multi-temporal, full-season historical street-view imagery and real-time mobile signal data to strengthen the results' robustness and representativeness. Moreover, COCO images differ from typical Chinese street view images in certain respects, such as the appearance of traffic lights, parking signs, or fire hydrants. Future work may fine-tune the model using local data to mitigate domain shift and further improve detection accuracy.

Moving forward, this research can be advanced through multi-dimensional data fusion, integrating urban morphology, socio-economic demographics, and behavioral patterns to decode why certain BE and NMT relationships emerge. Longitudinal studies may track post-intervention outcomes, such as health improvements from NMT adoption or economic shifts from reduced car dependency, and compare NMT across different demographic groups. Further, an adaptive computational framework can be implemented using AutoML platforms (e.g., H_2O AutoML, Auto-sklearn) to automatically benchmark multiple algorithms and hyperparameter configurations, simplifying model selection and reducing human bias, and using a modular framework that allows new built environment variables to be added or modified. By bridging predictive analytics with causal inference, this work aims to deliver adaptive, evidence-driven frameworks that evolve with urban dynamics, ensuring NMT initiatives not only enhance mobility but also foster inclusive, resilient cities.

Author Contributions: Conceptualization, Y.R. and X.Z.; methodology, X.Z.; software, X.Z.; validation, X.C. and Y.R.; formal analysis, Y.R.; investigation, X.C.; resources, X.Z.; data curation, Y.R. and X.Z.; writing—original draft preparation, Y.R.; writing—review and editing, Q.C. and S.W.; visualization, Y.R.; supervision, Q.C. and S.W.; visualization; project administration, Q.C.; funding acquisition, Y.R. and X.Z. All authors have read and agreed to the published version of the manuscript.

Funding: This research was supported by the Joint Funds of the Zhejiang Provincial Natural Science Foundation of China under Grant No. LHZY24A010001; and the Key R&D Program of Zhejiang (No.2024C03234); and the Talent introduction Program Youth Project of the Chinese Academy of Sciences (E43302020D, E2Z105010F).

Data Availability Statement: The data presented in this study are available on request from the corresponding author.

Conflicts of Interest: The authors declare no conflicts of interest.

Notes

1 The threshold of 100 stores per analysis unit corresponds to ≈2 stores per hectare (stores/ha), calculated based on the standard area of the analysis unit ($400^2\pi$ m^2 ≈ 0.0503 ha). Subsequent references to facility density will report values exclusively in stores/ha.

References

1. Balsas, C. Blending Individual Tenacity with Government's Responsibility in the Implementation of US Non-Motorized Transportation Planning (NMT). *Plan. Pract. Res.* **2017**, *32*, 197–211. [CrossRef]
2. Mansoor, U.; Kashifi, M.T.; Safi, F.R.; Rahman, S.M. A Review of Factors and Benefits of Non-Motorized Transport: A Way Forward for Developing Countries. *Environ. Dev. Sustain.* **2022**, *24*, 1560–1582. [CrossRef]
3. Næss, P. Built Environment, Causality and Travel. *Transp. Rev.* **2015**, *35*, 275–291. [CrossRef]
4. Nag, D.; Sen, J.; Goswami, A.K. Measuring Connectivity of Pedestrian Street Networks in the Built Environment for Walking: A Space-Syntax Approach. *Transp. Dev. Econ.* **2022**, *8*, 34. [CrossRef]
5. Miranda-Moreno, L.F.; Morency, P.; El-Geneidy, A.M. The Link between Built Environment, Pedestrian Activity and Pedestrian–Vehicle Collision Occurrence at Signalized Intersections. *Accid. Anal. Prev.* **2011**, *43*, 1624–1634. [CrossRef] [PubMed]
6. Mirzakhani, A.; Behzadfar, M.; Turró, M. Pedestrian Movement Patterns in Historical Urban Fabrics. *GeoJournal* **2023**, *88*, 6483–6506. [CrossRef]
7. Cao, X. (Jason); Mokhtarian, P.L.; Handy, S.L. The Relationship between the Built Environment and Nonwork Travel: A Case Study of Northern California. *Transp. Res. Part A Policy Pract.* **2009**, *43*, 548–559. [CrossRef]
8. Liu, C.; Shen, Q. An Empirical Analysis of the Influence of Urban Form on Household Travel and Energy Consumption. *Comput. Environ. Urban Syst.* **2011**, *35*, 347–357. [CrossRef]
9. Ewing, R.; Cervero, R. Travel and the Built Environment: A Meta-Analysis. *J. Am. Plan. Assoc.* **2010**, *76*, 265–294. [CrossRef]
10. Lanzini, P.; Stocchetti, A. From Techno-Centrism to Socio-Centrism: The Evolution of Principles for Urban Sustainable Mobility. *Int. J. Sustain. Transp.* **2021**, *15*, 815–825. [CrossRef]
11. Rahul, T.M.; Verma, A. Study of Impact of Various Influencing Factors on NMT Mode Choice. *Procedia—Soc. Behav. Sci.* **2013**, *104*, 1112–1119. [CrossRef]
12. Bautista-Hernández, D.A. Mode Choice in Commuting and the Built Environment in México City. Is There a Chance for Non-Motorized Travel? *J. Transp. Geogr.* **2021**, *92*, 103024. [CrossRef]
13. Gehrke, S.R.; Wang, L. Operationalizing the Neighborhood Effects of the Built Environment on Travel Behavior. *J. Transp. Geogr.* **2020**, *82*, 102561. [CrossRef]
14. Khan, M.; Kockelman, K.M.; Xiong, X. Models for Anticipating Non-Motorized Travel Choices, and the Role of the Built Environment. *Transp. Policy* **2014**, *35*, 117–126. [CrossRef]
15. Basu, N.; Oviedo-Trespalacios, O.; King, M.; Kamruzzaman, M.; Haque, M.M. The Influence of the Built Environment on Pedestrians' Perceptions of Attractiveness, Safety and Security. *Transp. Res. Part F Traffic Psychol. Behav.* **2022**, *87*, 203–218. [CrossRef]
16. Cheng, Y.-T.; Lin, J.-J. The Associations of Built Environments with Public Bike Use for Metro Commuters. *Transp. Plan. Technol.* **2024**, *47*, 903–921. [CrossRef]
17. Zhang, Y.; Thomas, T.; Brussel, M.; van Maarseveen, M. Exploring the Impact of Built Environment Factors on the Use of Public Bikes at Bike Stations: Case Study in Zhongshan, China. *J. Transp. Geogr.* **2017**, *58*, 59–70. [CrossRef]
18. Hu, Y.; Sobhani, A.; Ettema, D. To E-Bike or Not to e-Bike? A Study of the Impact of the Built Environment on Commute Mode Choice in Small Chinese City. *JTLU* **2021**, *14*, 479–497. [CrossRef]
19. Yu, Y.; Jiang, Y.; Qiu, N.; Guo, H.; Han, X.; Guo, Y. Exploring Built Environment Factors on E-Bike Travel Behavior in Urban China: A Case Study of Jinan. *Front. Public Health* **2022**, *10*, 1013421. [CrossRef]
20. Hankey, S.; Zhang, W.; Le, H.T.K.; Hystad, P.; James, P. Predicting Bicycling and Walking Traffic Using Street View Imagery and Destination Data. *Transp. Res. Part D Transp. Environ.* **2021**, *90*, 102651. [CrossRef]
21. Larkin, A.; Hystad, P. Evaluating Street View Exposure Measures of Visible Green Space for Health Research. *J. Expo. Sci. Environ. Epidemiol.* **2019**, *29*, 447–456. [CrossRef] [PubMed]
22. Ito, K.; Bansal, P.; Biljecki, F. Examining the Causal Impacts of the Built Environment on Cycling Activities Using Time-Series Street View Imagery. *Transp. Res. Part A Policy Pract.* **2024**, *190*, 104286. [CrossRef]
23. Ito, K.; Zhu, Y.; Abdelrahman, M.; Liang, X.; Fan, Z.; Hou, Y.; Zhao, T.; Ma, R.; Fujiwara, K.; Ouyang, J.; et al. ZenSVI: An Open-Source Software for the Integrated Acquisition, Processing and Analysis of Street View Imagery towards Scalable Urban Science. *Comput. Environ. Urban Syst.* **2025**, *119*, 102283. [CrossRef]
24. Barbierato, E.; Bernetti, I.; Capecchi, I.; Saragosa, C. Integrating Remote Sensing and Street View Images to Quantify Urban Forest Ecosystem Services. *Remote Sens.* **2020**, *12*, 329. [CrossRef]
25. Yue, H.; Xie, H.; Liu, L.; Chen, J. Detecting People on the Street and the Streetscape Physical Environment from Baidu Street View Images and Their Effects on Community-Level Street Crime in a Chinese City. *IJGI* **2022**, *11*, 151. [CrossRef]
26. Wu, F.; Li, W.; Qiu, W. Examining Non-Linear Relationship between Streetscape Features and Propensity of Walking to School in Hong Kong Using Machine Learning Techniques. *J. Transp. Geogr.* **2023**, *113*, 103698. [CrossRef]
27. Lei, Y.; Peng, D.; Zhang, P.; Ke, Q.; Li, H. Hierarchical Paired Channel Fusion Network for Street Scene Change Detection. *IEEE Trans. Image Process.* **2021**, *30*, 55–67. [CrossRef]

28. Wang, Y.; Wang, S.; Wang, X.; Huang, X.; Li, X.; Zhang, X.; Li, Z. Dynamic Visual Landscape Assessment of Disused Railway in Complex Terrain: A Study of Jingmen Railway. *IEEE J. Sel. Top. Appl. Earth Obs. Remote Sens.* **2024**, *17*, 10877–10892. [CrossRef]
29. Liu, C.; Huang, H.; Melack, J.M.; Tian, Y.; Jiang, J.; Fu, X.; Cao, Z.; Wang, S. Assessing Land Degradation and Restoration in Eastern China Grasslands from 1985 to 2018 Using Multitemporal Landsat Data. *IEEE J. Sel. Top. Appl. Earth Obs. Remote Sens.* **2024**, *17*, 19328–19342. [CrossRef]
30. Wu, C.; Kim, I.; Chung, H. The Effects of Built Environment Spatial Variation on Bike-Sharing Usage: A Case Study of Suzhou, China. *Cities* **2021**, *110*, 103063. [CrossRef]
31. Sharma, T.; Jain, S. Nexus between Built Environment, Travel Behaviour and Human Health: An Integrated Framework to Reinform Transport System. *J. Clean. Prod.* **2023**, *416*, 137744. [CrossRef]
32. Li, Y.; Zhang, C.; Wang, C.; Cheng, Z. Human Perception Evaluation System for Urban Streetscapes Based on Computer Vision Algorithms with Attention Mechanisms. *Trans. GIS* **2022**, *26*, 2440–2454. [CrossRef]
33. Feuillet, T.; Salze, P.; Charreire, H.; Menai, M.; Enaux, C.; Perchoux, C.; Hess, F.; Kesse-Guyot, E.; Hercberg, S.; Simon, C.; et al. Built Environment in Local Relation with Walking: Why Here and Not There? *J. Transp. Health* **2016**, *3*, 500–512. [CrossRef]
34. Tao, T.; Wu, X.; Cao, J.; Fan, Y.; Das, K.; Ramaswami, A. Exploring the Nonlinear Relationship between the Built Environment and Active Travel in the Twin Cities. *J. Plan. Educ. Res.* **2023**, *43*, 637–652. [CrossRef]
35. Qin, K.; Xu, Y.; Kang, C.; Kwan, M. A Graph Convolutional Network Model for Evaluating Potential Congestion Spots Based on Local Urban Built Environments. *Trans. GIS* **2020**, *24*, 1382–1401. [CrossRef]
36. Xiao, W.; Wei, Y.D. Assess the Non-Linear Relationship between Built Environment and Active Travel around Light-Rail Transit Stations. *Appl. Geogr.* **2023**, *151*, 102862. [CrossRef]
37. Ding, C.; Cao, X.; Dong, M.; Zhang, Y.; Yang, J. Non-Linear Relationships between Built Environment Characteristics and Electric-Bike Ownership in Zhongshan, China. *Transp. Res. Part D Transp. Environ.* **2019**, *75*, 286–296. [CrossRef]
38. Xu, Y.; Yan, X.; Liu, X.; Zhao, X. Identifying Key Factors Associated with Ridesplitting Adoption Rate and Modeling Their Nonlinear Relationships. *Transp. Res. Part A Policy Pract.* **2021**, *144*, 170–188. [CrossRef]
39. Rui, J.; Xu, Y. Beyond Built Environment: Unveiling the Interplay of Streetscape Perceptions and Cycling Behavior. *Sustain. Cities Soc.* **2024**, *109*, 105525. [CrossRef]
40. Hatami, F.; Rahman, M.M.; Nikparvar, B.; Thill, J.-C. Non-Linear Associations Between the Urban Built Environment and Commuting Modal Split: A Random Forest Approach and SHAP Evaluation. *IEEE Access* **2023**, *11*, 12649–12662. [CrossRef]
41. Bone, C.; Wulder, M.A.; White, J.C.; Robertson, C.; Nelson, T.A. A GIS-Based Risk Rating of Forest Insect Outbreaks Using Aerial Overview Surveys and the Local Moran's I Statistic. *Appl. Geogr.* **2013**, *40*, 161–170. [CrossRef]
42. Zhang, X.; Li, W.; Zhang, F.; Liu, R.; Du, Z. Identifying Urban Functional Zones Using Public Bicycle Rental Records and Point-of-Interest Data. *IJGI* **2018**, *7*, 459. [CrossRef]
43. Zhang, X.; Chen, Y.; Zhong, Y. Spatial and Temporal Characteristic Analysis of Imbalance Usage in the Hangzhou Public Bicycle System. *IJGI* **2021**, *10*, 637. [CrossRef]
44. Zhu, H.; Nan, X.; Yang, F.; Bao, Z. Utilizing the Green View Index to Improve the Urban Street Greenery Index System: A Statistical Study Using Road Patterns and Vegetation Structures as Entry Points. *Landsc. Urban Plan.* **2023**, *237*, 104780. [CrossRef]
45. Zhou, L.; Dang, X.; Sun, Q.; Wang, S. Multi-Scenario Simulation of Urban Land Change in Shanghai by Random Forest and CA-Markov Model. *Sustain. Cities Soc.* **2020**, *55*, 102045. [CrossRef]
46. Ewing, R.; Handy, S. Measuring the Unmeasurable: Urban Design Qualities Related to Walkability. *J. Urban Des.* **2009**, *14*, 65–84. [CrossRef]
47. Nasar, J.L. Urban Design Aesthetics: The Evaluative Qualities of Building Exteriors. *Environ. Behav.* **1994**, *26*, 377–401. [CrossRef]
48. Salleh, B.S.; Rahmat, R.A.A.O.; Ismail, A. Research Article A Study on Non-Motorised (NMT) Activities for Urban Environment. *Res. J. Appl. Sci. Eng. Technol.* **2014**, *7*, 290–295. [CrossRef]
49. Handy, S.L. Urban Form and Pedestrian Choices: Study of Austin Neighborhoods. *Transp. Res. Rec. J. Transp. Res. Board* **1996**, *1552*, 135–144. [CrossRef]
50. Ewing, R.; Dumbaugh, E. The Built Environment and Traffic Safety: A Review of Empirical Evidence. *J. Plan. Lit.* **2009**, *23*, 347–367. [CrossRef]
51. Arellana, J.; Saltarín, M.; Larrañaga, A.M.; Alvarez, V.; Henao, C.A. Urban Walkability Considering Pedestrians' Perceptions of the Built Environment: A 10-Year Review and a Case Study in a Medium-Sized City in Latin America. *Transp. Rev.* **2020**, *40*, 183–203. [CrossRef]
52. Alabi, R.O.; Elmusrati, M.; Leivo, I.; Almangush, A.; Mäkitie, A.A. Machine Learning Explainability in Nasopharyngeal Cancer Survival Using LIME and SHAP. *Sci. Rep.* **2023**, *13*, 8984. [CrossRef]
53. Lundberg, S.M.; Lee, S.-I. A Unified Approach to Interpreting Model Predictions. In *Proceedings of the Advances in Neural Information Processing Systems*; Guyon, I., Luxburg, U.V., Bengio, S., Wallach, H., Fergus, R., Vishwanathan, S., Garnett, R., Eds.; Curran Associates, Inc.: Red Hook, NY, USA, 2017; Volume 30, ISBN 9781510860964.

54. Rao, J.; Ma, J.; Dong, G. How Mobility-Based Exposure to Green Space and Environmental Pollution Influence Individuals' Wellbeing? A Structural Equation Analysis through the Lens of Environmental Justice. *Landsc. Urban Plan.* **2024**, *252*, 105199. [CrossRef]
55. Xiao, L.; Lo, S.; Liu, J.; Zhou, J.; Li, Q. Nonlinear and Synergistic Effects of TOD on Urban Vibrancy: Applying Local Explanations for Gradient Boosting Decision Tree. *Sustain. Cities Soc.* **2021**, *72*, 103063. [CrossRef]
56. Carlino, G.A.; Saiz, A. Beautiful City: Leisure Amenities and Urban Growth. *J. Reg. Sci.* **2019**, *59*, 369–408. [CrossRef]
57. Clark, T.N.; Lloyd, R.; Wong, K.K.; Jain, P. Amenities Drive Urban Growth. *J. Urban Aff.* **2002**, *24*, 493–515. [CrossRef]
58. Wang, M. Polycentric Urban Development and Urban Amenities: Evidence from Chinese Cities. *Environ. Plan. B Urban Anal. City Sci.* **2021**, *48*, 400–416. [CrossRef]

land

MDPI

Article

Spatial Gradient Effects of Landscape Pattern on Ecological Quality Along the Grand Canal

Yonggeng Xiong [1] and Aibo Jin [2,*]

[1] College of Landscape Architecture, Nanjing Forestry University, Nanjing 210037, China; tusiji@njfu.edu.cn
[2] School of Landscape Architecture, Beijing Forestry University, Beijing 100083, China
* Correspondence: jinaibo@bjfu.edu.cn

Abstract: The Grand Canal serves as a vital water transportation route, a UNESCO World Cultural Heritage site, and an ecological corridor. It is currently undergoing coordinated transformation through infrastructure development, heritage preservation, and ecological restoration. However, existing research has primarily focused on either cultural heritage conservation or localized ecological issues, with limited attention to the spatial relationship between landscape patterns and ecological quality along the entire corridor. To address this gap, this study examines eight sections of the Grand Canal and develops a gradient analysis framework based on equidistant buffer zones. The framework integrates the Remote Sensing Ecological Index (RSEI) with landscape pattern indices to assess ecological responses across spatial gradients. A Multi-scale Geographically Weighted Regression (MGWR) model is applied to reveal the spatially heterogeneous effects of landscape patterns on ecological quality. From 2013 to 2023, landscape patterns showed a trend toward increasing agglomeration and regularity. This is indicated by a rise in the Aggregation Index (AI) from 91.24 to 91.38 and declines in both patch density (PD) from 8.45 to 8.20 and Landscape Shape Index (LSI) from 199.74 to 196.72. During the same period, ecological quality slightly declined, with RSEI decreasing from 0.66 to 0.57. The effects of PD and Shannon's Diversity Index (SHDI) on ecological quality varied across canal sections. In highly urbanized areas such as the Tonghui River, these indices were positively correlated with ecological quality, whereas in less urbanized areas like the Huitong River, negative correlations were observed. Overall, the strength of these correlations tended to weaken with increasing buffer distance. This study provides a scientific foundation for the integrated development of ecological protection and spatial planning along the Grand Canal and offers theoretical insights for the refined management of other major inland waterways.

Keywords: territorial space coordination; landscape pattern; Remote Sensing Ecological Index; spatial gradient response; the Grand Canal

1. Introduction

The Grand Canal of China, the world's longest and largest artificial waterway, has served both as a critical transportation route and a cornerstone of Chinese cultural heritage for over 2500 years [1,2]. In 2014, it was inscribed on the World Heritage List, affirming its globally recognized cultural significance [3]. In recent decades, ecosystems along the canal have received increasing attention due to their marked spatial heterogeneity and heightened sensitivity to anthropogenic disturbances. These ecosystems are characterized by low ecological resilience and limited restoration capacity, which distinguish them from typical linear ecological corridors [4]. Amid global climate change, rapid urbanization, and the pursuit of ecological civilization, the Grand Canal is undergoing a functional transformation from a traditional navigation route to a multifunctional spatial system that integrates cultural preservation, ecological rehabilitation, and regional development [5]. Accordingly, a key challenge is how to systematically assess and improve ecological quality while maintaining the canal's cultural integrity [6,7].

In recent years, several ecological restoration initiatives have been launched along the Grand Canal [5,8,9], leading to initial improvements in ecological function and landscape structure in certain sections. Nevertheless, the canal still faces pressing issues, including intensified landscape fragmentation, increasing risks of ecological degradation, and an uneven spatial distribution of water resources [10,11]. More fundamentally, despite advances in heritage preservation, ecological rehabilitation, and urban planning, the spatial interplay between landscape pattern evolution and ecological quality remains insufficiently understood [12,13]. As a linear system with diverse functional demands, the Grand Canal requires a more integrated analytical approach that captures the relationship between landscape structure and ecosystem performance.

Landscape patterns, typically quantified using landscape pattern indices, describe the spatial arrangement and structural features of land-use types. These indices reflect the composition, configuration, and inter-relationships of landscape patches and are widely used in regional assessment and ecosystem restoration research [14,15]. In linear geographic units such as rivers and transportation corridors, they play a key role in revealing connections between land-use changes and ecological processes, including water quality, biodiversity, and ecosystem services [16–18]. For example, Han et al. identified spatial heterogeneity in ecological responses along the Qinhuai River using water quality metrics and landscape pattern indices [19]. Yang et al. examined the spatiotemporal evolution of ecosystem services along the Qinghai–Tibet Highway under the influence of climate factors [20]. Mohammadi et al. applied a multi-buffer method to explore spatial couplings between land-use patterns and pollution in the Talar watershed [21]. However, the Grand Canal presents additional complexities. As a highly artificial water transport system, it features frequent alterations between water bodies, green spaces, and urban construction areas, creating sharp landscape heterogeneity along its banks [22]. To effectively analyze the ecological responses of such fragmented spatial structures, a tailored, multi-scale analytical framework is required.

Traditional methods of ecological quality assessment, such as ecosystem service valuation [23,24], ecological risk indices [25], and carrying capacity evaluations [26,27] often rely on expert-defined weights. This subjectivity limits their objectivity and consistency, particularly in large-scale applications. In contrast, the Remote Sensing Ecological Index (RSEI), which integrates multi-source data using principal component analysis, offers a more objective, scalable, and computationally efficient approach [28]. It has been successfully applied in a variety of ecosystems [29–31] and, more recently, extended to linear geographic features. For instance, Zou et al. and Wang et al. applied RSEI to assess the ecological condition of the Qinghai–Tibet Railway and high-altitude highways [32,33], while Li et al. used it for the Grand Canal [4]. Despite its growing application, most studies focus on dynamic monitoring of ecological quality and lack systematic analysis of how landscape pattern transformations affect ecosystem conditions. Few have addressed the impacts of spatial changes, such as connectivity loss, habitat fragmentation, or intensified edge effects, on ecological succession [34,35]. Therefore, a comprehensive assessment framework is needed, one that integrates landscape structure indicators with ecological response mechanisms across multiple spatial scales.

Spatial heterogeneity is a defining feature of the relationship between landscape patterns and ecosystem dynamics. It provides a critical foundation for understanding localized ecological responses and for developing targeted management strategies [10]. Geographically Weighted Regression (GWR) is a commonly used spatial analysis method for identifying such local variations [36]. However, traditional GWR assumes a uniform spatial scale of all variables, making it less suitable for analyzing complex ecological systems with drivers operating at differing spatial extents [37]. The Multi-scale Geographically Weighted Regression (MGWR) model addresses this limitation by introducing variable-specific bandwidths, allowing for more precise detection of spatial non-stationarity [38]. MGWR has proven effective in studies related to urban heat islands, land-use transition, and ecosystem service distribution [39,40]. Nevertheless, MGWR does not fully account for distance sensitivity, which is essential in corridor studies. Therefore, combining MGWR with buffer-based gradient analysis allows for simultaneous integration of spatial location and ecological processes, helping to reveal how landscape patterns influence ecological quality at various distances.

In this context, this study takes the Grand Canal as a representative case and develops an integrated framework to examine the interactions among landscape patterns, ecological quality, and spatial responses from 2013 to 2023. The specific objectives are as follows: (1) To identify the spatial distribution and gradient evolution of landscape patterns using a buffer-based method; (2) to construct RSEI from multi-temporal remote sensing data to evaluate changes in ecological quality; (3) to apply the MGWR model to quantify spatial heterogeneity in the effects of landscape patterns on ecological quality; and (4) to assess spatial differentiation and effect intensity across buffer zones through gradient analysis. The results aim to clarify the spatiotemporal dynamics of ecological quality along the Grand Canal and provide scientific support for integrated ecological management and spatial planning in similar inland waterways.

2. Study Area and Data Sources

2.1. Study Area

The Grand Canal of China, extending nearly 3200 km, is the world's longest and oldest artificial waterway [16]. It comprises three main sections: the Beijing–Hangzhou Grand Canal, the Sui–Tang Grand Canal, and the Zhedong Canal. Due to historical and hydrological changes, much of the Sui–Tang section has dried up, making full restoration impractical. Therefore, this study focuses on the Beijing–Hangzhou and Zhedong segments. As of 2024, the Beijing–Hangzhou Canal has been completely rewatered and partially reopened for navigation, although certain sections remain restricted by local hydrological conditions [41]. The Zhedong Canal maintains overall navigability and relatively stable water levels, though some localized improvements are still needed. Based on differences in historical development and geographical location, the study area includes eight representative sections arranged from north to south: Tonghui River, North Canal, South Canal, Huitong River, Zhong Canal, Huaiyang Canal, Jiangnan Canal, and Zhedong Canal [7]. These sections span approximately 1947 km (Figure 1).

2.2. Data Sources and Pre-Processing

Canal and river data were obtained from the National Catalogue Service for Geographic Information and processed according to the General Plan for the Conservation and Management of the Grand Canal (2012–2030). Following the approach of Yang et al. [18], a series of equidistant buffer zones was created using the canal centerline as the baseline. These zones extend 5 km on both sides at 500 m intervals to capture spatial gradients in landscape pattern and ecological quality. Land-use data were obtained from the China Land Cover Dataset (CLCD), which has an overall classification accuracy of 79.31% [42]. Landsat 8 OLI imagery was obtained from Google Earth Engine (GEE), selecting images from May to October to correspond with peak vegetation growth and minimize seasonal variation [43]. All spatial data were pre-processed using ArcGIS Pro 3.0, including projection transformation and spatial alignment. A uniform resolution of 30 m was applied under the WGS_1984_UTM_Zone_50N coordinate system. Details of the data are provided in Table 1.

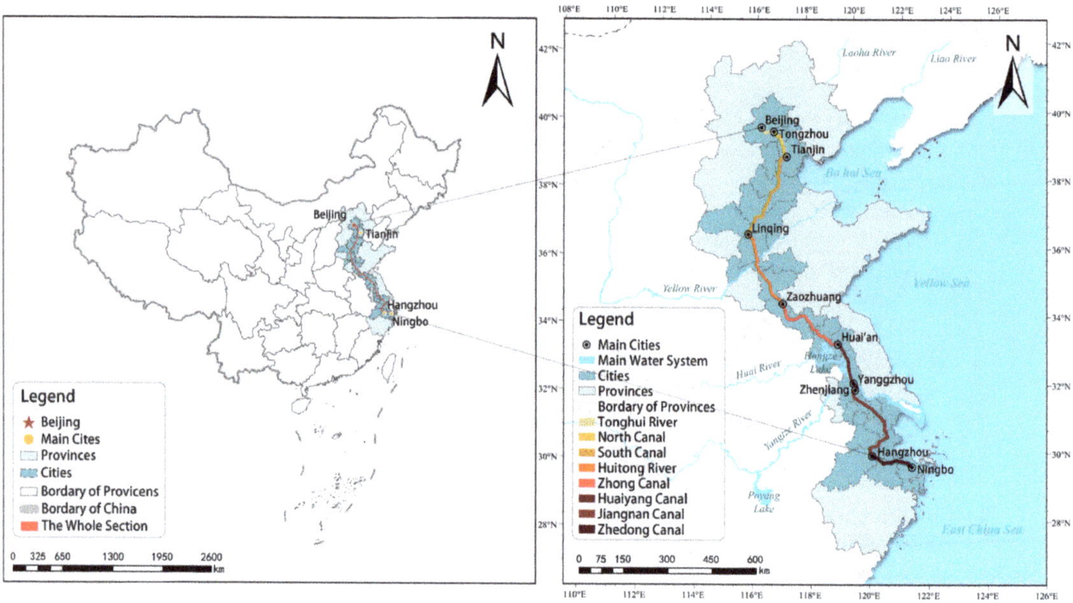

Figure 1. Location of this study area.

Table 1. Data sources.

Data	Resolution	Year	Data Source
Land-use data	30 m	2013–2023	https://doi.org/10.5281/zenodo.12779975 (accessed on 12 February 2025)
Landsat 8 OLI	30 m	2013–2023	https://www.usgs.gov/landsat (accessed on 13 February 2025)
Canals and rivers	–	2021	https://www.webmap.cn (accessed on 13 February 2025)

3. Methods

This study develops an integrated evaluation framework to examine the spatial interaction between landscape patterns and ecological quality along the Grand Canal (Figure 2). It first evaluates six key aspects of landscape configuration, including fragmentation, boundary complexity, shape irregularity, aggregation, connectivity, and diversity. Ecological quality is then quantified using the RSEI. A buffer-based gradient analysis is applied to identify landscape indices that are sensitive to distance from the canal and to reveal corresponding spatial response patterns. Finally, the MGWR model is employed to capture spatial heterogeneity and quantify the varying effects of landscape patterns on ecological quality across buffer gradients.

3.1. *Landscape Pattern Metrics*

To comprehensively characterize land-use structure along the Grand Canal, six representative landscape pattern indices were selected based on three core dimensions commonly used in landscape ecology: patch characteristics, spatial configuration and connectivity, and landscape composition and diversity [18,44]. Patch density (PD) and Edge Density (ED) measure fragmentation and boundary complexity. Landscape Shape Index (LSI),

Aggregation Index (AI), and Patch Cohesion Index (COHESION) capture spatial structure and connectivity. Shannon's Diversity Index (SHDI) represents the richness and evenness of land-use types. Collectively, these indices provide a comprehensive description of landscape patterns relevant to ecological processes and environmental quality [14,15]. All metrics were calculated at the landscape level using Fragstats 4.2, with detailed definitions presented in Table 2.

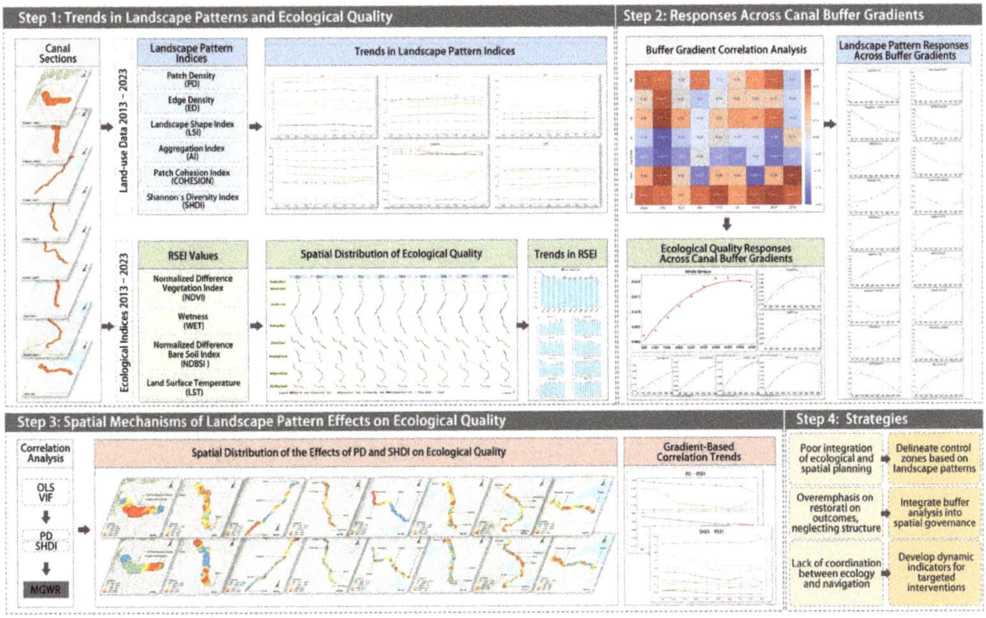

Figure 2. Framework of this study.

Table 2. Landscape pattern indices.

Index	Description
Patch Density (PD)	Quantifies landscape fragmentation; a higher number of patches indicates greater fragmentation.
Edge Density (ED)	Captures boundary complexity; longer total edge length suggests more intricate patch boundaries.
Landscape Shape Index (LSI)	Measures patch shape irregularity, normalized to remove area influence.
Aggregation Index (AI)	Indicates the degree of clustering among patches of the same type; higher values imply stronger aggregation.
Patch Cohesion Index (COHESION)	Reflects spatial connectivity of similar patches; higher values denote more continuous landscapes.
Shannon's Diversity Index (SHDI)	Represents landscape diversity in terms of richness and evenness of patch types, indicating overall heterogeneity.

3.2. Ecological Quality Assessment

The RSEI integrates four components: the Normalized Difference Vegetation Index (NDVI) for greenness, Land Surface Temperature (LST) for heat, Wetness Index (WET) for humidity, and the Normalized Difference Built-up and Soil Index (NDBSI) for dryness. These indices are combined using principal component analysis (PCA), enabling a comprehensive evaluation of surface vegetation, temperature, moisture, and aridity [45,46]. To reduce the influence of large water bodies on the calculation, the Modified Normalized Difference Water Index (MNDWI) was used to mask and exclude these areas before further processing [47,48]. Table 3 summarizes the formulas used for NDVI, LST, WET, and NDBSI [29].

Table 3. Formula of the four ecological indices.

Index	Formula
NDVI	$NDVI = (\rho_{NIR} - \rho_{Red})/(\rho_{NIR} + \rho_{Red})$
LST	$LST = 0.02LST_0 - 273.15$
WET	$WET = 0.1147\rho_{Red} + 0.2489\rho_{NIR1} + 0.2408\rho_{Blue} + 0.3132\rho_{Green} - 0.3122\rho_{NIR2} - 0.6416\rho_{SWIR1} - 0.5087\rho_{SWIR2}$
NDBSI	$NDBSI = (SI + IBI)/2$ $SI = [(\rho_{SWIR1} + \rho_{Red}) - (\rho_{Blue} + \rho_{NIR})]/[(\rho_{SWIR1} + \rho_{Red}) + (\rho_{Blue} + \rho_{NIR})]$ $IBI = \frac{2\rho_{SWIR1}/(\rho_{SWIR1}+\rho_{NIR})-[\rho_{NIR}/(\rho_{NIR}+\rho_{Red})+\rho_{Green}/(\rho_{Green}+\rho_{SWIR1})]}{2\rho_{SWIR1}/(\rho_{SWIR1}+\rho_{NIR})+[\rho_{NIR}/(\rho_{NIR}+\rho_{Red})+\rho_{Green}/(\rho_{Green}+\rho_{SWIR1})]}$

Note: In the formulas, ρ_i represents the surface reflectance of the corresponding band, and LST_0 refers to the original land surface temperature from the MOD11A2 product.

PCA was applied to the four normalized indices (NDVI, WET, LST, and NDBSI) to compute the initial RSEI ($RSEI_0$), with the first principal component (PC1) selected as the representative value [48,49]:

$$RSEI_0 = PC1[f(NDVI, WET, LST, NDBSI)] \tag{1}$$

The final RSEI was obtained through min–max normalization:

$$RSEI = (RSEI_0 - RSEI_{0\_min})/(RSEI_{0\_max} - RSEI_{0\_min}) \tag{2}$$

Based on prior studies [8,10], normalized RSEI values are classified into five ecological quality levels: excellent (0.8–1.0), good (0.6–0.8), moderate (0.4–0.6), fair (0.2–0.4), and poor (0.0–0.2).

3.3. Spatial Correlation Between Landscape Patterns and Ecological Quality

To evaluate the spatial relationship between landscape patterns and ecological quality, global spatial autocorrelation was first analyzed using Moran's I, which measures the degree of clustering among spatial units. The formula for Moran's I is as follows:

$$I = \frac{n}{\sum_{i=1}^{n}\sum_{j=1}^{n} w_{ij}} \cdot \frac{\sum_{i=1}^{n}\sum_{j=1}^{n} w_{ij}(x_i - \overline{x})(x_j - \overline{x})}{\sum_{i=1}^{n}(x_i - \overline{x})^2} \tag{3}$$

where n is the number of spatial units, x_i and x_j are the ecological quality values of units i and j, $\overline{x}$ is the global mean, and w_{ij} is the spatial weight. A positive value ($I > 0$) indicates clustering of similar values, a negative value ($I < 0$) indicates dispersion, and a value close to zero suggests a random distribution.

Following this, Ordinary Least Squares (OLS) regression was conducted to investigate the relationships between ecological quality and multiple landscape pattern indices. To mitigate multicollinearity and ensure the statistical validity of the model, variables with a Variance Inflation Factor (VIF) greater than 7.5 were excluded [38].

Considering the spatial resolution of remote sensing data, the heterogeneity of landscape patterns, and the perception scale of ecological processes, a 600 m regular grid was adopted as the spatial analysis unit [16]. The MGWR model was then constructed using MGWR 2.2 and is defined as follows:

$$Y_i = \beta_0(u_i, v_i) + \sum_{j=1}^{n} \beta_j(u_i, v_i)X_{ij} + \epsilon_i \tag{4}$$

where Y_i is the ecological quality of grid unit i, X_{ij} is the value of explanatory variable j in unit i, $\beta_j(u_i, v_i)$ is the spatially varying coefficient of variable j at location (u_i, v_i), $\beta_0(u_i, v_i)$ is the spatially varying intercept, and ϵ_i is the error term. The spatial scale of each explanatory variable is determined by its optimal bandwidth.

Finally, MGWR results were extracted based on equidistant buffer zones along the Grand Canal to assess the magnitude, direction, and spatial distribution of the effects of landscape pattern indices across buffer gradients.

4. Results

4.1. Trends in Landscape Pattern and Ecological Quality

4.1.1. Trends in Landscape Pattern Changes

To reveal temporal trends in landscape pattern evolution across canal sections, this study conducted a systematic analysis using six landscape pattern indices (Figure 3). At the whole-canal scale, the AI increased slightly from 91.24 to 91.38, indicating a general trend toward more clustered land-use. COHESION remained stable at 99.67, reflecting little change in structural connectivity. Meanwhile, ED decreased from 57.38 to 56.47 and PD declined from 8.45 to 8.20, reflecting a reduction in the number of patches and a corresponding increase in average patch size. The LSI fell from 199.74 to 196.72, suggesting a trend toward more regular patch configurations. SHDI increased slightly from 0.947 to 0.950 before declining to 0.945, implying a minor reduction in landscape type diversity.

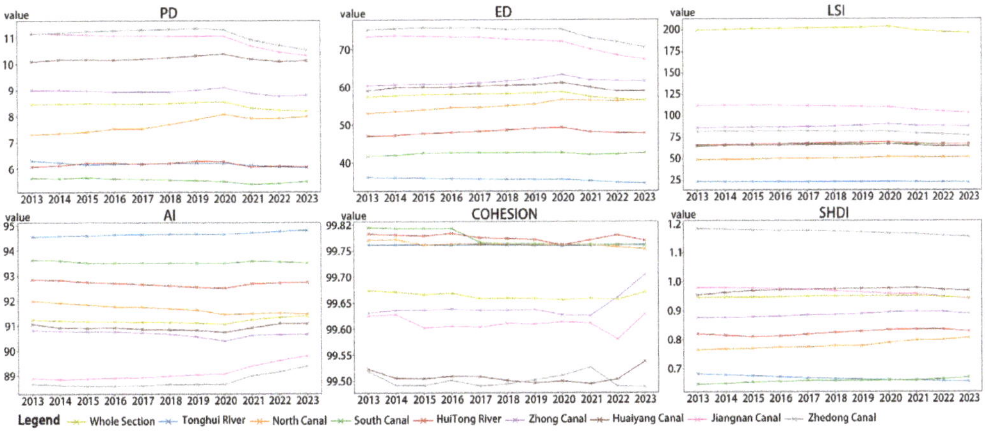

Figure 3. Changes in landscape pattern indices of the whole section and eight sections (2013–2023).

Notable variations in landscape pattern changes were observed across different canal sections. In the North Canal, AI declined from 91.98 to 91.46, while ED increased from 52.99 to 56.47, and LSI rose from 49.06 to 52.13, indicating more fragmented patterns with increased patch complexity. SHDI also increased, suggesting slightly improved landscape diversity. In contrast, the Jiangnan Canal and the Tonghui River both showed increases in AI, rising from 88.89 to 89.79 and from 94.56 to 94.80, respectively. ED decreased from 73.29 to 67.29 in the Jiangnan Canal and from 36.16 to 34.50 in the Tonghui River, indicating more aggregated patches with simpler boundaries. However, SHDI declined in both sections, decreasing from 0.98 to 0.94 in the Jiangnan Canal and from 0.68 to 0.66 in the Tonghui River, indicating a reduction in landscape heterogeneity. In the Huaiyang Canal, both AI and SHDI increased slightly, reflecting a modest rise in patch aggregation and landscape richness. The South Canal experienced a slight decline in AI, from 93.62 to 93.50, while SHDI increased from 0.65 to 0.67, reflecting enhanced landscape type diversity despite reduced aggregation. The Zhedong Canal exhibited an increase in AI to 89.36 and a significant decrease in LSI, alongside a drop in SHDI, indicating a trend toward simplification of the landscape structure. In the Zhong Canal, AI decreased slightly, while SHDI rose from 0.88 to 0.89, pointing to a marginal improvement in landscape diversity.

4.1.2. Trends in Ecological Quality Changes

Before analyzing ecological quality changes, PCA was conducted on four remotely sensed ecological indicators from 2013 to 2023 to assess the robustness of the RSEI model. As shown in Table 4, the first principal component (PC1) consistently accounted for the largest proportion of total variance each year, averaging 78.57% and never falling below 70%, confirming its effectiveness in representing ecological quality dynamics.

Table 4. Principal component analysis results (PC1).

Year	Loading of NDVI	Loading of LST	Loading of WET	Loading of NDBSI	Contribution Ratio of PC1 (%)
2013	−0.605	−0.381	0.410	0.566	77.14
2014	−0.617	−0.386	0.380	0.571	78.79
2015	−0.603	−0.404	0.370	0.580	78.98
2016	−0.600	−0.431	0.353	0.573	80.78
2017	−0.600	−0.416	0.336	0.595	77.36
2018	−0.600	−0.426	0.357	0.576	74.50
2019	−0.597	−0.432	0.358	0.573	79.59
2020	−0.597	−0.393	0.401	0.574	80.11
2021	−0.582	−0.417	0.388	0.580	80.05
2022	−0.616	−0.395	0.352	0.583	77.70
2023	−0.599	−0.414	0.381	0.569	79.31

To examine temporal trends in ecological quality, RSEI values from 2013 to 2023 were analyzed across eight canal sections and the whole-canal section (Figures 4 and 5). Overall, the average RSEI declined from 0.66 to 0.57, indicating a general degradation in ecological quality. However, clear spatial disparities were observed among sections.

The Tonghui River exhibited the most substantial improvement, with RSEI increasing from 0.33 to 0.49 and showing a steady upward trend beginning in 2016. The Huaiyang Canal, Jiangnan Canal, and Zhedong Canal also showed moderate but consistent improvement. In contrast, the North Canal and South Canal followed a U-shaped trajectory, reaching their lowest ecological quality around 2016, followed by gradual recovery after 2018. More concerning were the trends observed in the Huitong River and Zhong Canal, where RSEI values declined sharply from 0.78 to 0.66 and from 0.82 to 0.57, respectively, indicating sustained ecological degradation. These trends may reflect varying levels of anthropogenic pressure, restoration initiatives, and land-use changes across different sections of the Grand Canal.

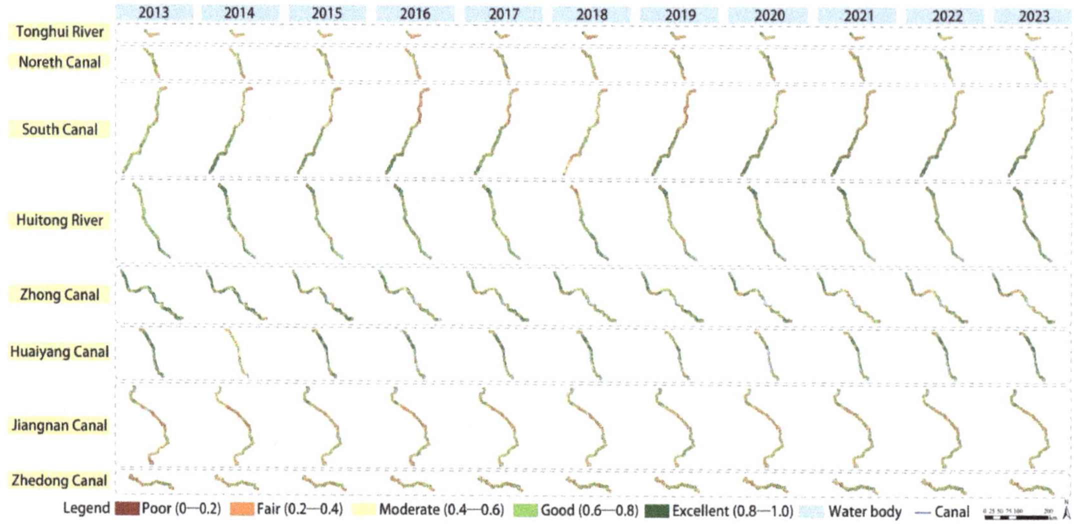

Figure 4. Spatial pattern of ecological quality in the eight sections (2013–2023).

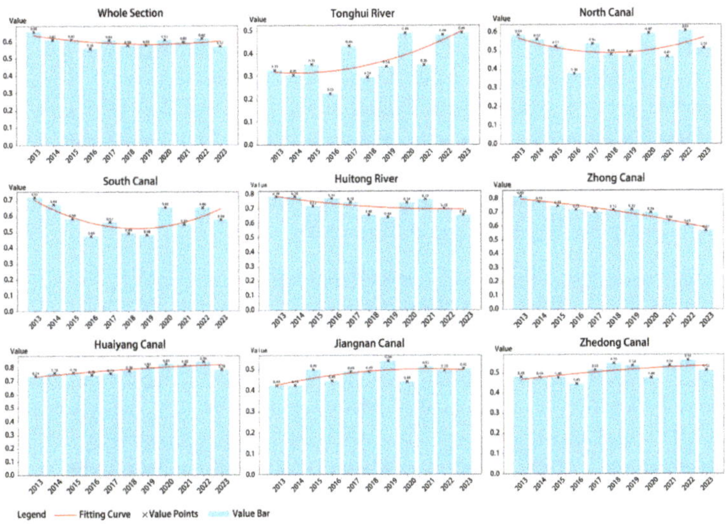

Figure 5. Changes in ecological quality in the whole section and eight sections (2013–2023).

4.2. Landscape Pattern and Ecological Quality Responses Across Canal Buffer Gradients

According to Spearman's correlation analysis (Figure 6), both landscape pattern indices and RSEI values exhibited significant spatial relationships with buffer distance from the canal. Among all indices, PD, SHDI, and RSEI showed the highest sensitivity to changes in distance. Across the entire study area, PD and RSEI were positively correlated with increasing distance from the canal, while COHESION showed a negative correlation. Other indices did not show statistically significant correlations at the overall level.

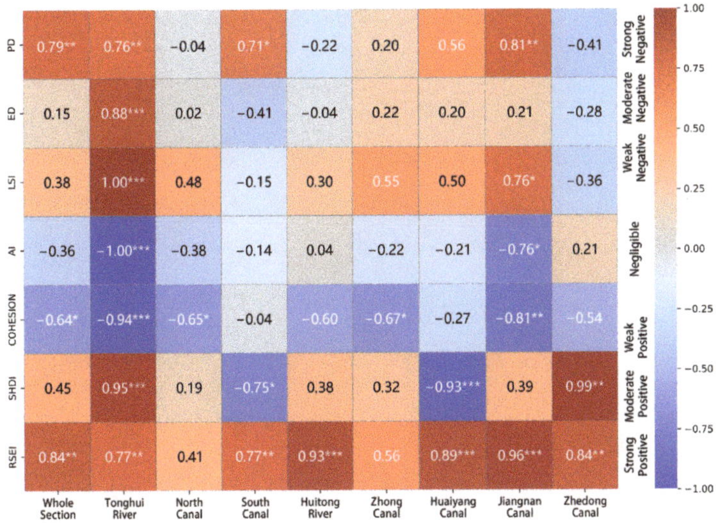

Figure 6. Correlation analysis between buffer gradients and landscape pattern and ecological quality across canal sections. Note: * indicates $p < 0.05$; ** indicates $p < 0.01$; *** indicates $p < 0.001$; values without asterisks are not significant ($p \geq 0.05$).

At the section level, the Tonghui River displayed the most pronounced gradient response: PD, ED, LSI, SHDI, and RSEI all increased with distance, whereas AI and

COHESION declined. Similarly, RSEI showed positive correlations with distance in the South Canal, Huitong River, Huaiyang Canal, Jiangnan Canal, and Zhedong Canal. LSI also tended to increase with distance in both the Tonghui and Jiangnan sections. Conversely, COHESION decreased with distance in the Tonghui River, North Canal, and Jiangnan Canal, indicating reduced landscape connectivity farther from the canal. SHDI exhibited a negative correlation with distance in the South Canal and Huaiyang Canal, suggesting a potential decline in landscape heterogeneity in more peripheral zones, possibly due to the dominance of homogenous land-use types in these areas.

4.2.1. Landscape Pattern Index Responses Across Canal Buffer Gradients

To examine how landscape patterns vary across different buffer distances, this study focused on indices significantly correlated with canal distance and analyzed their spatial trends (Figure 7). Across the entire study area, COHESION and PD exhibited clearly contrasting trends. COHESION continuously declined beyond 2000 m, reaching its lowest value in the outermost buffers, suggesting a weakening of patch connectivity. In contrast, PD peaked around 3500 m, indicating increased landscape fragmentation at greater distances from the canal.

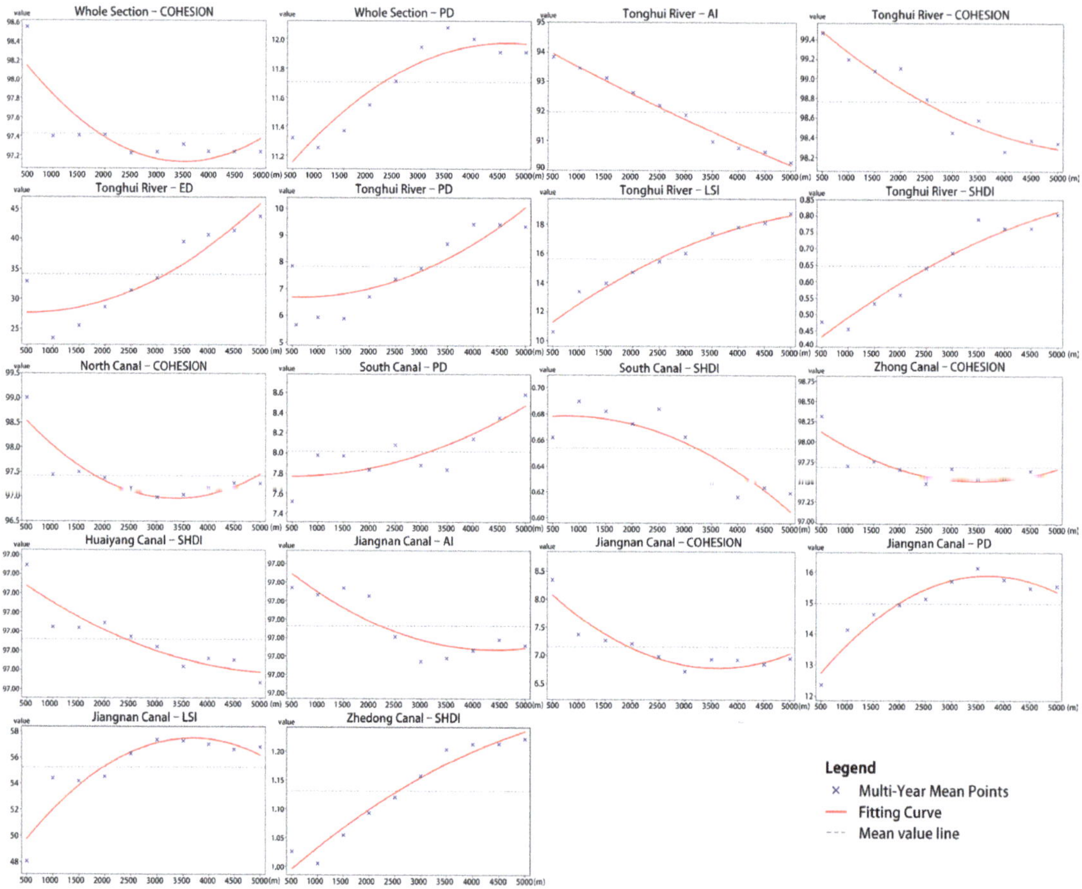

Figure 7. Fitted relationships between landscape patterns and canal buffer gradients across sections.

At the sectional level, COHESION showed a slight recovery following an initial decline in the Tonghui River, North Canal, Zhong Canal, and Jiangnan Canal, with minimum values observed at 4000 m, 3000 m, 2500 m, and 3000 m, respectively. In contrast, PD continued to rise in the Tonghui River and South Canal, while in the Jiangnan Canal, it peaked at 3500 m before declining slightly.

Further analysis revealed that ED, PD, LSI, and SHDI generally showed positive correlations with buffer distance in most sections, reflecting increasing fragmentation and landscape heterogeneity along the gradient. However, SHDI declined in the South Canal, Huaiyang Canal, and Zhedong Canal, indicating a reduction in landscape diversity with increasing distance in these regions. Notably, the South Canal exhibited the most pronounced increase in PD. The Jiangnan Canal, similar to the Tonghui River, exhibited a coordinated decline in AI and COHESION alongside increases in PD and LSI, reflecting a consistent spatial response pattern potentially driven by similar urban and ecological processes.

4.2.2. Ecological Quality Responses Across Canal Buffer Gradients

To further examine the spatial relationship between ecological quality and canal distance, this study analyzed variations in RSEI values across buffer zones along the study area (Figure 8). The results indicate a nonlinear trend: ecological quality generally increased with distance from the canal, peaked at mid-range buffers, and then declined. Specifically, the average RSEI across the full corridor peaked around 3500 m, suggesting that ecological conditions were most favorable in the middle buffer zones, possibly due to reduced development pressure and improved vegetation coverage.

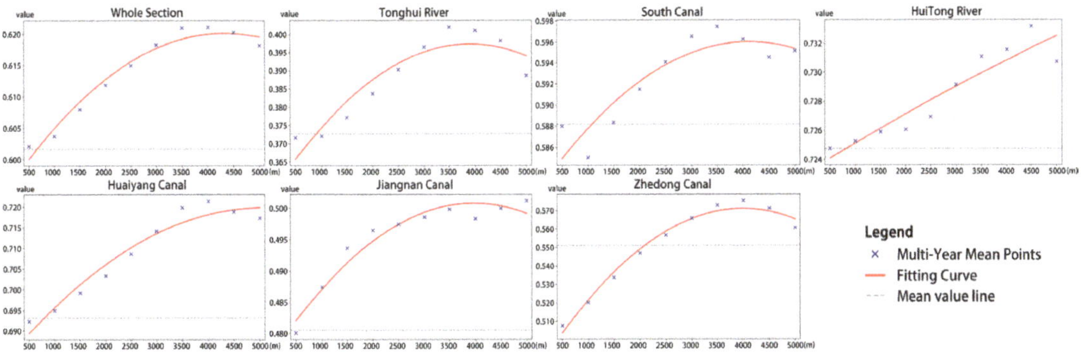

Figure 8. Fitted relationships between ecological quality and canal buffer gradients across sections.

Section-level analysis showed consistent yet nuanced trends. Both the Tonghui River and the South Canal followed the overall trend, with RSEI values peaking at 3500 m. The Huaiyang Canal and Zhedong Canal exhibited similar trends, with peaks occurring slightly farther out at 4000 m. In contrast, the Jiangnan Canal showed a delayed response, reaching its maximum ecological quality at 5000 m. Uniquely, the Huitong River showed a continuous increase in RSEI, peaking at 4500 m, suggesting a more sustained ecological benefit at greater distances from the canal.

4.3. Spatial Mechanisms of Landscape Pattern Effects on Ecological Quality

4.3.1. Spatial Effects of Landscape Pattern on Ecological Quality

To assess the spatial aggregation of ecological quality within each canal section, global spatial autocorrelation was measured using Moran's I. As shown in Table 5, all sections exhibited significant positive spatial autocorrelation, indicating clustered distributions of ecological quality. Z-scores exceeded the significance threshold substantially, and all *p*-values were 0.000, confirming the robustness of spatial clustering patterns.

Table 5. Global spatial autocorrelation of ecological quality in canal sections.

Canal Sections	Moran's I	Z-Score	*p*-Value
Tonghui River	0.756	42.803	0.000
North Canal	0.692	56.488	0.000
South Canal	0.762	105.278	0.000
Huitong River	0.696	116.113	0.000
Zhong Canal	0.750	95.144	0.000
Huaiyang Canal	0.726	70.513	0.000
Jiangnan Canal	0.732	71.388	0.000
Zhedong Canal	0.724	98.891	0.000

Following OLS regression and VIF screening, PD and SHDI emerged as the most influential landscape indices, showing strong correlations with ecological quality and acceptable multicollinearity. These two variables were retained in the MGWR model. As shown in Table 6, MGWR significantly improved model performance across all sections, with lower AICc values and greater explanatory power than the OLS model. This improvement highlights MGWR's capacity to capture spatially non-stationary and local variations relationships between landscape patterns and ecological quality.

Table 6. Comparison of OLS and MGWR models.

Canal Sections	OLS			MGWR		
	AICc	R^2	Adj.R^2	AICc	R^2	Adj.R^2
Tonghui River	3899.508	0.389	0.388	2114.404	0.826	0.810
North Canal	9726.510	0.020	0.019	5491.568	0.780	0.750
South Canal	27,821.192	0.042	0.042	12,619.346	0.839	0.819
Huitong River	21,105.191	0.095	0.095	10,451.657	0.828	0.804
Zhong Canal	21,406.759	0.060	0.060	11,500.445	0.789	0.767
Huaiyang Canal	10,757.009	0.057	0.057	5263.762	0.834	0.808
Jiangnan Canal	26,015.883	0.037	0.037	13,281.209	0.817	0.790
Zhedong Canal	13,292.110	0.091	0.090	7308.020	0.799	0.771

Regression coefficients derived from MGWR were further analyzed to interpret the spatial influence of landscape structure on ecological quality (Table 7). Notably, PD and SHDI exhibited marked heterogeneity across canal sections. In the Tonghui River and North Canal, average PD coefficients were positive, suggesting that a moderate level of fragmentation may enhance ecological quality, potentially by increasing patch-level diversity and microhabitat variability. Conversely, SHDI coefficients were mostly negative, implying that excessive diversity and irregularity could disrupt ecological continuity and reduce system stability. Coefficient ranges varied substantially, for example, PD in the North Canal ranged from −1.003 to 1.392, indicating strong spatial differentiation. Similarly, in the Huaiyang Canal and Zhedong Canal, SHDI coefficients fluctuated from negative to positive, further confirming the spatially non-stationary effects of landscape patterns on ecological quality.

Based on MGWR results, this study explored the spatial heterogeneity in the effects of PD and SHDI on RSEI across different canal sections (Figure 9). In the Tonghui River, South Canal, and Zhong Canal, PD generally showed a positive correlation with RSEI, although the strength of the association varied spatially. In the Tonghui River, the strongest positive effects of PD were concentrated within Beijing's Third Ring Road, while SHDI had its greatest influence in Tongzhou District. These findings suggest that high-density green patches and moderate landscape diversity may jointly enhance ecological quality, possibly by reinforcing habitat continuity and buffering microclimatic extremes.

Table 7. MGWR regression results.

Canal Sections	Landscape Pattern Indices	Average Value	Minimum Value	Median Value	Maximum Value
Tonghui River	PD	0.222	−0.198	0.282	0.453
	SHDI	0.324	0.080	0.265	0.618
North Canal	PD	0.125	−1.003	0.132	1.392
	SHDI	−0.186	−0.789	−0.307	0.545
South Canal	PD	0.057	−0.051	0.047	0.177
	SHDI	−0.262	−0.958	−0.350	0.741
Huitong River	PD	−0.002	−0.707	0.029	0.503
	SHDI	−0.313	−0.985	−0.373	0.768
Zhong Canal	PD	0.071	0.068	0.071	0.075
	SHDI	−0.204	−1.022	−0.251	0.634
Huaiyang Canal	PD	−0.032	−1.065	−0.025	0.880
	SHDI	−0.105	−0.934	−0.150	1.087
Jiangnan Canal	PD	−0.102	−1.255	−0.069	0.867
	SHDI	−0.014	−1.048	0.026	0.797
Zhedong Canal	PD	−0.138	−0.497	−0.135	0.089
	SHDI	0.075	−1.150	0.076	1.166

In the Zhong Canal, PD–RSEI correlations weakened downstream, and SHDI was largely negatively associated with RSEI, except in Suqian and Huai'an, where localized positive relationships were observed. This pattern indicates that landscape diversity shaped by urban spatial planning may enhance ecological quality under certain conditions. In the South Canal, PD–RSEI correlations intensified southward from Tianjin, peaking between Dezhou and Linqing. In the Jiangnan Canal, which spans highly urbanized zones, SHDI and RSEI were positively correlated in the Changzhou–Suzhou section, with PD also showing spatial continuity. This implies that even in highly developed areas, a well-structured landscape can sustain ecological quality.

Conversely, in the North Canal, Huaiyang Canal, and Huitong River sections, both PD and SHDI were negatively associated with RSEI, though the strength of these relationships varied across space. In the North Canal, the PD–RSEI relationship diverged sharply between urban and rural areas. SHDI exhibited negative correlations in most rural stretches, with exceptions in urban centers such as Beijing and Tianjin. In the Huaiyang Canal, positive effects were localized near Gaoyou Lake and the urban core of Yangzhou, while negative associations dominated the rest of the section. In the Huitong River, PD was negatively associated with RSEI across farmland, wetland, and forested areas. SHDI was positively correlated only in Liaocheng's urbanized areas, and negatively correlated elsewhere, highlighting the region's heightened sensitivity to landscape configuration in determining ecological quality.

4.3.2. Landscape Pattern Effects on Ecological Quality Across Buffer Gradients

To explore how landscape patterns influence ecological quality at varying distances from the canal, this study analyzed the correlation trends between PD and SHDI with RSEI across buffer gradients in each canal section (Figure 10). In the Tonghui River, North Canal, South Canal, and Zhong Canal, PD showed a consistently positive correlation with RSEI across the buffer distances, suggesting that increased patch density was generally associated with enhanced ecological quality. However, in the North Canal, the strength of this positive correlation declined notably beyond 1000 m, reaching a minimum at 4000 m before slightly recovering, a pattern indicative of spatial decay. In the Huitong River, PD was positively associated with RSEI within 2500 m but turned negative at greater distances, implying that the ecological benefits of patch density diminish with increasing distance from the canal. In contrast, PD remained negatively correlated with RSEI at all distances in the Huaiyang Canal, Jiangnan Canal, and Zhedong Canal, indicating that fragmentation in these areas may undermine ecological stability across the landscape.

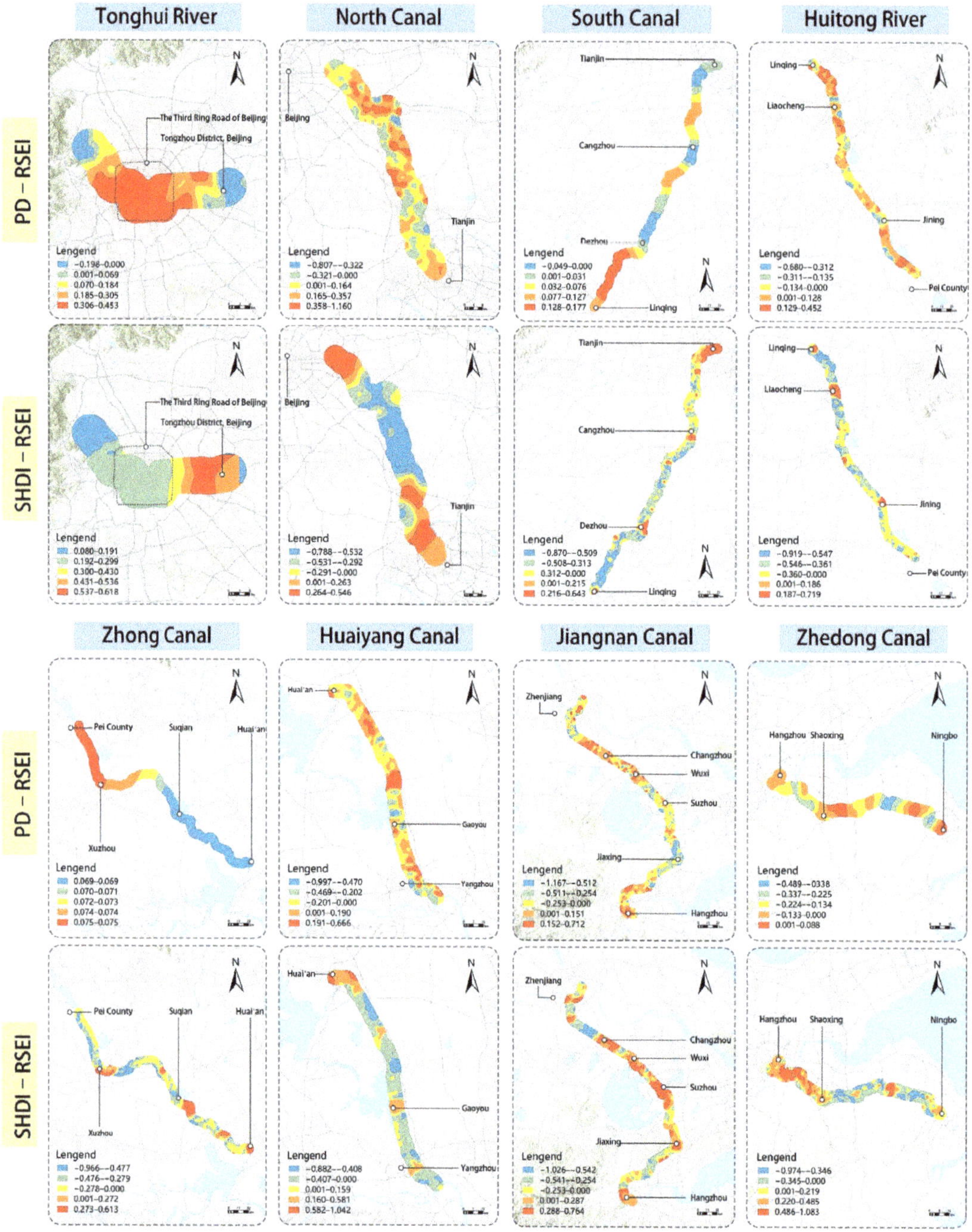

Figure 9. Spatial distribution of the effects of PD and SHDI on ecological quality across canal sections.

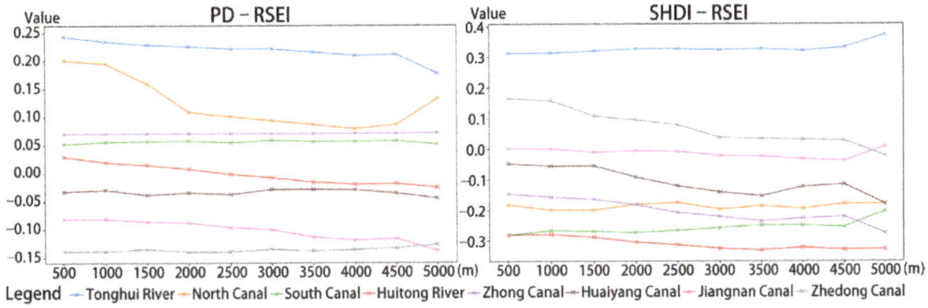

Figure 10. Buffer gradient variation in the correlation between landscape patterns and ecological quality across canal sections.

The correlation between SHDI and RSEI was more variable and generally negative. A relatively stable negative relationship was observed in the Zhong Canal, North Canal, South Canal, Huaiyang Canal, and Huitong River, implying that excessive landscape heterogeneity may impair ecological integrity in these sections. In the Zhedong Canal, this negative correlation intensified with distance, reinforcing the notion that uncontrolled diversity can lead to functional fragmentation. In contrast, the Tonghui River exhibited a distinct pattern: both PD and SHDI were consistently and positively correlated with RSEI. This finding suggests that moderate fragmentation and diversity, when guided by effective planning and ecological infrastructure, may jointly promote ecological quality. Meanwhile, in the Huaiyang Canal, the persistent negative correlation between both PD and SHDI with RSEI indicates that high levels of fragmentation and heterogeneity may interactively contribute to ecological degradation.

5. Discussion

5.1. Mechanistic Analysis of the Evolution of Landscape Pattern and Ecological Quality

From 2013 to 2023, the landscape pattern along the Grand Canal exhibited a clear trend toward spatial centralization and structural simplification. Land-use types became more concentrated, and patch configurations tended to be more regular. Section-level analysis revealed notable fragmentation in the North Canal, marked by increased patch density and edge complexity, a trend likely driven by intensive land development activities [50]. In contrast, both the Tonghui River and Jiangnan Canal showed rising AI values and declining SHDI, suggesting that urbanized artificial landscapes tend to become more homogeneous under high-intensity development [51,52]. This pattern may reflect a planned distribution and accessibility of green spaces [53]. Conversely, the South Canal, Huaiyang Canal, and Zhong Canal exhibited increasing SHDI values, indicating enhanced landscape heterogeneity likely resulting from localized ecological restoration efforts [8]. These spatial differences appear to be closely linked to regional variations in land-use functions, water infrastructure conditions, and the stringency of ecological policies. Notably, a decline in fragmentation was observed in sections such as the Tonghui River and Jiangnan Canal, contradicting the conventional assumption that urbanization invariably leads to fragmentation [54,55]. This suggests that intensive anthropogenic intervention may instead promote landscape agglomeration and structural regularization. This pattern is consistent with Carlier et al.'s findings on infrastructure-driven landscape simplification [56], emphasizing the need to incorporate spatial logic and policy context when analyzing landscape evolution. In particular, landscape homogenization in well-managed urban cores may not signal ecological degradation but rather reflect a strategic reorganization of spatial structure to improve systemic efficiency and connectivity under planning constraints.

In terms of ecological quality, pronounced spatial heterogeneity was observed across sections (Figure 4). Its evolution was influenced not only by ecological restoration measures but also by the effectiveness of land-use regulation and policy coordination. In the Yangtze River Delta, the Huaiyang Canal, Jiangnan Canal, and Zhedong Canal achieved steady improvements in ecological quality, supported by the integration of wetland restoration, ecological corridor construction, and urban boundary controls. These coordinated strategies effectively reconciled development pressures with ecological goals. In contrast, ecological conditions in the Zhong Canal and Huitong River, located in Shandong and northern Jiangsu, deteriorated during the study period due to a combination of intensified shipping, urban expansion, and agricultural encroachment. The underlying challenges included delayed ecological compensation mechanisms and fragmented spatial governance, which undermined restoration efforts. While previous studies reported general improvements in ecological quality across canal-adjacent provinces and cities [57,58], our results highlight the need for section-specific analysis, as macro-level improvements may mask local ecological decline. This supports Jiang et al.'s recommendation for differentiated restoration strategies tailored to the specific spatial and developmental contexts of each section [59].

Further insights emerged from the buffer zone analysis, which highlighted the spatial gradients in landscape–ecosystem interactions. Indices such as PD, COHESION, and SHDI showed distinct responses across distance zones, with more urbanized sections exhibiting stronger gradient sensitivity (Figure 7). Near-canal areas typically experience tighter ecological controls and enforcement, while outer zones were more susceptible to land-use conversion and development pressures, leading to increased fragmentation and reduced connectivity [60]. For example, the consistent increase in fragmentation in the Tonghui River section reflects the strong influence of urban expansion on landscape structure, consistent with Yang et al.'s findings from the Yongding River, where fringe areas showed reduced structural complexity and increased aggregation, with a notable transition in landscape configuration occurring around 3500 m from the river [18].

Ecological quality similarly exhibited a nonlinear spatial pattern, peaking around 3500 m from the canal (Figure 8). This mid-range zone likely represents a transitional area where development pressures and ecological processes temporarily balance. The result echoes the "core–edge" theory, in which built-up areas dominate the canal edge, while ecological functionality is better preserved in peripheral zones [61]. However, peak locations varied by section. In the Tonghui River and South Canal, ecological quality peaked around 3500 m, indicating a dual-function zone where development and ecological function coexist. In contrast, the Jiangnan and Huaiyang Canals exhibited delayed peaks, possibly reflecting interactions between urban expansion and water network restoration [51]. The Huitong River showed a continuous increase in ecological quality with increasing distance, reflecting a relatively natural shoreline and limited anthropogenic disturbance. These patterns suggest that the relationship between landscape structure and ecological quality is shaped by spatial thresholds and functional trade-offs. Moderate fragmentation in well-planned urban cores may enhance edge effects and functional diversity, while excessive heterogeneity in poorly managed zones may disrupt ecological continuity and reduce system resilience.

5.2. *Differential Responses of Ecological Quality to Landscape Pattern*

Based on the MGWR results, this study identified spatially heterogeneous effects of PD and SHDI on ecological quality across different canal sections (Figure 9). Positive correlations were commonly observed in well-planned urban areas, where structured green spaces and moderate landscape diversity enhance ecological resilience. In contrast, negative correlations emerged in fragmented or poorly managed zones, where excessive heterogeneity undermined connectivity and ecological stability. These differences reflect variations in land-use types, restoration effectiveness, and policy enforcement, underscoring the context-dependent mechanisms through which landscape patterns affect ecological quality.

For instance, in the Tonghui River, the southern segment of the South Canal, and the urban core of the Suzhou–Wuxi–Changzhou metropolitan area, PD showed a significant positive association with RSEI, while SHDI also contributed positively to selected locations. These results suggest that deliberate spatial planning, such as the concentration of high-density green patches can enhance landscape connectivity and integrate ecological functions. In the Tonghui River, within Beijing's Third Ring Road, long-standing implementation of green infrastructure plans and development boundaries, along with sustained ecological investment in Tongzhou District as Beijing's sub-center, have collectively strengthened the relationship between moderate fragmentation and improved ecological quality [62,63]. Such efforts have also contributed to a more comfortable walking environment for residents [64]. These findings suggest that the ecological impacts of human activity depend not only on development intensity, but also on spatial configuration, planning continuity, and institutional coherence. In historically well-regulated cities, anthropogenic interventions may even promote ecological recovery by enhancing spatial structure and connectivity.

Conversely, in the North Canal, Huaiyang Canal, and Huitong River, PD and SHDI generally showed negative correlations with RSEI, accompanied by significant spatial variation. These areas, primarily located in rural–urban transition zones of the Beijing–Tianjin–Hebei region and southwestern Shandong, have historically featured dispersed township development and extensive agricultural land-use, but limited ecological intervention. The Huitong River, which traverses multiple prefecture-level jurisdictions, lacks a unified ecological management framework, resulting in inconsistent policy enforcement and uneven construction standards. Consequently, ecological spaces have been fragmented, and landscape structures fail to achieve sustainable configurations, limiting the ability of landscape diversity to support ecological quality [65].

Interestingly, localized shifts from negative to positive correlations between PD or SHDI and RSEI were observed in urban sections of the Zhong Canal (Suqian–Huai'an) and Jiangnan Canal (Changzhou–Suzhou). This trend may be attributed to increasing spatial complexity and the reorganization of ecological function zones within cities. In recent years, under national spatial planning initiatives, cities have promoted the development of green wedges and waterfront ecological corridors, facilitating structural and functional integration of previously fragmented green spaces. These efforts have contributed to improvements in ecological quality. This contrasts with the findings by Aurora et al., who reported that urban expansion in Chiba Prefecture, Japan (1990–2021), generally led to ecological degradation [66]. In contrast, this study observed synchronous trends of urban growth and ecological improvement in certain Grand Canal sections, consistent with Yang et al.'s findings on internal spatial optimization in Kunming, Nanjing, and Guangzhou, where urban core restructuring supported ecological recovery from 1990 to 2020 [35]. These examples highlight the importance of considering multiple interacting landscape spaces in an integrated manner to optimize urban ecological performance [67].

Analysis of buffer gradients further revealed that the influence of patch density on ecological quality tends to weaken with increasing distance from the canal. In most sections, PD is positively correlated with RSEI in nearshore zones but declined in more distant areas (Figure 10). This pattern likely reflects stronger ecological governance, green investment, and spatial planning controls near the canal. In contrast, foreshore zones often face regulatory gaps and intensified land-use conversion, where abrupt transitions from forest to construction land reduce ecological connectivity and resilience. In these zones,

SHDI frequently showed negative correlations with RSEI, likely due to uncoordinated landscape diversity leading to function fragmentation. This scenario reflects an ecological risk profile characterized by high heterogeneity but low structural stability [14].

These spatial patterns must be interpreted within the broader historical and institutional context of the Grand Canal. As a major hydraulic and transportation corridor in ancient China, the canal profoundly shaped regional land-use and ecological processes. Historically, the construction and maintenance of the Grand Canal transformed natural hydrological systems, forming a linear landscape structure centered around waterways, docks, and settlement clusters [68]. These human interventions supported the expansion of irrigation networks, market towns, and transit hubs, driving long-term land-use changes [69]. Traditional uses such as irrigation and transport altered hydrological connectivity and aquatic ecosystems, while industrialization and modern urbanization introduced new pressures including water pollution and flow disruption [13]. These cumulative impacts fragmented ecosystems and impaired ecological function. In recent years, however, efforts in heritage conservation and ecological restoration have reshaped local landscape patterns and improved connectivity. This dynamic reflects a disturbance–response cycle, in which human activity affects hydrological flow, land-use intensity, and biodiversity, resulting in complex, nonlinear, and multi-scale feedback between landscape structure and ecological quality along the canal corridor [70].

5.3. Ecological Planning Strategies and Policy Implications

In recent years, policies promoting ecological protection and landscape construction along the Grand Canal have steadily advanced, gradually forming a development model that integrates cultural heritage preservation, ecological restoration, and functional coordination. The Outline of the Plan for the Protection, Inheritance, and Utilization of the Culture of the Grand Canal and the Plan for the Construction and Protection of the Grand Canal National Cultural Park have defined national-level implementation pathways for integrating culture and ecological values [71]. Provincial and municipal governments, including Jiangsu Province, Beijing, and Hangzhou, have actively implemented ecological restoration and spatial remediation efforts, offering valuable practical experience in landscape optimization and ecological reconstruction [72]. However, current policy frameworks still fall short in addressing the spatial coupling between landscape pattern evolution and ecological quality changes. First, special purpose planning often lacks alignment with the national territorial spatial planning system. Ecological governance frequently depends on fragmented, localized interventions that fail to incorporate spatial structure analysis. Second, many implementation efforts prioritize visible restoration outcomes, while neglecting the foundational role of spatial configuration in shaping ecological conditions. As a result, the relationships among ecological processes, spatial structure, and functional outcomes are often poorly captured. Third, canal navigation continues in many sections without effective coordination mechanisms to balance ecological protection with transport infrastructure. In certain sections, navigation facilities and heavy traffic still exert significant pressure on ecological spaces.

To address these challenges, spatial zoning strategies for regulating ecological quality should be refined based on landscape patterns' characteristics. On the one hand, ecological control units should be delineated, targeting areas with high fragmentation and low connectivity for prioritized regulation and resource allocation [73]. Restoration-prone zones should be incorporated into these units to guide targeted ecological enhancements. Zoning-specific and hierarchical regulation objectives should then be established, including measures such as restricting high-intensity development, improving ecological nodes through greening, and enhancing patch connectivity. On the other hand, a multi-scale spatial coordination mechanism should be developed across canal sections, buffer zones, and subregional units. Findings from buffer gradient analyses should be integrated into restoration plans and land-use boundary frameworks, thereby clarifying the spatial logic underlying coordination between ecological redlines and development activities [74]. This would enhance the precision and operational feasibility of ecological governance. International examples, such as the Rideau Canal in Canada and the Canal du Midi in France, demonstrate the benefits of functional zoning for ecological heritage management. These systems achieved synergistic integration of cultural preservation and ecological protection through refined landscape unit management and targeted interventions at the land–water interface [75,76]. In parallel, ecological impact factors should be systematically identified, and a dynamic evaluation index system should be developed to support zoning assessments, prioritize interventions, and guide integrated management. Such strategies will facilitate the scientific and adaptive advancement of ecological spatial governance.

5.4. Limitations and Future Research

Although this study offers meaningful insights into the landscape patterns and ecological quality along the Grand Canal, several limitations remain. First, the accuracy of available spatial data constrains the depth of analysis over large-scale regions. The current resolution of remote sensing data limits the ability to precisely detect fine-scale landscape units and complex ecosystem structures. Future research should incorporate higher-resolution imagery to enhance data accuracy and enable more detailed assessments of landscape and ecological changes. Second, the RSEI has intrinsic limitations when applied to ecological assessments at large watershed scales. Future studies should explore enhanced indices, such as an updated Water Body Ecological Quality Index (WQI), to provide a more comprehensive evaluation of ecological conditions. Third, in the buffer zone analysis, this study relied on multi-year mean values of landscape and ecological indices due to spatial and temporal constraints. This limits the ability to capture temporal dynamics and may obscure short-term or seasonal ecological changes. Future research should adopt year-by-year comparative analyses and integrate interannual trend evaluations to more accurately reflect ecosystem responses over time and under different development or restoration scenarios.

6. Conclusions

This study systematically examined the evolution of landscape patterns and their spatial impact on ecological quality along the Grand Canal from 2013 to 2023. The results revealed a general trend toward increased landscape aggregation and structural regularization, as indicated by an increase in the AI from 91.24 to 91.38 and decreases in PD and LSI from 8.45 to 8.20 and from 199.74 to 196.72, respectively. These changes reflect a shift toward more centralized land-use and more regular patch configurations. However, distinct spatial differences emerged among canal sections. The North Canal exhibited increased fragmentation, whereas landscape diversity declined significantly along the Tonghui River and Jiangnan Canal, highlighting the heterogeneous nature of landscape transformation across regions. In terms of ecological quality, the overall RSEI slightly declined from 0.66 to 0.57. Nevertheless, several river sections, including the Tonghui River, Huaiyang Canal, Jiangnan Canal, and Zhedong Canal, showed continuous improvement. In contrast, the North and South Canals followed a U-shaped trajectory, reflecting initial degradation followed by gradual recovery. Buffer gradient analysis further demonstrated spatial variations in ecological responses to landscape structure. Both PD and RSEI tended to increase with distance from the canal, while COHESION decreased, with all indicators peaking around 3500 m. This suggests that mid-distance zones represent key interfaces where landscape disturbance and ecological processes intersect. Moreover, the effects of PD and SHDI on ecological quality varied significantly across canal sections. In urbanized areas such as the Tonghui River and South Canal, patch density and landscape diversity jointly contributed to ecological enhancement. Conversely, the North Canal, Huaiyang Canal, and Huitong River exhibited predominantly negative correlations, implying that unregulated development and excessive heterogeneity may compromise ecological function. Overall, this study reveals the nonlinear and spatially heterogeneous relationships between landscape pattern changes and ecological quality. These insights provide a scientific foundation for the conservation and sustainable development of the Grand Canal and offer theoretical guidance for the ecological management of other major inland waterways.

Author Contributions: Conceptualization, Y.X. and A.J.; methodology, Y.X. and A.J.; software, Y.X.; validation, Y.X.; formal analysis, A.J.; investigation, Y.X. and A.J.; resources, A.J.; data curation, Y.X.; writing—original draft preparation, Y.X. and A.J.; writing—review and editing, A.J.; visualization, Y.X.; supervision, A.J.; project administration, A.J. All authors have read and agreed to the published version of the manuscript.

Funding: This research received no external funding.

Data Availability Statement: Data and materials are available from the authors upon request.

Acknowledgments: The authors thank the anonymous reviewers for their valuable comments and suggestions on this article.

Conflicts of Interest: The authors declare no conflicts of interest.

References

1. Luo, J.; Zhao, T.; Cao, L.; Bijecki, F. Semantic Riverscapes: Perception and evaluation of linear landscapes from oblique imagery using computer vision. *Landsc. Urban Plan.* **2022**, *228*, 104569. [CrossRef]
2. Cao, Y.; Chen, S. Rebel on the Canal: Disrupted Trade Access and Social Conflict in China. 1650–1911. *Am. Econ. Rev.* **2022**, *112*, 1555–1590. [CrossRef]
3. Tang, F.; Wang, L.; Guo, Y.; Fu, M.; Huang, N.; Duan, W.; Luo, M.; Zhang, J.; Li, W.; Song, W. Spatio-temporal variation and coupling coordination relationship between urbanisation and habitat quality in the Grand Canal, China. *Land Use Policy* **2022**, *117*, 106119. [CrossRef]
4. Li, Y.; Wu, L.; Han, Q.; Wang, X.; Zou, T.; Fan, C. Estimation of remote sensing based ecological index along the Grand Canal based on PCA-AHP-TOPSIS methodology. *Ecol. Indic.* **2021**, *122*, 107214. [CrossRef]
5. Zhou, S.; Ye, J.; Li, J.; Zhang, G.; Duan, Y. Identifying intrinsic drivers to changes in riparian ecosystem services by using PSR framework: A case study of the Grand Canal in Jiangsu, China. *Environ. Dev.* **2022**, *43*, 100728. [CrossRef]
6. Zhang, S.; Liu, J.; Pei, T.; Chan, C.; Wang, M.; Meng, B. Tourism value assessment of linear cultural heritage: The case of the Beijing-Hangzhou Grand Canal in China. *Curr. Issues Tour.* **2023**, *26*, 47–69. [CrossRef]
7. Yang, D.; Song, W. Ecological function regionalization of the core area of the Beijing-Hangzhou Grand Canal based on the leading ecological function perspective. *Ecol. Indic.* **2022**, *142*, 109247. [CrossRef]
8. Zhang, Y.; Tian, Q.; Wu, J. Coupling coordination degree and obstacle factors between the tourism industry and ecological environment in the Beijing-Hangzhou Grand Canal Basin, China. *Environ. Dev. Sustain.* **2024**, *26*, 2589–2613. [CrossRef]
9. Xie, Y.; Zhou, Q.; Xiao, X.; Chen, F.; Huang, Y.; Kang, J.; Wang, S.; Zhang, F.; Gao, M.; Du, Y. Satellite-Based Water Quality Assessment of the Beijing Section of the Grand Canal: Implications for SDG11.4 Evaluation. *Remote Sens.* **2024**, *16*, 909. [CrossRef]
10. Jiang, A.; Sun, F.; Zhang, B.; Wu, Q.; Cai, S.; Yang, Z.; Chang, Y.; Han, R.; Yu, S. Spatiotemporal dynamics and driving factors of vegetation coverage around linear cultural heritage: A case study of the Beijing-Hangzhou Grand Canal. *J. Environ. Manag.* **2024**, *349*, 119431. [CrossRef]
11. Du, L.; Peng, X.; Zhao, Y.; Lyu, J.; Pan, M. From nature-based to engineering-based: The interaction process and turning point of the city-water system relationship in the North China plain. *Appl. Geogr.* **2021**, *135*, 102556. [CrossRef]
12. Huang, Y.; Yang, S. Spatio-temporal evolution and distribution of cultural heritage sites along the Suzhou canal of China. *Herit. Sci.* **2023**, *11*, 188. [CrossRef]
13. Zhang, Y.; Jing, Z.; Huang, Q.; Wang, X.; Sun, W.; Zhang, C.; Wang, J.; Zhong, Y.; Wang, J.; Tan, L. On conservation of world heritage Beijing-Hangzhou grand canal for enhancing cultural ecosystem services. *Herit. Sci.* **2023**, *11*, 269. [CrossRef]

14. Gao, C.; Dang, Q.; Li, C.; Fan, Y. Analysis of Landscape Fragmentation Evolution Characteristics and Driving Factors in the Wei River Basin, China. *Land* **2025**, *14*, 538. [CrossRef]
15. Liu, J.; Jin, X.; Lin, J.; Liang, X.; Zhang, X.; Zhou, Y. Identification and characteristic analysis of semi-natural habitats in China's economically developed areas: New insights to inform cultivated land system ecological conservation. *J. Environ. Manag.* **2024**, *351*, 119804. [CrossRef]
16. Xie, Y.; Zhu, Q.; Bai, H.; He, H.; Zhang, Y. Combining ecosystem service value and landscape ecological risk to subdivide the riparian buffer zone of the Weihe River in Shaanxi. *Ecol. Indic.* **2024**, *166*, 112424. [CrossRef]
17. Fernandes, A.C.P.; Martins, L.M.D.; Pacheco, F.A.L.; Fernandes, L.F.S. The consequences for stream water quality of long-term changes in landscape patterns: Implications for land use management and policies. *Land Use Policy* **2021**, *109*, 105679. [CrossRef]
18. Yang, R.; Liu, Z.; Kang, W.; Su, J.; Zhang, R.; Xu, X. Historical Spatial Radiation Range of the Yongding River Corridor in Beijing Plain Section: Implications for Landscape Patterns and Ecological Restoration. *Land* **2023**, *12*, 1656. [CrossRef]
19. Han, H.; Yan, X.; Li, X.; Huang, Z.; Yan, X.; Xia, Y. Significant differences in optimal riparian buffer zone on water quality between different segments within the same river. *J. Environ. Manag.* **2025**, *381*, 125306. [CrossRef]
20. Yang, S.; Zhu, G.; Zhang, L.; Xu, H.; Cheng, J. Spatiotemporal Variations of Ecosystem Service Indicators and the Driving Factors Under Climate Change in the Qinghai-Tibet Highway Corridor. *Front. Environ. Sci.* **2022**, *10*, 935713. [CrossRef]
21. Mohammadi, M.; Darvishan, A.K.; Dinelli, E.; Bahramifar, N.; Alavi, S.J. How does land use configuration influence on sediment heavy metal pollution? Comparison between riparian zone and sub-watersheds. *Stoch. Environ. Res. Risk Assess.* **2022**, *36*, 719–734. [CrossRef]
22. Wen, H.; Xiao, Y.; Zhang, L. Spatial effect of river landscape on housing price: An empirical study on the Grand Canal in Hangzhou, China. *Habitat Int.* **2017**, *63*, 34–44. [CrossRef]
23. Xu, C.; Jing, Y.; Zhang, W.; Liu, Y.; Yin, W.; Zhao, S. Identification of Ecological Security Pattern and Ecological Restoration Zoning Strategy in the Shandong Section of the Beijing-Hangzhou Grand Canal. *Land* **2025**, *14*, 439. [CrossRef]
24. Yang, S.; Zhang, L.; Zhu, G. Effects of transport infrastructures and climate change on ecosystem services in the integrated transport corridor region of the Qinghai-Tibet Plateau. *Sci. Total Environ.* **2023**, *885*, 163961. [CrossRef]
25. Zhang, W.; Chang, W.; Zhu, Z.; Hui, Z. Landscape ecological risk assessment of Chinese coastal cities based on land use change. *Appl. Geogr.* **2020**, *117*, 102174. [CrossRef]
26. Ahmad, M.; Ahmed, Z.; Majeed, A.; Huang, B. An environmental impact assessment of economic complexity and energy consumption: Does institutional quality make a difference? *Environ. Impact Assess. Rev.* **2021**, *89*, 106603. [CrossRef]
27. Dogan, A.; Pata, U.K. The role of ICT, R&D spending and renewable energy consumption on environmental quality: Testing the LCC hypothesis for G7 countries. *J. Clean. Prod.* **2022**, *380*, 135038.
28. Hu, X.; Xu, H. A new remote sensing index for assessing the spatial heterogeneity in urban ecological quality: A case from Fuzhou City, China. *Ecol. Indic.* **2018**, *89*, 11–21. [CrossRef]
29. Aizizi, Y.; Kasimu, A.; Liang, H.; Zhang, X.; Zhao, Y.; Wei, B. Evaluation of ecological space and ecological quality changes in urban agglomeration on the northern slope of the Tianshan Mountains. *Ecol. Indic.* **2023**, *146*, 109896. [CrossRef]
30. Shan, W.; Jin, X.; Ren, J.; Wang, Y.; Xu, Z. Ecological environment quality assessment based on remote sensing data for land consolidation. *J. Clean. Prod.* **2019**, *239*, 118126. [CrossRef]
31. Firozjaei, M.K.; Fatholoumi, S.; Kiavvarz, M.; Biswas, A.; Homaee, M. Land Surface Ecological Status Composition Index (LSESCI): A novel remote sensing-based technique for modeling land surface ecological status. *Ecol. Indic.* **2021**, *123*, 107375. [CrossRef]
32. Zou, F.; Hu, Q.; Liu, Y.; Li, H.; Zhang, X.; Liu, Y. Spatiotemporal Changes and Driving Analysis of Ecological Environmental Quality along the Qinghai–Tibet Railway Using Google Earth Engine—A Case Study Covering Xining to Jianghe Stations. *Remote Sens.* **2024**, *16*, 951. [CrossRef]
33. Wang, X.; Liu, Q.; Jia, P.; Huang, X.; Yang, J.; Mao, Z.; Shen, S. Highway Ecological Environmental Assessment Based on Modified Remote Sensing Index—Taking the Lhasa–Nyingchi Motorway as an Example. *Remote Sens.* **2024**, *16*, 265. [CrossRef]
34. Liu, Q.; Qiao, J.; Li, M.; Dun, Y.; Zhu, X. Spatiotemporal evolution of ecological environmental quality and its dynamic relationships with landscape pattern in the Zhengzhou Metropolitan Area: A perspective based on nonlinear effects and spatiotemporal heterogeneity. *J. Clean. Prod.* **2024**, *480*, 144102. [CrossRef]
35. Yang, H.; Xu, W.; Yu, J.; Xie, X.; Xie, Z.; Lei, X.; Wu, Z.; Ding, Z. Exploring the impact of changing landscape patterns on ecological quality in different cities: A comparative study among three megacities in eastern and western China. *Ecol. Inform.* **2023**, *77*, 102255. [CrossRef]
36. Zhao, R.; Zhan, L.; Yao, M.; Yang, L. A geographically weighted regression model augmented by Geodetector analysis and principal component analysis for the spatial distribution of PM2.5. *Sustain. Cities Soc.* **2020**, *56*, 102106. [CrossRef]
37. Mansour, S.; Al Kindi, A.; Al-Said, A.; Al-Said, A.; Atkinson, P. Sociodemographic determinants of COVID-19 incidence rates in Oman: Geospatial modelling using multiscale geographically weighted regression (MGWR). *Sustain. Cities Soc.* **2021**, *65*, 102627. [CrossRef]

38. Roy, S.; Majumder, S.; Bose, A.; Chowdhury, I.R. The rich-poor divide: Unravelling the spatial complexities and determinants of wealth inequality in India. *Appl. Geogr.* **2024**, *166*, 103267. [CrossRef]
39. Cao, Y.; Liu, S.; Lu, Y.; Yang, H.; Yang, L. Multiscale impacts of urban nature on land surface temperature over two decades in a city with cloudy and foggy climates. *Urban Clim.* **2025**, *61*, 102389. [CrossRef]
40. Ren, D.; Qiu, A.; Cao, A.; Zhang, W.; Xu, M. Spatial Responses of Ecosystem Service Trade-offs and Synergies to Impact Factors in Liaoning Province. *Environ. Manag.* **2025**, *75*, 111–123. [CrossRef]
41. Du, J.; Zan, Y.; Meng, Q.; Zhang, H.; Sun, F.; Zhang, C.; Lu, C. Evaluation of Water Replenishment in the Northern Segment of the Yellow River Within the Beijing-Hangzhou Grand Canal, China. *Water* **2025**, *17*, 48. [CrossRef]
42. Yang, J.; Huang, X. The 30 m annual land cover dataset and its dynamics in China from 1990 to 2019. *Earth Syst. Sci. Data* **2021**, *13*, 3907–3925. [CrossRef]
43. Cui, R.; Hu, Z.; Wang, P.; Han, J.; Zhang, X.; Jiang, X.; Cao, Y. Crop Classification and Growth Monitoring in Coal Mining Subsidence Water Areas Based on Sentinel Satellite. *Remote Sens.* **2023**, *15*, 5095. [CrossRef]
44. Rakoto, P.Y.; Delami, K.; Hurley, J.; Amati, M.; Sun, Q. Revisiting the cooling effects of urban greening: Planning implications of vegetation types and spatial configuration. *Urban For. Urban Green.* **2021**, *64*, 127266. [CrossRef]
45. Yang, H.; Yu, j.; Xu, W.; Wu, Y.; Lei, X.; Ye, j.; Geng, J.; Ding, Z. Long-time series ecological environment quality monitoring and cause analysis in the Dianchi Lake Basin, China. *Ecol. Indic.* **2023**, *148*, 110084. [CrossRef]
46. Yuan, B.; Fu, L.; Zou, Y.; Zhang, S.; Chen, X.; Li, F.; Deng, Z.; Xie, Y. Spatiotemporal change detection of ecological quality and the associated affecting factors in Dongting Lake Basin, based on RSEI. *J. Clean. Prod.* **2021**, *302*, 126995. [CrossRef]
47. Zhang, Y.; She, J.; Long, X.; Zhang, M. Spatio-temporal evolution and driving factors of eco-environmental quality based on RSEI in Chang-Zhu-Tan metropolitan circle, central China. *Ecol. Indic.* **2022**, *144*, 109436. [CrossRef]
48. Xiong, Y.; Xu, W.; Lu, N.; Huang, S.; Wu, C.; Wang, L.; Dai, F.; Kou, W. Assessment of spatial–temporal changes of ecological environment quality based on RSEI and GEE: A case study in Erhai Lake Basin, Yunnan province, China. *Ecol. Indic.* **2021**, *125*, 107518. [CrossRef]
49. Zheng, Z.; Wu, Z.; Chen, Y.; Guo, C.; Marinello, F. Instability of remote sensing based ecological index (RSEI) and its improvement for time series analysis. *Sci. Total Environ.* **2022**, *814*, 152595. [CrossRef]
50. Zhu, L.; Wei, Y.; Wang, C.; Zhang, Y.; Zheng, H. Spatiotemporal variation of land use and its impact on ecosystem service value in the Beiyun River Basin from 1980 to 2015. *Acta Sci. Circumstantiae* **2021**, *41*, 301–310.
51. Zhang, L.; Zou, Y.; Liu, B. Evolution mechanisms and development strategies of local human residential environment landscape in the Jiangnan Canal Basin. *Chin. Landsc. Archit.* **2025**, *41*, 31–38.
52. Zhang, S.; Fu, B. Historical social connectivity of the Tonghui River system in Beijing from an urban spatial perspective. *Chin. Landsc. Archit.* **2022**, *38*, 32–37.
53. Yang, L.; Lu, Y.; Cao, M.; Wang, R.; Chen, J. Assessing accessibility to peri-urban parks considering supply, demand, and traffic conditions. *Landsc. Urban Plan.* **2025**, *257*, 105313. [CrossRef]
54. Huang, L.; Chen, X.; Yuan, Z.; Ye, C.; Liang, Y. Impact of landscape patterns on water quality in urbanized rivers at characteristic scale: A case of Pearl River Delta, China. *Environ. Manag.* **2024**, *74*, 715–728. [CrossRef]
55. Jatayu, A.; Rustiadi, E.; Juanda, B.; Pribadi, D.O. Metric-based approach for quantifying urban expansion impact on urban form changes in the JBMUR south conurbation corridor. *Land* **2023**, *12*, 864. [CrossRef]
56. Carlier, J.; Moran, J. Landscape typology and ecological connectivity assessment to inform Greenway design. *Sci. Total Environ.* **2019**, *651*, 3241–3252. [CrossRef]
57. Zhang, W.; Liu, Z.; Qin, K.; Dai, S.; Lu, H.; Lu, M.; Ji, J.; Yang, Z.; Chen, C.; Jia, P. Long-term dynamic monitoring and driving force analysis of eco-environmental quality in China. *Remote Sens.* **2024**, *16*, 1028. [CrossRef]
58. Xin, J.; Yang, J.; Yu, H.; Ren, J.; Yu, W.; Cong, N.; Xiao, X.; Xia, J.; Li, X.; Qiao, Z. Towards ecological civilization: Spatiotemporal heterogeneity and drivers of ecological quality transitions in China (2001–2020). *Appl. Geogr.* **2024**, *173*, 103439. [CrossRef]
59. Jiang, H.; Peng, J.; Zhao, Y.; Xu, D.; Dong, J. Zoning for ecosystem restoration based on ecological network in mountainous region. *Ecol. Indic.* **2022**, *142*, 109138. [CrossRef]
60. Xie, Y.; Yu, X.; Ng, N.; Li, K.; Fang, L. Exploring the dynamic correlation of landscape composition and habitat fragmentation with surface water quality in the Shenzhen river and deep bay cross-border watershed, China. *Ecol. Indic.* **2018**, *90*, 231–246. [CrossRef]
61. Li, W.; Cui, Y.; Qin, Y.; Qin, H.; Dong, J. Construction land allocation in Beijing-Tianjin-Hebei and Yangtze River Delta urban agglomerations. *J. Geogr. Sci.* **2025**, *35*, 501–520. [CrossRef]
62. Gao, X.; Qian, Y.; Fang, Y.; Shi, X.; Yao, S.; Dong, B.; Ji, K.; Wang, Z. Road network expansion and landscape dynamics in the Chaohu Lake wetland: A 20-year analysis. *Ecol. Indic.* **2025**, *173*, 113443. [CrossRef]
63. Pu, J.; Shen, A.; Liu, C.; Wen, B. Impacts of ecological land fragmentation on habitat quality in the Taihu Lake basin in Jiangsu Province, China. *Ecol. Indic.* **2024**, *158*, 111611. [CrossRef]
64. Yang, L.; Ao, Y.; Ke, J.; Lu, Y.; Liang, Y. To walk or not to walk? Examining non-linear effects of streetscape greenery on walking propensity of older adults. *J. Transp. Geogr.* **2021**, *94*, 103099. [CrossRef]

65. Wu, G.; Tan, L.; Yan, Y.; Tian, Y.; Shen, Y.; Cao, H.; Dong, M. Measures and planning for wetland restoration of Xianghe Segment of China's Grand Canal. *Int. J. Sustain. Dev. World Ecol.* **2016**, *23*, 326–332. [CrossRef]
66. Aurora, R.M.; Furuya, K. Spatiotemporal analysis of urban sprawl and ecological quality study case: Chiba Prefecture, Japan. *Land* **2023**, *12*, 2013. [CrossRef]
67. Yang, L.; Yang, H.; Yu, B.; Lu, Y.; Cui, J.; Lin, D. Exploring non-linear and synergistic effects of green spaces on active travel using crowdsourced data and interpretable machine learning. *Travel Behav. Soc.* **2024**, *34*, 100673. [CrossRef]
68. Bian, G.; Zhao, Y.; Yan, J. Exploring Characteristics and Patterns of In Situ Space Morphology: Perspective of the Historical and Cultural Canal Settlement. *Land* **2024**, *13*, 1119. [CrossRef]
69. Zhao, Y.; Bian, G.; Sun, T.; Yan, J. The In-Situ Spatial-Temporal Evolution of the Settlement Space along the Grand Canal Tianjin Section from the Perspective of Cultural Heritage. *Land* **2023**, *12*, 1023. [CrossRef]
70. Tang, F.; Zhou, X.; Wang, L.; Zhang, Y.; Fu, M.; Zhang, P. Linking Ecosystem Service and MSPA to Construct Landscape Ecological Network of the Huaiyang Section of the Grand Canal. *Land* **2021**, *10*, 919. [CrossRef]
71. Qin, J.; Wang, Z. Construction of recreation opportunity spectrum of Grand Canal National Cultural Park based on scene theory—Taking the Yuhe River section of Beijing as an example. *Chin. Landsc. Archit.* **2022**, *38*, 135–139.
72. Fu, J.; Dai, L. Exploring the construction path of the Grand Canal National Cultural Park based on territorial spatial planning. *J. Nat. Resour.* **2023**, *38*, 2312–2331. [CrossRef]
73. Franco, L.; Magalhães, M.R. Assessing the ecological suitability of land-use change. Lessons learned from a rural marginal area in southeast Portugal. *Land Use Policy* **2022**, *122*, 106381. [CrossRef]
74. Zhou, H.; Xiao, D.; Zhou, K. Corridor effect of the spatial changes of landscape patterns in and areas: A case study of the river corridor areas in the middle and lower reaches of Tarim River. *Chin. Sci. Bull.* **2006**, *51* (Suppl. 1), 82–91. [CrossRef]
75. Beaudoin, C.; Mistry, I.; Young, N. Collaborative knowledge mapping to inform environmental policy-making: The case of Canada's Rideau Canal National Historic Site. *Environ. Sci. Policy* **2022**, *128*, 299–309. [CrossRef]
76. Sun, P.; Bobbink, I.; Chouairi, A. Water narratives exploring the convergence of the Canal du Midi and its coastal landscape. *Shima* **2023**, *17*, 32–49. [CrossRef]

land

MDPI

Article

Route Selection for the Extension Line of Pinglu Canal: A Comparative Study

Honglin Feng [1], Zhefei Jin [1], Xiao Chu [2,*], Weihan Wang [3] and Haiyuan Yao [1]

1 Transport Planning and Research Institute, Ministry of Transport, Beijing 100028, China; fenghl@tpri.org.cn (H.F.); jinzf@tpri.org.cn (Z.J.); yaohy@tpri.org.cn (H.Y.)
2 School of Transportation and Logistics Engineering, Wuhan University of Technology, Wuhan 430063, China
3 State Key Laboratory of Coastal and Offshore Engineering, Dalian University of Technology, Dalian 116024, China; 1106270236@mail.dlut.edu.cn
* Correspondence: cx330@whut.edu.cn; Tel.: +86-13329724574

Abstract: The Pinglu Canal, China's first strategic river–sea transport canal, connects the inland waterway trunk line with the Beibu Gulf International Hub Port, while a diversion channel extension is needed for port development. This study proposed a comparative study of extension routes to Fangchenggang through qualitative analysis and quantitative modeling. Eventually, Route 4 is selected considering the engineering feasibility, ecological impact, and hydrodynamic modeling. It yields the lowest engineering cost, effectively bypasses the ecological protection red lines and cultural heritage areas, and only intersects the controllable ecological restoration zones in a limited manner. Hydrodynamic results show moderate tidal velocities, with maximum rising and falling speeds of 0.72–0.80 m/s and 0.72–0.86 m/s, respectively. The sediment deposition intensity (0.06 m/a) and total volume (58,600 m^3/a) are generally lower than the other alternatives. The comparative study shares useful insights into canal route selection and would support the sustainable development of the logistics transportation system in Southwest China.

Keywords: canal planning; route selection; hydrodynamic modeling; Pinglu Canal

1. Introduction

Canals, known as artificial waterways connecting different water systems, have played a critical role in promoting the shipping industry, regional economic development, and cultural exchange. From the perspective of shipping, on the one hand, canals are used to build efficient intermodal transportation networks by connecting inland natural waterways with other modes of transportation so as to regulate the longship transportation distances, insufficient connectivity, and high overall logistics costs, such as China's Grand Canal [1,2]; on the other hand, through the effective establishment of sea-crossing and ocean-crossing international channels, canals have effectively opened up the global trade network, greatly enhanced the efficiency of international logistics, and reorganized the new pattern of the world's maritime transportation, exemplified by the Suez Canal in Egypt [3] and the Panama Canal [4]. The canal project confronts a complex, multi-dimensional, comprehensive decision-making system of coordination and unity of transportation benefits, ecological protection, and regional development [5]. Common indicators of canal design include the geometric parameters (e.g., depth, width, slope, etc.), hydraulic characteristics (flow velocity, head loss, flow stability), investments (construction and operational costs), and the natural and social constraints of the region, such as terrain, geology, land-use restrictions, and ecological red lines [6].

Recently, great progress has been made in the canal route selection and engineering design [7,8]. Qualitative research commonly adopts methods such as fuzzy comprehensive evaluation and multi-objective decision-making methods (e.g., Technique for Order Preference by Similarity to an Ideal Solution), where expert scoring and hierarchical analysis are employed to assess the risks and feasibility of canal plans [9–11]. The identification of indicators and levels is generally achieved based on expert experience, and the evaluation results have advantages in determining the ranking of influencing factors [12]. Pham et al. [13] introduced the TOPSIS method to evaluate the benefits of route selection for the expansion of the Panama Canal and identified transportation cost as the most important factor, followed by transportation time, reliability, and route characteristics. Quantitative research mainly focuses on hydraulic modeling, optimization algorithms, and cost function construction to evaluate engineering parameters and their economic and ecological costs [14]. In terms of hydrodynamic modeling, scholars have developed coupled numerical models to simulate water flow and sediment dynamics. For example, Gad et al. [15] simulated the spatial variation of the seabed in navigation channels by developing a 2D depth-averaged finite element hydro-morphodynamical model. Coraci et al. [16] applied a two-dimensional finite element model to the Venice canals to analyze the intensity of sediment accumulation in the waterways. Meanwhile, hydrodynamic modeling has been extensively adopted in canal capacity evaluation and design optimization. Kamran et al. [17] and Mishra et al. [18] constructed hydrodynamic models based on HEC-RAS and MIKE11 software, respectively, to explore water flow behavior and assess conveyance capacity in canal systems. These models effectively capture transient flow dynamics and interactions with channel geometry, enabling precise evaluation of operational performance and flood risk analysis [19]. Tabari et al. [20] employed the SEEP/W software to establish saturated–unsaturated soil hydrodynamic models, aiming to support the optimized hydraulic design of canals for minimal water conveyance losses. Swamee et al. [21] proposed an explicit design equation as a representative and explored the quantitative relationship between excavation depth and project cost under different cross-sectional morphologies (triangular, rectangular, etc.) by introducing a shape coefficient. Ghazaw [22] optimized the seepage function and evapotranspiration model through the Lagrange multiplier method and systematically computed the economic cost of water loss in the channel cross-section. A comprehensive assessment combining qualitative and quantitative methods could enhance the credibility of engineering schemes in practical scenarios. Fan et al. [23] integrated fuzzy logic, Bayesian network, and utility–value theory to construct a multi-dimensional resilience assessment framework, which was successfully applied in the strait and canal projects.

As global economic development accelerates, canals are constructed to accommodate greater cargo transport demand and more frequent trade exchanges [24]. Specific measures include new canal construction, expansion, and operational optimization [25–27]. The expansion of the Panama Canal, which was completed in 2016, allowed for larger vessels with capacity increased by 26.3% (2018), and toll revenue is projected to rise by 56% (2025) [28]. The Pinglu Canal, which started construction in 2022, is the first canal project that connects rivers and estuaries since the founding of the People's Republic of China. Upon completion, 5000-ton river–maritime direct ships can navigate between China's coastal ports and major ports in Southeast Asia from the Xijiang ports, opening up a water trade channel between China and the Association of Southeast Asian Nations (ASEAN) and promoting the comprehensive development of economic and trade cooperation between China and ASEAN countries [29]. Liu et al. [30] compared the comprehensive benefits of the existing route and the Pinglu Canal and highlighted that the Pinglu Canal would yield better cargo volume, higher profits, and lower total carbon emissions. However, threats to aquatic habitats still exist, such as the increased risk of proliferation of harmful algal blooms during construction [31] and the reduction in habitat area for wildlife after canal construction [32]. In addition, the Pinglu Canal takes the Beibu Gulf seaport as an important international hub seaport, but the marine container transportation network may be significantly affected when the port encounters congestion, strikes, or natural disasters [33]. To promote the multimodal transport network and enhance shipping safety at the Beibu Gulf Port, it is necessary to extend the Pinglu Canal to Fangcheng Port to form a comprehensive and resilient transport system. The Fangchenggang Port Waterway extension is constructed on the basis of the ancient Huangchengao Canal, linking the Pinglu Canal with inland areas of southwest China via the Xijiang Golden Waterway. This project holds strategic value in expanding the Pinglu Canal's hinterland reach, restoring regional navigability, and optimizing land use in Fangchenggang and Qinzhou Bay. Moreover, the significant role of Beibu Gulf as an international hub port could be further strengthened.

Based on the construction scheme of the Pinglu Canal, preliminary routes of the extension line toward Fangchenggang East Bay are proposed first. A comparative analysis of alternative routes is then conducted from the perspectives of land resources, transportation planning, ecological protection, water conservancy, and flood control. In order to determine the route direction of the Fangchenggang waterway dredging corridor, a coupled two-dimensional hydrodynamic-sediment model is established to analyze the tidal wave phase relationship at the entrance and exit of the canal, as well as the sediment siltation intensity and the amount of siltation. The results share valuable insights into the selection of the optimal route.

2. Materials and Methods

2.1. Study Area

The new water evacuation canal of Fangchenggang (Ancient Huangchengao Canal) starts from the Tanyou Waterway in Dongwan of Fangchenggang in the south and connects to the Shajing channel of Pinglu Canal in Qinzhou Bay in the north, with a total length of about 16 km. The potential routes of the extension line are proposed based on the historical Huangchengao Canal, which is part of the ancient canal. The total length of the ancient canal is approximately 12 km, with natural sea trenches at its eastern and western ends. The central part of the Huangchengao Canal (about 5 km) shows clear features of artificial excavation. The canal is not navigable due to partial blockage and siltation, as shown in Figure 1. A new shipping canal is therefore planned to expand the hinterland of Pinglu Canal and Beibu Gulf Port. The canal is designed as an extension of a shipping line of Class I standard, enabling 5000-ton ships to navigate in both directions. The designed dimensions of land and marine excavation sections are 6.3 m $\times$ 80 m $\times$ 360 m (depth $\times$ bottom width $\times$ minimum bending radius, same as below) and 6.5 m $\times$ 130 m $\times$ 540 m, respectively. The minimum and maximum navigable water levels are -1.31 m and 3.76 m; the navigable clearance is 18 m. The net navigable width of the channel is 140 m and 240 m for the land and marine sections, respectively.

2.2. Route Planning

The canal construction aims to improve navigation conditions, increase channel capacity, and ensure ship navigation safety while also achieving comprehensive utilization of water resources (e.g., flood control, water usage, and environmental protection). Thus, the fundamental principles of route selection are proposed as follows:

- Optimizing overall benefits by focusing on the shipping development and the improvement of the integrated transport system of Beibu Gulf. Meanwhile, urban planning, land space, ecological protection, and comprehensive transportation also need to be considered to maximize the river–sea coordinated transport and social benefits.

- Following the national principle of land use strategy, priority is given to the use of existing ditches, ponds, and depressions, and arable land is protected as far as possible. Moreover, land acquisition and relocation should be minimized to reduce the amount of earth excavation work.
- The impact of canal construction on existing infrastructures (e.g., transportation, water conservancy, and utilities along the route) should be minimized. Coordination with existing road, rail, port, and water conservancy planning is essential to avoid mutual interference and redundant construction.
- Nature reserves, ecologically sensitive areas, and cultural heritage reserves should be avoided as much as possible to reduce the impact on the ecological environment and cultural heritage protection along the route.
- For canal route selection, a comprehensive evaluation and detailed comparison need to be conducted, including but not limited to natural conditions, terrain, geology, and engineering costs.

Figure 1. Route map of the ancient Huangchengao canal.

In the present study, four routes have been proposed based on the ship navigation requirements, natural conditions, and regional traffic conditions. The layout of four planned routes and their intersections with regional highways and railways are illustrated in Figure 2. It is observed that the main areas of intersection are Shajing Waterway, Maoling Waterway, and Tanyou Waterway, as well as major roads such as Qimao Road, Shaqi Road, and Rongmujiang Road. The specific routes of the extension line are presented in Figure 2:

Route 1: Starting from Maoling Channel on the east side of Zhongdun Island, sailing south through the coastal waterway, entering the inland waterway, and then turning southwest, crossing the highway, and then entering the section of the ancient canal, and then passing through the port area and crossing a number of traffic arteries before finally arriving at the final destination. The total length of Route 1 is 16.99 km, including a 4.20 km land excavation section and 12.79 km sea excavation section.

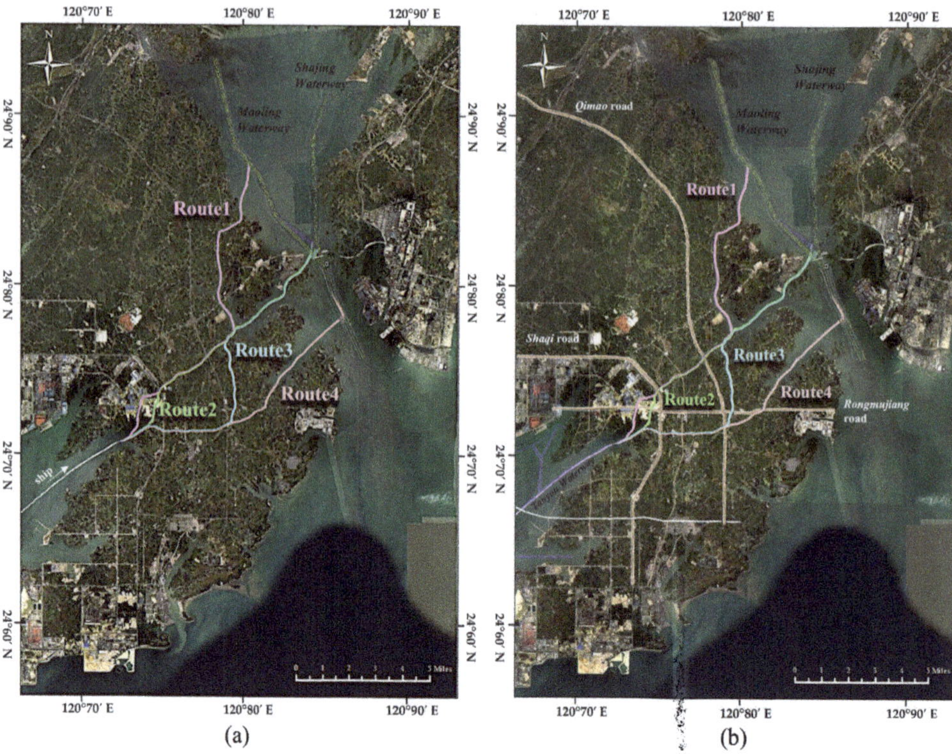

Figure 2. Geographical map of planned routes (**a**) and regional traffic conditions (**b**).

Route 2: Starting from the Shajing channel on the east side of Yagongshan, it sails southwestward to the inland waterway and then overlaps with Route 1, crosses the high-speed road and the ancient canal, and then turns southward through the harbor area, and finally arrives at the endpoint. The total length is approximately 14.90 km, including 4.80 km land excavation section and 10.10 km sea excavation section.

Route 3: Starting from the Shajing channel on the east side of Yagongshan, it sails southwestward into the inner area, then goes southward to the inland river, crosses the highway, passes through several villages, crosses the harbor area and the main traffic route, and finally arrives at the endpoint. The total length of Route 3 is about 17.34 km, including 4.20 km land excavation section and 13.14 km sea excavation section.

Route 4: Starting from the Qinzhou Bay channel in the southern sea area, it enters the inland waterway to the southwest, then crosses the highway to the south, then turns to the west, and crosses the harbor area and the main traffic line after passing through several villages, and finally arrives at the endpoint. The total length is 15.41 km, including 6.20 km land excavation section and 9.21 km sea excavation section.

2.3. Methods

2.3.1. Hydrodynamic and Sediment Modeling

This study develops a coupled two-dimensional hydrodynamic and sediment transport model based on MIKE21 to investigate the characteristics of tidal moments and waves on sedimentation for different routes. The coupled numerical model includes hydrodynamic module (HD), spectral wave module (SW), and sediment transport module (ST). The model was implemented using MIKE ZERO version 2011, and simulations were performed on a Windows 10 operating system running on a Lenovo desktop computer. The flowchart of the numerical model is illustrated in Figure 3, in which the hydrodynamic module provides water level and other elements for the wave module. The coupled hydrodynamic module and spectral wave module provide hydrodynamic fields for the sediment transport module.

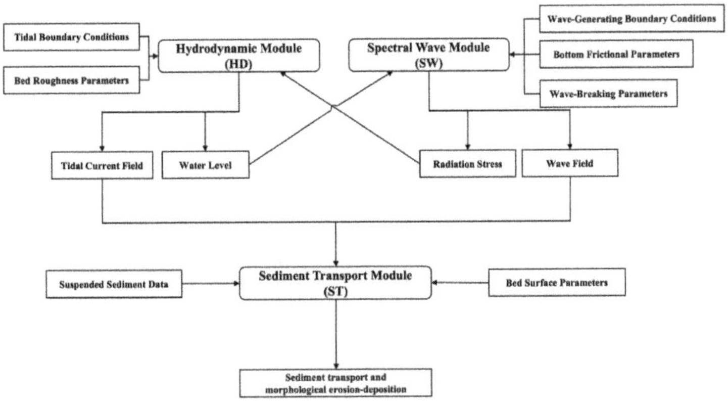

Figure 3. Flowchart of coupled hydrodynamic and sediment transport model based on MIKE21.

(1) Tidal Flow

A two-dimensional tidal flow model was built by using HD module of MIKE21 software package to compute and analyze the flow field. The governing equations are provided as follows.

Continuity equation:

$$\frac{\partial h}{\partial t}+\frac{\partial h\overline{u}}{\partial x}+\frac{\partial h\overline{v}}{\partial y}=hS \tag{1}$$

Momentum equation in X direction:

$$\begin{aligned}\frac{\partial h\overline{u}}{\partial t}+\frac{\partial h\overline{u}^2}{\partial x}+\frac{\partial h\overline{vu}}{\partial y} &= f\overline{v}h-gh\frac{\partial \eta}{\partial x}-\frac{h}{\rho_0}\frac{\partial p_a}{\partial x}-\frac{gh^2}{2\rho_0}\frac{\partial \rho}{\partial x}+\frac{\tau_{sx}}{\rho_0}-\frac{\tau_{bx}}{\rho_0}\\ &- \frac{1}{\rho_0}\left(\frac{\partial s_{xx}}{\partial x}+\frac{\partial s_{xy}}{\partial y}\right)+\frac{\partial}{\partial x}(hT_{xx})+\frac{\partial}{\partial x}(hT_{xy})+hu_sS\end{aligned} \tag{2}$$

Momentum equation in Y direction:

$$\begin{aligned}\frac{\partial h\overline{v}}{\partial t}+\frac{\partial h\overline{v}^2}{\partial y}+\frac{\partial h\overline{vu}}{\partial x} &= -f\overline{u}h-gh\frac{\partial \eta}{\partial y}-\frac{h}{\rho_0}\frac{\partial p_a}{\partial y}-\frac{gh^2}{2\rho_0}\frac{\partial \rho}{\partial y}+\frac{\tau_{sy}}{\rho_0}-\frac{\tau_{by}}{\rho_0}\\ &- \frac{1}{\rho_0}\left(\frac{\partial s_{yx}}{\partial x}+\frac{\partial s_{yy}}{\partial y}\right)+\frac{\partial}{\partial x}(hT_{xy})+\frac{\partial}{\partial y}(hT_{yy})+hv_sS\end{aligned} \tag{3}$$

where t is time and x and y are spatial coordinates in Cartesian coordinate system; η denotes the water surface elevation, d is the water depth, and h represents the total water depth $h = \eta + d$; u and v are the components of flow velocity in the direction of x and y; f is the Coriolis force, and g is the gravitational acceleration; ρ represents the water density. ρ_0 is the reference density; p_a is the atmospheric pressure; s_{xx}, s_{xy}, s_{yx}, s_{yy} are the radiational stress components; T_{xx}, T_{xy}, T_{yx}, T_{yy} are the horizontal viscous stresses; (τ_{sx}, τ_{sy}) and (τ_{bx}, τ_{by}) are the tangential stresses at the water surface and the bottom bed in the direction of X and Y; S is the flow rate of the source–sink term; u_s and v_s are the velocity components corresponding to the source–sink term.

(2) Waves

The SW module of MIKE21 software package has been adopted for sediment transport and siltation calculations. The triangular mesh is beneficial in dealing with complex boundaries in practical scenarios. Within the Cartesian coordinate system, the control equations of SW module are expressed as follows:

$$\frac{\partial N}{\partial t} + \nabla \cdot \left(\vec{v} N \right) = \frac{S}{\sigma} \tag{4}$$

$$\left(c_x, c_y \right) = \frac{d\vec{x}}{dt} = \vec{c_g} + \vec{U} \tag{5}$$

$$c_\sigma = \frac{d\sigma}{dt} = \frac{\partial \sigma}{\partial d} \left[\frac{\partial d}{\partial t} + \vec{U} \cdot \nabla_{\vec{x}} d \right] - c_g \vec{k} \cdot \frac{\partial \vec{U}}{\partial s} \tag{6}$$

$$c_\theta = \frac{d\theta}{dt} = -\frac{1}{k} \left[\frac{\partial \sigma}{\partial d} \frac{\partial d}{\partial m} + \vec{k} \cdot \frac{\partial \vec{U}}{\partial m} \right] \tag{7}$$

where N is the kinetic spectral density; t is time; $\vec{x}$ is the Cartesian coordinate system; $\vec{v}$ is the velocity of the wave group; S is the source term in the energy balance equation; ∇ is the differential operator; s is the propagation direction of the wave; θ and m are the directions perpendicular to s; $\nabla_{\vec{x}}$ is the two-dimensional differential operator on $\vec{x}$ space.

In the analysis of shoreline siltation characteristics, it is necessary to comprehensively consider the lifting and transporting effects of waves on the bottom bed sediment. Therefore, representative wave field needs to be selected to simulate the dynamic processes of sediment siltation. Following the Technical Specification for Simulation Tests of Water Transportation Engineering [34], the finite wave height data that play a weaker role in sediment movement have been omitted. The wave energy flow method is introduced to obtain the synthetic wave direction and synthetic wave height by weighted average. The synthetic wave direction is determined as follows:

$$\overline{\alpha} = \frac{1}{2} arcsin \frac{\sum P_j H_j^2 T_j sin2\alpha_j}{\sum P_j H_j^2 T_j} \tag{8}$$

where P_j, H_j, T_j, α_j are the effective wave height, period, frequency of occurrence, and wave direction corresponding to the wave of the j th magnitude. The synthesized wave height is calculated as below.

$$\overline{H} = \left(\frac{\sum H_i^2 P_j}{\sum P_j} \right)^{1/2} \tag{9}$$

(3) Sedimentation

On the basis of the tidal current calculations, the ST module in the MIKE21 software package is utilized to investigate the sediment transport, topographic scouring, and siltation of four routes, as demonstrated in Section 2.2. The fundamental equation of suspended sediment transport is expressed as follows:

$$\frac{\partial \overline{c}}{\partial t} + \overline{u} \frac{\partial \overline{c}}{\partial x} + \overline{v} \frac{\partial \overline{c}}{\partial y} = \frac{1}{h} \frac{\partial}{\partial x} \left(h D_x \frac{\partial \overline{c}}{\partial x} \right) + \frac{1}{h} \frac{\partial}{\partial y} \left(h D_y \frac{\partial \overline{c}}{\partial y} \right) + \sum_{i=1}^{n} \frac{S_i}{h} \tag{10}$$

where $\overline{c}$ denotes the mean suspended sediment concentration on the vertical line; D_x and D_y are the turbulent diffusion coefficients of suspended sediment along the x and y directions, and S_i is the siltation term. The main parameters determined in the ST module include bed shear stress, particle deposition velocity, suspended sand diffusion coefficient, and parameters related to the siltation term. The bed shear stress is calculated as below:

$$\tau_b = \frac{1}{2}\rho_w f_w \left(U_b^2 + U_\delta^2 + 2U_b U_\delta cos\beta\right) \tag{11}$$

where U_b is the horizontal orbital velocity of the wave water quality point at the bottom of the bed; U_δ is the flow velocity at the top of the wave boundary layer; β is the angle between the flow and wave directions; f_w is the wave bottom friction coefficient. Sediment settling velocity is the main parameter for calculating sediment siltation. For grain sizes less than 0.03 mm, sediment particles, which show flocculation state in seawater, have a settling velocity of 0.0004~0.0005 m/s. Sedimentation refers to the conversion process of the sediment from suspended sand to bottom-bed sediments. Siltation occurs when the bed shear stress τ_b is less than the critical siltation shear stress of sediment. The siltation rate is represented by a stochastic model of sediment–flow interaction.

$$S_D = w_s c_b p_d \tag{12}$$

where c_b is the suspended sediment concentration in the near-bottom layer; p_d is an expression for the probability of siltation, $p_d = 1 - \tau_b/\tau_{cd}$; the near-bottom sediment concentration c_b can be calculated by using the Pecklett number P_e and the plumbline mean suspended sand content [35].

$$c_b = \bar{c} \times \left(1 + \frac{P_e}{1.25 + 4.75p_d^{2.5}}\right) \tag{13}$$

where P_e is the Pecklett number defined as $P_e = 6w_s/\kappa U_f$, U_f is the drag flow rate, and κ is the von Karmen constant (0.4). Scouring refers to the process of transferring sediment from the bottom bed to the water column, and it occurs when the bed shear stress τ_b is greater than the critical scour shear stress τ_{ce}. The erosion rate can be expressed as the following formula [36].

$$S_E = e^{a(\tau_b - \tau_{ce})^{1/2}} \tag{14}$$

where E is the degree of erosion; τ_{ce} is the critical scour shear stress.

2.3.2. Model Computational Domain and Grid Division

The computational domain of the tidal current is shown in Figure 4, with a length of 372 km in the east–west direction and a width of 192 km in the north–south direction.

Part of Beibu Gulf north of the line connecting Dengloujiao of the Leizhou Peninsula (110°1′53.92″, 20°17′3.60″) and the mouth of the Honghe River (106°34′19.07″, 20°17′3.60″) is adopted as the computational area. The large-scale water depth is determined by the charts of Naval Hydrographic Bureau and the maritime authority. The bathymetry in the vicinity of the extension line of Pinglu Canal is further corrected according to the bathymetric charts of December 2019, and the shorelines are extracted following the satellite images of Google Earth.

In order to improve the computational efficiency and accuracy, a locally encrypted unstructured triangular grid is used in the present study. Larger grid sizes have been utilized in the sea area, e.g., roughly 2000 m at the open sea boundary, while finer grid sizes (about 5 m~10 m) are adopted in the route areas. Since Route 1 and Route 2 show a greater impact on the regional traffic system and the resulting railway expansion project is challenging, this paper evaluates the navigation conditions and siltation in the waterways for Route 3 and Route 4. The total number of computational nodes generated in the computational domain of Route 3 is 325,883, with a total of 640,035 grids. The total number of computational nodes generated for Route 4 is 318,856, with a total of 626,696 grids. The detailed grid division for Route 3 and Route 4 is illustrated in Figure 5.

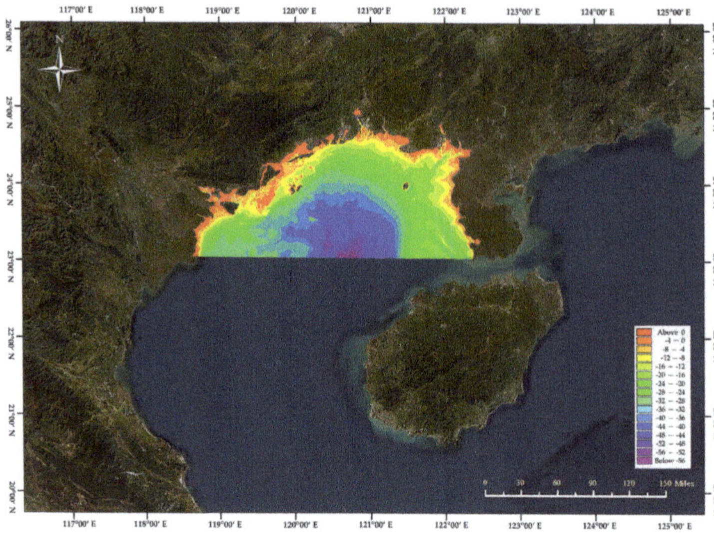

Figure 4. Schematic diagram of numerical simulation domain.

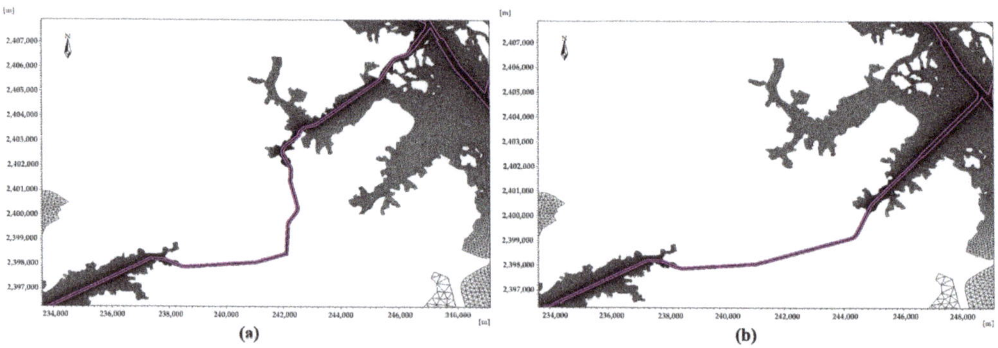

Figure 5. Grid division of coupled hydrodynamic-sediment model for Route 3 (a) and Route 4 (b).

2.3.3. Model Database

The hydrodynamic and sediment transport model was developed based on a comprehensive integration of various datasets for model setup, calibration, and validation. A detailed summary of the input data and parameter definitions is provided in Table 1.

Table 1. Data sources and applications.

Data Category	Data Source	Application
Tidal data	Time sequence dataset of tidal stations at Fangchenggang (November 2020) and Qinzhou Bay (August 2021)	Hydrodynamic (HD) module
Wave data	Time sequence wave dataset at Fangchenggang (November 2020) and Qinzhou Bay (August 2021)	Spectral Wave (SW) module

Table 1. Cont.

Data Category	Data Source	Application
Sediment parameters	Time sequence dataset of sediment concentration observations at Fangchenggang (November 2020) and Qinzhou Bay (August 2021)	Sediment Transport (ST) module
Boundary conditions	Electronic nautical charts provided by the Maritime Safety Administration; partial survey map of inner Qinzhou Bay (December 2019).	Model initialization and simulation

2.3.4. Model Calibration and Validation

The tide and wave measurements in Qinzhou Bay of 2021 have been obtained for model calibration and validation. In the present study, the hydrodynamic model is calibrated through the eddy viscosity coefficient (*Cs*) and Manning's coefficient (*M*), while the sediment transport module is calibrated through settling velocity and critical shear stress. Since detailed observation sequences of topographic siltation are absent, the historical nautical charts are derived from the maritime authorities to validate the topographic siltation features. The numerical model also takes into account the combined effect of tidal currents and waves (e.g., SES and SSSW) and is calibrated to conform with the macroscopic law of seabed evolution.

3. Results

3.1. Traffic Impacts Along the Route

As demonstrated in Section 2.2, four different routes have been proposed for the extension line between Pinglu Canal and Fangchenggang, of which Route 1 and Route 2 go down across the same highway and railway, and Route 3 and Route 4 go down across the same railway. Detailed information is provided in Figure 6.

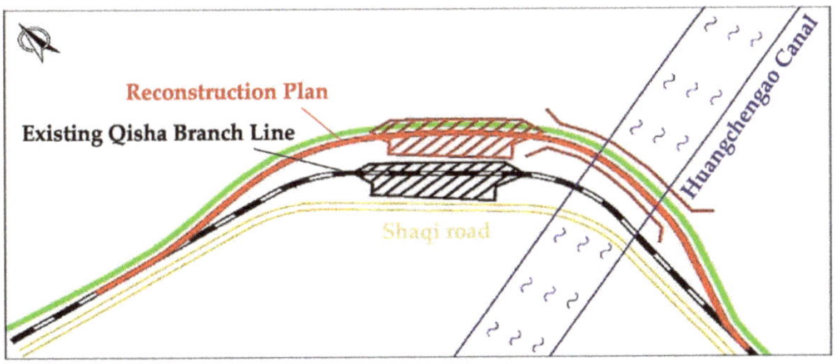

Figure 6. Schematic diagram of the railway reconstruction plan for Route 1 and Route 2.

(1) Route 1 and Route 2

Both routes bypass the highway and railway at the same location. The sufficient spatial availability along Shaqi Avenue and Qimao Road enables the potential construction of Route 1 and Route 2.

The planned canal waterway would pass under the railway about 400 m south of the Huangchengao Station. The rail elevation of Huangchengao Station is 22.26 m. To ensure the navigable clearance of the canal waterway, the rail elevation should not be lower than 25.26 m. Huangchengao Station and its freight yard need to be raised by 3.0 m. The total length of the reconstructed railway is approximately 6.8 km, as shown in Figure 6. In addition, Huangchengao Station is an intermediate railway station with six arrival and departure lines and a 6 million-ton comprehensive freight yard. From the perspective of engineering feasibility, the slope lifting height (i.e., 3.0 m) is too large, and the construction cannot transit smoothly. Moreover, the Huangchengao Station will be moved to the east by about 300 m, which has a great impact on the port area planning and is inconsistent with the current land planning scheme. To sum up, neither Route 1 nor Route 2 is recommended in the present study.

(2) Route 3

The planned canal would go under Shaqi Avenue and Qimao Road, where the geological conditions are favorable for the reconstruction project. However, the intersection of Canal Channel and Rongmujiang Avenue is located within the range of the hub interchange. The planned canal route would show a significant impact on its operation, and its feasibility is generally poor.

Route 3 passes under the railway between Huangchengao Station and Yunyue Station. The location of the underpass is approximately 2.7 km away from Huangchengao Station and 5.0 km away from Yunyue Station. To ensure navigable clearance, the rail surface elevation should not be lower than 25.26 m. Therefore, the existing Qisha Branch Line needs to be renovated by raising the slope. Considering the longitudinal section of the existing railway, the slope lifting requirements can be met without involving the reconstruction of the station. The railway reconstruction plan is overall feasible.

(3) Route 4

Consistent with the aforementioned routes, the geological conditions at the location of the underpass highway for Route 4 are relatively good, and its impact on highway reconstruction is overall limited. Route 4 and Route 3 underpass the railway at the same position. Based on the analysis results of Route 3, the railway reconstruction scheme is generally feasible for Route 4.

In summary, it is concluded that Route 4 can be recommended for the extension line.

3.2. Ecological Constraints and Cultural Relics Protection

According to the Fangchenggang City Territorial Spatial Master Plan (2020–2035), mangrove forests are distributed near the Fengliuling River, the Old Yangjiang River, and the Baisha River at both ends of the canal. As shown in Figure 7, the red area denotes "Prohibited development and construction area", the yellow area is "Restricted development and construction area", and the green area is "Ecological restoration planning area of mangrove". The four canal routes proposed in this paper do not intrude on the mangrove forests in the ecological protection red line area. Routes 1, 2, and 3 partially pass through the "Restricted development and construction area", while Routes 1, 3, and 4 partially pass through the "Ecological restoration planning area of mangrove".

In addition, Route 1 and Route 2 utilize the ancient Huangchengao Canal to cross the Huangchengao Ancient Site. It is necessary to carry out special impact assessments and achieve coordination of cultural relics protection. Routes 3 and 4 will open up new canal waterways by utilizing the tidal gullies on their southern sides without involving issues of cultural relic protection.

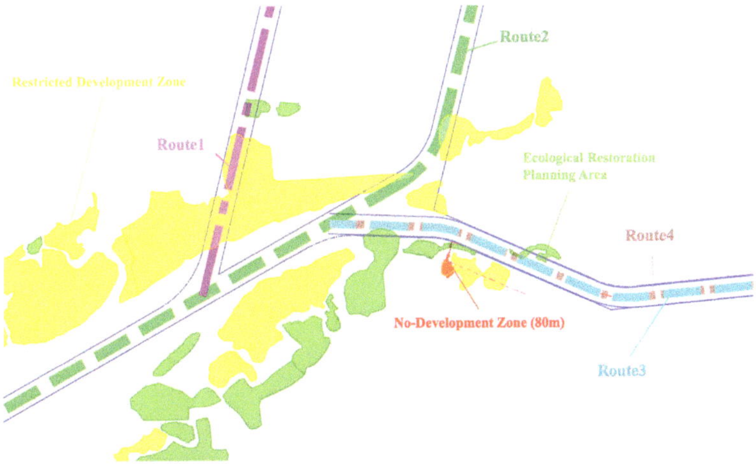

Figure 7. Relationship diagram between routes plan and mangrove conservation area.

3.3. Hydrodynamic Analysis

The high tide level data of 2021 is utilized to calculate tidal patterns and simulate the variations of the tide field of Qinzhou Bay. The land excavation section is reciprocating flow, which is jointly influenced by the tidal wave of Qinzhou Bay and Fangchenggang East Bay. Both Route 3 and Route 4 are characterized by the westward and eastward flows at high and low tides. The measured high tide initially flows westward at the beginning of high tide, and the flow rate gradually decreases, lasting about 7 h. Then it turns to eastward flow; the flow rate gradually increases and then decreases to turn to westward flow, lasting about 5 h. At the early stage, the falling tide is a westward flow and lasts about 5 h and then turns to the east direction. The maximum flow rate of the high tide occurs 2~4 h before the high tide, which is the eastward flow. The maximum flow rate at low tide occurred 3~5 h after high tide and is featured as westward flow. The high tide level occurred from 18:00 on 10 August to 12:00 on 11 August 2021. The maximum and average flow velocities of high tide and low tide at 56 characteristic points along Route 3 and 52 characteristic points along Route 4 are presented in Figures 8 and 9, respectively. Both figures illustrate the spatial distribution of key flow velocity parameters at dense characteristic points along the two selected routes. Figure 8 (for Route 3) and Figure 9 (for Route 4) visualize the tidal hydrodynamic variations in east–west flow directions during both flood and ebb tides. The tidal hydrodynamic features are highlighted in the terrestrial excavation sections as the projects are implemented, revealing the complex tidal responses within the modified areas. Using color gradients and flow velocity contours within a spatial coordinate system, the flow velocity patterns are demonstrated within the project waters. The color transition from blue to red enables quick identification of low-velocity regions to high-velocity regions. In general, the high and low tide velocities of Route 3 and Route 4 are basically similar in the sea-excavated section in Dongwan waters, with the maximum flow velocity at high tide ranging from 0.43 m/s to 1.18 m/s and the maximum flow velocity at low tide ranging from 0.33 m/s to 1.08 m/s, and the maximum flow velocity occurs in the waters near Tanyou Waterway. In the land excavation section, the maximum flow velocities of Route 3 and Route 4 decreased from west to east along the waterway with more uniform patterns.

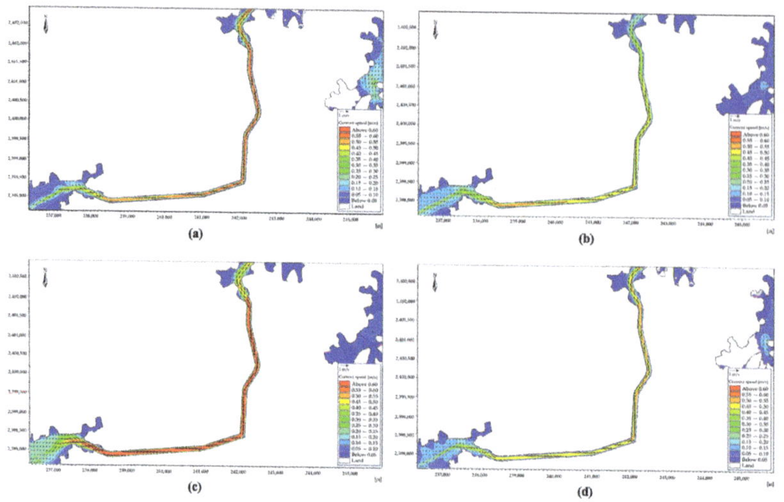

Figure 8. Calculation results of tidal field in land-excavated section as Route 3 is implemented. (**a**) Flood—Westward; (**b**) Flood—Eastward; (**c**) Ebb—Westward; (**d**) Ebb—Eastward.

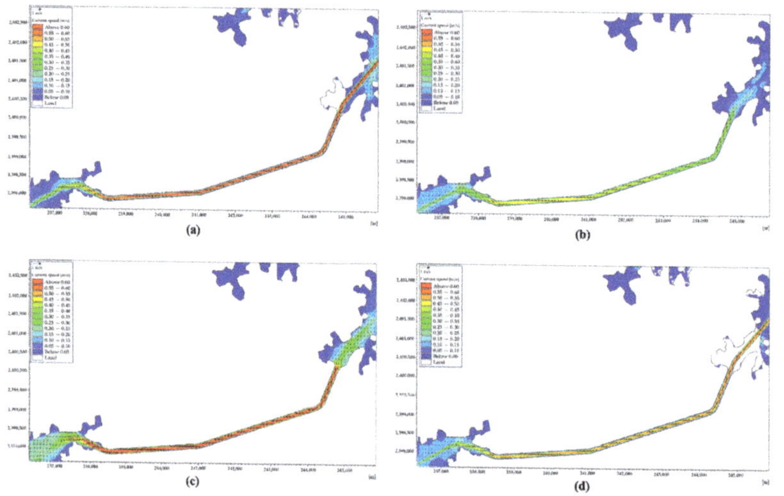

Figure 9. Calculation results of tidal field in land-excavated section as Route 4 is implemented. (**a**) Flood—Westward; (**b**) Flood—Eastward; (**c**) Ebb—Westward; (**d**) Ebb—Eastward.

The flow velocity of Route 3 was slightly smaller (i.e., 0.59 m/s~0.75 m/s) in high tide (as shown in Figure 8a,b). The flow velocity of Route 4 was 0.72 m/s~0.80 m/s, as shown in Figure 9a,b. During the low tide period, the flow velocity distributions of Route 3 and Route 4 are generally consistent. Specifically, the flow velocity fell in a range of 0.72 m/s~0.91 m/s (Figure 8c,d) for Route 3 and 0.72 m/s~0.86 m/s (Figure 9c,d) for Route 4. In the sea excavation section of Qinzhou Bay, Route 3 and Route 4 are connected to the main navigation channel near Longmen Harbor and the Legou operation area, respectively, after the widening of the cross-flow section. The maximum flow velocity shows a gradual decreasing and then increasing pattern, attributed to the mainstream of tides in Qinzhou Bay. The connecting section of Route 3 is close to the neck of Qinzhou Bay; therefore, its high tide flow velocity is relatively larger. The maximum flow velocity of high tide falls in a range of 0.50 m/s~1.29 m/s and 0.28 m/s~0.96 m/s for Route 3 and

Route 4, respectively. During the low tide period, the maximum flow velocity is noted as 0.38 m/s~1.07 m/s for Route 3 and 0.40 m/s~1.12 m/s for Route 4.

In summary, the land excavation sections of Route 3 and Route 4 are jointly influenced by the tidal waves of Qinzhou Bay and Fangchenggang East Bay. The results highlighted the changing characteristics of westward flow at high tide, eastward flow at high tide, westward flow at low tide, and eastward flow at low tide. The maximum transverse currents along two routes appear in the main channel connection area, of which Route 3 has a larger transverse current velocity due to the more zigzag patterns of the land excavation section, and the sea excavation section is affected by the flow of nearby shallow inlet channels. The flow patterns are relatively chaotic. The maximum transverse currents along the sea excavation section of Fangchenggang East Bay and the land excavation section are generally smaller. As the extension line of Pinglu Canal is implemented (e.g., Route 3 and Route 4), the tidal current field near the sea excavation section would be significantly altered, while the flow field does not present notable variations. In addition, the flow velocity distributions at the mouthwaters and the minimum water level for both routes meet the relevant requirements of coastal protection.

3.4. Wave Field

In the present study, the short-term wave measurement data from January to December 2008 in Fangchenggang and the statistical analysis of the European Center for Medium-Range Weather Forecasts (ECMWF) wave hour-by-hour hindcast data from 2016 to 2020 are adopted and analyzed. The waves of the project area are mainly in NNE and SE~SSW directions. Since Qinzhou Bay and Fangchenggang East Bay are surrounded by land in the east, north, and west directions, the wind area is limited, yielding small wave heights. As a result, the impact of wind waves mainly comes from the mouth of the bay to the south. The SE, SSE, S, and SSW waves are selected as representative waves for sediment transport simulation. Their wave energy accounts for about 45% of the total wave energy, and the annual action days are calculated as 133 days. Considering the shoreline direction of Qinzhou Bay (NNW~SSE) and Fangchenggang East Bay (NNE~SSW), the waves in the outer sea are synthesized according to the dominant wave directions of SE~S and S~SSW. The wave height data were pre-processed, and the small part of wave height less than 0.3 m was omitted to derive the representative wave field in SE~S and S~SSW directions, as shown in Figure 10. The spatial distribution of significant wave height is highlighted in both SE~S (Figure 10a) and S~SSW (Figure 10b) directions. Blue color denotes areas of lower significant wave height, while red color represents areas with elevated wave height. The results reveal substantial differences of wave energy intensity and spatial distributions.

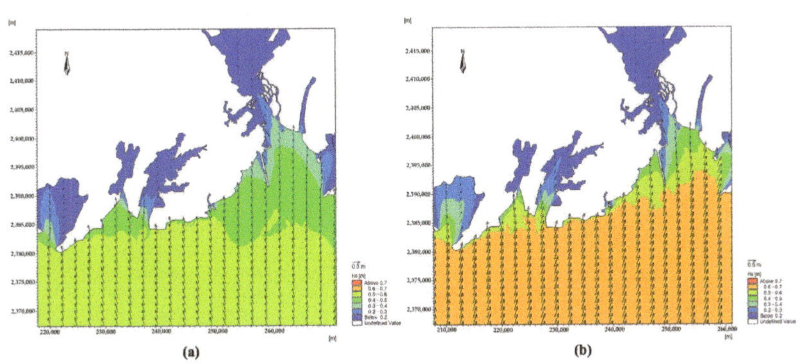

Figure 10. Numerical simulation of wave field. (**a**) SE~S direction; (**b**) S~SSW direction.

3.5. Sedimentation Features

The coupled hydrodynamic-sediment model based on MIKE21 has been utilized to investigate siltation processes as the extension line of Pinglu Canal is implemented. As demonstrated in Section 3.4, two representative wave directions (i.e., SE~S and S~SSW) have been selected to formulate three groups of power conditions, namely, pure tidal current, tidal current + SE~S wave, and tidal current + S~SSW wave. The upstream boundary condition of the numerical model is defined as the average river runoff and sand content, and the annual average siltation under normal weather is calculated by combining the frequency share of three groups of dynamic conditions. The results show that in the sea-excavated section of Dongwan waters, the same alignment of Route 3 and Route 4 leads to the same siltation intensities, with an annual average siltation intensity of 0.07 m/a~0.28 m/a. The volume of siltation would reach 304,600 m^3/a. In the excavated section, there is a minor difference in the annual average siltation intensities for Route 3′ (0.04 m/a~0.07 m/a) and Route 4 (0.04 m/a~0.09 m/a). The siltation volume is 59,300 m^3/a and 58,600 m^3/a for Route 3 and Route 4, respectively. In the excavation section of Qinzhou Bay, Route 3's annual average siltation intensity is approximately 0.07 m/a~0.32 m/a with a siltation volume of 165,700 m^3/a, while Route 4 has a smaller annual average siltation intensity (0.05 m/a~0.20 m/a) and siltation volume (134,100 m^3/a).

Following the substrate sampling in the project area, the sediment in the shallow water of Qinzhou Bay is mostly clayey chalk with a very fine grain size, which can be suspended and transported under certain wave currents. Considering the possibility of large waves and their driving mechanism on bed sediment transportation, there is a certain risk of sudden siltation in the project area. In this study, the representative wave with a return period of 10 years is adopted to represent the extreme wave conditions and explore the sudden siltation, and the representative wave direction is chosen to be the SSE direction wave selected for 24 h, which shows more significant influence on the project waters. The results demonstrate that the siltation intensity is 0.05 m and 0.04 m in the excavated section of the Fangchenggang East Bay and land area. The siltation intensity of Route 3 and Route 4 falls in a range of 0.05 m~0.07 m in Qinzhou Bay. A detailed comparison of four different routes is presented in Table 2, which also includes the results of the total engineering costs anticipated for the proposed routes. The engineering mainly includes waterway works, aids to navigation, river-crossing structures, bank protection, road construction, and water diversion facilities. Routes 1 and 2 involve exceptionally high costs for river-crossing structures, primarily due to the need for bridge construction. Route 3 incurs the highest cost in waterway works, owing to the large underwater excavation area and the substantial volume of dredging and rock-blasting operations. Overall, Route 4 demonstrates the highest economic efficiency.

Table 2. Route Comparison Results.

	Route 1	Route 2	Route 3	Route 4
Traffic impacts	Railway renovation is needed, not feasible.	Railway renovation is needed, not feasible.	Significant impact on highway interchange, low feasibility.	Limited impact on highways, high feasibility.
Ecological constraints	Partially through restricted development zones and ecological restoration areas, involving heritage protection zones, not feasible.	Partially through restricted development zones and ecological restoration areas, involving heritage protection zones, not feasible.	Partially through restricted development zones and ecological restoration areas, not feasible.	Bypassing ecological red line areas, occupies limited restoration areas, high feasibility.

Table 2. *Cont.*

	Route 1	Route 2	Route 3	Route 4
Hydrodynamics	/	/	Max flood velocity: 0.59–0.75 m/s; Max ebb velocity: 0.72–0.91 m/s; Higher transverse flow.	Max flood velocity: 0.72–0.80 m/s; Max ebb velocity: 0.72–0.86 m/s; More uniform flow distribution.
Sedimentation	/	/	Annual average intensity: 0.07–0.32 m/a; Total volume: 165,700 m^3/a.	Annual average intensity: 0.05–0.20 m/a; Total volume: 134,100 m^3/a.
Engineering cost (CNY)	4.35 billion	3.64 billion	3.24 billion	3.05 billion

Based on the comparative study of the traffic impact, ecological constraints, and hydrodynamic analysis, Route 4 is finally selected for the extension line of the Pinglu Canal in terms of engineering feasibility, environmental friendliness, and long-term stability. Specifically, Route 4 avoids the 3 m elevation of Huangchengao Railway Station and the reconstruction of the railway (e.g., Routes 1 and 2 for nearly 6.8 km). The total investment of the project could be reduced by about 40%. In terms of ecological protection, Route 4 bypasses the ecological red line protection area of mangroves, effectively reducing the degree of ecological and environmental impact. Following the hydrodynamic simulations, the flow velocity variations resulting from Route 4 are limited, with the maximum high tide flow velocity ranging from 0.72 m/s to 0.80 m/s and the maximum low tide flow velocity ranging from 0.72 m/s to 0.86 m/s. In contrast, the significant variations in flow velocity distribution by Route 3 would exert a greater impact on port operations and ship navigation safety. Moreover, both the average annual sedimentation intensity and siltation volume of Route 4 are smaller than those of Route 3. Therefore, Route 4 is finally recommended for the extension line of Pinglu Canal.

4. Discussion

4.1. Hydrodynamic Characteristics of Fangchenggang Waters

This study combines expert experience with numerical modeling to comprehensively evaluate the benefits of four routes in the extension line between Pinglu Canal and Fangchenggang. It is concluded that Route 4 satisfies the navigational function and yields stable hydrodynamic conditions (i.e., low cross-flow intensity). The operation and maintenance are expected to be convenient in the later stages, considering small siltation volumes. The present study highlights the significance of hydrodynamics on canal stability, which is consistent with the results reported by Mishra et al. [17]. It is worth noting that canals need to fulfill several different functions (e.g., irrigation, water conveyance and distribution, etc.) [37].

The present study confirms that the influence of transverse flow fields on navigation exists, especially in strongly tidal coastal zones. Huijts et al. [38] found that tidal variations would trigger a double-circle flow regime through a transverse density gradient and that the resulting tidal-phase-dependent transverse momentum transport mechanism significantly affects sediment distribution. Their findings further emphasize that the effects of transverse and longitudinal flow fields on coastal zones arise from their coupled interactions. The flow asymmetry between high and low tides and the difference in cross-flow duration in the results indirectly reveal the tidal asymmetry in the area of the Pinglu Canal extension line. In addition, the average annual siltation intensity (i.e., 0.18 m/a) is higher than that of the land excavation section, indicating that there is a tendency of sediment siltation in the sea section due to the influence of high tidal fluctuation and dynamical processes, which are closely related to the asymmetric push of tidal currents. Van et al. [39] also found that channel deepening interferes with the estuarine tidal asymmetry in the 3D numerical simulation, and channel deepening amplifies the vertical salinity gradient and the phase difference between near-bottom and surface flow velocities, thereby exacerbating the net transport of sediments upstream and leading to significantly higher sediment retention rates in localized areas. The accumulation of estuarine siltation is a great threat to ship navigation safety, and the dredging costs would increase accordingly (PION, 2018; Gar-CIA, 2024). The implementation of the extension line of the Pinglu Canal could improve the capability of port complexes to cope with natural disasters and set up an emergency plan from the perspective of stability.

4.2. Canal Route Design Under Ecological Constraints

Canal projects are usually characterized by cross-basin connectivity and high development intensity. Specific measures include artificial canal excavation, hub construction, waterway improvement, cross-basin water transfer, etc., which fundamentally alter the natural landscape, hydrological conditions, and ecological habitat [40]. Cao et al. [32] pointed out that the construction of the Pinglu Canal would lead to the destruction of regional wildlife habitat connectivity and trigger negative outcomes such as the "ecological isolation effect" and the "edge effect". The qualitative analysis by Cheng et al. [41] further revealed the underlying mechanisms: canal excavation increased the number of fragmented patches in the mangrove–tidal creek complex ecosystem (from 14 to 25), reduced the average patch size by 44.3%, and increased landscape fragmentation, confirming the costs of the "ecological island effect." In order to minimize the ecological disturbance, our study shifted from "active restoration" to "active avoidance" by choosing suitable routes to radically reduce the risk of habitat fragmentation. Comparative analysis shows that Route 1 and Route 2 cross the Huangchengao ancient site and occupy more mangrove-restricted development areas, which result in high ecological and cultural risks. In contrast, Route 4 bypasses the nature reserves, ecologically sensitive areas, and cultural relics protection areas as much as possible. The awareness of "active avoidance" has been a concern for a long time. Wells et al. [42] proposed a computerized graphic and map overlay methodology under the U.S. National Environmental Policy Act (1970) for selecting environmentally optimal canal routes with metrics of soil stability, drainage, wildlife migration, and historic site preservation. In the present study, hydrological features, ecological reserves, and historic site protection are combined to evaluate the suitability of different routes from the perspective of ecological benefits.

4.3. Limitations and Future Research Directions

The coupled hydrodynamic-sediment model based on MIKE21 has some shortcomings in both temporal and spatial scales. Only the typical high-tide hydrological conditions in 2021 were used, which may underestimate the temporal evolution mechanism of the flow regimes and the uncertainty of flow disturbances and siltation under extreme conditions. The conclusions of the flow field induced by tides may not be applicable to all estuarine waters, considering the unique geographical form of Qinzhou Bay. In the future, the hydrodynamic and morphological features could be investigated with the aid of a refined numerical model. Moreover, the ecological response mechanism of the extension line of the Pinglu Canal might be further explored through habitat suitability index (HSI) models. Therefore, the ecosystem can be better protected, and the development goal of sustainable transportation can be achieved. To validate the numerical model and habitat suitability calculations, it is necessary to carry out field measurements and ecological surveys to support a comprehensive evaluation of different route schemes.

5. Conclusions

In the context of the Pinglu Canal, the present study provides a comparative study of the route selection of its extension line to Fangchenggang. Four different routes are proposed according to the construction demand, natural conditions, and hydro-morphological features. A coupled two-dimensional hydrodynamic and sediment transport model has been established and validated. The comprehensive analysis demonstrates that Route 4 is recommended for the extension line of the Pinglu Canal. Some preliminary conclusions are drawn as follows:

(1) The elevation lifting (i.e., 3 m) of Huangchengao Railway Station and the reconstruction of the railway (approximately 6.8 km) are challenging. Route 1 and Route 2 are therefore excluded, considering the engineering feasibility. Route 3 imposes considerable adverse impacts on the hub interchange. Thus, Route 4 is eventually selected for the extension line of Pinglu Canal.

(2) The total length of Route 4 is approximately 15.41 km, including a 6.20 km land excavation section and a 9.21 km sea excavation section. It bypasses the mangrove ecological redline protection area and effectively minimizes the ecological environmental impact, whereas Routes 1, 2, and 3 intersect with ecological protection zones. The numerical modeling demonstrates that the canal waterway designed for Route 4 ensures stable hydrodynamic conditions, and the estimated siltation volume is generally acceptable.

(3) The implementation of the extension line of the Pinglu Canal would support the development of the logistics transportation system in Southwest China. It is worth noting that the coupled numerical model would be further improved to provide a refined simulation of hydro-morphological features in the future. Furthermore, the marine ecological response mechanism needs to be further investigated to achieve the sustainable development goal.

Author Contributions: Conceptualization, H.F. and X.C.; methodology, H.F.; software, Z.J.; validation, Z.J., W.W. and H.Y.; formal analysis, X.C.; investigation, H.F.; resources, H.Y.; data curation, W.W.; writing—original draft preparation, X.C.; writing—review and editing, H.F.; visualization, Z.J. All authors have read and agreed to the published version of the manuscript.

Funding: This research was funded by the Fangchenggang Supporting Special Project of the "Guangxi Beibu Gulf Port Master Plan" (No. 032207-001), and the National Natural Science Foundation of China (No. 52071250).

Data Availability Statement: The data that support the findings of this study are available from the corresponding author upon reasonable request.

Acknowledgments: The authors are grateful for the technical support provided by the Department of Transport of Guangxi Zhuang Autonomous Region.

Conflicts of Interest: The authors declare no conflicts of interest.

References

1. Yan, H. The making of the Grand Canal in China: Beyond knowledge and power. *Int. J. Herit. Stud.* **2021**, *27*, 584–600. [CrossRef]
2. Fang, L.; Cheng, X.; Yao, G. Research on the Strategy of Promoting the Coordinated Development of Grand Canal Water Transport and Other Transport Modes, Taking Jiangsu as an Example. *Sustainability* **2023**, *15*, 15314. [CrossRef]
3. Rusinov, I.; Gavrilova, I.; Sergeev, M. Features of sea freight through the Suez Canal. *Transp. Res. Procedia* **2021**, *54*, 719–725. [CrossRef]
4. Wang, M. The role of Panama Canal in global shipping. *Marit. Bus. Rev.* **2017**, *2*, 247–260. [CrossRef]

5. Dong, W.; Zhang, C.; Han, W.; Wang, J. Localized Canal Development Model Based on Titled Landscapes on the Grand Canal, Hangzhou Section, China. *Land* **2024**, *13*, 1178. [CrossRef]
6. Swamee, P.K.; Chahar, B.R. *Design of Canals*; Springer: New Delhi, India, 2015.
7. Williams, J.G.; Armstrong, G.; Katopodis, C.; Larinier, M.; Travade, F. Thinking like a fish: A key ingredient for development of effective fish passage facilities at river obstructions. *River Res. Appl.* **2012**, *28*, 407–417. [CrossRef]
8. Xiao, Z.; Zhang, X.; Fu, X.; Zhang, L.; Xu, H.; Liu, R.W.; Chong, C.S.; Qin, Z. Innovating Waterway Route Planning as a Service for Marine Traffic Applications. *IEEE Trans. Intell. Veh.* **2024**, 1–10. [CrossRef]
9. Jiang, M.; Wang, B.; Hao, Y.; Chen, S.; Lu, J. Vulnerability assessment of strait/canals in maritime transportation using fuzzy evidential reasoning approach. *Risk Anal.* **2023**, *43*, 1795–1810. [CrossRef] [PubMed]
10. Zhang, X.; Shen, J.; Sun, F.; Wang, S.; Zhang, S.; Chen, J. Allocation of flood drainage rights in watershed using a hybrid FBWM-Grey-TOPSIS method: A case study of the Jiangsu section of the Sunan Canal, China. *Int. J. Environ. Res. Public Health* **2022**, *19*, 8180. [CrossRef]
11. Li, Q.; Zhou, H.; Ma, Q.; Lu, L. Evaluation of Serviceability of Canal Lining Based on AHP–Simple Correlation Function Method–Cloud Model: A Case Study in Henan Province, China. *Sustainability* **2021**, *13*, 12314. [CrossRef]
12. Nakamura, P.; Tsukiyama, H. Analysis by an expert system of initial irrigation canal renovation project planning. *Water Resour. Manag.* **1992**, *6*, 223–233. [CrossRef]
13. Pham, T.Y.; Kim, K.Y.; Yeo, G.T. The Panama Canal expansion and its impact on east–west liner shipping route selection. *Sustainability* **2018**, *10*, 4353. [CrossRef]
14. Niazkar, M. Assessment of artificial intelligence models for calculating optimum properties of lined channels. *J. Hydroinform.* **2020**, *22*, 1410–1423. [CrossRef]
15. Gad, M.A.; Saad, A.; El-Fiky, A.; Khaled, M. Hydrodynamic modeling of sedimentation in the navigation channel of Damietta Harbor in Egypt. *Coast. Eng. J.* **2013**, *55*, 1350007-1–1350007-31. [CrossRef]
16. Coraci, E.; Umgiesser, G.; Zonta, R. Hydrodynamic and sediment transport modelling in the canals of Venice (Italy). *Estuar. Coast. Shelf Sci.* **2007**, *75*, 250–260. [CrossRef]
17. Kamran, M.; Yousaf, W.; Rajapakse, R.L.H.L.; Kareem Awan, W.; Riaz, M.; Asif, N.M.; Umar, M.; Shah, U.T. Innovative initiative for effective operation and monitoring using HEC-RAS modelling of Hakra Branch Canal System, Pakistan. *Irrig. Drain.* **2021**, *70*, 490–504. [CrossRef]
18. Mishra, A.; Anand, A.; Singh, R.; Raghuwanshi, N.S. Hydraulic modeling of Kangsabati main canal for performance assessment. *J. Irrig. Drain. Eng.* **2001**, *127*, 27–34. [CrossRef]
19. Zeng, F.; Jin, S.; Ye, L.; Gu, X.; Guo, J. Analysis of flood conveyance capacity of small-and medium-sized river and flood managements. *Nat. Hazards* **2023**, *116*, 447–467. [CrossRef]
20. Mohammad Rezapour Tabari, M.; Mazak Mari, M. The integrated approach of simulation and optimization in determining the optimum dimensions of canal for seepage control. *Water Resour. Manag.* **2016**, *30*, 1271–1292. [CrossRef]
21. Swamee, P.K.; Mishra, G.C.; Chahar, B.R. Minimum cost design of lined canal sections. *Water Resour. Manag.* **2000**, *14*, 1–12. [CrossRef]
22. Ghazaw, Y.M. Design and analysis of a canal section for minimum water loss. *Alex. Eng. J.* **2011**, *50*, 337–344. [CrossRef]
23. Fan, H.; Gong, X.; Lyu, J. Resilience assessment of strait/canal: A rule-based Bayesian network framework. *Transp. Res. Part D Transp. Environ.* **2023**, *124*, 103960. [CrossRef]
24. Miller, K.; Hyodo, T. Impact of the Panama Canal expansion on Latin American and Caribbean ports: Difference in difference (DID) method. *J. Shipp. Trade* **2021**, *6*, 8. [CrossRef]
25. Shahverdi, K.; Maestre, J.M. Holistic framework for canal modernization: Operation optimization, and economic and environmental analyses. *Water Resour. Manag.* **2023**, *37*, 6145–6164. [CrossRef]
26. Toprak, Z.F. Flow discharge modeling in open canals using a new fuzzy modeling technique (SMRGT). *CLEAN–Soil Air Water* **2009**, *37*, 742–752. [CrossRef]
27. Rezapour Tabari, M.M.; Tavakoli, S.; Mazak Mari, M. Optimal design of concrete canal section for minimizing costs of water loss, lining and earthworks. *Water Resour. Manag.* **2014**, *28*, 3019–3034. [CrossRef]
28. Pagano, A.M.; Light, M.K.; Sánchez, O.V.; Ungo, R.; Tapiero, E. Impact of the Panama Canal expansion on the Panamanian economy. *Marit. Policy Manag.* **2012**, *39*, 705–722. [CrossRef]
29. Yang, G.; Liao, L.; Chang, A. A preliminary study on the role of Pinglu canal in the new land-sea passage in the west. *J. Sustain. Urban. Plan. Prog.* **2024**, *8*, 28–34.
30. Liu, X.; Liu, Y.; Wang, S.; Dong, G. New canal construction and marine emissions strategy: A case of Pinglu. *Front. Mar. Sci.* **2024**, *10*, 1303879. [CrossRef]
31. Tan, K.; Huang, L.; Tan, K.; Kang, Z.; Kwan, K.Y. Metabarcoding analysis of potential HABs in the Maowei Sea, China: A baseline information for the "Pinglu Canal Waterway" mega-waterway project. *Estuar. Coast. Shelf Sci.* **2024**, *301*, 108752. [CrossRef]

32. Cao, Y.; Su, K.; You, Y.; Jiang, X.; Li, C.; Chen, J. Pinglu Canal Construction and Its Impact on Wildlife Habitat Connectivity: A Multi-Species Network Analysis and Ecological Restoration Strategy. *Land* **2025**, *14*, 150. [CrossRef]
33. Xia, M.; Chen, J.; Zhang, P.; Peng, P.; Claramunt, C. Spatial Structure and Vulnerability of Container Shipping Networks: A Case Study in the Beibu Gulf Sea Area. *J. Mar. Sci. Eng.* **2024**, *12*, 1307. [CrossRef]
34. *JTS/T 231—2021*; Code for Modelling Test for Port and Waterway Engineering. China Communications Press, Ministry of Transport of the People's Republic of China: Beijing, China, 2021. (In Chinese)
35. Teeter, A.M. Vertical Transport in Fine-Grained Suspension and Newly-Deposited Sediment. *Estuar. Cohesive Sediment Dyn.* **1986**, *14*, 170–191.
36. Parchure, T.M.; Mehta, A.J. Erosion of soft cohesive sediment deposits. *J. Hydraul. Eng.* **1985**, *111*, 1308–1326. [CrossRef]
37. Seyed Hoshiyar, S.M.; Pirmoradian, N.; Ashrafzadeh, A.; Parvaresh Rizi, A. Performance assessment of a water delivery canal to improve agricultural water distribution. *Water Resour. Manag.* **2021**, *35*, 2487–2501. [CrossRef]
38. Huijts, K.M.; de Swart, H.E.; Schramkowski, G.P.; Schuttelaars, H.M. Transverse structure of tidal and residual flow and sediment concentration in estuaries: Sensitivity to tidal forcing and water depth. *Ocean Dyn.* **2011**, *61*, 1067–1091. [CrossRef]
39. Van Maren, D.S.; van Kessel, T.; Cronin, K.; Sittoni, L. The impact of channel deepening and dredging on estuarine sediment concentration. *Cont. Shelf Res.* **2015**, *95*, 1–14. [CrossRef]
40. Zhang, N.; Bao, Z.; Zhang, L.; Gao, Y.; Xiao, Y.; Cheng, J.; Han, Z. Study on the Influence and Countermeasures of the Canal Projects on Aquatic Ecological Environment of Rivers along the Line. *E3S Web Conf.* **2023**, *393*, 01038. [CrossRef]
41. Cheng, Y.; Liu, J.; Xu, H.; Zhu, G. Impacts of Pinglu Canal Construction and Its Ecological Route Selection on Landscape Pattern and Suggestions on Environmental Protection. In *Advances in Transdisciplinary Engineering, Proceedings of the 3rd International Conference on Green Energy, Environment and Sustainable Development (GEESD2022), Beijing, China, 29 June 2022*; IOS Press: Amsterdam, The Netherlands, 2022; pp. 616–627.
42. Wells, D.M.; Mertes, J.D.; Allen, B.L.; Click, A.N. Procedure for selecting a minimal environmental impact routing for a water conveyance canal. *J. Am. Water Resour. Assoc.* **1975**, *11*, 714–733. [CrossRef]

land

Article

Research on the Coordination of Transportation Network and Ecological Corridors Based on Maxent Model and Circuit Theory in the Giant Panda National Park, China

Xinyu Li [1], Gaoru Zhu [1,2,*], Jiaqi Sun [1], Leyao Wu [1] and Yuting Peng [1]

1 School of Architecture and Design, Beijing Jiaotong University, Beijing 100044, China; 21311010@bjtu.edu.cn (X.L.); 21311042@bjtu.edu.cn (J.S.); 21311019@bjtu.edu.cn (L.W.); 23121606@bjtu.edu.cn (Y.P.)
2 Territorial Space and Transport Coordinated Development Institute, Beijing Jiaotong University, Beijing 100044, China
* Correspondence: grzhu@bjtu.edu.cn

Abstract: National parks serve as critical spatial units for conserving ecological baselines, maintaining genetic diversity, and delivering essential ecosystem services. However, accelerating socio-economic development has increasingly intensified the conflict between ecological protection and transportation infrastructure. Ecologically sustainable transportation planning is, therefore, essential to mitigate habitat fragmentation, facilitate species migration, and conserve biodiversity. This study examines the Giant Panda National Park and its buffer zone, focusing on six mammal species: giant panda, Sichuan snub-nosed monkey, leopard cat, forest musk deer, rock squirrel, and Sichuan takin. By integrating Maxent ecological niche modeling with circuit theory, it identified ecological source areas and potential corridors, and employed a two-step screening approach to design species-specific wildlife crossings. In total, 39 vegetated overpasses were proposed to serve all target species; 34 underpasses were integrated using existing bridge and culvert structures to minimize construction costs; and 27 canopy bridges, incorporating suspension cables and elevated pathways, were designed to connect forest canopies for arboreal species. This study established a multi-species and multi-scale conservation framework, providing both theoretical insights and practical strategies for ecologically integrated transportation planning in national parks, contributing to the synergy between biodiversity conservation and sustainable development goals.

Keywords: transportation network; habitat suitability assessment; minimum cumulative resistance; ecological source; ecological corridors; Giant Panda National Park; multi-species conservation

1. Introduction

National parks are critical to global biodiversity conservation and serve as exemplary models of human–nature coexistence [1]. However, with the rapid acceleration of global urbanization and the continuous expansion of infrastructure development, habitat fragmentation within national parks is intensifying, disrupting species migration and gene flow, and ultimately reducing habitat ranges, shrinking population sizes, and raising extinction risks [2,3]. As a flagship species endemic to China, the conservation of giant panda habitats is a crucial component of the country's ecological civilization initiatives. The official establishment of the Giant Panda National Park in 2021 marked a new milestone in China's ecological protection efforts. This park spans the transitional zone between the eastern edge

https://doi.org/10.3390/land14071465

of the Tibetan Plateau and the Sichuan Basin, encompassing diverse ecosystems and rich biodiversity. However, it simultaneously faces significant challenges, such as ecological fragmentation and transportation-induced disturbances. Ecological corridors, which link fragmented habitats and facilitate species movement, are essential for mitigating habitat fragmentation [4]. Meanwhile, wildlife crossings are a key strategy for incorporating ecological principles into transport infrastructure and minimizing their disruption to wildlife habitats. Therefore, accurately predicting the potential distribution of species habitats and scientifically optimizing ecological corridors and wildlife crossing designs are vital for achieving a balance between biodiversity conservation and transportation development [5].

In recent years, the integration of ecological conservation and transportation planning has emerged as a key topic in international academic research. Numerous studies have demonstrated the critical role of ecological corridors in mitigating habitat fragmentation and facilitating species movement and gene flow [6]. Reports from the International Union for Conservation of Nature (IUCN) have emphasized the ecological function of corridors in promoting landscape connectivity. A World Wildlife Fund report highlighted that large-scale infrastructure projects, such as those under the Belt and Road Initiative, may disrupt the habitats of many endangered species, further highlighting the urgency and importance of ecological corridor construction [7]. In Road Ecology: Science and Solutions, Forman et al. emphasized the theoretical foundation and practical strategies of road ecology, identifying wildlife crossings as an effective measure to mitigate the ecological impacts of transportation infrastructure [8,9]. Recent years have seen significant advances in corridor siting, structural design, habitat guidance, and performance monitoring. For example, Clevenger et al., in the *Wildlife Crossing Structure Handbook*, provided comprehensive guidelines for the design and evaluation of wildlife crossings in North America [10]. Moreover, advancements in technologies such as infrared camera monitoring, track analysis, and molecular techniques have greatly expanded research capacity in this field [11,12]. Despite these advancements, current research still faces several limitations. Most studies focus on single-species conservation, lacking a holistic understanding of multi-species ecological requirements, thus falling short of supporting comprehensive biodiversity goals. Moreover, the design of wildlife crossings often inadequately accounts for species-specific behavioral and ecological traits, limiting their functional effectiveness [13]. Therefore, strengthening research on multi-species collaborative protection, improving the design methods of wildlife crossings, and promoting the formulation of relevant standard systems are of great significance for enhancing the coordination of ecological corridors and transportation networks.

This study takes the Giant Panda National Park and its buffer zone as the study area and develops an ecological optimization approach for transportation networks based on the Maxent model and circuit theory [14], identifies ecological sources and corridors, optimizes transportation routes, and designs composite wildlife crossings to alleviate conflicts between ecological protection and transportation planning, thereby promoting regional ecological balance and sustainable development. The innovations of this study are reflected in the following three aspects: at the theoretical level, it proposes a four-stage ecological road network construction framework of 'analysis-evaluation-construction-optimization', which can systematically resolve the conflicts between the protection of the giant panda habitat and the traffic planning; at the methodological level, it innovatively uses the two-step screening method to realize the scientific siting of the composite wildlife crossings and the optimization of the ecological network; at the spatial level [15], the study breaks through the limitations of traditional small-scale research and builds a multi-species collaborative conservation planning system covering the Giant Panda National Park and its buffer zone. The findings offer theoretical insights and practical guidance for optimizing transportation networks in national parks, contributing to global biodiversity conservation and sustainable ecological infrastructure development.

2. Materials and Methods

2.1. Study Area

The Giant Panda National Park is located along the boundary between the first and second steps of China's topographic ladder, on the eastern edge of the Qinghai–Tibet Plateau. It lies within the alpine valley regions of the Min Mountains, Qionglai Mountains, and the Xiaoxiangling–Daxiangling ranges, which form the transitional zone between the Sichuan Basin and the Qinghai–Tibet Plateau. The region features highly complex terrain, with steep mountains, deep valleys, and a dense river network. The park's hydrological system drains into three major tributaries of the Yangtze River Basin: the Jialing River, Min River, and Tuo River. The total area of the park is approximately 22,000 square kilometers. The region experiences a continental monsoon climate that transitions from the northern subtropical zone to the warm temperate zone. It is generally humid and rainy, with a diverse and complex climatic pattern. Annual precipitation ranges between approximately 500 mm and 1200 mm [16].

Since the activity range of giant pandas cannot be confined within the boundaries of the national park, this study defines the research scope as encompassing both the Giant Panda National Park and a 10 km buffer zone surrounding it (Figure 1). The total study area, therefore, spans approximately 52,800 square kilometers.

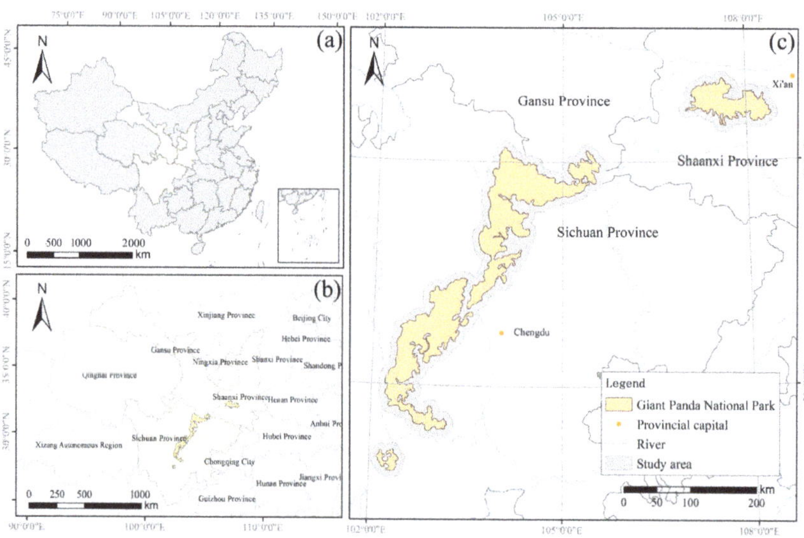

Figure 1. Study area boundary. (**a**) Administrative locations of Sichuan, Shaanxi, and Gansu Provinces in China; (**b**) spatial extent of the Giant Panda National Park; (**c**) river systems and the boundary of the study area.

2.2. Data Source and Pre-Processing

2.2.1. Distribution Data

Based on extensive literature review and comprehensive consideration of species records within Giant Panda National Park, national protection levels, species categories, and IUCN assessment statuses, this study selected six target terrestrial mammal species of varying sizes for analysis: giant panda (*Ailuropoda melanoleuca*), Sichuan snub-nosed monkey (*Rhinopithecus roxellana*), leopard cat (*Prionailurus bengalensis*), forest musk deer

(*Moschus berezovskii*), Sichuan takin (*Budorcas tibetana*), and rock squirrel (*Sciurotamias davidianus*) (Figure 2). Figure 3 presents the statistics of the selected species used in this study. Species distribution data were obtained from occurrence point datasets downloaded from the Global Biodiversity Information Facility (GBIF) database (https://www.gbif.org/) (accessed on 12 September 2024). The geographical distribution of occurrence points for each species is shown in Figure 4.

Figure 2. Photographs of the six target terrestrial mammal species. (**a**) Giant panda; (**b**) Sichuan snub-nosed monkey; (**c**) leopard cat; (**d**) forest musk deer; (**e**) Sichuan takin; (**f**) rock squirrel. (**a–e**) Source: National Geographic Media, 2025, available at https://www.natgeomedia.com (accessed on 30 June 2025); (**f**) source: Getty Images, 2025, available at https://www.gettyimages.com (accessed on 30 June 2025).

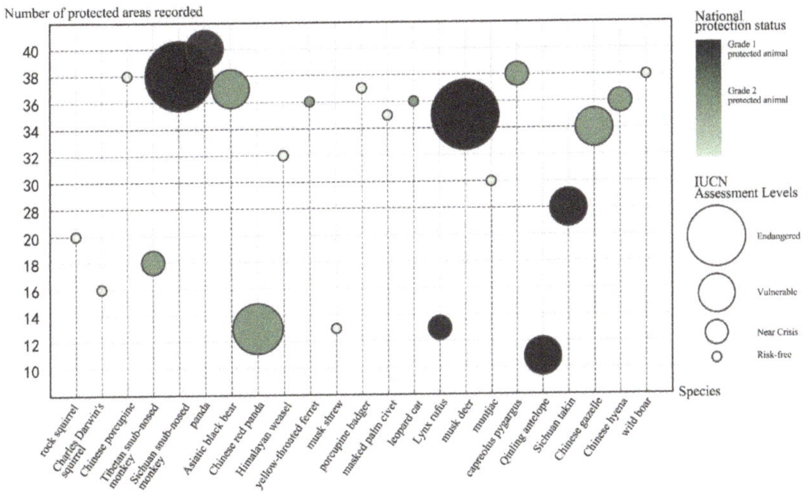

Figure 3. Statistics of selected species in Giant Panda National Park.

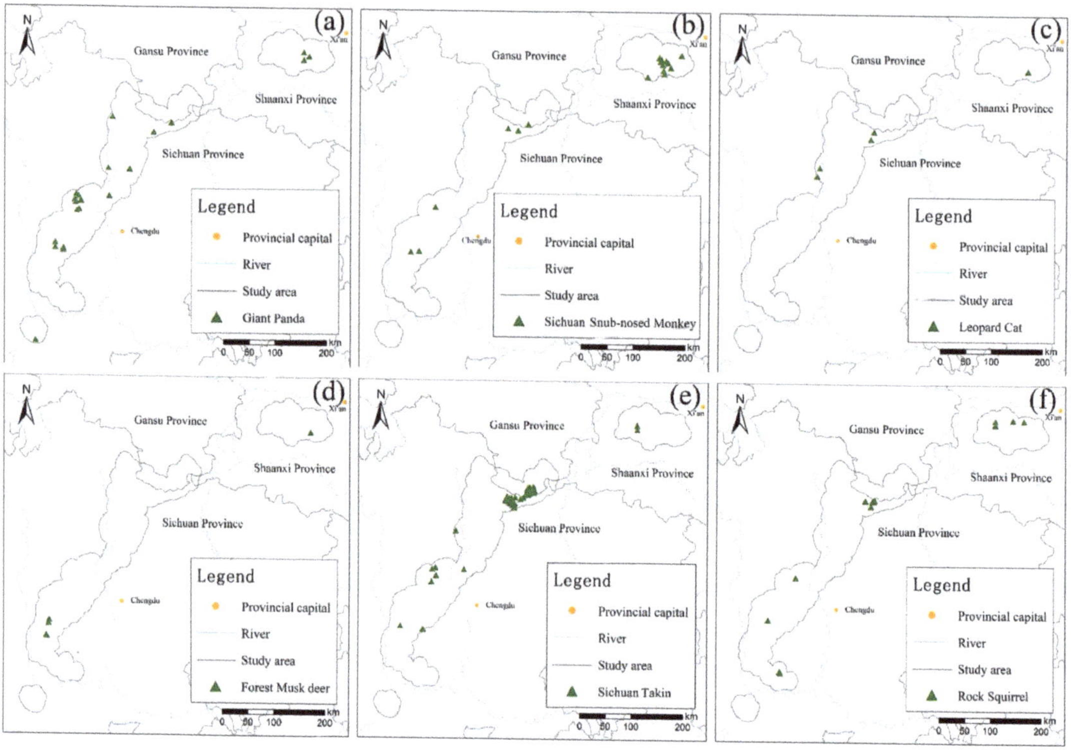

Figure 4. Distribution of species occurrence points. (**a**) Distribution points of the giant panda; (**b**) distribution points of the Sichuan snub-nosed monkey; (**c**) distribution points of the leopard cat; (**d**) distribution points of the forest musk deer; (**e**) distribution points of the Sichuan takin; (**f**) distribution points of the rock squirrel.

2.2.2. Road Data

The current railway and highway data of Sichuan, Shaanxi, and Gansu provinces are obtained from Open Street Map; the planned railway, highway, and provincial highway data are obtained from the traffic planning maps of Sichuan, Gansu, and Shaanxi provinces and are geographically calibrated by ArcGIS 10.8.

2.2.3. Environmental Data

Based on the habitat selection characteristics, living habits and literature studies of the selected wildlife, a comprehensive habitat factor index system was constructed by considering and determining the influencing factors from four aspects, namely, geographic environment, climatic factors, biological factors, and disturbance factors (Table 1). From the Geospatial Data Cloud Platform (GDCP), 30 m precision DEM data were obtained, and elevation, slope, and direction were calculated from DEM data by the spatial analysis tool of ArcGIS 10.8. The water source data were obtained from Open Street Map, annual precipitation and mean annual temperature were obtained from the official website of World Clim, canopy height data were obtained from https://www.3decology.org/ (accessed on 19 September 2024), the Normalized Difference Vegetation Index (NDVI) is obtained from the Resource and Environment Science and Data Center of the Chinese Academy of Sciences, and 1:1 million vegetation type distribution data were obtained from the National Glacier Frost and Frost Service Platform. The type distribution data is from the National Glacial and Permafrost Desert Science Data Centre, the settlement data is from the National Bureau of Statistics, the 10 m accuracy land cover data is from Esri | Sentinel-2 Land Cover Explorer, and the natural disaster point data is from the Centre for Resource and Environmental Science and Data (Table 2).

Table 1. Integrated habitat factor indicator system.

Variable Category	Sub Variable
Geography	Slope Elevation Aspect Distance to rivers
Climatic factors	Annual precipitation Mean annual temperature
Biological factors	Canopy height Vegetation type NDVI
Disturbances	Land cover type Distance to settlements Distance to roads Distance to disaster sites

Table 2. Data sources.

Data	Sources	Spatial Accuracy
Species distribution data	GBIF Global Biodiversity Information Facility https://www.gbif.org/ (accessed on 12 September 2024)	——
DEM data	Geospatial Data Cloud	30 m
Water data	Open Street Map	——
Precipitation data	World Clim Global Climate Data official website	——
Temperature data	World Clim Global Climate Data official website	——
Vegetation canopy height	Neural network guided interpolation for mapping canopy height of China's forests by integrating GEDI and ICE-Sat-2 data. [17]	30 m
NDVI data	http://www.nesdc.org.cn/ (accessed on 19 September 2024)	30 m
Vegetation type	National Cryosphere Desert Data Center	1:100 million
Settlement data	National Bureau of Statistics	——
Land cover data	Esri | Sentinel-2 Land Cover Explorer	10 m
Natural disaster sites	Resource and Environmental Science Data Platform	——
Roads data	Open Street Map	——

2.2.4. Data Processing

In ArcGIS 10.8, all data required for the research were unified into the GCS_Beijing_1954 projection coordinate system, and the point data of six species were unified into CSV format. All environmental variable data were converted into raster format and resampled to a resolution of 30 m $\times$ 30 m (Figure 5). Subsequently, the raster files of various environmental variables were converted into ASCII format required for the Maxent model and then input into Maxent for operation.

2.3. Methods

2.3.1. Maxent Modeling and Circuit Theory

This study employed the Maxent model and circuit theory to identify species ecological corridors. The Maxent 3.4.4 [18], based on Maximum Entropy Machine Learning (MEML), uses ecological niche modeling and species occurrence data to analyze the relationship between species distribution and environmental variables. By iteratively fitting the model, it calculates the probability distribution with maximum entropy, thereby predicting the potential geographic distribution of the species [19]. The result can visually represent suitable habitat distribution and facilitate habitat suitability assessment.

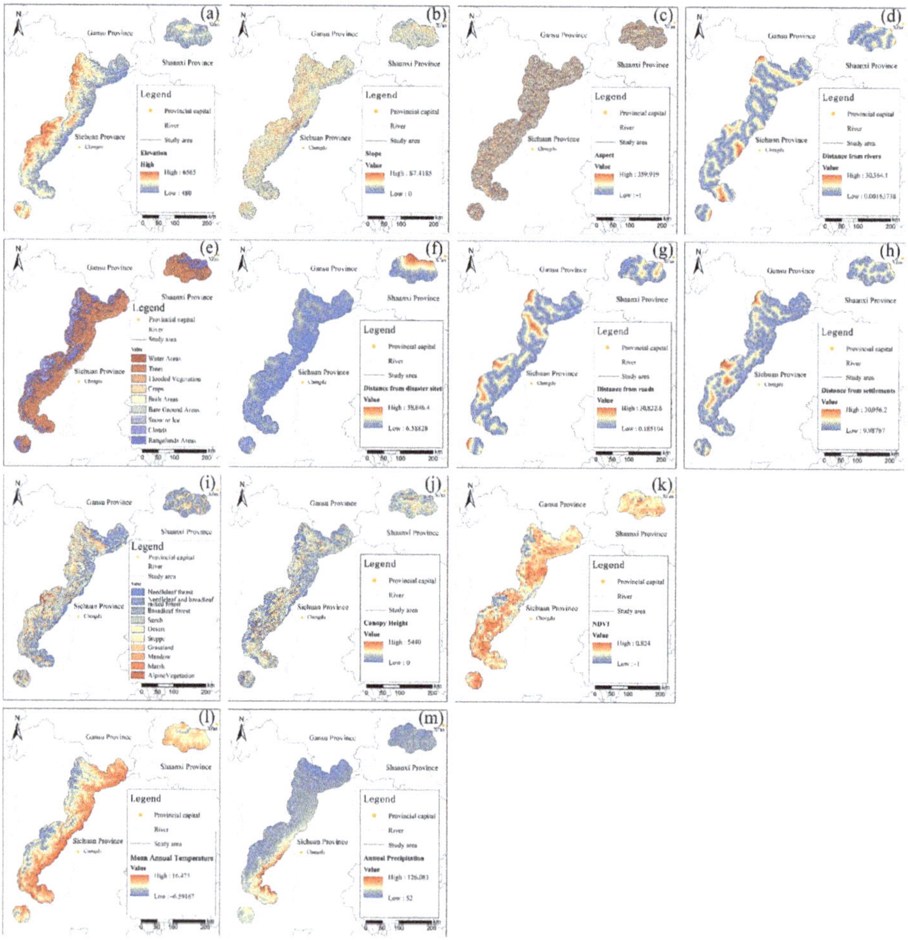

Figure 5. Habitat factors. (**a**) Elevation; (**b**) slope; (**c**) aspect; (**d**) distance to river; (**e**) land cover type; (**f**) distance to disaster sites; (**g**) distance to road; (**h**) distance to settlements; (**i**) vegetation type; (**j**) canopy height; (**k**) NDVI; (**l**) mean annual temperature; (**m**) annual precipitation.

Circuit theory, applied in landscape ecology, is grounded in the random walk model from probability theory. The Linkage Mapper 3.0.0 (The Nature Conservancy, https://linkagemapper.org/) (accessed on 8 October 2024), based on circuit theory, identifies adjacent ecological sources through neighborhood and distance data, constructs a regional network between ecological sources, calculates the cost-weighted distance from each grid on the resistance surface to the ecological sources, and superimposes it on the ecological sources to find out the path with the lowest cumulative cost between the ecological sources [20,21]. The ecological corridor is the path that minimizes the cumulative cost between ecological sources.

Species distribution data and corresponding environmental variables were input into the Maxent model and iteratively fitted to estimate the probability distribution at maximum entropy, thus characterizing species distribution patterns. The relative importance of variables was evaluated using the Jackknife procedure, and 10-fold cross-validation was performed to minimize random errors in model predictions [22].

The habitat suitability outputs (TIF files) from the Maxent model were reclassified using the Reclassify tool in ArcGIS 10.8 (Esri Inc., Redlands, CA, USA), Habitat suitability was divided into five levels based on value thresholds: 0–0.2 as unsuitable areas, 0.2–0.4 as low-suitability areas, 0.4–0.6 as moderate-suitability areas, 0.6–0.8 as high-suitability areas, and 0.8–1.0 as very high-suitability areas [23]. Areas classified as moderate suitability and above ($\geq$0.4) were identified as ecological core areas. Considering differences in species habits, as well as the size and number of core patches, patches larger than 8–30 km^2 were selected as ecological source areas.

Based on the Jackknife test results provided by the Maxent 3.4.4, the importance of each environmental variable to species distribution was evaluated. A higher training gain under the condition of a single variable indicates greater independent explanatory power of that variable, and thus a more significant impact on the species' distribution. According to this principle, the variable contribution rates and Jackknife gain values were extracted for each species during the modeling process, and the five most influential variables were selected as key resistance factors [24]. Given the variability in species' ecological responses to environmental factors, a unified weighting scheme would be insufficient to capture their specific behavioral and habitat preferences. Therefore, species-specific resistance factors were identified, and differentiated weights were assigned based on the variable contributions of each species. The contribution values of the five factors were then normalized to ensure their sum equals 1 and were further adjusted with reference to the relevant literature and expert opinions, thereby generating the final weighting coefficients used in the construction of the integrated resistance surface.

In ArcGIS 10.8, each resistance factor was reclassified into five resistance levels (1 representing the lowest resistance and 5 representing the highest) using the Natural Breaks (Jenks) method. The Raster Calculator was then used to apply weighted overlay of the five factors based on their corresponding weights, resulting in the integrated resistance surface for each species [25]. This surface was subsequently used for least-cost path analysis in the construction of ecological corridors.

2.3.2. Two-Step Screening Method

In this study, intersections between planned and existing railways and highways with ecological corridors and source areas of multiple species within Giant Panda National Park were defined as conflict points. These points were categorized into multiple sets based on the type of wildlife crossing. For each road, the number of conflict points was counted, and roads containing more than five conflict points were selected for analysis using the two-step screening method.

The two-step screening method was developed to identify suitable locations for constructing composite wildlife corridors. In the first step, all ecological conflict points were sequentially numbered, and a fixed-radius buffer zone was established around each point. Then, the number of surrounding conflict points within each buffer was counted to assess the degree of spatial clustering. In the second step, points located at the peak of a clustering interval—representing local hotspots of ecological conflict—were selected as priority locations for corridor construction. The overall screening framework is illustrated in Figure 6.

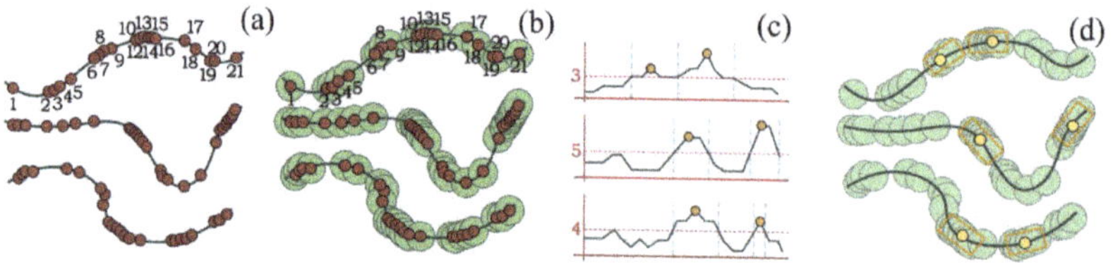

Figure 6. Two-step screening approach. (**a**) Numbering the ecological conflict points in order (red points); (**b**) creating a 5 km buffer for each conflict point (green circles); (**c**) selection of points where the number of aggregated points in the buffer zone is at its peak (orange points); (**d**) screening of completed points to be optimized (yellow points).

2.4. Study Framework

This study proposes a systematic framework for the ecological development of the transportation network within the Giant Panda National Park. Based on the Maxent model, we conducted an integrated analysis of species distribution data and environmental variables to evaluate habitat suitability and identify ecological sources. A minimum cumulative resistance model was constructed by quantifying the impact of environmental variables on animal habitats. By integrating circuit theory with the Linkage Mapper tool and the minimum resistance surface, we delineated ecological corridors by identifying least-cost paths between ecological sources. Multi-species corridor overlay analysis and a two-step screening method were used to locate spatial conflict points. In response to the ecological impacts of existing roads, three types of composite wildlife crossings were designed based on the biological characteristics of different species, optimizing the synergy between ecological corridors and the transportation network. This provides scientific basis and practical guidance for balancing biodiversity conservation and infrastructure development in national parks. The overall technical framework of the study is illustrated in Figure 7.

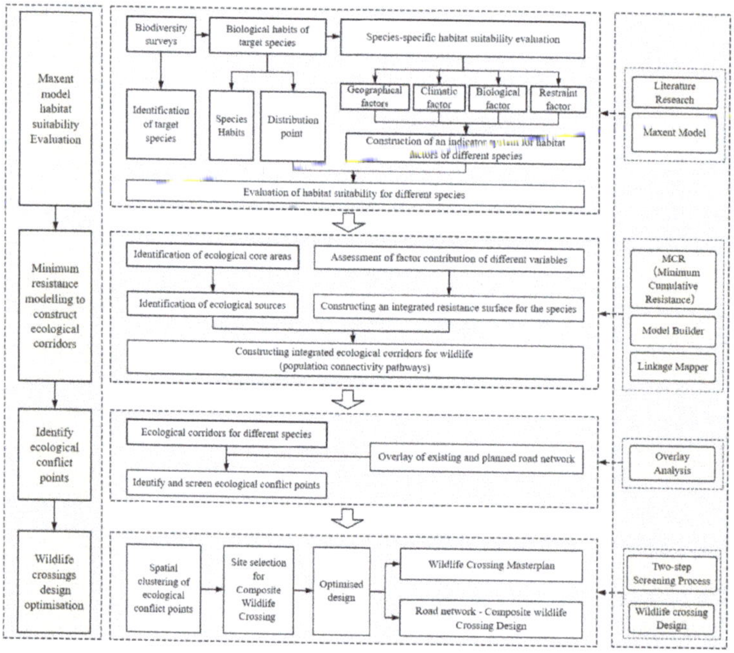

Figure 7. Technical framework.

3. Results

3.1. Habitat Suitability Assessment Based on the Maxent Model

3.1.1. Habitat Suitability Distribution

Based on the results of the habitat suitability evaluation using the Maxent model, the average AUC value across the six species was 0.915 (with values above 0.9 indicating high accuracy), as shown in Figure 8 [26,27]. This demonstrates that the Maxent model reliably predicted suitable habitats, successfully identifying the potential distribution ranges for each species.

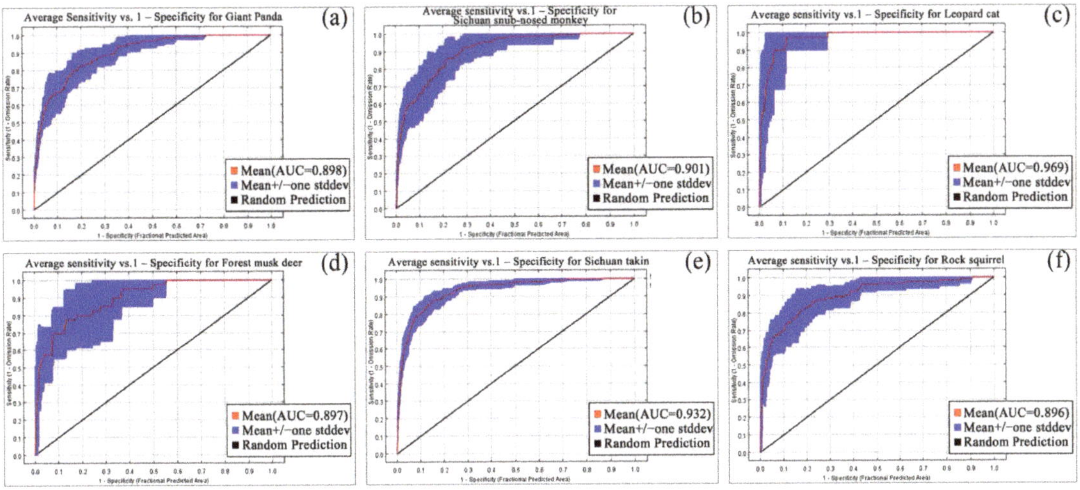

Figure 8. Accuracy verification ROC Curves of each species. (**a**) ROC Curves of the giant panda; (**b**) ROC Curves of the Sichuan snub-nosed monkey; (**c**) ROC Curves of the leopard cat; (**d**) ROC Curves of the forest musk deer; (**e**) ROC Curves of the Sichuan takin; (**f**) ROC Curves of the rock squirrel.

Figure 9 shows the spatial distribution of habitat suitability for each target species. The results indicate that suitable habitats for the giant panda are primarily located in the central and western regions of the study area within Sichuan Province, as well as in the northern portion of Shaanxi Province. Suitable areas for the Sichuan snub-nosed monkey are mainly located in the northern part of the study area in Shaanxi, with some distribution in the northern part of Sichuan. The leopard cat exhibits a broader yet more fragmented distribution, extending from Sichuan and Shaanxi into southern Gansu Province. Suitable habitats for the forest musk deer are mainly found in the central and southern parts of the study area in Sichuan, with limited distribution in the mountainous regions of Shaanxi. The Sichuan takin is primarily distributed in the northern part of the study area in Sichuan and the southern mountainous areas of Shaanxi. The rock squirrel's habitat is relatively widespread, spanning the central mountainous areas of both Shaanxi and Sichuan, and extending into southern Gansu. The scale and proportion of habitat suitability zones for each species are summarized in Table 3.

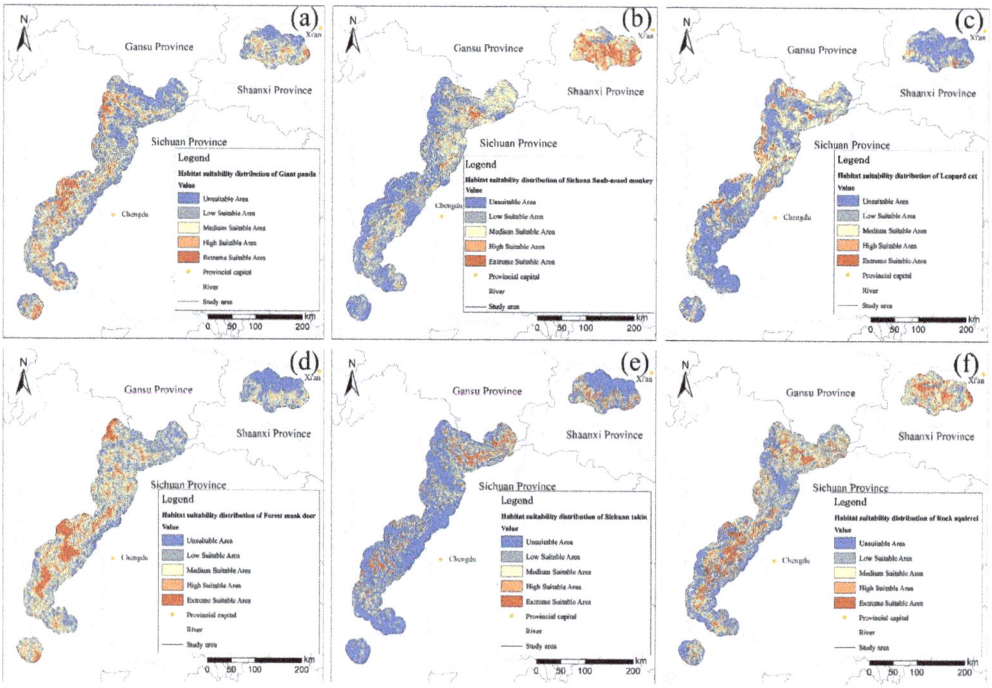

Figure 9. Habitat suitability evaluation of species. (**a**) Habitat suitability evaluation of the giant panda; (**b**) habitat suitability evaluation of the Sichuan snub-nosed monkey; (**c**) habitat suitability evaluation of the leopard cat; (**d**) habitat suitability evaluation of the forest musk deer; (**e**) habitat suitability evaluation of the Sichuan takin; (**f**) habitat suitability evaluation of the rock squirrel.

Table 3. Scale of habitat suitability areas for each species.

Habitat Suitability Zone	Medium-Suitability Area (km^2)	Medium-Suitability Percentage (%)	High-Suitability Area (km^2)	High-Suitability Percentage (%)	Extreme High-Suitability Area (km^2)	Extreme High-Suitability Percentage (%)
Giant panda	11,779	5.51	7674	14.57	2902	22.36
Leopard cat	9035	17.15	3865	7.34	1723	3.27
Sichuan snub-nosed monkey	10,967	20.82	618	11.73	269	5.11
Forest musk deer	1370	26.01	776	14.73	3879	7.36
Rock squirrel	11,966	22.72	8468	16.08	4745	9.01
Sichuan takin	6095	11.57	3660	6.95	2272	4.31

3.1.2. Ecological Source Distribution of Species

Based on the extraction results of ecological source areas for each species in ArcGIS 10.8, the ecological sources of the giant panda are relatively dispersed yet evenly distributed, covering multiple mountain ranges. Table 4 summarizes the number, total area, and proportional coverage of ecological source regions for each species. The ecological sources of the Sichuan snub-nosed monkey are predominantly located in the Minshan region of Shaanxi Province, with a tendency to concentrate at higher latitudes. The ecological sources of the leopard cat are primarily located in the Minshan and Baishuijiang regions, with a slightly higher concentration in the northern part of Sichuan Province. The forest musk deer's sources are mainly distributed in the central and southern sections of the Qionglai Mountains. The rock squirrel's ecological sources are highly fragmented and limited in size, with moderate clustering in the central and northern Qinling Mountains. The ecological sources of the Sichuan takin are mainly distributed in the Minshan and Baishuijiang regions, near the border between Sichuan and Gansu Provinces (Figure 10).

Table 4. Ecological source area for each species.

Species	Ecological Source Area (km²)	Number (Pcs)	Percentage of Study Area Occupied (%)
Giant panda	5319	86	10.06
Leopard cat	7045	69	13.33
Sichuan snub-nosed monkey	9871	86	18.67
Forest musk deer	12,023	50	22.75
Rock squirrel	3338	90	6.31
Sichuan takin	3956	75	7.48

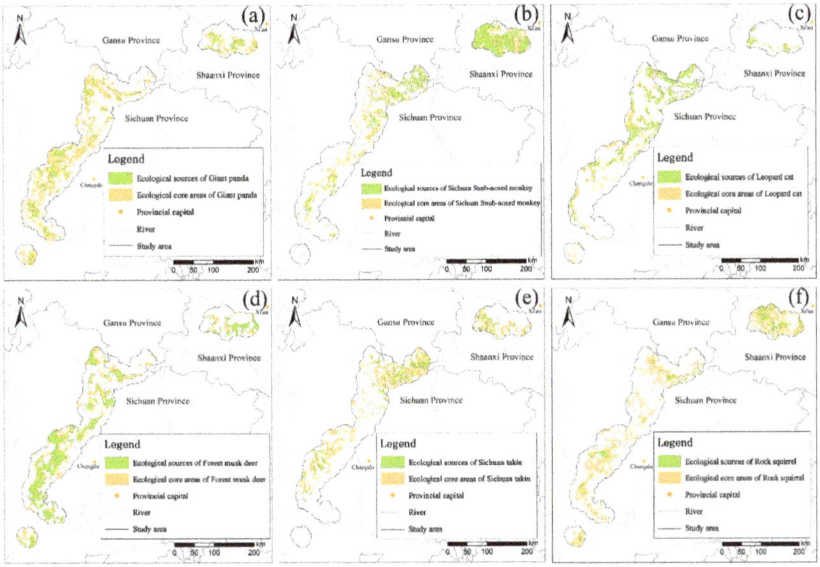

Figure 10. Ecological sources of each species. (**a**) Ecological sources of the Giant panda; (**b**) ecological sources of the Sichuan snub-nosed monkey; (**c**) ecological sources of the leopard cat; (**d**) ecological sources of the forest musk deer; (**e**) ecological sources of the Sichuan takin; (**f**) ecological sources of the rock squirrel.

3.2. Construction of Wildlife Ecological Corridors Based on Circuit Theory

3.2.1. Integrated Species Resistance Surface Construction

In the standard training gain model [28], a higher score indicates that the environmental variable has greater importance in predicting species distribution and contributes more significantly to the model's performance. The results demonstrated that the primary environmental factors influencing habitat suitability varied considerably among the six target species, reflecting distinct ecological adaptations. The contribution rates of all environmental variables to the predicted habitat suitability of each species are presented in Figure 11.

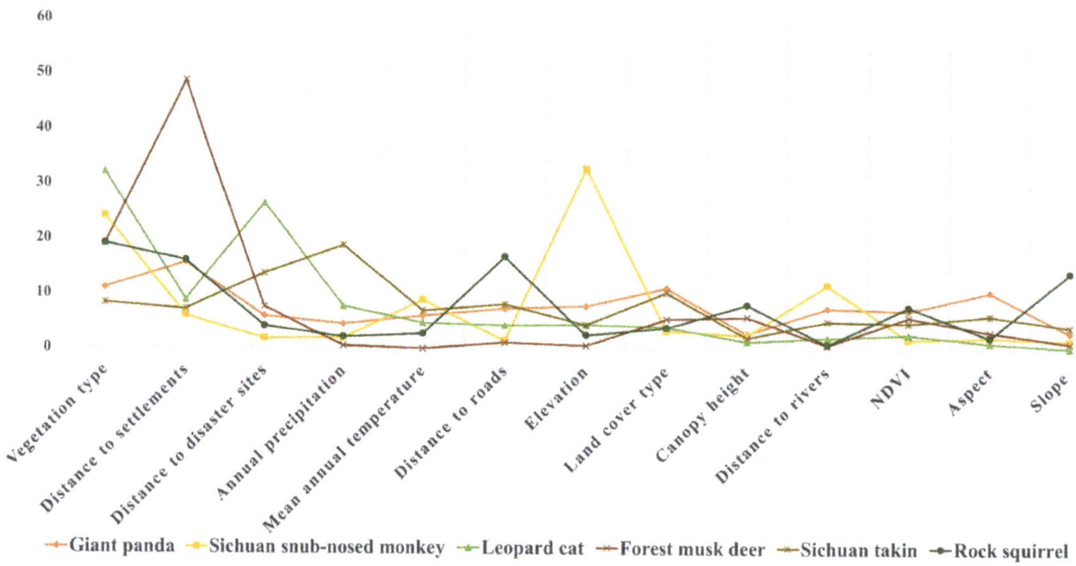

Figure 11. The Jackknife test results for the variable importance of each species.

The weight values of the top five most influential environmental variables for each species are presented in Table 5. For the giant panda, the most influential variable was distance to settlements, followed by land cover type, vegetation type, aspect, and elevation. This suggests a strong avoidance of human disturbance and a preference for complex mountainous environments far from populated areas. The habitat suitability of the Sichuan snub-nosed monkey was primarily affected by elevation, followed by vegetation type, distance to rivers, distance to settlements, and mean annual temperature, indicating a tendency to inhabit mid- to high-elevation forested areas with abundant vegetation and water availability. For the leopard cat, the most important factor was vegetation type, followed by distance to disaster sites, annual precipitation, distance to settlements, and mean annual temperature, highlighting its dependence on concealed habitats and specific climatic conditions. The forest musk deer showed a similar pattern, with distance to settlements being the dominant factor, followed by vegetation type, distance to disaster sites, canopy height, and vegetation cover, indicating a preference for less disturbed montane mixed-forest environments. The habitat suitability of the Sichuan takin was most influenced by distance to roads, vegetation type, land cover type, distance to disaster sites, and annual precipitation, reflecting its sensitivity to infrastructure and preference for forest–meadow ecotones. For the rock squirrel, vegetation type was again the most influential factor, followed by distance to roads, distance to settlements, slope, and canopy height, indicating its preference for needleleaf–broadleaf mixed forests and environments buffered from human activity.

Based on the top five most influential environmental variables for each species, integrated resistance surface models were constructed accordingly. The results are presented in Figure 12.

Table 5. Habitat suitability of species.

Species	Weight	Environment Variable	Grading Criteria				
			1	2	3	4	5
Giant panda	0.30	Distance to settlements	15.8–30.9 km	10.2–15.8 km	6.4–10.2 km	3.1–6.4 km	≤3.1 km
	0.20	Land cover type	trees	water, rangelands	snow or ice	crops	built, bare ground
	0.20	Vegetation type	needleleaf and broadleaf mixed forest	scrub, grassland	desert	needleleaf, broadleaf forest	steppe, meadow, alpine vegetation
	0.20	Aspect	−1–70.9°	70.9–141.4°	141.4–211.9°	211.9–283.8°	283.8–359.9°
	0.10	Elevation	454–1466 m	1466–2136 m	2136–2877 m	2877–3724 m	3724–7107 m
Sichuan snub-nosed monkey	0.40	Elevation	454–1466 m	1466–2136 m	2136–2877 m	2877–3724 m	3724–7107 m
	0.30	Vegetation type	grassland	desert	needleleaf and broadleaf mixed forest, needleleaf forest, alpine vegetation	broadleaf forest, steppe	scrub
	0.15	Distance to rivers	17,328–30,593 km	11,472–17,328 km	7289–11,472 km	3466–7289 km	≤3466 km
	0.075	Distance to settlements	15.8–30.9 km	10.2–15.8 km	6.4–10.2 km	3.1–6.4 km	≤3.1 km
	0.075	Mean annual temperature	11.9–16.6 °C	8.8–11.9 °C	5.5–8.8 °C	1.5–5.5 °C	−6.5–1.5 °C
Leopard cat	0.40	Vegetation type	grassland	scrub	needleleaf and broadleaf mixed forest, needleleaf, broadleaf forest	desert, steppe, meadow	alpine vegetation
	0.35	Distance to disaster sites	0.71–4142 km	4142–9665 km	9665–19,560 km	19,560–33,828 km	33,828–58,911 km
	0.10	Annual precipitation	52–64.7 mm	64.7–72.5 mm	72.5–83.1 mm	83.1–97.0 mm	97.0–126.0 mm
	0.10	Distance to settlements	15.8–30.9 km	10.2–15.8 km	6.4–10.2 km	3.1–6.4 km	≤3.1 km
	0.05	Mean annual temperature	11.92–16.69 °C	8.8–11.92 °C	5.5–8.8 °C	1.53–5.5 °C	−6.59–1.53 °C
Forest musk deer	0.50	Distance to settlements	15.8–30.9 km	10.2–15.8 km	6.4–10.2 km	3.1–6.4 km	≤3.1 km
	0.20	Vegetation type	grassland	needleleaf and broadleaf mixed forest	scrub, alpine vegetation	needleleaf, broadleaf forest, desert, steppe	marsh
	0.15	Distance to disaster sites	33.83–58.91 km	19.56–33.83 km	9.67–19.56 km	4.14–9.67 km	≤4.14 km
	0.075	Canopy height	≤4.82 cm	4.82–12.12 cm	12.12–17.92 cm	17.92–25 cm	25–62.7 cm
	0.075	NDVI	0.615–0.824	0.521–0.615	0.425–0.521	0.293–0.425	−1.293
Sichuan takin	0.35	Annual precipitation	52–64.7 mm	64.7–72.5 mm	72.5–83.1 mm	83.1–97.0 mm	97.0–126.0 mm
	0.25	Distance to disaster sites	17.9–30.8 km	12.0–17.9 km	7.3–12.0 km	3.3–7.3 km	≤4.1 km
	0.20	Land cover type	flooded vegetation	rangelands	water, trees, crops, bare ground, clouds	snow or ice	built
	0.10	Vegetation type	scrub, steppe	grassland	alpine vegetation, needleleaf and broadleaf mixed forest	broadleaf forest, meadow	desert
	0.10	Distance to roads	15.8–30.9 km	10.2–15.8 km	6.4–10.2 km	3.1–6.4 km	≤3.3 km
Rock squirrel	0.25	Vegetation type	needleleaf and broadleaf mixed forest	cultivated vegetation	needleleaf forest, meadow, alpine vegetation	desert, grassland	scrub
	0.20	Distance to roads	15.8–30.9 km	10.2–15.8 km	6.4–10.2 km	3.1–6.4 km	≤3.3 km
	0.20	Distance to settlements	15.8–30.9 km	10.2–15.8 km	6.4–10.2 km	3.1–6.4 km	≤3.1 km
	0.20	Slope	0–14.4°	14.4–24.1°	24.1–32.6°	32.6–42.9°	42.9–87.8°
	0.15	Canopy height	25–62.7 cm	17.9–2 cm	12.1–17.9 cm	4.8–12.1 cm	≤4.8 cm

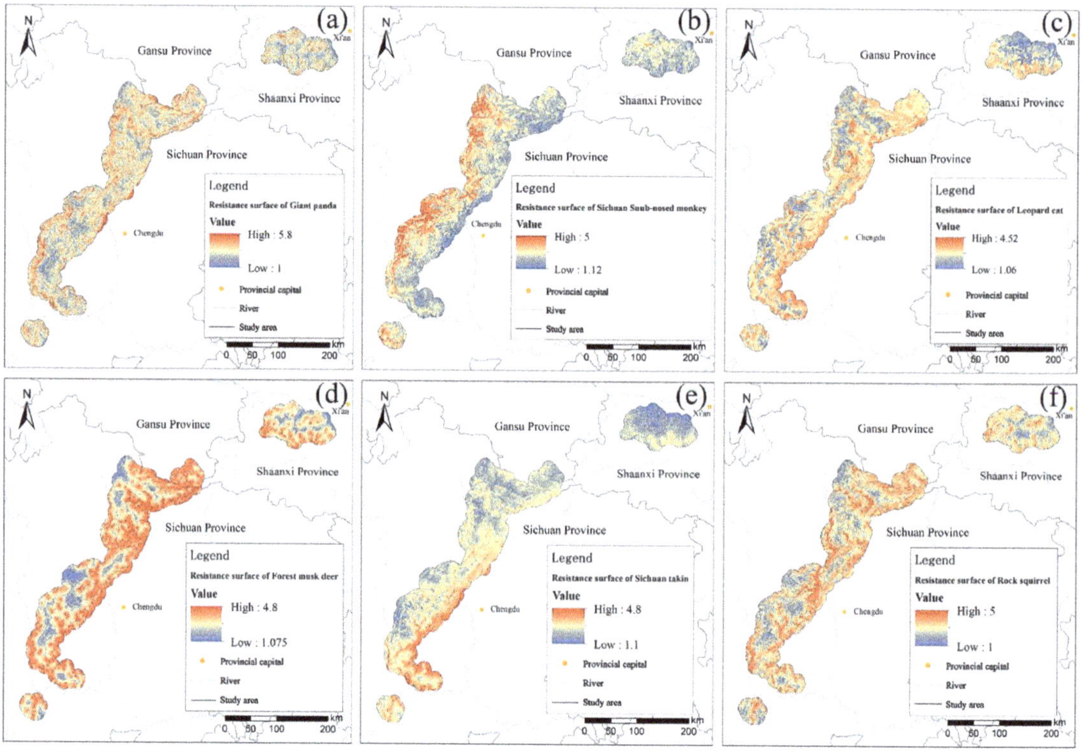

Figure 12. Integrated resistance surfaces of each species. (**a**) Integrated resistance surfaces of the Giant panda; (**b**) integrated resistance surfaces of the Sichuan snub-nosed monkey; (**c**) integrated resistance surfaces of the leopard cat; (**d**) integrated resistance surfaces of the forest musk deer; (**e**) integrated resistance surfaces of the Sichuan takin; (**f**) integrated resistance surfaces of the rock squirrel.

3.2.2. Construction of Species Ecological Corridors and Identification of Ecological Conflict Points

Based on the ecological source areas and integrated resistance surfaces for each species [29,30], ecological corridors were constructed separately using the Linkage Mapper tool. As shown in Figure 13, the results yielded 172 ecological corridors connecting 86 ecological sources for giant pandas; 180 ecological corridors for Sichuan snub-nosed monkeys, connecting 86 ecological sources; 136 ecological corridors for leopard cats, connecting 69 ecological sources; 78 ecological corridors for forest musk deer, connecting 50 ecological sources; 192 ecological corridors for rock squirrels, connecting 90 ecological sources; and 154 ecological corridors for Sichuan takins, connecting 75 ecological sources. 75 ecological source areas.

The intersections of planned railways, highways, provincial highways, and existing railways and roads in the Giant Panda National Park with the ecological corridors of the six species and the ecological sources of each species were designated as ecological conflict points (Figure 14). Finally, we obtained 138 ecological conflict points between giant pandas and roads, 100 ecological conflict points between Sichuan snub-nosed monkeys and roads, 54 ecological conflict points between leopard cats and roads, 37 ecological conflict points between forest musk deer and roads, 205 ecological conflict points between rock squirrels and roads, and 129 ecological conflict points between Sichuan takins and roads.

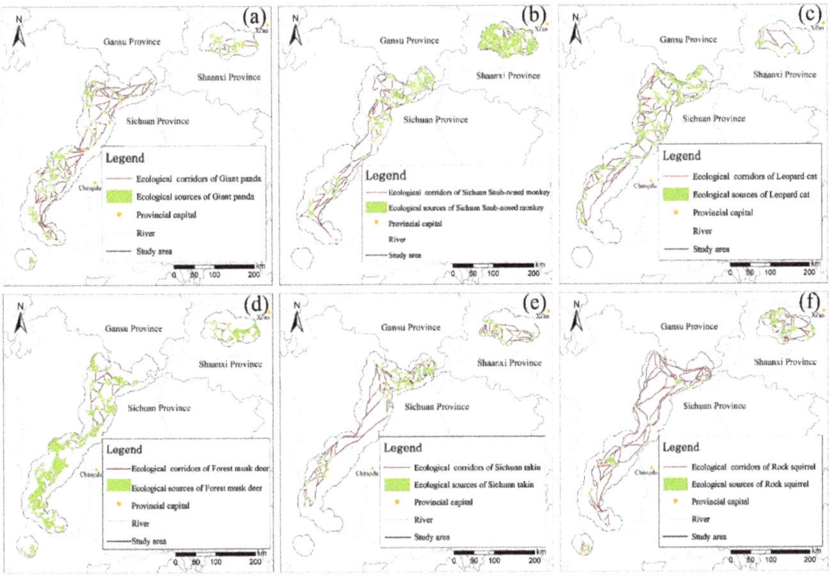

Figure 13. Ecological corridors of each species. (**a**) Ecological corridors of the Giant panda; (**b**) ecological corridors of the Sichuan snub-nosed monkey; (**c**) ecological corridors of the leopard cat; (**d**) ecological corridors of the forest musk deer; (**e**) ecological corridors of the Sichuan takin; (**f**) ecological corridors of the rock squirrel.

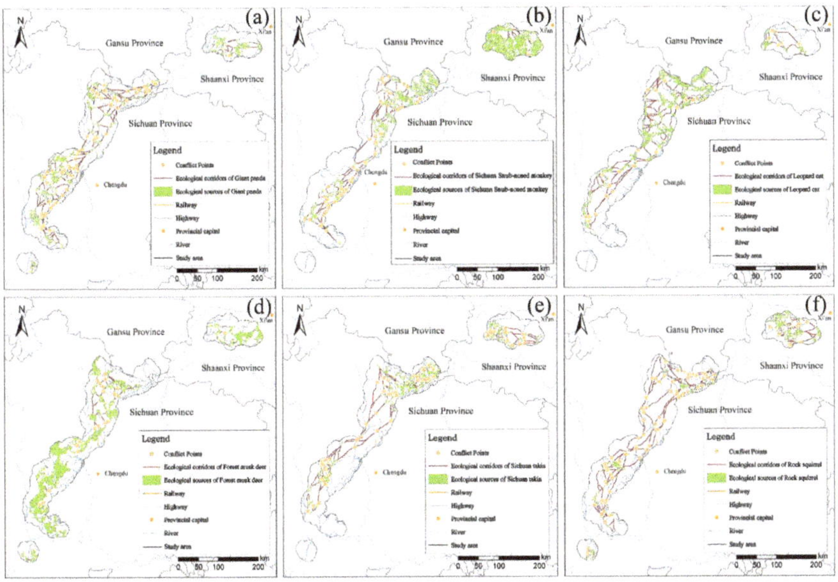

Figure 14. Ecological conflict points of species. (**a**) Ecological conflict points of the Giant panda; (**b**) ecological conflict points of the Sichuan snub-nosed monkey; (**c**) ecological conflict points of the leopard cat; (**d**) ecological conflict points of the forest musk deer; (**e**) ecological conflict points of the Sichuan takin; (**f**) ecological conflict points of the rock squirrel.

3.3. Site Selection for Multifunctional Wildlife Crossings

The intersections of planned railways, expressways, provincial roads, and existing railways and highways in the Giant Panda National Park with the ecological corridors and ecological sources of six target species were identified as ecological conflict points, and the conflict points of the species involved in the under-penetrating wildlife crossings applicable to non-arboreal wildlife, the canopy wildlife crossings of arboreal wildlife, and the wildlife overpasses of large mammals can be extracted through the two-step screening method, and the conflict point of the wildlife involved in the wildlife crossings can be selected as the construction site location of wildlife crossings.

A 5 km buffer zone was established around each conflict point. The selection of the buffer radius was mainly based on empirical research on the ecological impact range of roads and related infrastructure on mammals. Studies have shown that the negative effects of roads and infrastructure on wildlife are most significant within the 0–5 km range [31]. Therefore, the 5 km buffer provides a reasonable balance and adaptability for different species. Based on this framework, the number of conflict points within each buffer zone was counted. Line graphs were then plotted based on the data of each point, revealing fluctuations that indicate the presence of clustering. Taking the G5 Expressway as an example, the graph shows clustering between points 5 to 17 and points 25 to 33. Peak points 12 and 27 were selected as the preferred locations for wildlife crossings within these clusters. The same method was applied to 30 roads to determine the final set of priority crossing sites. The clustering analysis and site prioritization process along the G5 Expressway is exemplified in Figure 15.

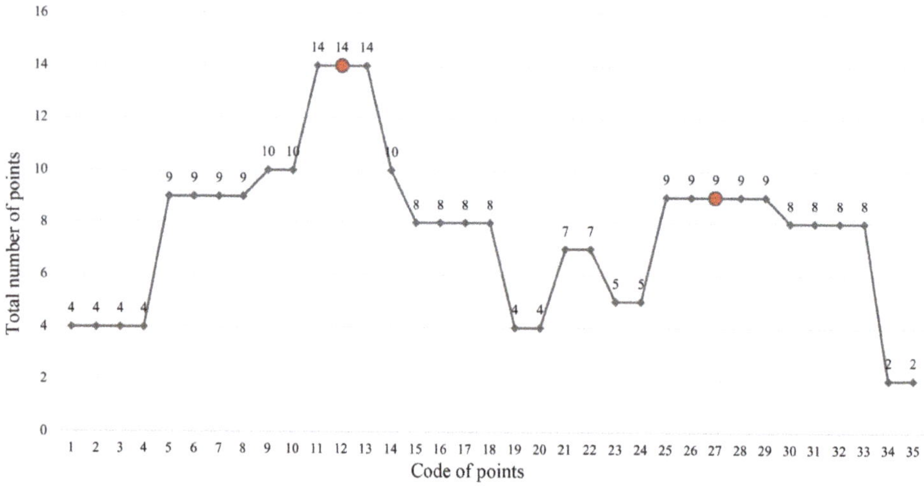

Figure 15. Example of G5 high-speed point selection method; red dots indicate the peak points selected as priority wildlife crossing sites.

Based on these three types of conflict points, a total of 39 overpasses, 34 underpasses, and 27 canopy bridges were identified as sites for the construction of wildlife crossings (Figure 16).

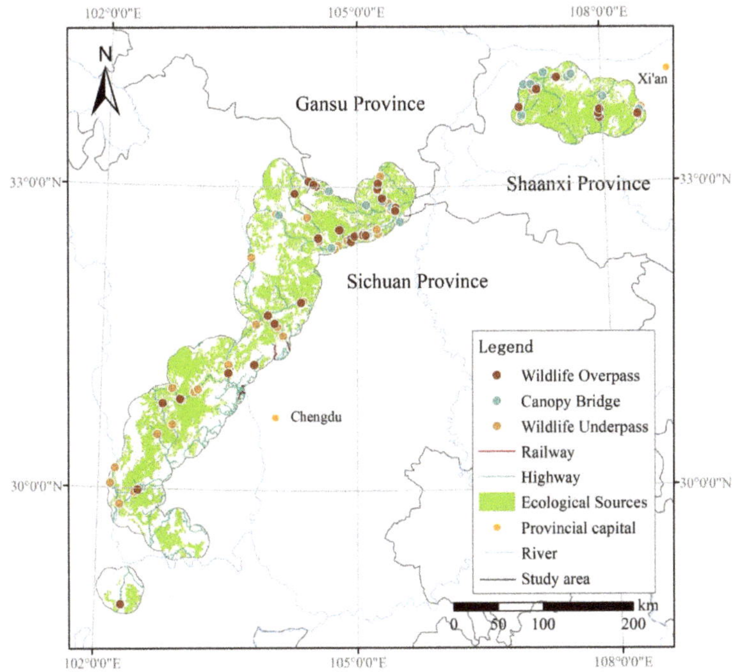

Figure 16. Wildlife crossing masterplan.

3.4. Design of Multifunctional Wildlife Crossings

The design and selection of wildlife crossing types are influenced by species-specific ecological traits, topographic conditions, engineering feasibility, and construction costs. This study designs three types of composite wildlife crossings—overpasses, underpasses, and canopy bridges—tailored to species assemblages, with the aim of minimizing movement disruption and population fragmentation while enhancing habitat connectivity. Figure 17 displays the specific locations for each type of wildlife crossing.

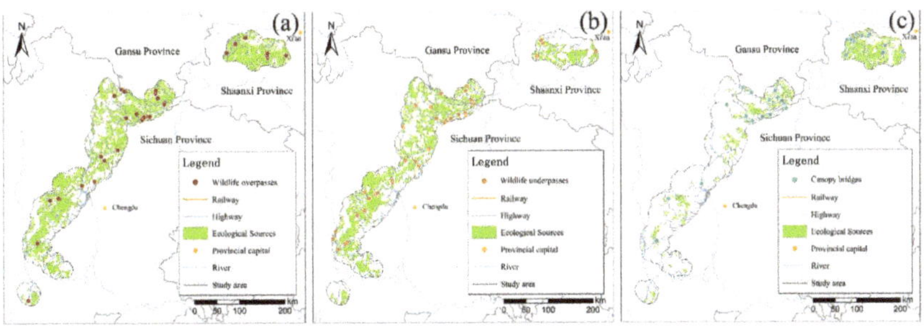

Figure 17. Locations of different types of wildlife crossings. (**a**) Locations of overpasses; (**b**) locations of underpasses; (**c**) locations of canopy bridges.

Overpasses are primarily designed for large mammals and are applicable to all six target species in this study. The soil above the pass is planted with locally dominant plants, such as alpine rhododendron and fir trees, to simulate the natural environment and provide hiding and foraging sites, while bamboos and shrubs are planted around the fences to reduce the interference, and fallen logs are installed inside the pass to guide the wildlife to traverse the crossing, which is a good solution to meet the crossing needs of various wildlife types, despite the high cost of construction [32]. Although the construction cost is high, it can meet the needs of various types of wildlife.

Underpasses are suitable for species of varying body sizes. In this study, they are mainly designed for four target species: the giant panda, leopard cat, forest musk deer, and Sichuan takin. These structures can be integrated into traffic bridges and culverts, which are low-cost and convenient to set up. The design of the crossings takes into account geological stability and ecological consistency and makes use of natural paths combined with drainage facilities to ensure safe passage in the rainy season. This design also sets up attractive vegetation at the entrances and exits and increases the natural elements in the interior to attract wildlife to use them [33,34], which has cross-border and cross-strait advantages. Other advantages include its large span, small area, and low cost.

Canopy bridges are specifically designed for arboreal species, including the Sichuan snub-nosed monkey and rock squirrel. These structures mimic natural canopy pathways using slings and rattan bridges anchored to stable trees, with bionic materials employed to minimize landscape disturbance. They enable arboreal species to move between canopy layers, reducing their dependence on ground-level travel [35].

Each crossing type offers distinct features that address the specific needs of different species, achieving a balance between ecological conservation and engineering feasibility.

4. Discussion

4.1. Comparative Analysis of Existing Studies

In recent years, the planning and construction of wildlife crossing structures within national parks have become a central topic in ecological conservation research. North America and Europe were early pioneers in this field and have developed relatively mature systems [36]. For instance, Banff National Park in Canada has established 44 wildlife crossing structures—comprising 6 vegetated overpasses and 38 underpasses—along an 82 km section of the Trans-Canada Highway, achieving a corridor density of 0.54 crossings per kilometer [37,38]. These crossings primarily serve large mammals such as grizzly bears, American black bears, and elk, forming one of the most representative road-ecological connectivity systems worldwide. In the western United States and countries like the Netherlands [39], wildlife crossings have been adapted to species size and topography, including multifunctional overpasses and canopy bridges for smaller species, with continuous monitoring to optimize design performance [40].

By contrast, although China entered this field relatively late, it has made significant progress in recent years with the advancement of ecological civilization policies. National-level standards, such as the *Technical Specification for Terrestrial Wildlife Corridor Design*, have been issued, and key demonstration projects—such as those along the Qinghai–Tibet Railway and the Beijing–Xinjiang Expressway—have been implemented [41]. Nevertheless, most existing practices still focus on single species and rely heavily on underpasses, which are insufficient for ungulates that prefer open environments [42].

There remains substantial room for optimization in terms of corridor density and structural diversity.

This study takes the Giant Panda National Park as the research area and establishes an integrated ecological network of composite wildlife corridors, considering six representative terrestrial mammal species with varied ecological traits. A total of 100 candidate corridors were identified, including 39 vegetated overpasses, 34 underpasses, and 27 canopy bridges, each matched to specific species groups. Although the overall corridor density (0.017 crossings per km) is lower than that of Banff National Park, the study area features greater species diversity, more complex terrain, and a higher degree of ecological fragmentation. Unlike the traditional site-based layout methods used in Banff, this study employed a data-driven approach combining Maxent modeling and Linkage Mapper analysis to identify optimal corridor locations based on resistance surfaces and habitat suitability, offering greater reproducibility and regional adaptability.

The corridor identification process established a complete spatial analysis chain from species distribution modeling and resistance surface construction to least-cost path identification. Compared to expert-based or field-oriented methods, this approach improves methodological transparency and transferability. The proposed method for identifying composite wildlife corridors enhances the scientific basis for corridor design and better addresses the conservation needs of multiple species in complex national park ecosystems, providing a quantitative foundation for future decision-making in corridor planning.

4.2. Policy Recommendations

While China has made significant strides in the design and implementation of wildlife crossings, several critical challenges still hinder their widespread application and long-term effectiveness. To further promote the ecological development of transportation networks within national parks and enhance the effectiveness of wildlife corridors for biodiversity conservation, the following policy recommendations are proposed.

(1) Improve the legal and regulatory framework and standard system.

At present, China lacks a comprehensive legal and regulatory system specifically dedicated to the construction of wildlife crossings. To better facilitate the ecological development of transportation infrastructure in national parks, it is imperative to formulate and refine relevant laws, regulations, and technical standards [43]. These should clearly define key aspects such as design criteria, construction standards, and monitoring and evaluation mechanisms, thereby providing a legal foundation for the scientific planning and effective implementation of wildlife corridors.

(2) Enhance cross-sectoral coordination and collaboration. A routine coordination mechanism should be established among the departments of transportation, ecological environment, forestry, and natural resources. Regular joint meetings should be convened to discuss issues related to ecological protection in transportation projects. During project implementation, these departments should collaborate on construction oversight to minimize ecological damage [44]. Additionally, an efficient information-sharing platform should be developed to facilitate the timely exchange of project updates and ecological monitoring data, enabling early identification and resolution of emerging issues.

(3) Increase financial investment and policy support.

The government should establish a dedicated wildlife crossing construction fund to support the planning, construction, and maintenance of wildlife corridors in national parks. Policies such as tax incentives and financial subsidies should be introduced to encourage private enterprises and social capital to participate in these projects [45]. For companies engaged in ecologically friendly business practices, targeted fiscal support should be provided to stimulate broader societal involvement in the ecological development of transportation networks.

(4) Promote international cooperation and academic exchange.

China should actively participate in activities organized by international wildlife conservation organizations and establish long-term partnerships with countries and regions that have advanced experience in this field. Regular international conferences should be held to explore new technologies and innovative approaches in wildlife corridor construction. Through international collaboration, China can enhance its global influence and voice in this domain, contributing to the advancement of global biodiversity conservation efforts [46,47].

5. Conclusions

As core areas for biodiversity conservation, national parks are increasingly challenged by habitat fragmentation caused by the expansion of transportation networks. This study focused on six terrestrial mammal species within the Giant Panda National Park, analyzing their distribution and behavioral patterns. A total of 13 variables across four categories were selected to construct a comprehensive evaluation framework. The Maxent model was employed to assess habitat suitability for each target species and to identify ecological source areas. Through weighted integration of geographic, biological, and anthropogenic factors, potential dispersal pathways were identified, and a composite ecological corridor network was established.

Building upon the integration of ecological corridors and transportation infrastructure, this study identified ecological conflict zones and proposed multifunctional wildlife crossings benefiting multiple species. A two-step selection method was adopted to determine optimal locations for these crossings, resulting in the design of three types of wildlife crossing structures—overpasses, underpasses, and canopy bridges—tailored to the needs of Giant Panda National Park.

This study contributes both theoretical insights and practical tools for the ecological design of transportation networks. Theoretically, this study enriches the conceptual framework of wildlife crossing design and provides a comprehensive and in-depth theoretical basis. Practically, it offers design solutions for three types of wildlife crossings that mitigate traffic-induced disturbances and enhance biodiversity conservation, setting a reference model for other protected areas and advancing the ecological integration of transportation networks.

Although this study has made significant progress in advancing the ecological integration of transportation networks within national parks, several limitations warrant further improvement.

One notable limitation is that the identification of habitat suitability factors in this study primarily focused on environmental variables for individual species, without fully accounting for interspecific ecological interactions. Future research could benefit from integrating food web or ecological network perspectives to better capture the combined effects of predation, competition, and niche overlap on species distribution.

Moreover, the predictive accuracy of the Maxent model remains limited by the spatial precision and reliability of species occurrence data. To reduce uncertainty and improve model robustness, future studies should consider incorporating higher-quality datasets, such as those derived from remote sensing, infrared camera monitoring, and UAV observations.

In addition, the current process of wildlife crossing design and site selection warrants further refinement. Although this study identified ecological conflict zones based on species distribution patterns, practical implementation should also address terrain complexity, land-use heterogeneity, and socioeconomic constraints. A more adaptive planning framework is recommended—one that integrates factors such as accessibility, human disturbance intensity, and structural safety. Furthermore, the use of a fixed buffer distance in the two-step screening method may oversimplify species-specific behavioral variation. Buffer zones should instead be adjusted based on species' movement ranges and ecological dependencies, while also considering the coupled effects of topography and anthropogenic activity. Field-based validation will be essential to ensure the ecological functionality and operational feasibility of the proposed strategies.

Author Contributions: Conceptualization, G.Z.; methodology, X.L. and Y.P.; software, X.L., L.W. and J.S.; formal analysis, X.L., J.S. and L.W.; investigation, X.L., J.S. and L.W.; resources, X.L. and G.Z.; data curation, X.L., Y.P. and J.S.; validation, G.Z. and Y.P.; visualization, X.L.; writing—original draft preparation, X.L., J.S. and L.W.; writing—review and editing, G.Z. and Y.P.; supervision, G.Z.; project administration, G.Z.; funding acquisition, G.Z. All authors have read and agreed to the published version of the manuscript.

Funding: This work was supported by the National Natural Science Foundation of China (No. 42471119) and the Hainan Provincial Transportation Technology Project (HNJTT-KXC-2024-2-27-01).

Data Availability Statement: The original contributions presented in the study are included in the article, further inquiries can be directed to the corresponding author.

Acknowledgments: We thank the reviewers for their constructive comments on the original manuscript.

Conflicts of Interest: The authors declare no conflicts of interest.

References

1. Watson, J.E.; Dudley, N.; Segan, D.B.; Hockings, M. The performance and potential of protected areas. *Nature* **2014**, *515*, 67–73. [CrossRef]
2. Tucker, M.A.; Böhning-Gaese, K.; Fagan, W.F.; Fryxell, J.M.; Van Moorter, B.; Alberts, S.C.; Ali, A.H.; Allen, A.M.; Attias, N.; Avgar, T.; et al. Moving in the Anthropocene: Global reductions in terrestrial mammalian movements. *Science* **2018**, *359*, 466–469. [CrossRef] [PubMed]
3. Haddad, N.M.; Brudvig, L.A.; Clobert, J.; Davies, K.F.; Gonzalez, A.; Holt, R.D.; Lovejoy, T.E.; Sexton, J.O.; Austin, M.P.; Collins, C.D.; et al. Habitat fragmentation and its lasting impact on Earth's ecosystems. *Sci. Adv.* **2015**, *1*, e1500052. [CrossRef]
4. Dickson, B.G.; Albano, C.M.; Anantharaman, R.; Beier, P.; Fargione, J.; Graves, T.A.; Gray, M.E.; Hall, K.R.; Lawler, J.J.; Leonard, P.B.; et al. Circuit-theory applications to connectivity science and conservation. *Conserv. Biol.* **2019**, *33*, 239–249. [CrossRef] [PubMed]
5. Phillips, S.J.; Anderson, R.P.; Dudík, M.; Schapire, R.E.; Blair, M.E. Opening the black box: An open-source release of Maxent. *Ecography* **2017**, *40*, 887–893. [CrossRef]
6. LaPoint, S.; Balkenhol, N.; Hale, J.; Sadler, J.; van der Ree, R. Ecological connectivity research in urban areas. *Funct. Ecol.* **2015**, *29*, 868–878. [CrossRef]
7. Hughes, A.C. Understanding and minimizing environmental impacts of the Belt and Road Initiative. *Conserv. Biol.* **2019**, *33*, 883–894. [CrossRef]
8. Das, T.K. Road ecology: Science and solutions by Richard T. T. Forman, Daniel Sperling, John A. Bissonette, Anthony P. Clevenger, Carol D. Cutshall, Virginia H. Dale, Lenore Fahrig, Robert France, Charles R. Goldman, Kevin Heanue, Julia A. Jones, Frederick J. Swanson, Thomas Turrentine, and Thomas C. Winter Island Press Covelo, WA, and London, UK (2002) 481 pages ISBN 1-55963-932-6 (cloth) U.S. List Price: $55.00 ISBN 1-55963-933-4 (paper) U.S. List Price: $27.50. *Environ. Prog.* **2003**, *22*, O16.
9. Forman, R.; Alexander, L.E. Roads and their major ecological effects. *Annu. Rev. Ecol. Syst.* **1998**, *29*, 207–231. [CrossRef]
10. Glista, D.J.; DeVault, T.L.; DeWoody, J.A. A review of mitigation measures for reducing wildlife mortality on roadways. *Landsc. Urban Plan.* **2009**, *91*, 1–7. [CrossRef]
11. Soanes, K.; Taylor, A.C.; Sunnucks, P.; Vesk, P.A.; Cesarini, S.; van der Ree, R. Evaluating the success of wildlife crossing structures using genetic approaches and an experimental design: Lessons from a gliding mammal. *J. Appl. Ecol.* **2018**, *55*, 129–138. [CrossRef]
12. Steenweg, R.; Hebblewhite, M.; Kays, R.; Ahumada, J.; Fisher, J.T.; Burton, C.; Townsend, S.E.; Carbone, C.; Rowcliffe, J.M.; Whittington, J.; et al. Scaling-up camera traps: Monitoring the planet's biodiversity with networks of remote sensors. *Front. Ecol. Environ.* **2017**, *15*, 26–34. [CrossRef]
13. Ascensão, F.; Fahrig, L.; Clevenger, A.P.; Corlett, R.T.; Jaeger, J.A.; Laurance, W.F.; Pereira, H.M. Environmental challenges for the Belt and Road Initiative. *Nat. Sustain.* **2018**, *1*, 206–209. [CrossRef]
14. Zeller, K.A.; McGarigal, K.; Whiteley, A.R. Estimating landscape resistance to movement: A review. *Landsc. Ecol.* **2012**, *27*, 777–797. [CrossRef]
15. Beier, P.; Noss, R.F. Do Habitat Corridors Provide Connectivity? *Conserv. Biol.* **1998**, *12*, 1241–1252. [CrossRef]
16. Shen, W.; Kang, Y.; Zhang, X.; Liu, Y.; Zhou, X.; Li, J. Spatio-temporal variation of vegetaton and its topographic diferentiation in the Sichuan area of Giant Panda National Park. *Acta Ecol. Sin.* **2024**, *44*, 9081–9093.
17. Liu, X.; Su, Y.; Hu, T.; Yang, Q.; Liu, B.; Deng, Y.; Tang, H.; Tang, Z.; Fang, J.; Guo, Q. Neural network guided interpolation for mapping canopy height of China's forests by integrating GEDI and ICE-Sat-2 data. *Remote Sens. Environ.* **2022**, *269*, 112844. [CrossRef]

18. Phillips, S.J.; Anderson, R.P.; Schapire, R.E. Maximum entropy modeling of species geographic distributions. *Ecol. Model.* **2006**, *190*, 231–259. [CrossRef]
19. Elith, J.; Phillips, S.J.; Hastie, T.; Dudík, M.; Chee, Y.E.; Yates, C.J. A statistical explanation of MaxEnt for ecologists. *Divers. Distrib.* **2011**, *17*, 43–57. [CrossRef]
20. McRae, B.H.; Beier, P. Circuit theory predicts gene flow in plant and animal populations. *Proc. Natl. Acad. Sci. USA* **2007**, *104*, 19885–19890. [CrossRef]
21. Peterman, W.E. ResistanceGA: An R package for the optimization of resistance surfaces using genetic algorithms. *Methods Ecol. Evol.* **2018**, *9*, 1638–1647. [CrossRef]
22. Naimi, B.; Araújo, M.B. sdm: A reproducible and extensible R platform for species distribution modelling. *Ecography* **2016**, *39*, 368–375. [CrossRef]
23. Torres, A.; Jaeger, J.A.G.; Alonso, J.C. Assessing large-scale wildlife responses to human infrastructure development. *Proc. Natl. Acad. Sci. USA* **2016**, *113*, 8472–8477. [CrossRef] [PubMed]
24. Gueta, T.; Carmel, Y. Quantifying the value of user-level data cleaning for big data: A case study using mammal distribution models. *Ecol. Inform.* **2016**, *34*, 139–145. [CrossRef]
25. Koen, E.L.; Garroway, C.J.; Wilson, P.J.; Bowman, J. The Effect of Map Boundary on Estimates of Landscape Resistance to Animal Movement. *PLoS ONE* **2010**, *5*, e11785. [CrossRef]
26. Shcheglovitova, M.; Anderson, R.P. Estimating optimal complexity for ecological niche models: A jackknife approach for species with small sample sizes. *Ecol. Model.* **2013**, *269*, 9–17. [CrossRef]
27. Wisz, M.S.; Hijmans, R.J.; Li, J.; Peterson, A.T.; Graham, C.H.; Guisan, A. Effects of sample size on the performance of species distribution models. *Divers. Distrib.* **2008**, *14*, 763–773. [CrossRef]
28. Adriaensen, F.; Chardon, J.P.; De Blust, G.; Swinnen, E.; Villalba, S.; Gulinck, H.; Matthysen, E. The application of 'least-cost' modelling as a functional landscape model. *Landsc. Urban Plan.* **2003**, *64*, 233–247. [CrossRef]
29. Rayfield, B.; Fortin, M.; Fall, A. Connectivity for conservation: A framework to classify network measures. *Ecology* **2011**, *92*, 847–858. [CrossRef]
30. Pinto, N.; Keitt, T.H. Beyond the least-cost path: Evaluating corridor redundancy using a graph-theoretic approach. *Landsc. Ecol.* **2009**, *24*, 253–266. [CrossRef]
31. Benítez-López, A.; Alkemade, R.; Verweij, P.A. The impacts of roads and other infrastructure on mammal and bird populations: A meta-analysis. *Biol. Conserv.* **2010**, *143*, 1307–1316. [CrossRef]
32. Mata, C.; Hervás, I.; Herranz, J.; Suárez, F.; Malo, J.E. Are motorway wildlife passages worth building? Vertebrate use of road-crossing structures on a Spanish motorway. *J. Environ. Manag.* **2008**, *88*, 407–415. [CrossRef] [PubMed]
33. Lesbarrères, D.; Fahrig, L. Measures to reduce population fragmentation by roads: What has worked and how do we know? *Trends Ecol. Evol.* **2012**, *27*, 374–380. [CrossRef] [PubMed]
34. Bissonette, J.A.; Cramer, P.C. *Evaluation of the Use and Effectiveness of Wildlife Crossings*; The National Academies Press: Washington, DC, USA, 2008; p. 161.
35. Teixeira, F.Z.; Coelho, A.V.; Esperandio, I.B.; Kindel, A. Vertebrate road mortality estimates: Effects of sampling methods and carcass removal. *Biol. Conserv.* **2013**, *157*, 317–323. [CrossRef]
36. Bissonette, J.A.; Adair, W. Restoring habitat permeability to roaded landscapes with isometrically-scaled wildlife crossings. *Biol. Conserv.* **2008**, *141*, 482–488. [CrossRef]
37. Stewart, L.; Russell, B.; Zelig, E.; Patel, G.; Whitney, K.S. Wildlife Crossing Design Influences Effectiveness for Small and Large Mammals in Banff National Park. *Case Stud. Environ.* **2020**, *4*, 1231752. [CrossRef]
38. Clevenger, A.P.; Chruszcz, B.; Gunson, K.E. Highway mitigation fencing reduces wildlife-vehicle collisions. *Wildl. Soc. Bull.* **2001**, *29*, 646–653.
39. Clevenger, A.P.; Waltho, N. Performance indices to identify attributes of highway crossing structures facilitating movement of large mammals. *Biol. Conserv.* **2005**, *121*, 453–464. [CrossRef]
40. Clevenger, A.P.; Huijser, M.P. *Wildlife Crossing Structure Handbook: Design and Evaluation in North America*; Central Federal Lands Highway Division: Lakewood, CO, USA, 2011; FHWA-CFL-TD-11-003.
41. Wang, Y.; Guan, L.; Yang, Y.; Zhou, H.; Wang, Y.; Cao, G.; Kong, Y. A Review of Highway Wildlife Crossing Structures. *Transp. Res.* **2019**, *5*, 79–87+109.
42. Cushman, S.A.; Landguth, E.L. Multi-taxa population connectivity in the Northern Rocky Mountains. *Ecol. Model.* **2012**, *231*, 101–112. [CrossRef]
43. Trombulak, S.C.; Frissell, C.A. Review of ecological effects of roads on terrestrial and aquatic communities. *Conserv. Biol.* **2000**, *14*, 18–30. [CrossRef]
44. Cash, D.W.; Adger, W.N.; Berkes, F.; Garden, P.; Lebel, L.; Olsson, P.; Pritchard, L.; Young, O. Scale and Cross-Scale Dynamics: Governance and Information in a Multilevel World. *Ecol. Soc.* **2006**, *11*, 8. [CrossRef]

45. Baylis, K.; Honey-Rosés, J.; Börner, J.; Corbera, E.; Ezzine-de-Blas, D.; Ferraro, P.J.; Lapeyre, R.; Persson, U.M.; Pfaff, A.; Wunder, S. Mainstreaming Impact Evaluation in Nature Conservation. *Conserv. Lett.* **2016**, *9*, 58–64. [CrossRef]
46. Laurance, W.F.; Clements, G.R.; Sloan, S.; O'connell, C.S.; Mueller, N.D.; Goosem, M.; Venter, O.; Edwards, D.P.; Phalan, B.; Balmford, A.; et al. A global strategy for road building. *Nature* **2014**, *513*, 229–232. [CrossRef]
47. Hilty, J.; Worboys, G.L.; Keeley, A.; Woodley, S.; Lausche, B.J.; Locke, H.; Carr, M.; Pulsford, I.; Pittock, J.; White, J.W.; et al. *Guidelines for Conserving Connectivity Through Ecological Networks and Corridors*; IUCN: Gland, Switzerland, 2020.

MDPI AG
Grosspeteranlage 5
4052 Basel
Switzerland
Tel.: +41 61 683 77 34

Land Editorial Office
E-mail: land@mdpi.com
www.mdpi.com/journal/land

www.ingramcontent.com/pod-product-compliance
Lightning Source LLC
Chambersburg PA
CBHW061556220926
51850CB00005B/352
* 9 7 8 3 7 2 5 8 4 7 7 5 4 *